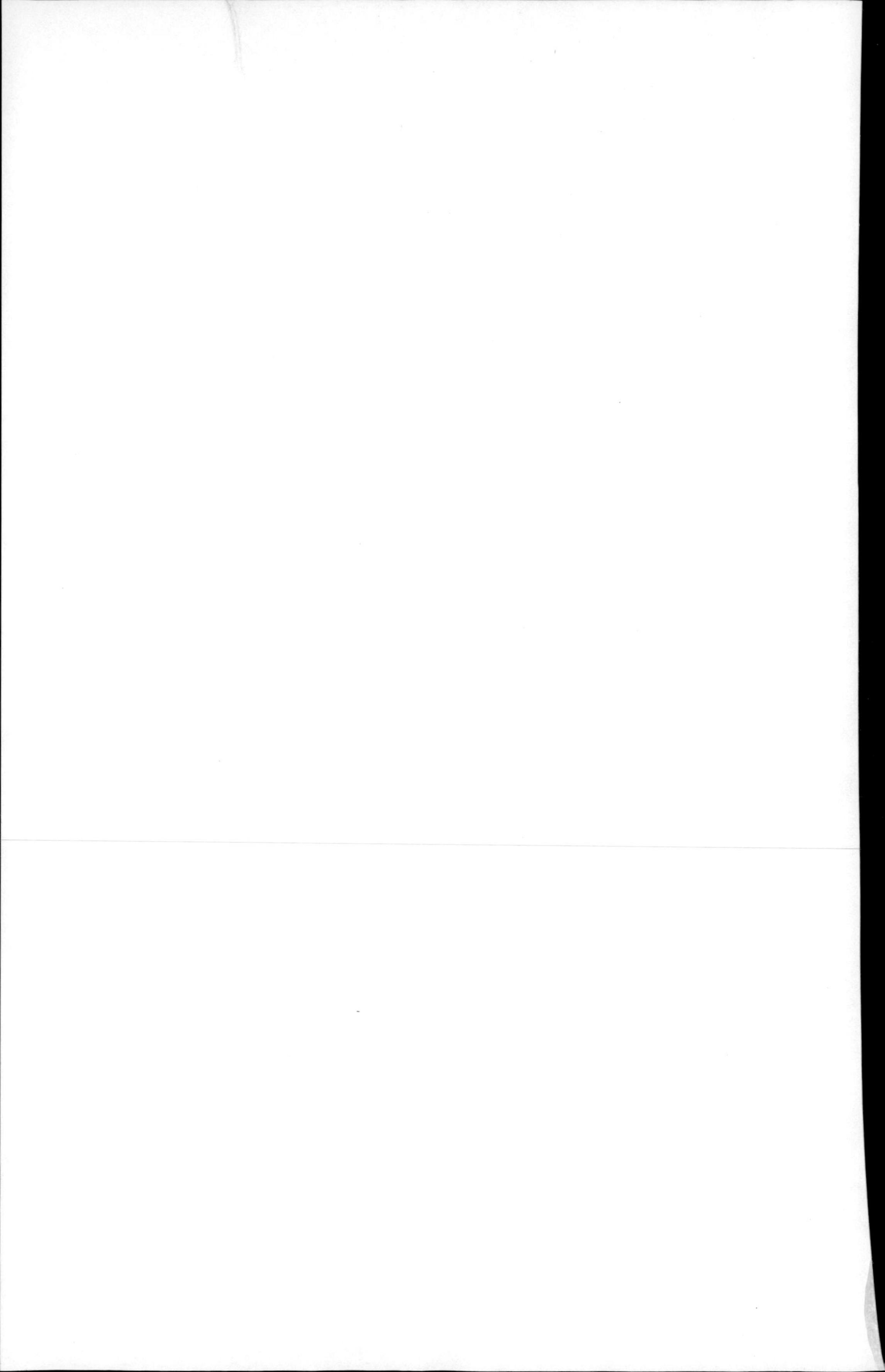

Achieving the Paris Climate Agreement Goals

Sven Teske
Editor

Achieving the Paris Climate Agreement Goals

Global and Regional 100% Renewable Energy Scenarios with Non-energy GHG Pathways for +1.5°C and +2°C

Editor
Sven Teske
Institute for Sustainable Futures
University of Technology Sydney
Sydney, NSW, Australia

Additional material to this book can be downloaded from http://extras.springer.com.

ISBN 978-3-030-05842-5 ISBN 978-3-030-05843-2 (eBook)
https://doi.org/10.1007/978-3-030-05843-2

Library of Congress Control Number: 2018966518

This Springer imprint is published by the registered company Springer Nature Switzerland AG.
The registered company address is: Gewerbestrasse 11, 6330 Cham, Switzerland

For the next generation.
For my son, Travis.

Climate Model: Foreword

In October of 2018, the Intergovernmental Panel on Climate Change issued its starkest warning yet: we have around 12 years to avoid the worst effects of anthropogenic climate change. The consumption of fossil fuels, the reckless destruction of forests and other natural ecosystems, and the release of powerful greenhouse gases have already caused around 1.0 °C of warming above pre-industrial levels.

Continuing at the current rate, we are likely to reach 1.5 °C by 2030 – and all the evidence suggests that a world beyond 1.5 °C is not one we want to live in.

While making the 2016 documentary film, *Before the Flood*, I witnessed first-hand the impacts of an already-changing climate: the rapid melting of ice in the Arctic Circle, massive bleaching of coral reefs in the Bahamas, and rampant deforestation in Indonesia and the Amazon. Better than ever, we understand the heart-breaking impact of human activity on our natural world. It is estimated, for example, that 60% of animals have been wiped out since 1970.

Higher temperatures and extreme weather events will cause ever more severe harm to biodiversity and ecosystems and even greater species loss and extinction. And when we lose biodiversity, we lose resilience. Currently, natural ecosystems absorb about half of human-caused carbon dioxide emissions. If we continue to degrade the natural world, we could lose completely the Earth's ability to adapt to climate change.

A passion for nature conservation and animal protection has driven much of my foundation's work over the past 20 years. Ultimately, however, the climate crisis is a humanitarian one. If business-as-usual continues, the impact on human beings will be immeasurable. Water supplies will become more insecure. Sea level rise will profoundly impact islands, low-lying coastal areas, and river deltas. Small island communities like those I visited in the South Pacific are already preparing for migration to safer lands. Fatal floods, droughts, hurricanes, and wildfires are the new normal, and happening closer to home. An estimated 41 million Americans live within a 100-year flood zone. Texas saw its third 500-year flood 3 years in a row.

Poor air quality is a public health emergency across the world and now the fourth-highest cause of death – contributing to strokes, heart attacks, and lung cancer – causing public unrest in countries like China and India, where the poorest find themselves at the mercy of pollution from industrial facilities and the burning of biomass. In states like Texas, Colorado, and North Dakota, communities are fighting back against gas drilling operations near playgrounds or soccer fields, where children breathe in poisonous gases.

These health impacts are only part of the story. Climate change, as the US Pentagon notes, is a national security threat. In a 2017 report by the Environmental Justice Foundation, senior US military experts pointed to the likelihood of tens of millions of climate refugees displaced by extreme weather – in a world already struggling with a refugee crisis. We already know that many conflicts are driven by environmental factors and access to natural resources. The truth is that, where ecosystems collapse, societies collapse too.

Politically, there has been a monumental failure to grasp the scale of this problem. Climate scientists still face disinformation campaigns and a press corps that often draws a false equivalence between those who support the scientific consensus for human-caused climate change and those who do not. Surveys suggest that most Americans do not know a scientific consensus exists, and scientists like Michael Mann, who spoke to me for *Before the Flood*, face abuse for exposing the truth. As a result, scientific research programs, critical to better understanding and addressing climate change, are often attacked or defunded.

Nevertheless, in the face of these challenges, some progress is being made. With the growth of the environmental movement, public awareness of the climate crisis has increased significantly. Governments and the private sector are beginning to ramp up their efforts. Renewable energy is booming. And the UN Sustainable Development Goals, ratified by 193 countries, now call for a halt to deforestation and land degradation by 2030. After decades of climate negotiations, the Paris Agreement now calls upon the world's governments to keep warming "well below 2°C" while striving for 1.5°C.

While we are beginning to move in the right direction, the reality is that these efforts are simply not ambitious enough to address the climate crisis at scale. The IPCC warns that to avoid the worst consequences of climate change, we must stay below the 1.5 °C limit. But what does that mean in practical terms?

Determined to find solutions, my foundation supported a 2-year research program led by a team of international climate and energy experts to develop a roadmap for how we can actually stay below this critical climate threshold. The findings, outlined in this book, give cause for optimism. With a transition to 100% renewable energy by mid-century and a major land conservation and restoration effort, it is possible to stay below the 1.5 °C limit with technologies that are available right now. It will be a lot of work, but the costs will be far less than the $5 trillion per year governments currently spend subsidizing the fossil fuel industries responsible for climate change.

The climate model and energy transition pathways compiled in this book offer an exciting, positive, and achievable vision of a better world in which we are no longer dependent on fossil fuels and where the conservation and restoration of nature is treated as indispensable to our survival. This is not fantasy. This is science.

Science is showing us the way forward, but you do not need to be a scientist to understand that climate change is the defining issue of our time. If our world warms past 1.5 °C, our way of life will profoundly change for the worse. Why not manage the transition in a way that is orderly and equitable? Human beings caused this problem, but with our vast knowledge and ingenuity, we can also fix it.

We are resilient. We can adapt. We can change.

Chairman of the Leonardo DiCaprio Foundation Leonardo DiCaprio

Contact Information

Lead Author: Dr Sven Teske
University of Technology Sydney – Institute for Sustainable Futures (UTS-ISF)
Address: Building 10, 235 Jones Street, Sydney, NSW, Australia 2007/Telephone: +61 2 9514 4786
https://www.uts.edu.au/research-and-teaching/our-research/institute-sustainable-futures

Author: Dr. Sven Teske E-mail: sven.teske@uts.edu.au	Chapters: 1, 2.2, 3.1, 3.2, 3.5, 3.6, 7, 8 (Power Sector analysis), 9, 10, 13
Author: Prof. Dr. Damien Giurco E-mail: damien.giurco@uts.edu.au	Chapters: 11, 13
Author: Tom Morris E-mail: tom.morris@uts.edu.au	Chapters: 3.2, 3.5, 3.6, 7
Author: Kriti Nagrath E-mail: kriti.nagrath@uts.edu.au	Chapters: 3.2, 7
Author: Franziska Mey E-mail: franziska.mey@uts.edu.au	Chapter: 10
Author: Dr Chris Briggs E-mail: chris.briggs@uts.edu.au	Chapter: 10
Author: Elsa Dominish E-mail: elsa.dominish@uts.edu.au	Chapter: 10, 11
Author: Dr Nick Florin E-mail: nick.florin@uts.edu.au	Chapter 11

Graduate School of Energy Science, Kyoto University – for Chapter 11
Author: Takuma Watari,
Author: Benjamin Mclellan

German Aerospace Center (DLR), Institute for Engineering Thermodynamics (TT),
Department of Energy Systems Analysis
Address: Pfaffenwaldring 38-40, Germany D-70569/Telephone: +49-711 6862 355
http://www.dlr.de/tt/en/desktopdefault.aspx/tabid-2904/4394_read-6500/

Author: Dr. Thomas Pregger Chapters: 3.1, 3.4, 5, 8
E-mail: thomas.pregger@dlr.de (Long-term energy model), 13
Author: Dr. Tobias Naegler Chapters: 3.1, 3.4, 5, 8
E-mail: tobias.naegler@dlr.de (Long-term energy model), 13
Author: Dr. Sonja Simon Chapters: 3.1, 3.4, 5, 8
E-mail: sonja.simon@dlr.de (Long-term energy model), 13

German Aerospace Center (DLR), Institute of Vehicle Concepts (FK), Department of Vehicle Systems and Technology Assessment
Address: Pfaffenwaldring 38-40, Germany D-70569/Telephone: +49-711 6862 533
https://www.dlr.de/fk/en/desktopdefault.aspx/

Author: Johannes Pagenkopf, Chapters: 3.3, 6, 13
E-mail: johannes.pagenkopf@dlr.de
Author: Bent van den Adel Chapters: 3.3, 6, 13
E-mail: Bent.vandenAdel@dlr.de
Author: Özcan Deniz Chapters: 3.3, 6, 13
E-mail: oezcan.deniz@dlr.de
Author: Dr. Stephan Schmid Chapters: 3.3, 6, 13
E-mail: stephan.schmid@dlr.de

University of Melbourne
Address: Australian-German Climate and Energy College, Level 1, 187 Grattan Street, University of Melbourne, Parkville, Victoria, Australia 3010
www.energy-transition-hub.org

Author: A/Prof. Dr. Malte Meinshausen Chapters: 2.1, 3.8, 4, 12, 13
Affiliation: University of Melbourne
E-mail: malte.meinshausen@unimelb.edu.au/
Telephone: +61 3 90356760
Author: Dr. Kate Dooley Chapters: 3.8, 4.1, 7
Affiliation: University of Melbourne
E-mail: kate.dooley@unimelb.edu.au/
Telephone: +61 3 90356760

Editor: Janine Miller

Executive Summary

Abstract An overview of the motivations behind the writing of this book, the scientific background and context of the research. Brief outline of all methodologies used, followed by assumptions and the storyline of each scenario. Presentation of main results of the renewable energy resources assessment, transport scenario, long-term energy pathway, the power sector analysis, employment analysis and an assessment for required metals for renewable energy and storage technologies. Key results of non-energy greenhouse mitigation scenarios which are developed in support of the energy scenario in order to achieve the 1.5 °C target. Concluding remarks and policy recommendations including graphs and tables.

Introduction The Paris Climate Agreement aims to hold global warming to well below 2 degrees Celsius (°C) and to "pursue efforts" to limit it to 1.5 °C. To accomplish this, countries have submitted *Intended Nationally Determined Contributions* (INDCs) outlining their post-2020 climate actions (Rogelj 2016). This research aimed to develop practical pathways to achieve the Paris climate goals based on a detailed bottom-up examination of the potential of the energy sector, in order to avoid reliance on net negative emissions later on.

The study described in this book focuses on the ways in which humans produce energy, because energy-related carbon dioxide (CO_2) emissions are the main drivers of climate change. The analysis also considers the development pathways for non-energy-related emissions and mitigation measures for them because it is essential to address their contributions if we are to achieve the Paris climate change targets.

State of Research—Climate Beyond reasonable doubt, climate change over the last 250 years has been driven by anthropogenic activities. In fact, the human-induced release of greenhouse gas emissions into the atmosphere warms the planet even more than is currently observed as climate change, but some of that greenhouse-gas-induced warming is masked by the effect of aerosol emissions.

Carbon dioxide emissions are so large that they are the dominant driver of human-induced climate change. A single kilogram of CO_2 emitted will increase the atmospheric CO_2 concentration over hundreds or even thousands of years. Since the Intergovernmental Panel on Climate Change (IPCC) Fifth Assessment Report, the finding that cumulative CO_2 emissions are roughly linearly related to temperature has shaped scientific and political debate. The remaining permissible CO_2 emissions that are consistent with a target temperature increase of 2 °C or 1.5 °C and their comparison with remaining fossil fuel resources are of key interest.

The IPCC Fifth Assessment Report concluded that beyond 2011, cumulative CO_2 emissions of roughly 1000 $GtCO_2$ are permissible for a "likely below 2.0 °C" target change, and approximately 400 $GtCO_2$ are permissible for a 1.5 °C target change. However, the recently published IPCC Special Report on the 1.5 °C target suggests substantially higher carbon emissions of 1600 $GtCO_2$ will achieve a 2.0 °C change and 860 GtCO2 will achieve a 1.5 °C change, which must be reduced by a further 100 $GtCO_2$ to account for additional Earth system feedback over the twenty-first century. One of the key reasons behind this difference is definitional: how far do we consider that we are away from 1.5 °C warming? While that question seems simple, it is surprisingly complex when the observational data on coverage, the internal variability and the pre-industrial to early-industrial temperature differences are considered.

This study does not resolve the differences in opinions about carbon budgets, but it does provide emission pathways that are consistent with the 1.5 °C target increase in the 1.5 °C Scenario, or with the "well below 2.0 °C" target increase in the 2.0 °C Scenario consistent with other scenarios in the literature and classified as such by the IPCC Special Report on 1.5 °C.

Global Trends in the Energy Sector In 2017, the ongoing trends continued: solar photovoltaics (PV) and wind power dominated the global market for new power plants; the price of renewable energy technologies continued to decline; and fossil fuel prices remained low. A new benchmark was reached, in that the new renewable capacity began to compete favourably with existing fossil fuel power plants in some markets. Electrification of the transport and heating sectors is gaining attention, and although the amount of electrification is currently small, the use of renewable technologies is expected to increase significantly.

The growth of solar PV has been remarkable and is nearly double that of the second-ranking wind power. The capacity of new solar PV in 2017 was greater than the combined increases in the coal, gas and nuclear capacities. Renewable energy technologies achieved a global average generation share of 23% in the year 2015, compared with 18% in the year 2005. Storage is increasingly used in combination with variable renewables as battery costs decline, and solar PV plus storage has started to compete with gas peaking plants. However, bioenergy (including traditional biomass) remains the leading renewable energy source in the heating (buildings and industry) and transport sectors.

Since 2013, global energy-related carbon dioxide (CO_2) emissions from fossil fuels have remained relatively flat. Early estimates based on preliminary data suggest that this changed in 2017, with global CO_2 emissions increasing by around 1.4% (REN21-GSR 2018). These increased emissions were primarily attributable to increased coal consumption in China, which grew by 3.7% in 2017 after a 3-year decline. The increased Chinese consumption, as well as a steady growth of around 4% in India, is expected to lead to an upturn in global coal use, reversing the annual global decline from 2013 to 2016.

In 2017, as in previous years, renewables saw the greatest increases in capacity in the power sector, whereas the growth of renewables in the heating, cooling and transport sectors was comparatively slow. Sector coupling—the interconnection of power, heating and transport and particularly the electrification of heating and transport—is gaining increasing attention as a means of increasing the uptake of renewables in the transport and thermal sectors. Sector coupling also allows the integration of large proportions of variable renewable energy, although this is still at an early stage. For example, China is specifically encouraging the electrification of heating, manufacturing and transport in high-renewable areas, including promoting the use of renewable electricity for heating to reduce the curtailment of wind, solar PV and hydropower. Several US states are examining options for electrification, specifically to increase the overall renewable energy share.

Methodology for Developing Emission Pathways The complete decarbonisation of the global energy supply requires entirely new technical, economic and policy frameworks for the electricity, heating and cooling sectors as well as for the transport system. To develop a global plan, the authors combined various established computer models:

- *Generalized Equal Quantile Walk (GQW)*: This statistical method is used to complement the CO_2 pathways with non-CO_2 regional emissions for relevant greenhouse gases (GHGs) and aerosols, based on a statistical analysis of the large number (~700) of multi-gas emission pathways underlying the recent IPCC Fifth Assessment Report and the recently published IPCC Special Report on 1.5 °C. The GQW method calculates the median non-CO_2 gas emission levels every 5 years—conditional on the energy-related CO_2 emission level percentile of the "source" pathway. This method is a further development under this project—building on an earlier Equal Quantile Walk method—and is now better able to capture the emission dynamics of low-mitigation pathways.
- *Land-based sequestration design*: A Monte Carlo analysis across temperate, boreal, subtropical and tropical regions has been performed based on various literature-based estimates of sequestration rates, sequestration periods and areas available for a number of sequestration options. This approach can be seen as a quantified literature-based synthesis of the potential for land-based CO_2 sequestration, which is not reliant on biomass plus sequestration and storage (bioenergy with carbon capture and storage, BECCS).
- *Carbon cycle and climate modelling (Model for the Assessment of Greenhouse Gas-Induced Climate Change, MAGICC)*: This study uses the MAGICC climate model, which also underlies the classification used by both the IPCC Fifth Assessment Report and the IPCC Special Report on 1.5 °C in terms of the abilities of various scenarios to maintain the temperature change below 2 °C or 1.5 °C. MAGICC is constantly evolving, but its core goes back to the 1980s, and

it represents one of the most established reduced-complexity climate models in the international community.

- *Renewable Resource Assessment [R]E-SPACE*: RE-SPACE is based on a Geographic Information Systems (GIS) approach and provides maps of the solar and wind potentials in space-constrained environments. GIS attempts to emulate processes in the real world at a single point in time or over an extended period (Goodchild 2005). The primary purpose of GIS mapping is to ascertain the renewable energy resources (primarily solar and wind) available in each region. It also provides an overview of the existing electricity infrastructures for fossil fuel and renewable sources.
- *Transport model (TRAEM)*: The transport scenario model allows the representation of long-term transport developments in a consistent and transparent way. The model disaggregates transport into a set of different modes and calculates the final energy demand by multiplying each transport mode's specific transport demand with powertrain-specific energy demands, using a passenger km (pkm) and tonne km (tkm) activity-based bottom-up approach.
- *Energy system model (EM)*: The energy system model (a long-term energy scenario model) is used as a mathematical accounting system for the energy sector. It helps to model the development of energy demands and supply according to the development of drivers and energy intensities, energy potentials, future costs, emission targets, specific fuel consumption and the physical flow between processes. The data available significantly influence the model architecture and approach. The energy system model is used in this study to develop long-term scenarios for the energy system across all sectors (power, heat, transport and industry), without applying cost-optimization based on uncertain cost assumptions. However, an ex-post analysis of costs and investments shows the main economic effects of the pathways.
- *Power system models [R]E 24/7*: Power system models simulate electricity systems on an hourly basis with geographic resolution to assess the requirements for infrastructure, such as the grid connections between different regions and electricity storage, depending on the demand profiles and power-generation characteristics (Teske 2015). High-penetration or renewable energy-only scenarios will contain significant proportions of variable solar PV and wind power because they are inexpensive. Therefore, power system models are required to assess the demand and supply patterns, the efficiency of power generation and the resulting infrastructural needs. Meteorological data, typically in 1 h steps, are required for the power-generation model, and historical solar and wind data were used to calculate the possible renewable power generation. In terms of demand, either historical demand curves were used, or if unavailable, demand curves were calculated based on assumptions of consumer behaviour in the use of electrical equipment and common electrical appliances. Figure 1 provides an overview of the interaction between the energy- and GIS-based models. The climate model is not directly linked with it but provided the carbon budgets for the 2.0 °C and the 1.5 °C Scenarios.

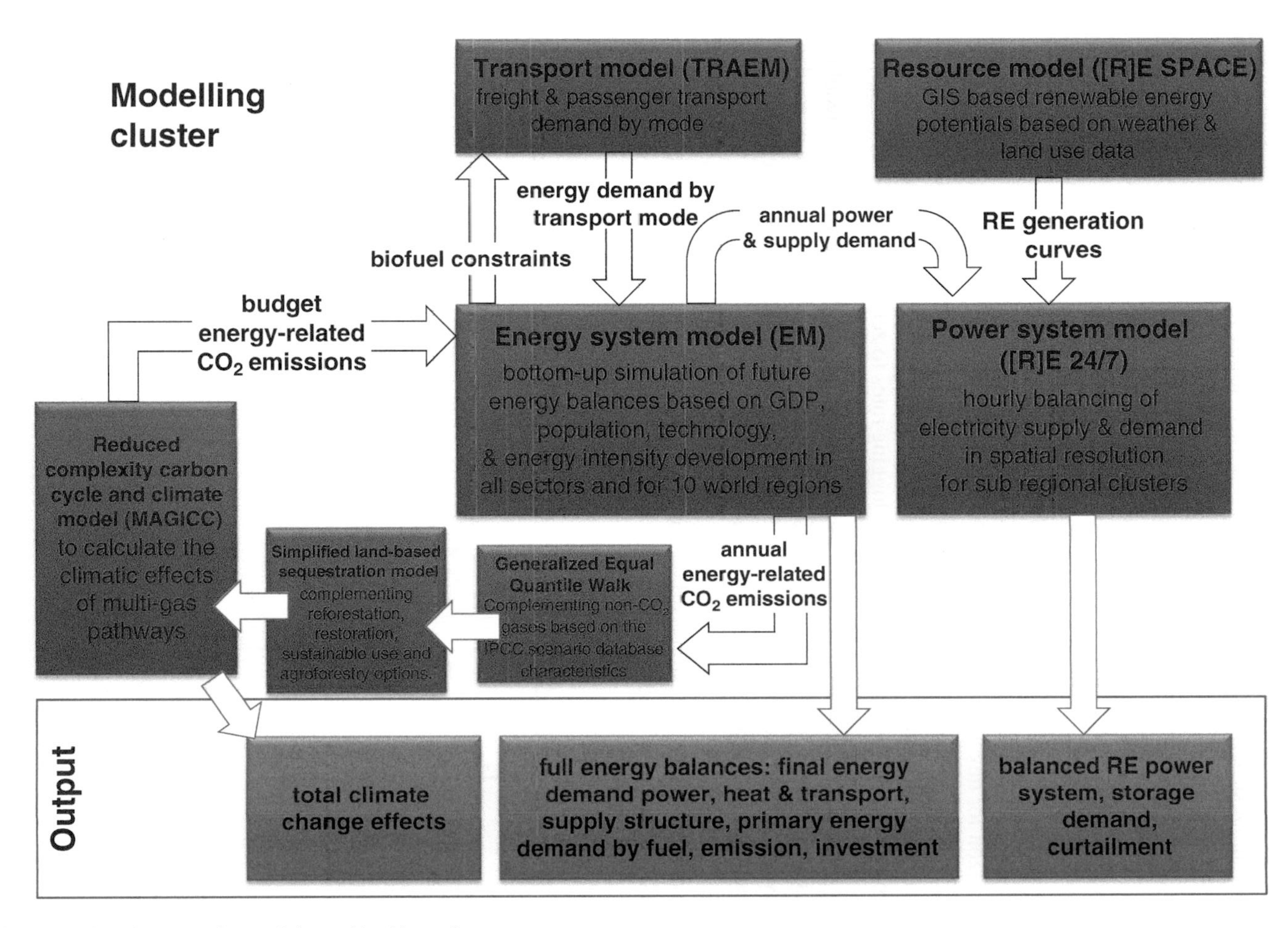

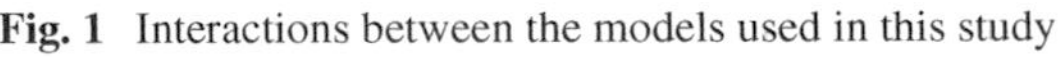
Fig. 1 Interactions between the models used in this study

Besides the climate and energy models, employment effects and the metal resource requirements for selected materials have been calculated. Now that the methodology has been outlined, the next sections present the results and assumptions for the nonenergy GHG mitigation scenarios, followed by the energy sector scenarios

Nonenergy-GHG Mitigation Scenarios The most important sequestration measure could be large-scale reforestation, particularly in the subtropics and tropics (see yellow pathways in Fig. 2). The second most important pathway in terms of the amount of CO_2 sequestered is the sustainable use of existing forests, which basically means reduced logging within those forests. In subtropical, temperate and boreal regions, this could provide substantial additional carbon uptake over time. The time horizon for this sequestration option is assumed to be slightly longer in temperate and boreal regions, consistent with the longer time it takes for these forest ecosystems to reach equilibrium. The "forest ecosystem restoration" pathway is also important, which basically assumes a reduction in logging rates to zero in a fraction of forests.

Overall, the median assumed sequestration pathways, shown in Fig. 2, would result in the sequestration of 151.9 GtC. This is approximately equivalent to all historical land-use-related CO_2 emissions and indicates the substantial challenges that accompany these sequestration pathways.
Given the competing forms of land use throughout the world today, the challenge of reversing overall terrestrial carbon stocks back to pre-industrial levels cannot be underestimated. There would be significant benefits, but also risks, if this

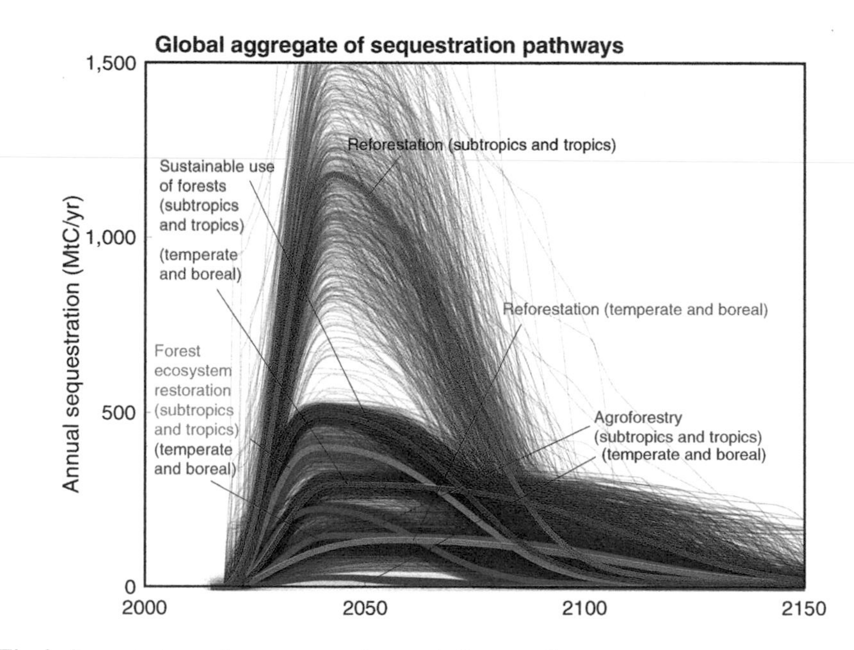

Fig. 2 Sequestration pathways—annual sequestration over time

sequestration option were to be used *instead* of mitigation. However, the benefits are clearly manifold, ranging from biodiversity protection, reduced erosion, improved local climates, protection from wind and potentially reduced air pollution.

Assumptions for Scenarios Scenario studies cannot predict the future. Instead, scenarios describe what is required for a pathway that will limit warming to a certain level and that is feasible in terms of technology implementation and investment. Scenarios also allow us to explore the possible effects of transition processes, such as supply costs and emissions. The energy demand and supply scenarios described in this study have been constructed based on information about current energy structures and today's knowledge of energy resources and the costs involved in deploying them. As far as possible, the study also takes into account potential regional constraints and preferences.

The energy modelling used primarily aims to generate transparent and coherent scenarios, ambitious but still plausible storylines, out of several possible techno-economic pathways. Knowledge integration is the core of this approach because we must consider different technical, economic, environmental and societal factors. Scenario modelling follows a hybrid bottom-up/top-down approach, with no objective cost-optimization functions. The analysis considers key technologies for successful energy transition and focuses on the role and potential utility of efficiency measures and renewable energies. Wind and solar energies have the highest economic potential and dominate the pathways on the supply side. However, the variable renewable power from wind and PV remains limited to a maximum of 65%, because sufficient secured capacity must always be maintained in the electricity system. Therefore, we also consider concentrating solar power (CSP) with high-temperature heat storage as a solar option that promises large-scale dispatchable and secured power generation.

The 5.0 °C Scenario (Reference Scenario): The reference scenario only takes into account existing international energy and environmental policies and is based on the International Energy Agency (IEA) World Energy Outlook (IEA 2017). Its assumptions include, for example, continuing progress in electricity and gas market reforms, the liberalization of cross-border energy trade and recent policies designed to combat environmental pollution. The scenario does not include additional policies to reduce GHG emissions. Because the IEA's projections only extend to 2040, we extrapolate their key macroeconomic and energy indicators forward to 2050. This provides a baseline for comparison with the 2.0 °C and 1.5 °C Scenarios.

The 2.0 °C Scenario: The first alternative scenario aims for an ambitious reduction in GHG emissions to zero by 2050 and a global energy-related CO_2 emission budget of around 590 Gt between 2015 and 2050. This scenario is close to the assumptions and results of the Advanced E[R] scenario published in 2015 by Greenpeace (Teske et al. 2015). However, it includes an updated base year, more coherent regional developments in energy intensity, and reconsidered trajectories and shares of the deployment of renewable energy systems. Compared with the 1.5 °C Scenario, the

2.0 °C Scenario allows for some delays due to political, economic and societal processes and stakeholders.

The 1.5 °C Scenario: The second alternative scenario aims to achieve a global energy-related CO_2 emission budget of around 450 Gt, accumulated between 2015 and 2050. The 1.5 °C Scenario requires immediate action to realize all available options. It is a technical pathway, not a political prognosis. It refers to technically possible measures and options without taking into account societal barriers. Efficiency and renewable potentials need to be deployed even more quickly than in the 2.0 °C Scenario, and avoiding inefficient technologies and behaviours is an essential strategy for developing regions in this scenario.

Global Transport Transport emissions have increased at a rapid rate in recent decades and accounted for 21% of total anthropogenic CO_2 emissions in 2015. The reason for this steady increase in emissions is that passenger and freight transport activities are increasing in all world regions, and there is currently no sign that these increases will slow in the near future. The increasing demand for energy for transport has so far been predominantly met by GHG-emitting fossil fuels. Although (battery) electric mobility has recently surged considerably, it has done so from a very low base, which is why in terms of total numbers, electricity remains an energy carrier with a relatively minor role in the transport sector.

The key results of our transport modelling demonstrate that meeting the 2.0 °C Scenario, and especially the 1.5 °C Scenario, will require profound measures in terms of rapid powertrain electrification and the use of biofuels and synthetically produced fuels to shift transport performance to more efficient modes. This must be accompanied by a general limitation of further pkm and tkm growth in the OECD countries.

The *5.0 °C Scenario* follows the IEA World Energy Outlook (WEO) scenario until 2040, with extrapolation to 2050. Only a minor increase in electrification over all transport modes is assumed, with passenger cars and buses increasing their electric vehicle (EV) shares. For example, this study projects a share of 30% for battery electric vehicles (BEVs) in China by 2050 in response to the foreseeable legislation and technological advancement in that country, whereas for the world car fleet, the share of BEVs is projected to increase to only around 10%. Growth in the shares of electric powertrains and two- and three-wheel vehicles in the commercial road vehicle fleet will be small, as will the rise in further rail electrification. Aviation and navigation (shipping) are assumed to remain fully dependent on conventional kerosene and diesel, respectively.

In the *2.0 °C Scenario* minimal progress in electrification until 2020 will occur, whereas a significant increase in electrification of the transport sector between 2020 and 2030 is projected. This will occur first in OECD regions, followed by emerging economies and finally in developing countries. Battery-driven electric passenger cars are projected to achieve shares of between 21% and 30%, whereas heavy commercial electric vehicles and buses could achieve even higher shares of between 28% and 52% by 2030. This uptake will require a massive build-up of battery

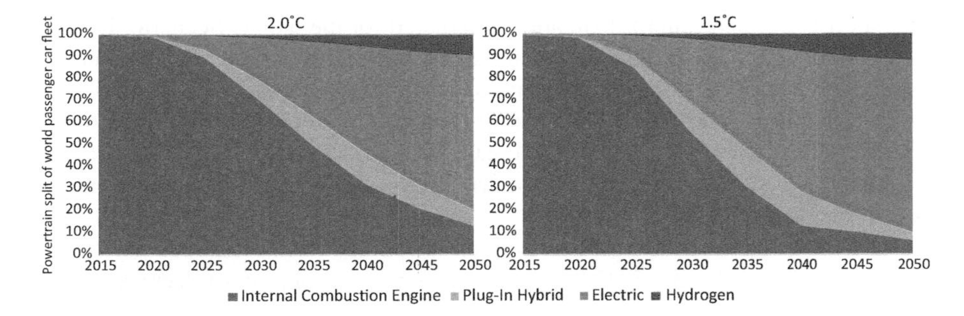

Fig. 3 Powertrain split of the world passenger car fleet in the 2 °C Scenario (left) and 1.5 °C Scenario (right)

production capacity in coming years. Two- and three-wheel vehicles—mainly used in Asia and Africa—will be nearly completely electrified (batteries and fuel cells) by 2030. Looking ahead to 2050, 60–70% of buses and heavy trucks will become (battery-driven) electric, and fuel-cell electric vehicles will increase their market share to around 37%. In the 2.0 °C Scenario, developing countries in Africa and countries in the oil-producing countries of the Middle East will remain predominantly dependent on internal combustion engines, using bio- or synthetic-based fuels.

In the *1.5 °C Scenario*, an earlier and more rapid increase in electric powertrain penetration is required, with the OECD regions at the forefront. The emerging economic regions must also electrify more rapidly than in the 2.0 °C Scenario. On a global level, internal combustion engines will be almost entirely phased out by 2050 in both the 2.0 °C and 1.5 °C Scenarios. In OECD regions, cars with internal combustion engines (using oil-based fuels) will be phased out by 2040, whereas in Latin America or Africa, for example, a small share of internal combustion engine internal combustion engine (ICE)-powered cars, fuelled with biofuels or synthetic fuels, will still be on the road but will be constantly replaced by electric drivetrains (Fig. 3).

Efficiency improvements are modelled across all transport modes until 2050, resulting in improved energy intensity over time. We project an increase in annual efficiency of 0.5–1% in terms of MJ/tonnes km or MJ/passenger km, depending on the transport mode and region. Regardless of the types of powertrains and fuels, increasing the efficiency at the MJ/pkm or MJ/tkm level will result from the following measures:

- Reductions in powertrain losses through more efficient motors, gears, power electronics, etc.
- Reductions in aerodynamic drag
- Reductions in vehicle mass through lightweighting
- The use of smaller vehicles
- Operational improvements (e.g. through automatic train operation, load factor improvements)

Transport performance will increase in all scenarios on a global scale but with different speeds and intensities across modes and world regions. Current trends in transport performance until 2050 are extrapolated for the 5.0 °C Scenario. In relative terms, all transport carriers will increase their performance from the current levels, and in particular, energy-intensive aviation, passenger car transport and commercial road transport are projected to grow strongly. In the 2.0 °C Scenario and 1.5 °C Scenario, we project a strong increase in rail traffic (starting from a relatively low base) and slower growth or even a decline in the use of the other modes in all world regions (Fig. 4).

The modal shifts from domestic aviation to rail and from road to rail are modelled. In the 2.0 °C and 1.5 °C Scenarios, passenger car pkm must decrease in the OECD countries (but increase in the developing world regions) after 2020 in order to maintain the carbon budget. The passenger car pkm decline will be partly compensated by an increase in the performances of other transport modes, specifically public transport rail and bus systems.

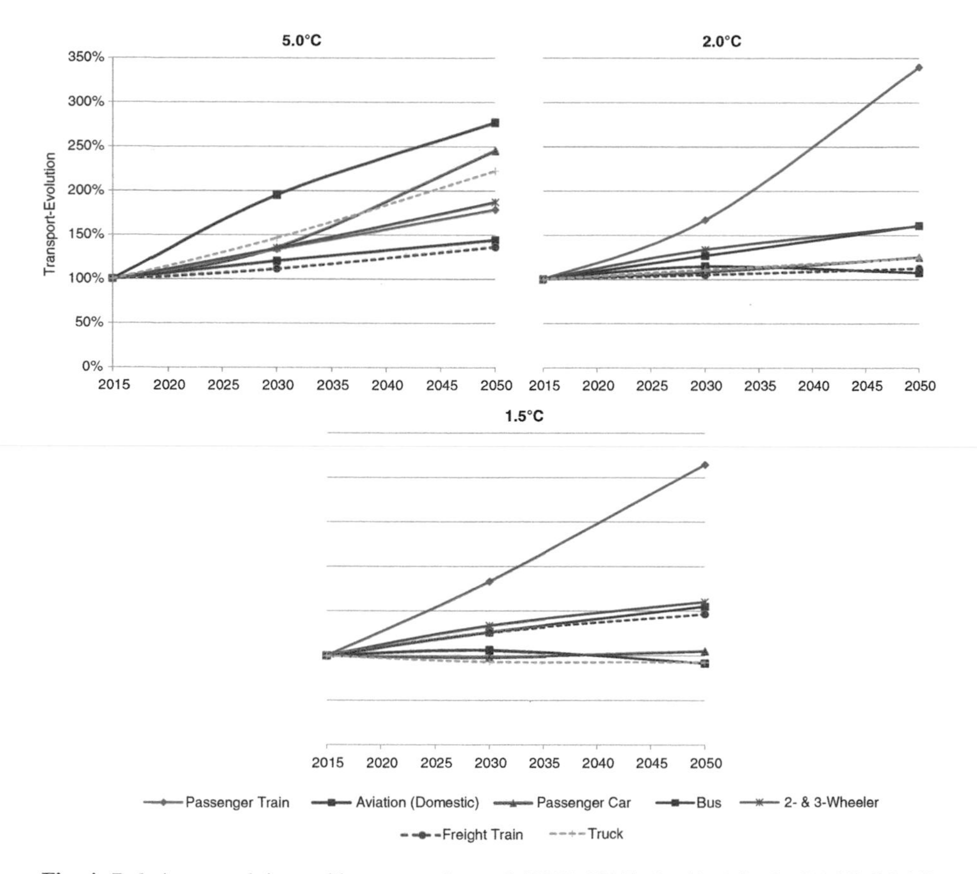

Fig. 4 Relative growth in world transport demand (2015, 100% pkm/tkm) in the 5.0 °C, 2.0 °C and 1.5 °C Scenarios

Global Renewable Energy Potential To develop the 2.0 °C and 1.5 °C Scenarios, the economic renewable energy potential in a space-constrained environment was analysed. Land is a scarce resource. The use of land for nature conservation, agricultural production, residential areas and industry, as well as for infrastructure such as roads and all aspects of human settlement, limits the amount of land available for utility-scale solar and wind projects. Furthermore, solar and wind generation requires favourable climatic conditions, so not all available areas are suitable for renewable power generation. To assess the renewable energy potential based on the area available, all scenario-relevant regions and subregions were analysed with the [R]E-SPACE methodology to quantify the available land area in square kilometres with a defined set of constraints:

- Residential and urban settlements
- Infrastructure for transport (e.g. rail, roads)
- Industrial areas
- Intensive agricultural production land
- Nature conservation areas and national parks
- Wetlands and swamps
- Closed grasslands (as the land-use type)

In addition to this spatial analysis, the remaining available land areas were correlated with the available solar and wind resources. For CSP, a minimum solar radiation of 2000 kilowatt hours per square meter and year (kWh/m^2 year) is assumed to be the minimum deployment criterion, whereas the onshore wind potential under an average annual wind speed of 5 m/s has been omitted.

The 2.0 °C Scenario utilizes only a fraction of the available economic potential of the assumed suitable land for utility-scale solar PV and CSP plants. This estimate does not include solar PV rooftop systems, which have significant additional potential. India has the highest solar utilization rate of 8.5%, followed by Europe and the Middle East, each of which utilizes around 5%. Onshore wind potential has been utilized to a larger extent than solar potential. In the 2.0 °C Scenario, space-constrained India will utilize about half of all the onshore wind energy utilized, followed by Europe, which will utilize one fifth. This wind potential excludes offshore wind, which has significant potential, but mapping the offshore wind potential was beyond the scope of this analysis.

The 1.5 °C Scenario is based on the accelerated deployment of all renewables and the more ambitions implementation of efficiency measures. Thus, the total installed capacity of solar and wind power plants by 2050 is not necessarily larger than it is in the 2.0 °C Scenario, and the utilization rate is in the same order of magnitude. The increased deployment of renewable capacity in the OECD Pacific (Australia), the Middle East and Africa will be due to the production of synthetic bunker fuels based on hydrogen or synthetic fuels (*synfuels*) to supply the global transport energy for international shipping and aviation.

Key results of the global long-term energy scenarios show that the efficiency and uptake of renewable energy are two sides of the same coin. All sectors, including transport, industry and all commercial and residential buildings, must use energy efficiently and from a huge range of renewable energy technologies. Compared with the 5.0 °C Scenario, which was defined using assumptions from the IEA, the alternative scenarios require more stringent efficiency levels. The 1.5 °C Scenario involves the even faster implementation of efficiency measures than in the 2.0 °C Scenario and the decelerated growth of energy services in all regions, in order to avoid a further strong increase in fossil fuel use after 2020.

Global energy intensity will decline from 2.4 MJ/US$GDP in 2015 to 1.25 MJ/US$GDP in 2050 in the 5.0 °C Scenario compared with 0.65 MJ/US$GDP in the 2.0 °C Scenario and 0.59 MJ/US$GDP in the 1.5 °C Scenario. This is a result of the estimated power, heat and fuel demands for all sectors, with more stringent efficiency levels in the alternative scenarios than in the 5.0 °C case. It reflects a further decoupling of the energy demand and gross domestic product (GDP) growth as a prerequisite for the rapid decarbonisation of the global energy system.

Total final energy demand is estimated based on assumptions about the demand drivers, specific energy consumption and the development of energy services in each region. In the 5.0 °C Scenario, the global energy demand will increase by 57% from 342 EJ/year in 2015 to 537 EJ/year in 2050. In the 2.0 °C Scenario, the final energy will be 19% lower than the current consumption and will reach 278 EJ/year by 2050. The final energy demand in the 1.5 °C Scenario will be 253 EJ, 26% below the 2015 demand, and, in 2050, will be 9% lower than in the 2.0 °C Scenario.

Global electricity demand will significantly increase in the alternative scenarios due to the electrification of the transport and heating sectors, which will replace fuels, but will also be due to a moderate increase in the electricity demand of "classical" electrical devices on a global level. In the 2.0 °C Scenario, the electricity demand for heating will be about 12,600 TWh/year from electric heaters and heat pumps, and, in the transport sector, there will be an increase of about 23,400 TWh/year due to electric mobility. The generation of hydrogen (for transport and high-temperature process heat) and the manufacture of synthetic fuels for transport will add an additional power demand of 18,800 TWh/year. The gross power demand will thus rise from 24,300 TWh/year in 2015 to 65,900 TWh/year in 2050 in the 2.0 °C Scenario, 34% higher than in the 5.0 °C Scenario. In the 1.5 °C Scenario, the gross electricity demand will increase to a maximum of 65,300 TWh/year in 2050.

Global electricity generation from renewable energy sources will reach 100% by 2050 in the alternative scenarios. "New" renewables—mainly wind, solar and geothermal energy—will contribute 83% of the total electricity generated. The contribution of renewable electricity to total production will be 62% by 2030 and 88% by 2040. The installed capacity of renewables will reach about 9500 GW by 2030 and 25,600 GW by 2050. The proportion of electricity generated from renewables in 2030 in the 1.5 °C Scenario is assumed to be 73%. The 1.5 °C Scenario will have a generation capacity of renewable energy of about 25,700 GW in 2050.

From 2020 onwards, the continuing growth of wind and PV to 7850 GW and 12,300 GW, respectively, will be complemented by the generation of up to 2060 GW of solar thermal energy as well as limited biomass-derived (770 GW), geothermal

(560 GW) and ocean-derived energy (around 500 GW) in the 2.0 °C Scenario. Both the 2.0 °C and 1.5 °C Scenarios will lead to the generation of high proportions (38% and 46%, respectively) of energy from variable power sources (PV, wind and ocean) by 2030, which will increase to 64% and 65%, respectively, by 2050. This will require a significant change in how the power system is operated. The main findings of the power sector analysis are summarized in the section below.

Calculated average electricity-generation costs in 2015 (referring to full costs) were around 6 ct/kWh. In the 5.0 °C Scenario, these generation costs will increase, assuming rising CO_2 emission costs in the future, until 2050, when they reach 10.6 ct/kWh. The generation costs will increase in the 2.0 °C and 1.5 °C Scenarios until 2030, when they will reach 9 ct/kWh, and then drop to 7 ct/kWh by 2050. In both alternative scenarios, the generation costs will be around 3.5 ct/kWh lower than in the 5.0 °C Scenario by 2050. Note that these estimates of generation costs do not take into account integration costs such as power grid expansion, storage and other load-balancing measures.

Total electricity supply costs in the 5.0 °C Scenario will increase from today's $1560 billion/year to more than $5 500 billion/year in 2050, due to the growth in demand and increasing fossil fuel prices. In both alternative scenarios, the total supply costs will be $5050 billion/year in 2050, about 8% lower than in the 5.0 °C Scenario.

Global investment in power generation between 2015 and 2050 in the 2.0 °C Scenario will be around $49,000 billion, which will include additional power plants to produce hydrogen and synthetic fuels and the plant replacement costs at the end of their economic lifetimes. This value is equivalent to approximately $1360 billion per year on average, which is $28,600 billion more than in the 5.0 °C Scenario ($20,400 billion). An investment of around $51,000 billion for power generation will be required between 2015 and 2050 in the 1.5 °C Scenario ($1420 billion per year on average). In both alternative scenarios, the world will shift almost 95% of its total energy investment to renewables and cogeneration.

Fuel Cost Savings Because renewable energy has no fuel costs other than biomass, the cumulative savings in fuel cost in the 2.0 °C Scenario will reach a total of $26,300 billion in 2050, equivalent to $730 billion per year. Therefore, the total fuel costs in the 2.0 °C Scenario will be equivalent to 90% of the energy investments in the 5.0 °C Scenario. The fuel cost savings in the 1.5 °C Scenario will sum to $28,800 billion or $800 billion per year.

Final energy demand for heating will increase by 59% in the 5.0 °C Scenario from 151 EJ/year in 2015 to around 240 EJ/year in 2050. Energy efficiency measures will help to reduce the energy demand for heating by 36% in 2050 in the 2.0 °C Scenario, relative to that in the 5.0 °C case, and by 40% in the 1.5 °C Scenario.

Global Heat Supply In 2015, renewables supplied around 20% of the final global energy demand for heating, mainly from biomass. Renewable energy will provide 42% of the world's total heat demand in 2030 in the 2.0 °C Scenario and 56% in the 1.5 °C Scenario. In both scenarios, renewables will provide 100% of the total heat demand in 2050. This will include the direct use of electricity for heating, which will increase by a factor of 4.2–4.5 between 2015 and 2050 and will constitute a final share of 26% in 2050 in the 2.0 °C Scenario and 30% in the 1.5 °C Scenario.

Estimated investments in renewable heating technologies to 2050 will amount to more than $13,200 billion in the 2.0 °C Scenario (including investments for plant replacement after their economic lifetimes)—approximately $368 billion per year. The largest share of investment is assumed to be for heat pumps (around $5700 billion), followed by solar collectors and geothermal heat use. The 1.5 °C Scenario assumes an even faster expansion of renewable technologies. However, the lower heat demand (compared with the 2.0 °C Scenario) will result in a lower average annual investment of around $344 billion per year.

Energy demand in the transport sector will increase in the 5.0 °C Scenario from around 97 EJ/year in 2015 by 50% to 146 EJ/year in 2050. In the 2.0 °C Scenario, assumed changes in technical, structural and behavioural factors will reduce this by 66% (96 EJ/year) by 2050 compared with the 5.0 °C Scenario. Additional modal shifts, technological changes and a reduction in the transport demand will lead to even higher energy savings in the 1.5 °C Scenario of 74% (or 108 EJ/year) in 2050 compared with the 5.0 °C case.

Transport Energy Supply By 2030, electricity will provide 12% (2700 TWh/year) of the transport sector's total energy demand in the 2.0 °C Scenario, and, in 2050, this share will be 47% (6500 TWh/year). In 2050, around 8430 PJ/year of hydrogen will be used as a complementary renewable option in the transport sector. In the 1.5 °C Scenario, the annual electricity demand will be about 5200 TWh in 2050. The 1.5 °C Scenario also assumes a hydrogen demand of 6850 PJ/year by 2050. Biofuel use will be limited to a maximum of around 12,000 PJ/year in the 2.0 °C Scenario. Therefore, around 2030, synthetic fuels based on power-to-liquid will be introduced, with a maximum amount of 5820 PJ/year in 2050. Because of the lower overall energy demand in transport, biofuel use will decrease in the 1.5 °C Scenario to a maximum of 10,000 PJ/year. The maximum synthetic fuel demand will amount to 6300 PJ/year.

Global primary energy demand in the 2.0 °C Scenario will decrease by 21% from around 556 EJ/year in 2015 to 439 EJ/year. Compared with the 5.0 °C Scenario, the overall primary energy demand will decrease by 48% by 2050 in the 2.0 °C Scenario (5.0 °C, 837 EJ in 2050). In the 1.5 °C Scenario, the primary energy demand will be even lower (412 EJ) in 2050 because the final energy demand and conversion losses will be lower.

Global Primary Energy Supply Both the 2.0 °C and 1.5 °C Scenarios aim to rapidly phase out coal and oil, after which renewable energy will have a primary energy share of 35% in 2030 and 92% in 2050 in the 2.0 °C Scenario. In the 1.5 °C Scenario, renewables will have a primary share of more than 92% in 2050 (this will include nonenergy consumption, which will still include fossil fuels). Nuclear energy is phased out in both the 2.0 °C and 1.5 °C Scenarios. The cumulative primary energy consumption of natural gas in the 5.0 °C Scenario will sum to 5580 EJ, the cumulative coal consumption will be about 6360 EJ, and the crude oil consumption to 6380 EJ. In the 2.0 °C Scenario, the cumulative gas demand is 3140 EJ, the cumulative coal demand 2340 EJ and the cumulative oil demand 2960 EJ. Even lower fossil fuel use will be achieved under the 1.5 °C Scenario: 2710 EJ for natural gas, 1570 EJ for coal and 2230 EJ for oil. In both alternative scenarios, the primary energy supply in 2050 will be based on 100% renewable energy (Fig. 5).

Bunker Fuels In 2015, the annual bunker fuel consumption was in the order of 16,000 PJ, of which 7400 PJ was for aviation and 8600 PJ for navigation. Annual CO_2 emissions from bunker fuels accounted for 1.3 Gt in 2015, approximately 4% of the global energy-related CO_2 emissions. In the 5.0 °C case, we assume the development of the final energy demand for bunkers according to the IEA World Energy Outlook 2017, Current Policies scenario. This will lead to a further increase in the demand for bunker fuels by 120% until 2050 compared with the base year 2015. Because no substitution with "green" fuels is assumed, CO_2 emissions will rise by the same order of magnitude. Although the use of hydrogen and electricity in aviation is technically feasible (at least for regional transport) and synthetic gas use in navigation is an additional option under discussion, this analysis adopts a conservative approach and assumes that bunker fuels are only replaced by biofuels and synthetic liquid fuels. In the 2.0 °C and 1.5 °C Scenarios, we assume the limited use of sustainable biomass potentials and the complementary central production of power-to-liquid synfuels.

In the 2.0 °C Scenario, this production is assumed to take place in three world regions: Africa, the Middle East and OECD Pacific (especially Australia), where synfuel generation for export is expected to be most economic. The 1.5 °C Scenario requires even faster decarbonisation, so it follows a more ambitious low-energy pathway. The production of synthetic fuels will cause significant additional electricity demand and a corresponding expansion of renewable power-generation capacities. In the case of liquid bunker fuels, these additional renewable

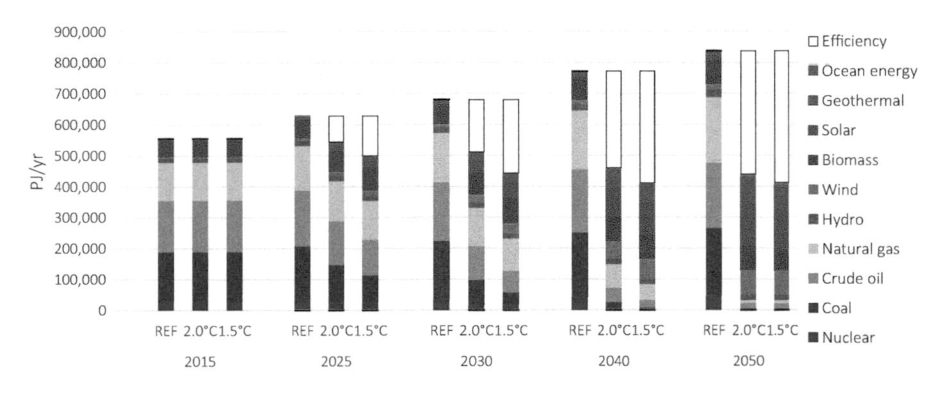

Fig. 5 Global projections of total primary energy demand (PED) by energy carrier in the various scenarios

power-generation capacities could amount to 1100 GW in the 2.0 °C Scenario and more than 1200 GW in the 1.5 °C Scenario if the flexible utilization of 4000 full-load hours per year can be achieved. However, such a scenario requires high electrolyser capacities and high-volume hydrogen storage to ensure not only flexibility in the power system but also high utilization rates by downstream synthesis processes (e.g. via Fischer-Tropsch plants).

Annual global energy-related CO_2 emissions will increase by 40% in the 5.0 °C Scenario, from 31,180 Mt in 2015 to more than 43,500 Mt in 2050. The stringent mitigation measures in both alternative scenarios will cause annual emissions to fall to 7070 Mt in 2040 in the 2.0 °C Scenario and to 2650 Mt in the 1.5 °C Scenario, with further reductions to almost zero by 2050. In the 5.0 °C Scenario, the cumulative CO_2 emissions from 2015 until 2050 will sum to 1388 Gt. In contrast, in the 2.0 °C and 1.5 °C Scenarios, the cumulative emissions for the period from 2015 until 2050 will be 587 Gt and 450 Gt, respectively. Therefore, the cumulative CO_2 emissions will decrease by 58% in the 2.0 °C Scenario and by 68% in the 1.5 °C Scenario compared with the 5.0 °C case. Thus, a rapid reduction in annual emissions will occur in both alternative scenarios.

Global Power Sector Analysis

Global and regional long-term energy results were used to conduct a detailed power sector analysis with the methodology described in Sect. 1.7 of Chap. 3. Both the 2.0 °C and 1.5 °C Scenarios rely on high proportions of variable solar and wind generation. The aim of the power sector analysis was to gain insight into the stability of the power system in each region—subdivided into up to eight subregions—and to gauge the extent to which power grid interconnections, dispatch generation services and storage technologies are required. The results presented in this chapter are projections calculated based on publicly available data. Detailed load curves for some subregions and countries were not available, or, in some cases, the relevant information is classified. Therefore, the outcomes of the [R]E 24/7 model are estimates and require further research with more detailed localized data, especially regarding the available power grid infrastructures. The power sector projections for developing countries, especially in Africa and Asia, assume unilateral access to energy services by the residential sector by 2050 and require transmission and distribution grids in regions where there are none at the time of writing. Further research, in cooperation with local utilities and government representatives, is required to develop a more detailed understanding of the power infrastructure needs.

Development of Global Power Plant Capacities The size of the global market for renewable power plants will increase significantly under the 2.0 °C Scenario. The annual market for solar PV power must increase by a factor of 4.5, from close to 100 GW in 2017 to an average of 454 GW by 2030. The annual onshore wind market must expand to 172 GW by 2025, about three times higher than in 2017. The offshore wind market will continue to increase in importance within the renewable power sector. By 2050, offshore wind installations will increase to 32 GW annually—11 times higher than in 2017. Concentrated solar power (CSP) plants will play an important role in the generation of dispatchable solar electricity to supply bulk power, especially for industry, and to provide secured capacities to power systems. By 2030, the annual CSP market must increase to 78 GW, compared with 3 GW in 2020 and only 0.1 GW in 2017.

In the 1.5 °C Scenario, the phase-out of coal and lignite power plants is accelerated, and a total capacity of 618 GW—equivalent to approximately 515 power stations (1.2 GW on average)—must end operation by 2025. This will mean a phase-out of two coal power plants per week from 2020 onwards, on average. The replacement power will come from a variety of renewable power generators, both variable and dispatchable. The annual market for solar PV energy must be around 30% higher than it was in 2025, as under the 2.0 °C Scenario. The onshore wind market also has an accelerated trajectory under the 1.5 °C Scenario, whereas the offshore wind market is assumed to be almost identical to that in the 2.0 °C Scenario, because of long lead times for these projects. The same is assumed for CSP plants, which are utility-scale projects, and significantly higher deployment seems unlikely in the time remaining until 2025.

Utilization of Power Plant Capacities On a global scale, in the 2.0 °C and 1.5 °C Scenarios, the shares of variable renewable power generation will increase from 4% in 2015 to 38% and 46%, respectively, by 2030, and will increase to 64% and 65%, respectively, by 2050. The reason for the variations in the two cases is the different assumptions made regarding efficiency measures, which may lead to lower overall demand in the 1.5 °C Scenario than in the 2.0 °C Scenario. During the same period, dispatchable renewables—CSP plants, bioenergy generation, geothermal energy and hydropower—will remain around 32% until 2030 on a global average and then decrease slightly to 29% under the 2.0 °C Scenario (and to 27% under the 1.5 °C Scenario) by 2050. The system share of dispatchable conventional generation capacities—mainly coal, oil, gas and nuclear energy—will decrease from a global average of 60% in 2015 to only 14% in 2040. By 2050, the remaining dispatchable conventional gas power plants will be converted to operate on hydrogen as a synthetic fuel, to avoid stranded investments and to achieve higher dispatch power capacities. Increased variable shares—mainly in the USA, the Middle East region and Australia—will produce hydrogen for local and the export markets, as fuel for both renewable power plants and the transport sector.

Development of Maximum and Residual Loads for the Ten World Regions The maximum load will increase in all regions and within similar ranges under both the 2.0 °C and 1.5 °C Scenarios. The load in OECD countries will rise most strongly in response to increased electrification, mainly in the transport sector, whereas the load in developing countries will increase as the overall electricity demand increases in all sectors.

The most significant increase will be in Africa, where the maximum load will surge by 534% over the entire modelling period due to favourable economic development and increased access to energy services by households. In OECD Pacific (South Korea, Japan, Australia and New Zealand), efficiency measures will reduce the maximum load to 87% by 2030 relative to that in the base year, and it will increase to 116% by 2050 with the expansion of electric mobility and the increased electrification of the process heat supply in the industry sector. The 1.5 °C Scenario calculates slightly higher loads in 2030 due to the accelerated electrification of the industry, heating and business sectors, except in three regions (the Middle East, India and Non-OECD Asia Other Asia), where the early application of efficiency measures will lead to an overall lower demand at the end of the modelling period, for the same GDP and population growth rates.

In this analysis, the residual load is the load remaining after the variable renewable power generation. Negative values indicate that the energy generated from solar and wind exceeds the actual load and must be exported to other regions, stored or curtailed. In each region, the average generation should be consistent with the average load. However, maximum loads and maximum generations do not usually occur at the same time, so surplus electricity can be produced and must be exported or stored as far as possible. In rare individual cases, solar- or wind-based generation plants can also temporarily reduce their output to a lower load, or some plants can be shut down. Any reduction in energy generation from solar and wind sources in response to low demands is defined as "curtailment". In this analysis, curtailment rates of up to 5% by 2030 and 10% by 2050 are assumed to have no substantial negative economic impact on the operation of power plants and therefore will not trigger an increase in storage capacities. Figure 6 illustrates the development of maximum loads across all ten world regions under the 2.0 °C and 1.5 °C Scenarios.

Global Storage and Dispatch Capacities The world market for storage and dispatch technologies and services will increase significantly in the 2.0 °C Scenario. The annual market for new hydro-pumped storage plants will grow on average by 6 GW per year to a total capacity of 244 GW in 2030. During the same period, the total installed capacity of batteries will increase to 12 GW, requiring an annual market of 1 GW. Between 2030 and 2050, the energy service sector for storage and storage technologies must accelerate further. The battery market must grow by an annual installation rate of 22 GW and, as a result, will overtake the global cumulative capacity of pumped hydro between 2040 and 2050. The conversion of gas infrastructure from natural gas to hydrogen and synthetic fuels will start slowly between 2020 and 2030, with the conversion of power plants with annual capacities of around 2 GW. However, after 2030, the transformation of the global gas industry to hydrogen will accelerate significantly, with the conversion of a total of 197 GW gas power plants and gas cogeneration facilities each year. In parallel, the average capacity of gas and hydrogen plants will decrease from 29% (2578 h/year) in 2030 to 11% (975 h/year) by 2050, converting the gas sector from a supply-driven to a service-driven industry.

At around 2030, the 1.5 °C Scenario will require more storage throughput than the 2.0 °C Scenario, but the storage demands for the two scenarios will be equal at the end of the modelling period. It is assumed that the higher throughput can be managed with equally higher installed capacities, leading to full-load hours of up to 200 h per year for batteries and hydro-pumped storage.

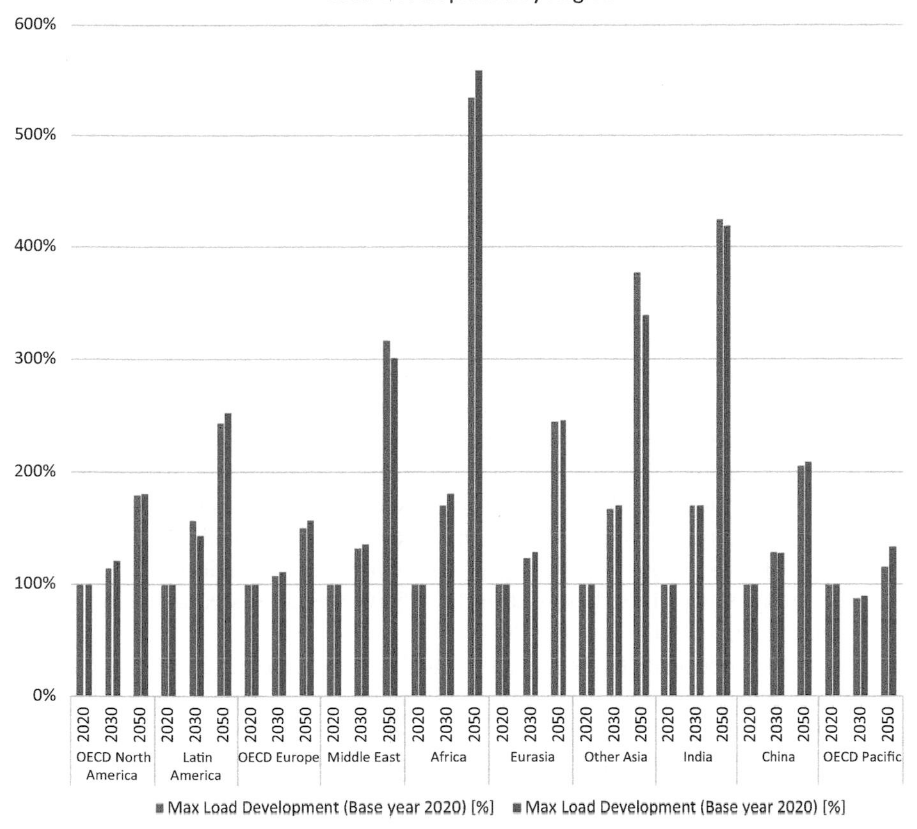

Fig. 6 Development of maximum loads in ten world regions in 2020, 2030 and 2050 under the 2.0 °C and 1.5 °C Scenarios

Trajectories for a Just Transition of the Fossil Fuel Industry The implementation of the 2.0 °C and 1.5 °C Scenarios will have a significant impact on the global fossil fuel industry. While this may appear to be stating the obvious, current climate debates have not yet led to an open debate about the orderly withdrawal from the coal, oil and gas extraction industries. Instead, the political debate about coal, oil and gas is focused on the security of supply and price security. However, mitigating climate change is only possible when fossil fuels are phased out.

Coal: Under the 5.0 °C Scenario, the required production of thermal coal—excluding coal for nonenergy uses, such as steel production—will remain at 2015 levels, with an annual increase of around 1% per year until 2050. Under the 2.0 °C Scenario, coal production will decline sharply between 2020 and 2030 at a rate of around 6% per year. By 2030, global coal production will be equal to China's annual production in 2017, at 3.7 billion tonnes, whereas that volume will be reached in 2025 under the 1.5 °C Scenario.

Oil: Oil production in the 5.0 °C Scenario will grow steadily by 1% annually until the end of the modelling period in 2050. Under the 2.0 °C Scenario, oil production will decline by 3% annually until 2025 and then by 5% per year until 2030. After 2030, oil production will decline by around 7% per year on average, until the oil produced for energy use is phased out entirely by 2050. The oil production capacity of the USA, Saudi Arabia and Russia in 2017 would be sufficient to supply the global demand in 2035 calculated under the 2.0 °C Scenario. The 1.5 °C Scenario reduces the required production volume by half by 2030, reducing it further to the equivalent of the 2017 production volume of just one of the three largest oil producers (USA, Saudi Arabia or Russia) by 2040.

Gas: In the 5.0 °C Scenario, gas production will increase steadily by 2% a year for the next two decades, leading to an overall production increase of about 50% by 2050. Compared with coal and oil, the gas phase-out will be significantly slower in the 2.0 °C and 1.5 °C Scenarios. These scenarios also assume that the gas infrastructure, such as gas pipelines and power plants, will be used afterwards for the hydrogen and/or renewable methane produced with electricity from renewable sources. Under the 2.0 °C Scenario, gas production will only decrease by 0.2% per year until 2025, by 1% until 2030 and, on average, by 4% annually until 2040. This represents a rather slow phase-out and will allow the gas industry to gradually transfer to hydrogen. The phase-out in the 1.5 °C Scenario is equally slow, and a 4%/year reduction will occur after 2025.

The trajectories predicted by the 2.0 °C and 1.5 °C Scenarios for global coal, oil and gas production are consistent with the Paris Agreement targets and can be used to calculate possible employment effects, in terms of job losses in the fossil fuel industry, job gains in the renewable energy industry and options for transitioning the gas industry into an industry based on renewably produced hydrogen.

Employment The transition to a 100% renewable energy system is not just a technical task, it is also a socially and economically challenging process. It is imperative that this transition is managed in a fair and equitable way. One of the key concerns is the employment of workers in the affected industries. However, it should be noted that the "just transition" concept is concerned not only with workers' rights but also with the broader community. This includes considering, for example, community participation in decision-making processes, public dialogue and policy mechanisms that create an enabling environment for new industries to ensure local economic development. Although it is acknowledged that a just transition is important, there are limited data on the effects that this transition will have on employment. There is even less information on the types of occupations that will be affected by the transition, either by project growth or declines in employment. This study provides projections for jobs in construction, manufacturing, operations and maintenance and fuel and heat supply across 12 technologies and 10 world regions, based on the 5.0 °C, 2.0 °C and 1.5 °C Scenarios. Projected employment is calculated regionally, but the results are presented at the global level.

Employment—Quantitative Results The 2.0 °C and 1.5 °C Scenarios will generate more energy-sector jobs in the world as a whole at every stage of the projection. The 1.5 °C Scenario will increase renewable energy capacities faster than the 2.0 °C Scenario, and, therefore, employment will increase faster. By 2050, both scenarios will create around 47 million jobs, so employment will be within similar ranges.

- In 2025, there will be 30.9 million energy-sector jobs under the 5.0 °C Scenario, 45.5 million under the 2.0 °C Scenario and 52.3 million under the 1.5 °C.
- In 2030, there will be 31.7 million energy-sector jobs under the 5.0 °C Scenario, 52.9 million under the 2.0 °C Scenario and 58.5 million under the 1.5 °C Scenario.
- In 2050, there will be 29.9 million energy-sector jobs under the 5.0 °C Scenario, 48.7 million under the 2.0 °C Scenario and 46.3 million under the 1.5 °C Scenario.

Under the 5.0 °C Scenario, job will drop to 4% below the 2015 levels by 2020 and then remain quite stable until 2030. Strong growth in renewable energy will lead to an increase of 44% in total energy-sector jobs by 2025 under the 2.0 °C Scenario and 66% under the 1.5 °C Scenario. In the 2.0 °C (1.5 °C) Scenario, renewable energy jobs will account for 81% (86%) in 2025 and 87% (89%) in 2030, with PV having the greatest share of 24% (26%), followed by biomass, wind and solar heating.

Employment—Occupational Calculations Jobs will increase across all occupations between 2015 and 2025, except in metal trades, which display a minor decline of 2%, as shown in Fig. 7. However, these results are not uniform across regions. For example, China and India will both experience a reduction in the number of jobs for managers and clerical and administrative workers between 2015 and 2025.

Mineral and Metal Requirements Under the 2.0 °C and 1.5 °C Scenarios Within the context of the increasing requirements for metal resources by renewable energy and storage technologies, the rapid increases in demands for both cobalt and lithium are of greatest concern. The demands for both metals will exceed the current production rates by 2023 and 2022, respectively. The demands for these metals will increase more rapidly than will that for silver, partly because solar PV is a more established technology and silver use has become very efficient, whereas the electrification of the transport system and the rapid expansion in lithium battery use have only begun to accelerate in the last few years. The potential to offset primary demand is different depending on the technology. Offsetting demand through secondary sources of cobalt and lithium has the most potential to reduce total primary demand, as these technologies have a shorter lifetime of approximately 10 years. The cumulative demands for both metals will exceed current reserves, but with high recycling rates, they can remain below the resource levels. However, there is a delay in the period during which recycling can offset demand, because there must be sufficient batteries in use and they must exhaust their current purpose before they can be collected and recycled. This delay could be further extended by strategies that reuse vehicular batteries as stationary storage, which might reduce costs in the

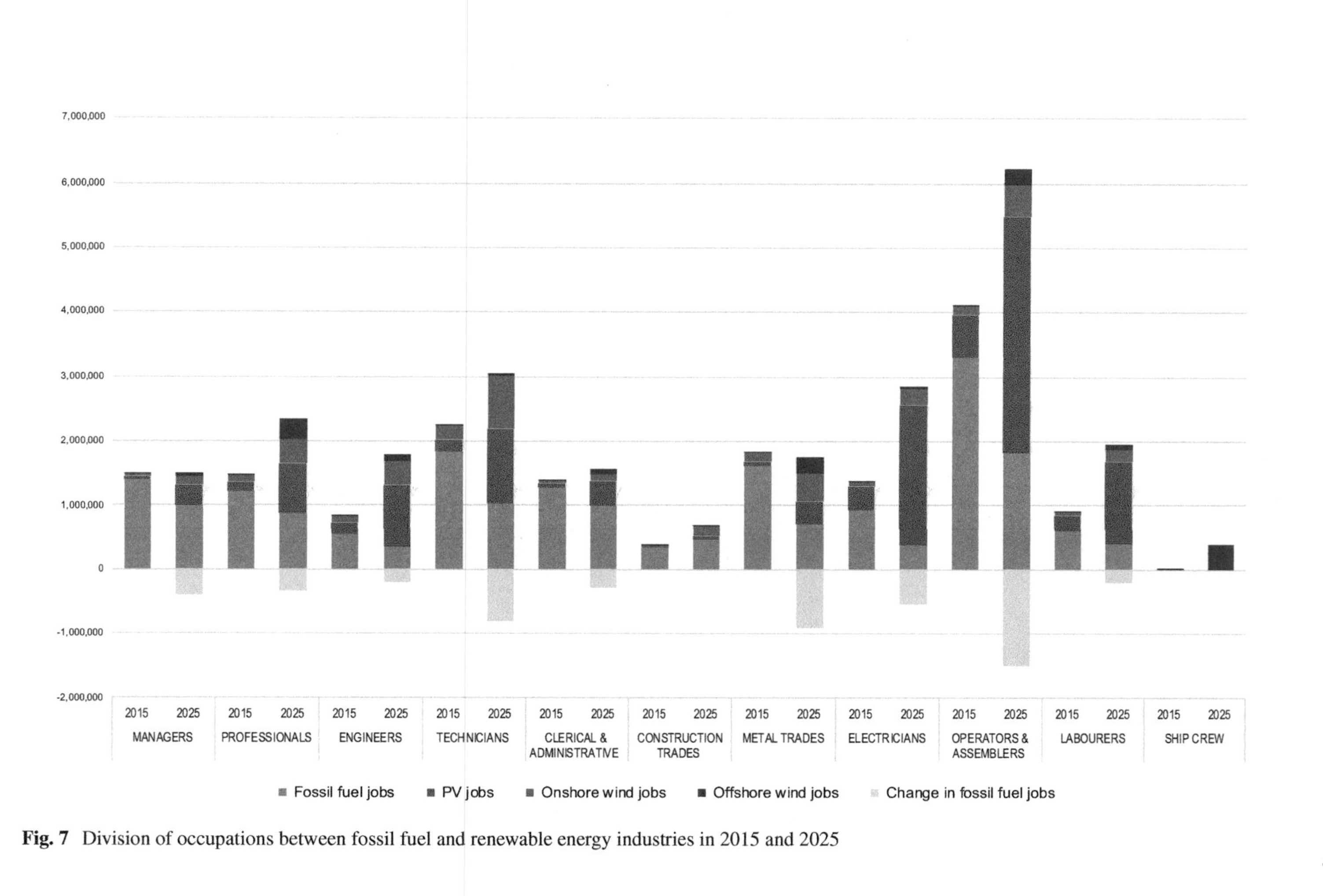

Fig. 7 Division of occupations between fossil fuel and renewable energy industries in 2015 and 2025

short term and increase the uptake of PV. The efficiency of cobalt in batteries also significantly reduces its demand, and this reduction is already happening as manufacturers move towards lower cobalt chemistries.

Increasing the efficiency of the material used is potentially the most successful strategy to offset the demand for PV metals, and recycling will have a smaller impact on demand because the lifespan of solar PV panels is long and their potential for recycling is low. Although the increased demand for silver by 2050 will not be as extreme as that for cobalt or lithium, it will still be considerable. This is important, especially when considering that solar PV currently consumes approximately 9% of end-use silver. It is possible to create silver-less solar panels, but these panels are not expected to be on the market in the near future.

Examination of the Climate Implications of Our Scenarios

One of the Paris Agreement's most outstanding achievements has been the consensus by 195 countries to limit climate change to well below 2 °C and to pursue their best efforts to limit it to 1.5 °C. Together with the goal to reduce emissions to net zero levels, the international agreement clearly sets a framework in which regional and national emission trajectories can be designed and evaluated. The strong focus on a <2.0 °C temperature increase is partly driven by the knowledge that 2 °C warming does not equate to a safe climate: not for small islands that are threatened by rising seas, not for farmers dependent on rainfall in drought-stricken areas and not for communities that are threatened by extreme rainfall events or more intense cyclones.

Here, we use probabilistic methods to examine the scenarios that have been developed to evaluate their implications for long-term temperature and sea-level rises, using models and settings that are also used in the recent IPCC Special Report on 1.5 °C warming. Our lowest scenario has—by design—an approximate 50% or higher chance of a 2100 temperature level that is below 1.5 °C—after a slight overshoot. In contrast to the SSP1_19 scenario, which is the main 1.5 °C-compliant scenario in the next IPCC Assessment Report, our 1.5 °C Scenario does not rely on massive net negative emissions. Even the most stringent mitigation scenarios developed in this study are unable to halt sea-level rise. In fact, a 30 cm rise in sea level by 2100, which will continue thereafter, seems to be the unavoidable legacy of our past use of fossil fuels, unless we remove this CO_2 from the atmosphere in much larger amounts than even the complete reforestation of the planet would permit.

Faced with the grim challenge of ongoing climate risks on the one side and the many positive effects and economic benefits of switching from fossil fuels to renewables on the other, the path is clear. A rapid shift towards a new era of smart, renewable and sector-coupled energy supply, combined with clever demand-side measures and adaptations to the impacts of climate change, will allow us and our children to address the legacy of our past reliance on fossil fuels.

Acknowledgement

The authors would like to thank the Leonardo DiCaprio Foundation (https://www.leonardodicaprio.org) which funded the research for Chaps 1, 2, 3, 4, 5, 6, 7, 8, and 9 and Chap. 12. Their ongoing support and dedication to this project was key and kept all researchers highly motivated.

Furthermore, we thank the German Greenpeace Foundation "Umweltstiftung Greenpeace" (https://umweltstiftung-greenpeace.de/die-stiftung) for funding the employment calculation research documented in Chap. 10. Last but not least, our thanks also to Earthworks (earthworks.org) for funding the research about metal requirements presented in Chap. 11.

This project has been supported by numerous people between July 2017 and November 2018, and our thanks go to each of them. A special thanks to Karl Burkart (Leonardo DiCaprio Foundation); Melanie Stoehr and Claudia Voigt (Umweltstiftung Greenpeace); Payal Sampat (Earthworks); Anna Leidreiter, Anna Skowron and Rob van Riet from the World Future Council (https://www.worldfuturecouncil.org/); Dr. Joachim Fuenfgelt from Bread for the World (https://www.brot-fuer-die-welt.de/en/bread-for-the-world); and Stefan Schurig from F20—Foundations 20 (http://www.foundations-20.org/) who provided initial support to make this project possible. Finally, we would like to thank Greenpeace International and Greenpeace Germany for their ongoing support of the Energy [R]evolution energy scenario research series between 2004 and 2015 which resulted in the development of the long-term energy scenario model, the basis for the long-term energy pathways.

Contents

List of Figures

List of Tables

Chapter 1
Introduction

Sven Teske and Thomas Pregger

Abstract Brief introduction to the UNFCCC Paris Agreement and its main goals, followed by the project background, motivation and objectives. Presentation of the specific research questions for the energy and climate scenario development. Short overview of published 100% renewable energy scenarios and the main differences between those scenarios and the newly developed 1.5 °C and 2.0 °C scenarios presented in the book. Overview about the basic assumptions in regard to technology preferences in future energy pathways. Discussion of the advantages and limitations of scenarios in the energy and climate debate.

UNFCCC Paris Agreement, Article 2:

1. *This Agreement, in enhancing the implementation of the Convention, including its objective, aims to strengthen the global response to the threat of climate change,in the context of sustainable development and efforts to eradicate poverty, including by:*
   (a) *Holding the increase in the global average temperature to well below 2.0 °C above pre-industrial levels and pursuing efforts to limit the temperature increase to 1.5 °C above pre-industrial levels, recognizing that this would significantly reduce the risks and impacts of climate change;*
   (b) *Increasing the ability to adapt to the adverse impacts of climate change and foster climate resilience and low greenhouse gas emission development, in a manner that does not threaten food production; and*

(continued)

S. Teske (✉)
Institute for Sustainable Futures, University of Technology Sydney, Sydney, NSW, Australia
e-mail: sven.teske@uts.edu.au

T. Pregger
Department of Energy Systems Analysis, German Aerospace Center (DLR), Institute for Engineering Thermodynamics (TT), Pfaffenwaldring, Germany
e-mail: thomas.pregger@dlr.de

S. Teske (ed.), *Achieving the Paris Climate Agreement Goals*,
https://doi.org/10.1007/978-3-030-05843-2_1

(c) *Making finance flows consistent with a pathway towards low greenhouse gas emissions and climate-resilient development.*

2. *This Agreement will be implemented to reflect equity and the principle of common but differentiated responsibilities and respective capabilities, in the light of different national circumstances.*

The Paris Climate Agreement aims to hold global warming to well below 2.0 °C and to "pursue efforts" to limit it to 1.5 °C. To accomplish this, countries have submitted Intended Nationally Determined Contributions (INDCs) outlining their post-2020 climate actions (Rogelj et al. 2016). The aim of this research is to develop practical pathways to achieve the Paris climate goals in an economically feasible and sustainable matter.

The study described in this book focuses on changing the ways in which humans produce energy, because energy-related CO_2 emissions are the main driver of climate change. The analysis also considers the developmental pathways of non-energy-related emissions and mitigation measures because it is essential to address their contributions if we are to achieve the Paris climate change targets. The analysis considers options or 'scenarios' for the transition to net zero emissions across all sectors that allow unnecessary techno-economic, societal, and environmental risks to be avoided.

Scenario studies are an important way of linking expected or assumed anthropogenic activities and their resulting emissions with environmental effects, such as global warming. They also provide important insights into these techno-economic, societal, and political options and their various effects. Therefore, they are widely used to analyse possible carbon emission pathways, to guide decision-makers, and to motivate or justify interventions and developments. However, comprehensive, transparent, and robust results and conclusions are required as the bases for such decision-making. Ideally, this information will come from scenario studies that investigate a broad range of possible conditions and available options. Such studies must adopt a holistic approach and integrate comprehensive state-of-the-art background knowledge, including about the impacts of sectoral and technological changes, the influence of market developments, and the effects of certain pathways.

Existing global scenario studies do not provide a comprehensive view of the possible development pathways and technological options required to achieve these ambitious climate targets. Each study usually provides a few selected pathways, representing a narrow range of possible energy futures. One reason for this is that most scenario models are based on objective cost-optimizing functions, which overemphasize the cost efficiency based on uncertain cost assumptions. Another reason is that disruptive developments are not usually considered in scenario narratives. The history of scenario-based systems analysis is littered with many examples of misleading and fallacious 'optimized' scenario pathways and derived policy recommendations (see e.g., Mai et al. 2013; Mohn 2016).

Furthermore, in most existing 2.0 °C and 1.5 °C scenarios, achievement of the climate targets is based on technologies that have significant, and to some extent unknown, disadvantages. These technologies include nuclear power generation, carbon capture and sequestration, and geoengineering (see e.g., Rogelj et al. 2018; Kriegler et al. 2015). Such scenarios involve considerable risk. Moreover, the reader is usually given only limited access to the model assumptions and results, and therefore has limited information about the transparency and traceability of the factors that influence these model-based analyses and the conclusions drawn from them.

The primary objective of this report is to provide a holistic picture of what will be involved in the transition to 100% renewable energy. This report examines power, heat, and fuel supplies on a global scale. Its main focus is on the role of efficiency and renewable energies. We aim to contribute a different and complementary view of the global transition to renewable energy. We provide two exemplary development pathways for each of the 10 regions of the world. We consider both pathways to be achievable, based on the current state of knowledge, and both are consistent with the "well below 2.0 °C" climate target.

In addition to scenario building, we assess the major economic and infrastructural implications of the two pathways in comparison with a 5.0 °C 'reference' scenario based on the International Energy Agency (IEA)'s Current Policies scenario published in the *Word Energy Outlook 2017* (IEA-WEO 2017). We do not claim that our scenarios are optimal with regard to the economy or society. We want to provide a transparent basis for the further concretization and development of energy system transformation, and to demonstrate the enormous challenges we face and the need for action. In contrast to most other studies, we have excluded options with large uncertainties about the economic, societal, and environmental risks associated with technologies such as nuclear power, unsustainable biomass use, CCS, and geoengineering.

Another important objective is to combine bottom-up energy scenarios with non-energy greenhouse gas (GHG)-mitigation scenarios to construct a complete picture of possible climate mitigation pathways and the contributions of the illustrated strategies to achieving the Paris targets. Land-use changes and emissions of other GHGs and aerosols are the focus of this analysis. Finally, GHG concentrations, radiative forcing, and the implications of global mean temperature and sea-level rises are modelled by applying a feasible model with reduced complexity, which is frequently used for integrated assessment models, as the climate model.

Any scenario building on a global scale must severely simplify the complex transition processes and their interrelations. The introduction of different new technologies occurs under very different conditions and at different scales, and an in-depth analysis is required in each case to identify the optimal or feasible solutions. Global governance will also be required for the fast and deep decarbonisation of the world's energy systems, especially in relation to carbon pricing and efficiency standards.

However, all perspectives need a common understanding of what is required to meet the ambitious Paris climate targets. We believe that the results of this study will contribute to such a common understanding and will demonstrate how urgent is the need to act.

The 2.0 °C Scenario represents a far more likely pathway than the 1.5 °C Scenario. Whereas the 2.0 °C Scenario takes into account unavoidable delays due to political, economic, and societal processes and stakeholders, the 1.5 °C Scenario requires immediate action. Under the 1.5 °C Scenario, efficiency measures and renewable energy options must be deployed, and the further development of energy services must be limited and constrained. Furthermore, for the 1.5 °C Scenario to be achievable, it will be essential for developing countries to avoid inefficient technologies and behaviours.

References

IEA-WEO (2017), International Energy Agency, World Energy Outlook 2017, OECD Publishing, Paris/International Energy Agency, Paris, https://doi.org/10.1787/weo-2017-en.

Mai, T.; Logan, J.; Blair, N.; Sullivan, P.; Bazilian, M. (2013): RE-ASSUME- A Decision Maker's Guide to Evaluating Energy Scenarios, Modeling, and Assumptions. National Renewable Energy Laboratory, Golden, CO (USA). URL: http://iea-retd.org/wp-content/uploads/2013/07/RE-ASSUME_IEA-RETD_2013.pdf (accessed on July 27th, 2015).

Mohn, K. (2016) Undressing the emperor: A critical review of IEA's WEO. University of Stavanger Business School, Norwegian School of Economics. Available at: http://www1.uis.no/ansatt/odegaard/uis_wps_econ_fin/uis_wps_2016_06_mohn.pdf

Rogelj J, den Elzen M, Höhne M, Franzen T, Fekete H, Winkler H, Schaeffer R, Sha F, et al (2016) Paris Agreement climate proposals need a boost to keep warming well below 2 °C. Nature 534: 631–639. DOI:https://doi.org/10.1038/nature18307

Rogelj J, Popp A, Calvin KV, Luderer G, Emmerling J, Gernaat D, Fujimori S, Strefler J, Hasegawa T, Marangoni G, Krey V, Kriegler E, Riahi K, van Vuuren DP, Doelman J, Drouet L, Edmonds J, Fricko O, Harmsen M, Havlik P, Humpenöder F, Stehfest E, Tavoni M (2018) Scenarios towards limiting global mean temperature increase below 1.5 °C. Nature Climate Change 8 (4): 325–332. DOI:https://doi.org/10.1038/s41558-018-0091-3

Kriegler E, Riahi K, Bauer N, Schwanitz VJ, Petermann N, Bosetti V, et al (2015) Making or breaking climate targets: The AMPERE study on staged accession scenarios for climate policy. Technological Forecasting & Social Change 90 (2015) 24–44

Chapter 2
State of Research

Sven Teske, Malte Meinshausen, and Kate Dooley

Abstract This chapter sets the context for the climate and energy scenario development. The first part summarizes the scientific status quo of climate change research and explains how the global climate has changed over recent decades and the likely outcomes if we continue with business as usual and fail to drastically reduce GHG emissions.

The second part reviews the development of the global energy markets during the past decade. Trends in the power-, transport- and heating sector in regard to technologies and investments are provided for the year of writing (2018). The developments put the energy scenarios presented in the following chapters into a global context.

2.1 Scientific Status Quo of Climate Change Research

A summary of the latest scientific publications explains how the global climate has changed in recent decades and the likely outcomes if we continue with 'business as usual' and fail to drastically reduce greenhouse gas emissions.

2.1.1 Basics of Climate Change and Radiative Forcing

The Earth's current climate is the result of a delicate balance between incoming short-wave solar radiation and outgoing long-wave radiation that moves back to space. Roughly half (165 W/m^2) the incoming short-wave radiation (340 W/m^2) reaches the

S. Teske (✉)
Institute for Sustainable Futures, University of Technology Sydney, Sydney, NSW, Australia
e-mail: sven.teske@uts.edu.au

M. Meinshausen · K. Dooley
Australian-German Climate and Energy College, University of Melbourne,
Parkville, Victoria, Australia
e-mail: malte.meinshausen@unimelb.edu.au; kate.dooley@unimelb.edu.au

S. Teske (ed.), *Achieving the Paris Climate Agreement Goals*,
https://doi.org/10.1007/978-3-030-05843-2_2

surface of the Earth. The rest is reflected back to space by clouds, aerosol particles, or scattering gases, such as N_2 and O_2 (100 W/m^2) or is absorbed by the troposphere (75 W/m^2) (Stephens et al. 2012). With the exception of clouds and aerosols, this window to incoming solar radiation onto Earth's surface is relatively transparent, so that most of the Sun's energy that comes towards Earth is absorbed either in the atmosphere or on the Earth's surface. This window is our basic heating engine, and the incoming radiation that is our energy source is somewhat dimmed by aerosol emissions and by changes in the amount of sunlight reflected back to space by changes in land use. However, these are not humanity's greatest influences on the Earth's climate.

There is a second window for radiation, through which outgoing long-wave radiation passes, and we are much more effectively closing this window than the one for incoming solar radiation. We can imagine that this second atmospheric window already has a thick curtain across it, largely formed by water vapour. That curtain prevents surface long-wave radiation from going directly to space. In other words, that curtain acts like a thick blanket. The long-wave radiation that the Earth loses back into space originates not from its surface but from much higher atmospheric layers, which tend to be colder. Without absorbers of long-wave radiation in the atmosphere, the Earth's surface would be much colder and inhospitable to humans.

To the parts of the outgoing radiation window that are not yet covered by water vapour, humanity is now adding more layers of absorbers of other long-wave radiation in the form of GHGs. By adding an assortment of GHGs, we are thickening the existing curtain and closing the curtain further across the long-wave radiation window. Compared with the overall incoming solar radiation at 340 W/m^2, the 'curtain' generated by human-induced increases in the concentrations of long-lived greenhouse gases (CO_2, CH_4, halocarbons, N_2O and fluorinated gases) appears to be of little importance, as it "only" amounts to 2.83 W/m^2. The addition and subtraction of many other smaller human influences results in a slightly reduced net current (year 2011) forcing of 2.29 W/m^2.

2.1.1.1 Anthropogenic Contribution

Beyond reasonable doubt, climate change over the last 250 years has been driven by anthropogenic activities. In fact, the human-induced release of GHGs into the atmosphere has the potential to cause more than 100% of the currently observed climate change. The reason that climate change is not even greater than it is that some human-induced changes mask some of the warming attributable to elevated GHG concentrations. This masking effect arises from the emission of cooling aerosols and changes in land use that increase the reflectivity of the Earth's surface.

2.1.1.2 Carbon Budget and Future Warming

Although the anthropogenic contribution to climate change occurs via a large set of GHG emissions (the current emission scenarios that feed into CMIP6 include 43 GHG emission species), and multiple aerosols and land-use changes, there is one dominant influence: carbon dioxide (CO_2) emissions.

It is not only the magnitude of the anthropogenic emissions of CO_2 that makes it such a significant driver of human-induced climate change. There is also an inherent difference between CO_2 and almost all other GHGs and aerosols. Over the time scales of interest here, CO_2 does not have a finite lifetime in the atmosphere. All other gases react chemically, become photo-dissociated in the stratosphere, or are, for example, consumed by the bacteria in soils. However, once CO_2 is released from the near-permanent carbon pools of fossil fuel reservoirs, it only travels between a set of 'active' carbon pools. These active pools are the land biosphere, the ocean, and the atmosphere. Therefore, if CO_2 is added to one, the level in all three pools will rise and over time, a new, higher equilibrium concentration is reached. For example, whereas CO_2 is consumed by plants during the photosynthesis process and then built into plant tissue as carbon, this same carbon is released again as CO_2 when forests burn, when organic matter in the soil decomposes, and when humans and other animals oxidize the food they eat. Therefore, a kilogram of CO_2 emissions will increase the atmospheric CO_2 concentration for hundreds or even thousands of years. Initially, the average CO_2 concentration will shoot up by that kilogram, and then drop relatively quickly again before a new equilibrium is slowly re-established by the redistribution of carbon into the land biosphere and the ocean.

The IPCC Fifth Assessment Report highlighted this key difference between CO_2 and other GHGs. The airborne fraction of CO_2 emissions diminishes over time, as for other GHGs. However, the airborne CO_2 fraction does not decline to zero over 100 years, 1000 years, or even longer periods. Furthermore, carbon-cycle feedback mechanisms mean that higher CO_2 concentrations cause more carbon to remain in the atmosphere. Acting in the other direction, any extra amount of CO_2 in the atmosphere will have less and less effect on radiative forcing, i.e., how much each CO_2 molecule contributes to the warming of the planet. These factors act in concert with another feature of the climate system: the Earth's inertia to warming. When the thermostat of your kitchen oven is set to 220 °C, the oven will take a while to heat to that level. The situation is much the same with the Earth's climate. The IPCC Fifth Assessment Report notes that three effects (the carbon cycle and its feedbacks, the saturation effect of forcing, and the delayed response of the atmosphere to warming) combine to create what is almost a stepwise function in the warming caused by CO_2 emissions. In other words, every extra kilogram of CO_2 produces a slightly greater increase in temperature than the preceding kilogram, and the warming effect is much the same 10 years after the emission of that kilogram as it is after 100 or 500 years. Over time, less of the CO_2 will remain in the atmosphere, but the Earth's inertia will still cause the temperature to reflect the extra warmth arising from the initial emission.

This feature of the Earth's warming and the carbon cycle can be exploited to derive a very simple linear relationship between cumulative carbon emissions and warming. In fact, the resultant warming is a simple function of the sum of all the CO_2 that has ever been emitted, largely independent of when a certain amount of CO_2 was emitted in the past. Based on this understanding, we can compute the carbon budgets for specific levels of warming. As a complication, of course, an unknown amount of warming arises in response to other GHG emissions and aerosols. When deriving carbon budgets, this extra level of warming is normally derived from a range of future emission scenarios. Therefore, the ultimate level of warming is the sum of the linear CO_2-induced warming level (often described as the 'transient climate response to cumulative emissions of carbon') and a smaller and somewhat uncertain contribution that depends on the other GHGs and aerosols.

2.1.2 Carbon Budgets for 1.5 °C and 2.0 °C Warming

The IPCC Fifth Assessment Report used the results of earth system models to derive its carbon budgets. Earth system models are the most complex computer models we have of how the Earth, its atmosphere, oceans, and vegetation are interacting with each other. The IPCC investigated the amounts of cumulative carbon emissions (in a multi-gas world) that would be consistent with, for example, temperatures maintained below either 1.5 °C or 2.0 °C higher than to pre-industrial levels. The recently published IPCC Special Report on the 1.5 °C degree target cites different carbon budget numbers, depending on whether a low estimate of historical temperatures is assumed or surface air temperatures are consistently applied. Therefore, there is some complexity and uncertainty around the carbon budget, which is related to the fact that different interpretations can be made about how far we are still away from the 1.5 °C target (for example). If we assume that we are still 0.63 °C away from 1.5 °C warming (a very optimistic estimate, which is unfortunately based on a too optimistic account of historical emissions), from January 2018 onwards, we can still emit 1320 $GtCO_2$ before reaching 2.0 °C warming (66% chance) and 770 $GtCO_2$ before reaching 1.5 °C warming (50% chance). These figures must be reduced by a further 100 $GtCO_2$ to account for the additional Earth system feedback that has occurred over the twenty-first century. However, when a more realistic measure of historical temperature evolution is used (i.e., calculated by consistently using proxies for surface air temperatures over the land and ocean, rather than by mixing ocean surface water temperatures with air temperatures over land), the carbon budgets are no longer very high. Specifically, the carbon budget required to maintain the Earth's temperature below 2.0 °C, with 66% probability, then decreases to 1170 GtCO2 from 1 January 2018 onwards and to 560 GtCO2 for a 50% chance of staying below 1.5 °C (before the extra 100 GtCO2 that must be subtracted for additional Earth system feedbacks is taken into account; see Table 2.2 in the IPCC Special Report on 1.5 °C warming).

The substantial difference between these two sets of figures (and also their differences from earlier IPCC estimates of the carbon budget) is the focus of a current intense scientific debate, which is unlikely to be settled in the next few months (see e.g., Schurer et al. 2017, 2018; Hawkins et al. 2017). As already mentioned, one of the key factors in deriving a carbon budget is the estimated current level of warming. Among other reasons, the recent warming 'hiatus' in temperature may explain why the recently derived carbon budgets are more relaxed than expected. When the 'hiatus' period substantially influences the level of warming that is taken as a starting point, the ‘distance’ from 1.5 °C and 2.0 °C might seem larger than it is. The recent upswing in the global mean temperatures gives us an idea of how much natural variability is superimposed on the long-term warming trend. Other points for discussion in the determination of carbon budgets include the pre-industrial warming level and the already-mentioned amount of warming induced by non-CO_2 gases.

This study does not aim to resolve the differences in opinions about carbon budgets, but it does provide emissions pathways that can be considered to be consistent with both a target level of 1.5 °C warming in the case of the 1.5 °C scenario, and with a “well-below 2.0 °C” target level in the case of the 2.0 °C scenario.

Whatever the precise carbon budget, recent effects of climate change provide another set of stark reminders that it is more urgent than ever to replace fossil fuels. If we wait for the wild-fire seasons that will occur at global warming levels of 1.5 °C or 2.0 °C, with intensified droughts or ever-more intense hurricane, it might be much too late to avoid their widespread catastrophic impacts. Even at 1.5 °C warming, there is a risk that the continuous melting of the Greenland ice sheet will cause sea levels to rise by meters over the coming centuries. Fossil fuels have undoubtedly allowed great growth in prosperity across the globe, but their replacement with the cleaner, cheaper and emission-free technologies that are available today is overdue.

2.2 Development of Energy Markets—Past and Present

Renewable energy technologies have been developing rapidly since the beginning of the century, and they have emerged from niche markets to become mainstream. This section provides an overview of the development of renewable energy in the power, heating, and transport sectors up to the year of writing (2018). These developments will put the energy scenarios presented in the following chapters into a global context. The research and data in Sect. 2.2.1 are based on the REN21 Renewables 2018—Global Status Report Renewables.

2.2.1 Global Trends in Renewable Energy in 2018

In 2017, ongoing trends continued: solar photovoltaics (PV) and wind power dominated the global market for new power plants, the price of renewable energy technologies continued to decline, and fossil fuel prices remained low. A new benchmark was reached, in that the new renewable capacity began to compete favourably with existing fossil fuel power plants in some markets (Malik 2017). Electrification of the transport and heating sectors is gaining attention, and although the amount of electrification is currently small, the use of renewable technologies is expected to increase significantly.

The growth in solar PV has been remarkable, nearly double that of second-ranking wind power, and the capacity of new solar PV is greater than the combined increases in the coal, gas, and nuclear capacities (FS-UNEP 2018). Storage is increasingly used in combination with variable renewables as battery costs decline, and solar PV plus storage has started to compete with gas peaking plants (Carroll 2018). However, bioenergy (including traditional biomass) remains the leading renewable energy source in the heating (buildings and industry) and transport sectors.

Renewable energy's share of the total final energy consumption has increased only modestly in recent years, despite tremendous growth in the modern renewable energy sector. There are two main reasons for this. One is that the growth in the overall energy demand (except for the drop in 2009 after the global economic recession) has counteracted the strong forward momentum of modern renewable energy technologies. The other is the declining share of traditional biomass, as people switch to other forms of energy. Traditional biomass makes up nearly half of all renewable energy used, and its use has increased at a rate lower than the growth in total energy consumption.

Since 2013, the global energy-related CO_2 emissions from fossil fuels have remained relatively flat. Early estimates based on preliminary data suggest that this changed in 2017, with global CO_2 emissions growing by around 1.4% (REN21-GSR 2018). These increased emissions were primarily due to increased coal consumption in China, which grew by 3.7% in 2017 after a 3-year decline (ENERDATA 2018). This increased Chinese consumption, as well as steady growth of around 4% in India, is expected to lead to an upturn in global coal use, reversing the annual global decline observed from 2013 to 2016 (ENERDATA 2018).

In contrast to the upturn in global coal use, in 2017, 26 countries joined the Powering Past Coal Alliance, which is committed to phasing-out coal power by 2030, with new pledges from Angola, Denmark, Italy, Mexico, New Zealand, and the United Kingdom (Carrington 2017). An increasing number of companies who owned, developed or operated coal power plants have moved away from the coal business (Shearer 2017). Also in 2017, 26 of 28 European Union member states signed an agreement to build no more coal-fired power plants from 2020 onwards, and the Port of Amsterdam, which currently handles 16 million tonnes of coal per year, announced plans to become coal-free by 2030 (Campbell 2017).

The global oil price averaged USD 52.5 per barrel in 2017, equivalent to about half the record high prices that occurred between 2011 and 2014, although it was still almost double the prices from 1996 to 2005 (Statista 2018). Natural gas prices fell from 2013 to 2016, and early indicators suggest that prices remained low or decreased further in 2017 (BP 2017). Low fossil fuel prices have challenged renewable energy markets, especially in the heating and transport sectors (IEA-RE 2016).

The value of direct global fossil fuel consumption subsidies in 2016 was estimated to be about USD 360 billion, a 15% reduction since 2015—but still more than 20% higher than the total renewable industry turnover in 2017 (IEA-WEB

2018). The value of fossil fuel subsidies also increases by an order of magnitude if externalities are considered (IMF 2015). Although the Group of Twenty (G20) reaffirmed their 2009 commitment to phasing-out inefficient fossil fuel subsidies in 2017, progress has been slow and there are calls from large investors, insurers, and civil society to both increase transparency and accelerate the process (G20-2017). The main problems identified include that the G20 has not defined 'inefficient subsidies'; there is no mandatory reporting; and there are no timelines for phase-out commitments (Asmelash 2017).

At the global policy level, international climate negotiations have continued to influence energy markets. Following the 2015 Paris Agreement of the United Nations Framework Convention on Climate Change (UNFCCC), a technical meeting on its implementation took place in Bonn, Germany, in November 2017 at the 23rd Conference of the Parties (COP23) (UNFCCC 2017). Although renewable energy figured prominently in a large proportion of the Nationally Determined Contributions (NDCs) that countries submitted in the lead-up to COP22 in 2016, the climate negotiations in 2017 were unable to resolve the question of how NDCs should be organized, delivered, and updated, leaving uncertainty about how national renewable energy commitments could be ramped up (Timberley 2017).

Despite these uncertainties, an increasing number of communities, cities, and regions have introduced 100% renewable energy targets. The number of cities powered by at least 70% renewable electricity has more than doubled in 2 years, from 42 in 2015 to 101 in 2017. These cities now include Auckland, Brasilia, Nairobi, and Oslo (CDP-WEB 2018).

Carbon pricing policies, which include carbon taxes and emission trading schemes, were in place in 64 jurisdictions around the world in 2017, up from 61 in 2016. In December 2017 (REN21-GSR 2018), China launched the first phase of its long-awaited nationwide carbon emissions trading scheme, which will focus on the power sector. Carbon trading will be based in Shanghai and will include about 1700 power companies emitting more than 3 billion tonnes of CO_2 annually (Xu and Mason 2017). For comparison, the emissions trading scheme of the European Union included around 1.7 billion tonnes of CO_2 in 2016 (EC 2017). Reforms to the European Union scheme were agreed upon at the end of 2017, which will reduce the number of emission certificates issued and accelerate the cancellation of surplus certificates (Agora 2018).

The global investment in renewable energy in 2017 (excluding hydropower plants larger than 50 megawatts [MW]) was USD 280 billion (REN21-GSR 2018), up by 2% from 2016, but 13% below the all-time high, which occurred in 2015. It is noteworthy that each dollar represents more capacity on the ground as prices per GW decrease. Nearly all the investment was in either solar PV (58%) or wind power (38%). Developing countries accounted for the largest share of investment for the third consecutive year, at 63% of the total investment. China alone accounted for 45% of global investment, with a 30% increase since 2016. The United States was next, with 14%, followed by Japan (5%) and India (4%). Investment remained steady or trended upwards in Latin America and the USA, but has been falling in Europe since about 2010, with a drop of 30% from 2016 to 2017 (UNEP-FS 2018).

Pressure to diversify and stable growth in the renewables sector over the past decade has increased the interest of the fossil fuel industry in renewables. Large oil corporations more than doubled their acquisitions, project investments, and venture capital stakes in renewable energy in 2016 relative to those in 2015. This increased the investment in clean energy companies to USD 6.2 billion over the past 15 years, with more than 70% of deals involving solar PV or wind, and 16% involving biofuels (Bloomberg 2017). However, this is dwarfed by the spending of these companies on fossil fuels. One estimate is that renewables account for about 3% of the total annual spending (around USD 100 billion) by the world's five biggest oil companies (Schneyer and Bousso 2018). Bank finance for fossil fuels increased in 2017 by 11% relative to that in 2016, after a significant decline in 2016 (RAN 2018).

In 2017, as in previous years, renewables saw the greatest increases in capacity in the power sector, whereas the growth of renewables in the heating, cooling, and transport sectors was comparatively slow. Sector coupling—the interconnection of power, heating, and transport, and particularly the electrification of heating and transport—is gaining increasing attention as a way to increase the uptake of renewables by the transport and thermal sectors. Sector coupling also allows the integration of large shares of variable renewable energy, although this is still at an early stage. For example, China is specifically encouraging the electrification of heating, manufacturing, and transport in high-renewable areas, including promoting the use of renewable electricity for heating to reduce the curtailment of wind, solar PV, and hydropower. Several USA states are examining options for electrification, specifically to increase the overall renewable energy share (NEEP 2017).

2.2.1.1 Trends in the Renewable Power Sector

The capacity for generating renewable power saw its largest annual increase ever in 2017, with an estimated 178 GW of capacity added. The total global renewable power capacity increased by almost 9% relative to that in 2016. Solar PV additions reached a record high and represented about 55% of newly installed renewable power capacity in 2017. The increase in the solar PV capacity was greater than the combined net additions to the fossil fuel and nuclear capacities. For the first time, the installed solar PV capacity surpassed the global operating capacity of nuclear power. Wind and hydropower installations were in second and third positions, contributing about 29% and 11% of the increase in renewable generation capacity, respectively (REN21-GSR 2018).

In 2017, renewables accounted for an estimated 70% of net additions to the global power-generating capacity, up from 63% in 2016 (REN21-GSR 2018). The cost-competitiveness of renewable power generation continued to improve. Wind power and solar PV are now competitive with the generation of new fossil fuel energy in many markets, and even with existing fossil fuel generation in some markets. The costs of bio-electricity, hydropower, and geothermal power projects commissioned in 2017 were mostly within the range of the cost of fossil-fuel-fired electricity generation. Offshore wind prices also fell significantly in 2017, as competitive tenders in Germany, the UK, and the Netherlands resulted in bids that were competitive with new conventional power generation.

By the end of 2017, the countries with the greatest total installed renewable electric capacities were China, the USA, Brazil, Germany, and Canada. When only solar and wind capacities are considered, the top countries were China, the USA, and Germany, followed by Japan, India, and Italy, and then by Spain and the UK, which had about equal amounts of capacity by the year's end.

Seventeen countries have more than 90% renewable electricity, with the majority supplied almost completely by hydropower. However, three of these, Uruguay, Costa Rica, and Ethiopia, also have significant contributions from wind power (32%, 10%, and 7%, respectively) (REN21-GSR 2018). Increasing proportions of variable renewable electricity (VRE) must be integrated into electricity systems, and VRE penetration reached locally significant levels in 2017. The countries leading the way with wind and solar penetration are Denmark (52%), Uruguay (32%), and Cape Verde (31%), with another three countries at or above the 25% VRE penetration mark. Several countries and regions integrated much higher shares of VRE into their energy systems as instantaneous shares of the total demand for short periods during 2017, e.g., South Australia (more than 100% of load from wind power alone, and 44% of load from solar PV alone on another occasion), Germany (66% of load from wind and solar combined), Texas (54% of load from wind alone), and Ireland (60% of load from wind alone) (Parkinson 2017).

Curtailment—the forced reduction of wind and solar generation—is an indicator of grid integration challenges. In six of the jurisdictions with the highest VRE penetration, the curtailment rates were low (0–6%) in 2016 (Wynn 2018). Although curtailment initially tends to increase as the VRE share increases, some jurisdictions have successfully introduced measures, such as transmission upgrades, that have significantly reduced curtailment (Wynn 2018). Integration challenges have led to high curtailment rates in China, the world's largest wind and solar PV market (ECNS.CN 2018). These were reduced in 2017, with the average curtailment of wind power for the year at around 12%, down from 17% in 2016, and the average curtailment of solar PV was 6–7%, down 4.3% relative to that in 2016 (Haugwitz 2018).

The ongoing increase in the growth and geographic expansion of renewable energy was driven by the continued decline in the prices for renewable energy technologies (in particular, for solar PV and wind power), caused by the increasing power demand in some countries and by targeted renewable energy support mechanisms.[44] Solar PV and onshore wind power are now competitive with new fossil fuel generation in an increasing number of locations, due in part to declines in system component prices and to improvements in generation efficiency. The bid prices for offshore wind power also dropped significantly in Europe during 2016 (FT 12.9.2017).

Such declines in cost are particularly important in developing and emerging economies, and in isolated electric systems (such as on islands or in isolated rural communities) where electricity prices tend to be high (if they are not heavily subsidized), where there is a shortage of generation, and where renewable energy resources are particularly plentiful, making renewable electricity more competitive relative to other options. Many developing countries are racing to bring new power-generating capacities online to meet rapidly increasing electricity demands, often turning to renewable technologies (which may be grid-connected or off-grid) through policies such as tendering or feed-in tariffs, in order to achieve the desired growth quickly.

Approximately 1.06 billion people, most in sub-Saharan Africa, lived without electricity in 2016, 223 million fewer than in 2012 (IEA-WEO 2016; IEA-EAO 2017). Distributed renewables for energy access (DREA) systems were serving an estimated 300 million people at the end of 2016, and they comprised about 6% of new electricity connections worldwide between 2012 and 2016 (IRENA-P 2017). In places where the electricity grid does not reach or is unreliable, DREA technologies provide a cost-effective option to improve energy access. For example, about 13% of the population of Bangladesh gained access to electricity through solar home systems (SHS), and more than 50% of the off-grid population in Kenya is served by DREA systems (Dahlberg 2018). Off-grid solar devices, such as solar lanterns and SHS, displayed annual growth rates of 60% between 2013 and 2017 (Dahlberg 2018).

2.2.1.2 Heating and Cooling

Energy use for heating and cooling is estimated to have accounted for just over half of the total world final energy consumption in 2017, with about half of that used for industrial process heat (IEA-RE 2017). Around 27% of this was supplied by renewables. The largest share of renewable heating was from traditional biomass, which continued to supply about 16.4% of the global heat demand, predominantly for cooking in the developing world (IEA-RE 2017). Renewable energy—excluding traditional biomass—supplied approximately 9% of the total global heat production in 2017, up from about 6% in 2008 (REN21-GSR 2018).

Renewable heating and cooling receives much less attention than renewable power generation, and has been identified as the 'sleeping giant of energy policy' for the past decade (IEA-Collier 2018). The use of modern renewable heat has increased at an average rate close to 3% per year since 2008, gradually increasing its share of heat supply, but it lags well behind the average annual increase of 17% in modern renewables for electricity (IEA-RE 2017). Renewable energy technologies for heating and cooling include a variety of solar thermal collectors for different temperature levels; geothermal and air-sourced heat pumps; bioenergy used in traditional combustion applications or converted to gaseous, liquid, or solid fuels and subsequently used for heat; and any type of renewable electricity used for heating. Heat may be supplied by on-site equipment or by a district heating network.

A wide range of temperature requirements exist, from temperatures of around 40–70 °C for space and water heating in buildings, to steam at several hundred degrees Celsius for some industrial processes (Averfalk et al. 2017; USA-EPA 2017). The variety in renewable heating systems and applications is much greater than in the renewable power sector, which makes standardization to reduce costs by economies of scale more challenging, and makes it difficult for policy makers to find effective mechanisms to increase the renewable share. Trends in the use of modern renewable energy for heating vary according to the technology, although the relative shares of the main renewable heat technologies have remained stable for the past few years. In 2017, bioenergy (excluding traditional biomass) accounted for the greatest share, providing an estimated 68% of renewable heat, followed by renewable electricity at 18%, solar thermal at around 7%, district heating at 4% (which was nearly all bioenergy), and geothermal at 2% (REN21-GSR 2018). Although additional bio-heat and solar thermal capacities were added in 2017, the growth in both markets continued to slow. The trends in direct geothermal heating are unclear.

Bioenergy systems provide individual heating in residential and medium-sized office buildings, either as stand-alone systems or in addition to an existing central heating system, and bioenergy also accounts for 95% of district heating (IEA-RE 2017). District heating systems are suitable for use in densely populated regions with an annual heating demand during ≥4 months of the year, such as in the northern part of China, Denmark, Germany, Japan, Poland, Russia, Sweden, the UK, and the northern United States (IRENA-RE-H 2017). However, district heating supplies a very small proportion of global heating needs. The majority of district heating systems are fuelled by either coal or gas, with the share of renewables ranging from 0% to 42% (IRENA-RE-H 2017). Switching existing districting heating systems from fossil fuels to renewables has considerable potential (IRENA-RE-H 2017). Since the 1980s, Sweden has progressively switched from an almost entirely fossil-fuelled heating supply to systems supplied by 90% renewables and recycled heat (Brown 2018). District heating can combine different sources of heat, and can play a positive role in the integration of VRE through the use of electric heat pumps.

Solar thermal collector installations continued to decline globally in 2017, with a reduction of 3% (REN21-GSR 2018) compared with 2016, but the markets in China and India remained strong. In Europe, hybrid systems, in which solar thermal systems are used in combination with gas-fired central heating or bioenergy, are becoming more common, with specialized companies offering standardized technology.

Electricity accounts for an estimated 6% of the total heating and cooling consumption in buildings and industry, with about half of that electricity estimated to be renewable (IEA-RE 2017). The further electrification of heating and cooling drew increasing attention in 2017, particularly in the United States and China. Residential solar PV systems are also increasingly connected to electricity-using heat pumps in buildings rather than feeding the energy into the public electricity grid, especially when feed-in tariffs for solar electricity are reduced or have been entirely phased out.

Space cooling accounts for about 2% (REN21-GSR 2018) of the total world final energy consumption, and is supplied almost entirely by electricity (IEA-RE 2017). Solar-based space-cooling systems are still in the minority compared with conventional air-conditioning systems.

2.2.1.3 Transport

The global energy demand for transport increased by an average of 2.1% between 2000 and 2016, and is responsible for approximately 29% of the final global energy use and 24% of GHG emissions (IEA-WEO 2017). The vast majority (92%) of global transport energy needs are met by oil, with small proportions met by biofuels (2.9%) and electricity (1.4%) (IEA-WEO 2017).

There are three main entry points for renewable energy in the transport sector: the use of 100% liquid biofuels or biofuels blended with conventional fuels; natural gas vehicles and infrastructure (these can run on upgraded biogas); and the electrification of transport (if the electricity is itself renewable), which can be via batteries or hydrogen in fuel cells.

Biofuels (bioethanol and biodiesel) make by far the greatest contribution to renewable transport. The overall renewable share of road transport energy use was estimated to be 4.2% in 2016, with nearly all of that from biofuels (IEA-RE 2017). In 2017, global bioethanol production increased by 2.5% relative to that in 2016, with a slight decline in Brazil offset by increases in the USA, Europe, and China (IEA OIL 2018). Biodiesel production remained relatively stable in 2017, following a 9% increase in 2016 relative to 2015 (IEA OIL 2018).

The technology for producing, purifying, and upgrading biogas for use in transport is relatively mature, and the numbers of natural gas vehicles (NGVs) and the associated infrastructure are increasing slowly but steadily internationally. Many countries have relatively well-developed NGV infrastructures, and NGVs provide a good entry point for biogas in the transport sector (IRENA-RV-2017). The largest producers of biogas for vehicle fuel in 2016 were Germany, Sweden, Switzerland, the UK, and the USA (IRENA-RV-2017). The main barriers to the further expansion of biogas for transport are economic, with supply costs of USD 0.22–1.55 per cubic metre (m^3), compared with natural gas prices, which are as low as USD 0.13 per m^3. However, the lack of consistent regulation and access to gas grids are also significant barriers (IRENA-RV-2017).

Historically, the electrification of the transport sector has been limited to trains, light rail, and some buses. In 2017, there were signs that the entire sector would open to electrification. Fully electric passenger cars, scooters, and bicycles are rapidly becoming common-place, and prototypes for heavy-duty trucks, planes, and ships were released in 2017 (Hawkins 2017).

The number of electric vehicles (EVs) on the road passed the three million mark in 2017 (Guardian 25.12.2017). Annual sales are still only a very small proportion of the global total (1%), but this is set to change. In 2017, partly influenced by the scandal over diesel emissions cheating, five countries announced their intention to ban sales of new diesel and petrol cars from 2030 (India, the Netherlands, and Slovenia) or 2040 (France and the UK) (Bloomberg 11.2017). The announcement of electric product lines by car manufacturers in 2017 was another breakthrough. However, the number of EVs on the road is dwarfed by the number of electric bikes.

The global fleet was estimated to be around 200 million at the end of 2016, most of them in China, and 30% of bicycles sold in the Netherlands were e-bikes in 2017 (Wang 2017). Electric two- and three-wheel vehicles account for less than 0.5% of all transport energy use, but they account for most of the remaining renewable share after biofuels (IEA-RE 2017).

Further electrification of the transport sector will potentially create a new market for renewable energy and ease the integration of variable renewable energy, if market and policy settings ensure that the charging patterns are effectively harmonized with the requirements of electricity systems. There are already examples of countries and cities supplying electricity for both heavy and light rail from renewable electricity, including the Netherlands (BI 2017), Delhi (Times of India 2017), and Santiago de Chile (CT 2017).

Road transport accounts for 67% of global transport energy use, and two-thirds of that is used for passenger transport.

Aviation accounts for around 11% of the total energy used in transport (US-EIA-2017). In October 2016, the International Civil Aviation Organization (ICAO), a UN agency, announced a landmark agreement to mitigate GHG emissions in the aviation sector. By the end of 2017, 106 states representing 90.8% of global air traffic had settled on a global emissions reduction scheme (Guardian 6.10.2016). Together with technical and operational improvements, this agreement will support the production and use of sustainable aviation fuels, specifically drop-in fuels produced from biomass and different types of waste (ICAO 2018). In 2017, Norway announced a target of 100% electric short-haul flights by 2040 (Guardian 18.1.2018).

Shipping consumes around 12% of the global energy used in transport (US-EIA-2017) and is responsible for approximately 2.0% of global CO_2 emissions. There are multiple entry points for renewable energy: ships can incorporate wind (sails) and solar energy directly, and can use biofuels, synthetic fuels, or hydrogen produced with renewable electricity for propulsion. China saw the launch of the world's first all-electric cargo ship in 2017, and in Sweden, two large ferries were converted from diesel to electricity in 2017 (China Daily 14.11.2017). In 2017, the International Maritime Organization's (IMO's) Marine Environment Protection Committee (MEPC) approved a roadmap (2017–2023) to develop a strategy for reducing GHG emissions from ships. The roadmap includes plans for an initial GHG strategy to be adopted in 2018 (IMO 2017).

Rail accounts for around 1.9% of the total energy used in transport and is the most highly electrified transport sector. The share of rail transport powered by electricity was estimated to be 39% in 2015, up from 29% in 2005 (IEA-UIC 2017). Just over a third of the electricity (9% of rail energy) is estimated to be derived from renewable sources (IEA-UIC 2017). Some jurisdictions are opting to ensure that the proportion of energy from renewable sources in their transport sectors is well above the share of renewable energy in their power sectors. For example, the Dutch railway company NS announced that its target to power all electric trains with 100% renewable electricity was achieved ahead of schedule in 2017 (Caughill 2017), and the New South Wales Government in Australia announced a renewable tender for the Sydney's light rail system.

Following the historic Paris Climate Agreement in December 2015, the international community has focused increasing attention on the decarbonisation of the transport sector. At the climate conference in November 2017 in Bonn, Germany, a multi-stakeholder alliance launched the Transport Decarbonisation Alliance (UN-P 2018). France, the Netherlands, Portugal, Costa Rica, and the Paris Process on Mobility and Climate (PPMC) are members of the Alliance, which includes countries, cities, regions, and private-sector companies committed to ambitious action on transport and climate change (UN-P 2018).

References

Agora (2018), Agora Energiewende and Sandbag. The European Power Sector in 2017. State of Affairs and Review of Current Developments. (London, 2018) p.37. https://sandbag.org.uk/project/european-energy-transition-power-sector-2017/

Asmelash, H.B. (2017). "Phasing Out Fossil Fuel Subsidies in the G20: Progress, Challenges, and Ways Forward." (Geneva, Switzerland: International Centre for Trade and Sustainable Development, 2017), Page vii. https://www.ictsd.org/sites/default/files/research/phasing_out_fossil_fuel_subsidies_in_the_g20-henok_birhanu_asmelash.pdf

Averfalk, H., Werner, S. et al. (2017), Averfalk, H., Werner, S. et al. "Transformation Roadmap from High to Low Temperature District Heating Systems". Annex XI final report. International Energy Agency Technology Collaboration on District Heating and Cooling including Combined Heat and Power (2017), page 5. http://www.iea-dhc.org/index.php?eID=tx_nawsecuredl&u=1440&g=3&t=1521974500&hash=312c2176feeabc6fcf75a735f3031f91e33940f1&file=fileadmin/documents/Annex_XI/IEA-DHC-Annex_XI_Transformation_Roadmap_Final_Report_April_30-2017.pdf ,

Bloomberg (2017) Deals increased from 21 in 2015 to 44 in 2016 (still below the high point in 2011): Anna Hiretnstein, "Big Oil Is Investing Billions to Gain a Foothold in Clean Energy", Bloomberg Markets, 25 October 2017, https://www.bloomberg.com/news/articles/2017-10-24/big-oil-is-investing-billions-to-gain-a-foothold-in-clean-energy

(Bloomberg 11.2017) Anna Hirtenstein, "Global Electric Car Sales Jump 63 Percent", Bloomberg, 21 November 2017, https://www.bloomberg.com/news/articles/2017-11-21/global-electric-car-sales-jump-63-percent-as-china-demand-surges,

BP (2017), Average prices to 2016 from: BP Statistical Review of World Energy "Natural gas—2016 in review", presentation, (BP,June 2017) https://www.bp.com/content/dam/bp/en/corporate/pdf/energy-economics/statistical-review-2017/bp-statistical-review-of-world-energy-2017-natural-gas.pdf According to https://tradingeconomics.com/commodity/natural-gas, gas prices fell by 14% between 2 Jan 2017 and 1 January 2018 (viewed 21 March 2018)

BI(2017)PatrickCaughill,"AllDutchTrainsNowRunOn100%WindPower",BusinessInsider.3June 2017. http://uk.businessinsider.com/wind-power-trains-in-netherlands-2017-6?r=US&IR=TT

Brown, P (2018), Paul Brown, "District Heating Warms Cities Without Fossil Fuels ", Ecowatch, 15 January 2018, https://www.ecowatch.com/district-heating-energy-efficiency-2525685749.html

Carroll (2018), Phil Carroll, "are gas-fired peaking plants on the way out?" , (Missouri, USA, Finley Engineering 28 Feb 2018). http://finleyusa.com/industries/energy/whitepapers/are-gas-fired-peaking-plants-on-the-way-out/are-gas-fired-peaking-plants-on-the-way-out/

Campbell (2017), Shaun Campbell, , "EU nations pledge no more coal plants from 2020", Windpower Monthly, 6 April 2017. https://www.windpowermonthly.com/article/1429875/eu-nations-pledge-no-coal-plants-2020; Megan Darby. "Port of Amsterdam set to be coal-free by 2030". Climate Home News, 16 March 2017. http://www.climatechangenews.com/2017/03/16/port-amsterdam-set-coal-free-2030/

Caughill (2017), Patrick Caughill "All Dutch Trains Now Run On 100% Wind Power", Business Insider. 3 June 2017. http://uk.businessinsider.com/wind-power-trains-in-netherlands-2017-6?r=US&IR=T

Carrington (2017) Damian Carrington, "Political watershed' as 19 countries pledge to phase out coal", The Guardian, 16 Nov 2017 (Bonn). https://www.theguardian.com/environment/2017/nov/16/political-watershed-as-19-countries-pledge-to-phase-out-coal

CDP-WEB (2018), Carbon Disclosure Project (CDP), "The World's Renewable Energy Cities" https://www.cdp.net/en/cities/world-renewable-energy-cities website, September 2018

China Daily (14.11.2017) Qiu Quanlin, "Fully electric cargo ship launched in Guangzhou", China Daily, 14 November 2017, http://www.chinadaily.com.cn/business/2017-11/14/content_34511312.htm.; ABB, "HH Ferries electrified by ABB win prestigious Baltic Sea Clean Maritime Award 2017", press release (Zurich: 14 June 2017), http://new.abb.com/news/detail/1688/HH-ferries-electrified-by-ABB-win-prestigious-baltic-sea-clean-maritime-award-2017; Fred Lambert, "Two massive ferries are about to become the biggest all-electric ships in the world", Electrek, 24 August 2017, https://electrek.co/2017/08/24/all-electric-ferries-abb/; Daniel Boffey,

CT (2017) Chile's Santiago Metro Will Meet 60% Of Its Energy Demand From Renewables", Clean Technica, 8 July 2017. https://cleantechnica.com/2017/07/08/chiles-santiago-metro-will-meet-60-energy-demand-renewables/

Dahlberg (2018), Dahlberg Advisors and Lighting Global, Off-Grid Solar Market Trends Report 2018 (Washington, DC: International Finance Corporation, 2018), p. 70, https://www.gogla.org/sites/default/files/resource_docs/2018_mtr_full_report_low-res_2018.01.15_final.pdf..

ECNS.CN (2018), ECNS.CN, People Daily Online, Wang Zihao,China's clean power waste continues to drop, 8th June 2017, viewed 13th March 2018, http://www.ecns.cn/2017/06-08/260610.shtml

EC (2017), European Commission, Report From The Commission To The European Parliament And The Council - Report on the functioning of the European carbon market, COM(2017) 693 final, (Brussels,Belgium 23 November 2017), page 23 Table 8 provides an overview to all verified emissions (in million tonnes CO2 equivalents), https://ec.europa.eu/commission/sites/beta-political/files/report-functioning-carbon-market_en.pdf

ENERDATA (2018), ENERDATA, Rise in global energy-related CO2 emissions in 2017, 29 January 2018, https://www.enerdata.net/publications/executive-briefing/global-increase-co2-emissions-2017.html

FS-UNEP (2018), Frankfurt School - UNEP Collaborating Centre for Climate & Sustainable Energy Finance (FS-UNEP) in co-operation with Bloomberg New Energy Finance. (2018) Global Trends in Renewable Energy Investment 2018 (Frankfurt: 2018). Page 34.

FT (12.9.2017), Financial Times, 12 September 2017, Natalie Thomas, Powerful turbines slash price of offshore wind farms, https://www.ft.com/content/28b0eb2e-96f1-11e7-a652-cde3f882dd7b

G20 (2017), G20 reaffirm commitment to phasing out subsidies: G20. G20 Hamburg Climate and Energy Action Plan for Growth: Annex to G20 Leaders' Declaration. (Hamburg, Germany: 2017) Page 11 https://www.g20germany.de/Content/DE/_Anlagen/G7_G20/2017-g20-climate-and-energy-en.pdf?__blob=publicationFile&v=6;

Guardian (18.1.2018) The Guardian, Norway aims for all short-haul flights to be 100% electric by 2040, AFP, 18th January 2018, viewed 16th March 2018, https://www.theguardian.com/world/2018/jan/18/norway-aims-for-all-short-haul-flights-to-be-100-electric-by-2040

Guardian (6.10.2016) Oliver Milman. "First deal to curb aviation emissions agreed in landmark UN accord", 6 October 2016, https://www.theguardian.com/environment/2016/oct/06/aviation-emissions-agreement-united-nations; ICAO Website, Environment, September 2018, https://www.icao.int/environmental-protection/Pages/ClimateChange_ActionPlan.aspx

Haugwitz (2018), 2017 data from: AECEA, "China 2017—what a year with 53 GW of added solar PV! What's in for 2018!" Briefing Paper—China Solar PV development, January 2018 ,Frank Haugwitz, AECEA.

Hawkins (2017) Andrew Hawkins. "This Electric Truck Startup Thinks It Can Beat Tesla To Market" The Verge. 15 December 2017. https://www.theverge.com/2017/12/15/16773226/thor-trucks-electric-truck-etone-tesla; Miquel Ros. "7 Electric Aircraft You Could Be Flying In Soon." CNN Travel. 21 November 2017. https://edition.cnn.com/travel/article/electric-aircraft/index.html.

Hawkins, Ed, Pablo Ortega, Emma Suckling, Andrew Schurer, Gabi Hegerl, Phil Jones, Manoj Joshi et al. "Estimating changes in global temperature since the preindustrial period." Bulletin of the American Meteorological Society 98, no. 9 (2017): 1841–1856.

Guadian (25.12.2017) Adam Vaughan. "Electric and Plug-In Hybrid Cars Whiz Past 3m Mark Worldwide". The Guardian, 25 December 2017. https://www.theguardian.com/environment/2017/dec/25/electric-and-plug-in-hybrid-cars-3m-worldwide

ICAO (2018) ICAO Website, Environment, September 2018,https://www.icao.int/environmental-protection/GFAAF/Pages/default.aspx

(IEA- Collier 2018), Ute Collier ,Commentary: More policy attention is needed for renewable heat, (Paris, France: International Energy Agency, 25 January 2018), http://www.iea.org/newsroom/news/2018/january/commentary-more-policy-attention-is-needed-for-renewable-heat.html

IEA-EAO (2017) International Energy Agency, Energy Access Outlook 2017, Table 2.1, https://www.iea.org/publications/freepublications/publication/WEO2017SpecialReport_EnergyAccessOutlook.pdfhttp://www.iea.org/publications/freepublications/publication/WEO2017_Special_Report_Energy_Access_Outlook_ExecutiveSummary_English.pdf.

IEA-RE (2016), International Energy Agency (IEA), MediumTerm Renewable Energy Market Report 2016 (Paris: 2016),Page 214 https://www.iea.org/newsroom/news/2016/october/mediumterm-renewable-energy-market-report-2016.html ;

IEA-WEB (2018) International Energy Agency (IEA),, Energy Subsidies; website, September 2018, https://www.iea.org/statistics/resources/energysubsidies/

IEA-WEO (2016), 2014 total from IEA, "World Energy Outlook 2016—Electricity Access Database", http://www.worldenergyoutlook.org/media/weowebsite/2015/WEO2016Electricity.xlsx, 2016; 2016 data from

IEA-RE (2017),International Energy Agency Renewables 2017. Analysis and Forecasts to 2022. (Paris, France: 2017), Page 121. http://www.iea.org/bookshop/761-Market_Report_Series:_Renewables_2017

IEA-WEO (2017), IEA, World Energy Outlook 2017 (Paris: 2017) p. 648. https://www.iea.org/weo2017/ ; GHG emissions share for 2015 from IEA, CO2 Emissions from Fuel Combustion 2017, (Paris: 2017) p. 12. http://www.iea.org/bookshop/757-CO2_Emissions_from_Fuel_Combustion_2017

IEA OIL (2018), International Energy Agency: Analysis and Forecasts to 2023, (Paris, France: 2018), p.77 & p.134 http://www.oecd.org/publications/market-report-series-oil-25202707.htm

IEA-UIC (2017), IEA and International Union of Railways, Railway Handbook 2017, Energy Consumption and CO2 emissions, (Paris: 2017). P. 26, https://uic.org/IMG/pdf/handbook_iea-uic_2017_web2-2.pdf

IMF (2015), The International Monetary Fund (IMF) includes externalities in fossil fuel subsidies and arrives at US$5.3 trillion for 2015, 6.5% of the world's GDP; Damian Carrington, "Fossil fuels subsidized by $10m a minute, says IMF", The Guardian, 18 May 2015, https://www.theguardian.com/environment/2015/may/18/fossil-fuel-companies-getting-10m-a-minute-in-subsidies-says-imf,

IMO (2017) International Maritime Organization (IMO) Website, September 2018, http://www.imo.org/en/MediaCentre/HotTopics/GHG/Pages/default.aspx

IRENA-P (2017)Estimate of 300 million from International Renewable Energy Agency (IRENA), "2016 a record year for renewables, latest IRENA data reveals", press release (Abu Dhabi: 30 March 2017), http://irena.org/newsroom/pressreleases/2017/Mar/2016-a-Record-Year-for-Renewables-Latest-IRENA-Data-Reveals;

IRENA-RE-H (2017), International Renewable Energy Agency (IRENA), Renewable Energy in District Heating and Cooling: A Sector Roadmap for REmap, , (Abu Dhabi, UAE 2017) page 17, table 2, http://www.irena.org/-/media/Files/IRENA/Agency/Publication/2017/Mar/IRENA_REmap_DHC_Report_2017.pdf

IRENA-RV (2017),IRENA, Biogas for Road Vehicles Technology Brief (Abu Dhabi: March 2017), p. 2, http://www.irena.org/DocumentDownloads/Publications/IRENA_Biogas_for_Road_Vehicles_2017.pdf.

Naureen S Malik, (2017) "Renewables Are Starting to Crush Aging U.S. Nukes, Coal Plants". Bloomberg. 2 November 2017 https://www.bloomberg.com/news/articles/2017-11-02/renewables-are-starting-to-crush-aging-u-s-nukes-coal-plants

NEEP (2017), Northeast Energy Efficiency Partnerships Northeastern Regional Assessment of Strategic Electrification: Summary Report. (Massachusetts, USA 2017), http://www.neep.org/reports/strategic-electrification-assessment; National Development and Reform Commission, National Energy Administration. Notice on publication of the "Measures for resolving curtailment of hydro, wind and PV power generation",

Parkinson (2017), Giles Parkinson, "Wind Output Hits Record In July, Wind And Solar 59% In S.A.", Reneweconomy, 31 August 2017. http://reneweconomy.com.au/wind-output-hits-record-in-july-wind-and-solar-59-in-s-a-45242/ , Giles Parkinson, "Rooftop Solar Provides 48% Of South Australia Power, Pushing Grid Demand To Record Low", Reneweconomy, 18 September 2017, http://reneweconomy.com.au/rooftop-solar-provides-48-of-south-australia-power-pushing-grid-demand-to-record-low-47695/;

RAN (2018), Overall finance went USD 126 billion in 2015, to USD 104 billion in 2016, to USD 115 billion in 2017, from: Rainforest Action Network, BankTrack, Sierra Club, and Oil Change International, Banking On Climate Change: Fossil Fuel Finance Report Card 2018, (2018). p. 3. https://www.ran.org/bankingonclimatechange2018

REN21-GSR (2018), REN21; Renewables 2018 Global Status Report, Paris: REN21 Secretariat, ISBN 978-3-9818911-3-3, http://www.ren21.net/wp-content/uploads/2018/06/17-8652_GSR2018_FullReport_web_final_.pdf

Schurer, A. P., Cowtan, K., Hawkins, E., Mann, M. E., Scott, V., & Tett, S. F. B. (2018). Interpretations of the Paris climate target. Nature Geoscience, 11(4), 220.

Shearer (2017), Christine Shearer, "Coal Phase-Out: First Global Survey of Companies and Political Entities Exiting Coal", Endcoal. 18 October 2017. https://endcoal.org/2017/10/coal-phase-out-first-global-survey-of-companies-and-political-entities-exiting-coal/

Schurer, A. P., Mann, M. E., Hawkins, E., Tett, S. F., & Hegerl, G. C. (2017). Importance of the pre-industrial baseline for likelihood of exceeding Paris goals. Nature climate change, 7(8), 563

Statista (2018), Statista: "Average annual Brent crude oil price from 1976 to 2018 (in U.S. dollars per barrel)",https://www.statista.com/statistics/262860/uk-brent-crude-oil-price-changes-since-1976/ , updated 2018, Viewed 21 March 2018.; "Average annual OPEC crude oil price from 1960 to 2018 (in U.S. dollars per barrel)", https://www.statista.com/statistics/262858/change-in-opec-crude-oil-prices-since-1960/. viewed September 2018.

Schneyer, Bousso (2018), Ernest Scheyder, Ron Bousso, "Peak Oil? Majors Aren't Buying Into The Threat From Renewables", Reuters, 8 November 2018. https://uk.reuters.com/article/us-oil-majors-strategy-insight/peak-oil-majors-arent-buying-into-the-threat-from-renewables-idUKKBN1D80GA

Stephens G.L., Wild M., Stackhouse Jr P.W., L'Ecuyer T., Kato S., Henderson D.S. (2012) The Global Character of the Flux of Downward Longwave Radiation, American Meteorological Society, 25, pp:2329–2340. DOI: https://doi.org/10.1175/JCLI-D-11-00262.1

Timberley (2017), Jocelyn Timperley, COP23: Key outcomes agreed at the UN climate talks in Bonn, CarbonBrief, 19 November 2017, https://www.carbonbrief.org/cop23-key-outcomes-agreed-un-climate-talks-bonn

Times of India (2017) "Solar Energy To Power Delhi Metro's Phase Iii". The Times of India, Saurabh Mahapatra.. 24 April 2017. https://timesofindia.indiatimes.com/city/delhi/solar-energy-to-power-metro-ph-iii/articleshow/58332740.cms;

UNEP-FS (2018), Frankfurt School - UNEP Collaborating Centre for Climate & Sustainable Energy Finance (FS-UNEP) in co-operation with Bloomberg New Energy Finance. Global Trends in Renewable Energy Investment 2018 (Frankfurt, Germany: 2018). Pages 14, 15 & 26.

UNFCCC (2017), United Nation—Climate Change, UNFCCC—The Paris Agreement, website viewed 12th March 2018, http://unfccc.int/paris_agreement/items/9485.php; Jocelyn Timperley, COP23: Key outcomes agreed at the UN climate talks in Bonn, CarbonBrief, 19 November 2017, https://www.carbonbrief.org/cop23-key-outcomes-agreed-un-climate-talks-bonn

UN-P (2018), United Nations, Press release, New Transport Decarbonisation Alliance for Faster Climate Action, 11th November 2018, https://cop23.unfccc.int/news/new-transport-decarbonisation-alliance-for-faster-climate-action

USA-EPA (2017), United States of America, Environmental Protection Agency (EPA), Renewable Heating and Cooling, Renewable Industrial Process Heat, https://www.epa.gov/rhc/renewable-industrial-process-heat, updated 26 October 2017,

US-EIA (2017) U.S. Energy Information Administration. International Energy Outlook 2017. Transportation sector passenger transport and energy consumption by region and mode: https://www.eia.gov/outlooks/aeo/data/browser/#/?id=50-IEO2017®ion=0-0&cases=Reference&start=2010&end=2020&f=A&linechart=Reference-d082317.2-50-IEO2017&sourcekey=0

Wynn (2018), Wynn, G. Power-Industry Transition, Here and Now: Wind and Solar Won't Break the Grid: Nine Case Studies. Institute for Energy Economics and Financial Analysis. (Cleveland, US, 2018) p.15. http://ieefa.org/wp-content/uploads/2018/02/Power-Industry-Transition-Here-and-Now_February-2018.pdf

Wang (2017) Brian Wang. "Electric bikes could grow from 200 million today to 2 billion in 2050", Next Big Future, 27t April 2017. https://www.nextbigfuture.com/2017/04/electric-bikes-could-grow-from-200-million-today-to-2-billion-in-2050.html; Sales and Trends. "E-Bike Puts Dutch Market Back on Growth Track". Bike Europe. 6 March 2018. http://www.bike-eu.com/sales-trends/nieuws/2018/3/e-bike-puts-dutch-market-back-on-growth-track-10133083

XU, Mason (2017), Muyu Xu, Josephine Mason, "China aims for emission trading scheme in big step vs. global warming", Reuters, 19 December 2017, https://www.reuters.com/article/us-china-carbon/china-aims-for-emission-trading-scheme-in-big-step-vs-global-warming-idUSKBN1ED0R6

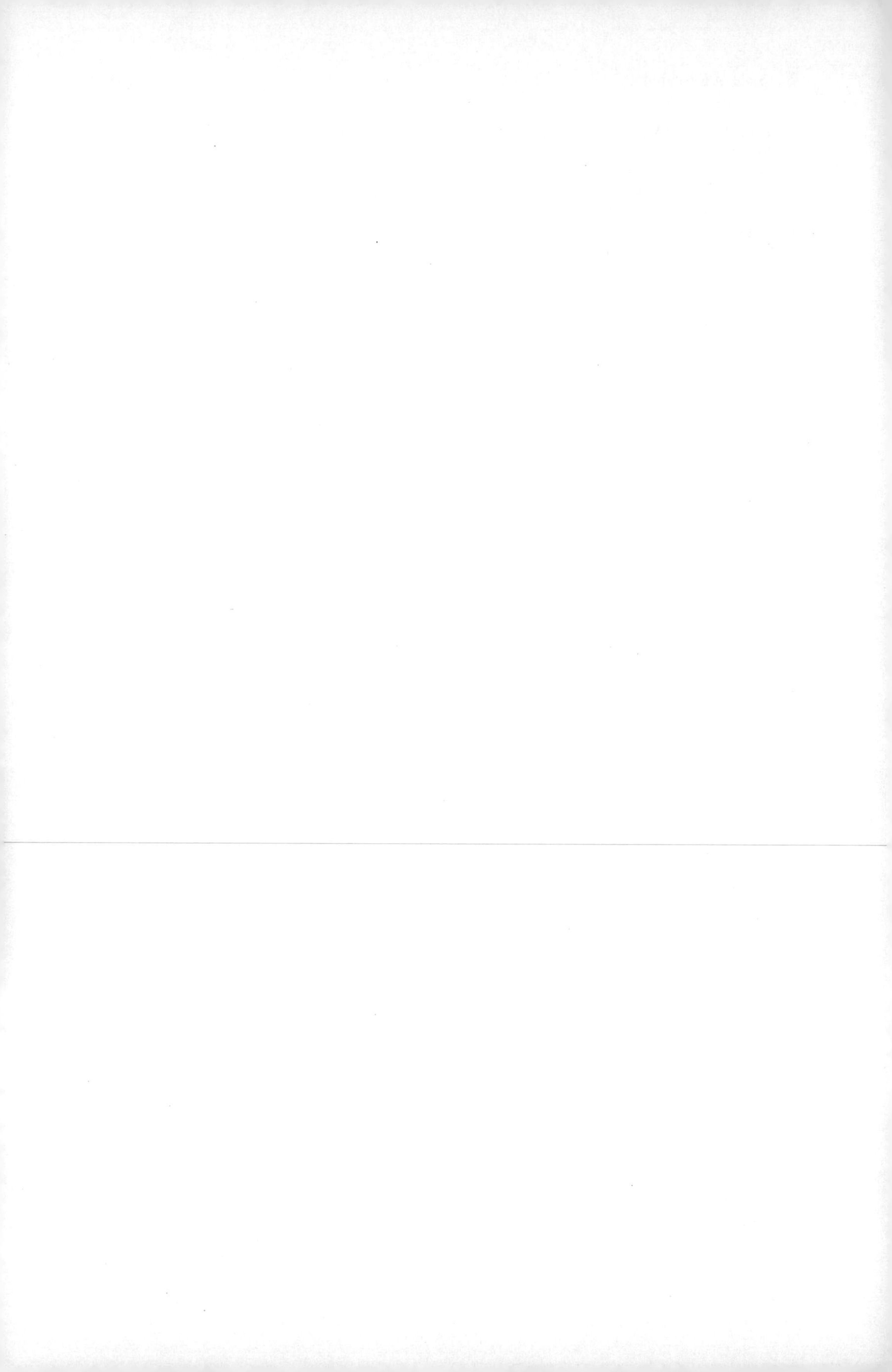

Chapter 3
Methodology

Sven Teske, Thomas Pregger, Sonja Simon, Tobias Naegler, Johannes Pagenkopf, Bent van den Adel, Malte Meinshausen, Kate Dooley, C. Briggs, E. Dominish, D. Giurco, Nick Florin, Tom Morris, and Kriti Nagrath

Abstract A detailed overview of the methodologies used to develop the 2.0 °C and 1.5 °C scenario presented in this book. Starting with the overall modelling approach, the interaction of seven different models is explained which are used to calculate and developed detailed scenarios for greenhouse gas emission and energy pathways to stay within a 2.0 °C and 1.5 °C global warming limit. The following models are presented:

- For the non-energy GHG emission pathways, the *Generalized Equal Quantile Walk (GQW)* method, the land-based sequestration design method and the Carbon cycle and climate (MAGICC) model.
- For the energy pathways, a renewable energy resources assessment for space constrained environments ([R]E-SPACE, the transport scenario model (TRAEM), the Energy System Model (EM) and the power system model [R]E 24/7.

The methodologies of an employment analysis model, and a metal resource assessment tool are outlined. These models have been used to examine the analysis of the energy scenario results.

S. Teske (✉) · C. Briggs · E. Dominish · D. Giurco · N. Florin · T. Morris · K. Nagrath
Institute for Sustainable Futures, University of Technology Sydney, Sydney, NSW, Australia
e-mail: sven.teske@uts.edu.au; chris.briggs@uts.edu.au; elsa.dominish@uts.edu.au; damien.giurco@uts.edu.au; nick.florin@uts.edu.au; tom.morris@uts.edu.au; kriti.nagrath@uts.edu.au

T. Pregger · S. Simon · T. Naegler
Department of Energy Systems Analysis, German Aerospace Center (DLR), Institute for Engineering Thermodynamics (TT), Pfaffenwaldring, Germany
e-mail: thomas.pregger@dlr.de; sonja.simon@dlr.de; tobias.naegler@dlr.de

J. Pagenkopf · B. van den Adel
Department of Vehicle Systems and Technology Assessment, German Aerospace Center (DLR), Institute of Vehicle Concepts (FK), Pfaffenwaldring, Germany
e-mail: johannes.pagenkopf@dlr.de; Bent.vandenAdel@dlr.de

M. Meinshausen · K. Dooley
Australian-German Climate and Energy College, University of Melbourne, Parkville, Victoria, Australia
e-mail: malte.meinshausen@unimelb.edu.au; kate.dooley@unimelb.edu.au

S. Teske (ed.), *Achieving the Paris Climate Agreement Goals*,
https://doi.org/10.1007/978-3-030-05843-2_3

Achieving the goals of the Paris Climate Agreement (UNFCCC 2015) will require the total decarbonisation of the energy system by 2050, with a global emissions peak no later than 2020 (Hare and Roming 2016) and a drastic reduction in non-energy-related greenhouse gases (GHGs), including land-use-related emissions (Rogelj and den Elzen 2016). Over the past decades, numerous computer models have been developed to analyse different emissions pathways and to investigate the effects of changes in policy and technology and adjustments in global and regional economies. A wide range of climate models is used to calculate non-energy-related GHG emissions pathways and their impacts on the global climate. The Intergovernmental Panel on Climate Change (IPCC) states that "Climate models have continued to be developed and improved since the AR4 [published in 2007-author], and many models have been extended into Earth System models by including the representation of biogeochemical cycles important to climate change" (Flato and Marotzke 2013). Whereas climate models analyse the effects of a variety of GHG emissions, energy scenarios only cover energy-related CO_2. Their purpose is to investigate future energy systems to identify feasible technological and/or economic pathways. Like climate models, energy models are diverse and vary significantly in their methodologies. The IPCC's Special Report on Renewable Energy Sources and Climate Change Mitigation states that there is "enormous variation in the detail and structure of the models used to construct the scenarios" (Fischedick and Schaeffer 2011). Energy scenarios with high penetrations of variable renewable power generation—solar photovoltaic (PV) and wind power—require a higher degree of time resolution to assess the security of 24/7 electricity supplies than those with mainly dispatchable power generation.

Modelling the energy system involves a variety of methodological requirements, which pose specific challenges when addressed on the global level: the quantitative projection of developments in (future) technologies and potential markets; a consistent database of renewable energy potentials and their temporal and spatial distributions; reliable data on the current situations in all regions; an assessment of energy flows and emissions across all energy subsectors, such as industry, transport, residential, etc.; and a comprehensive assessment of all CO_2 emissions, in order to assess the impact of the energy system on climate change. Finally, analysing and assessing the energy transition require a long-term perspective on future developments.

Changes to energy markets require long-term decisions to be made because infrastructural changes are potentially required, and are therefore independent of short-term market developments. The power market cannot function optimally without long-term infrastructure planning. Grid modifications and the roll-out of smart metering infrastructure, for example, require several years to implement. These technologies form the basis of the energy market and allow energy trading. Therefore, the time required for infrastructure planning and other substantial transformation processes must be considered in the scenario-building approach.

Although numerous energy scenarios that provide 100% renewable energy at the community, state, and national levels have been published in the past decade (Elliston and MacGill 2014; Teske and Dominish 2016; Klaus et al. 2010; Teske and Brown 2012), only a handful of analyses have been performed on a global level. The main research projects on 100% renewable energy supplies published between 2015 and 2018 were:

- A Road Map to 100 Percent Renewable Energy in 139 Countries by 2050, Mark Jacobson, Charles Q. Choi, Stanford Engineering, Stanford University, USA, 2017 (Jacobson and Choi 2017);
- Internet of Energy, A 100% Renewable Electricity System, Christian Breyer, Neo Carbon Energy, Lappeenranta University of Technology, Finland, 2016 (Breyer 2016; Breyer and Bogdanov 2018);
- Energy [R]evolution—A sustainable World Energy Outlook 2015, Greenpeace International with the German Aerospace Centre (DLR), Institute of Engineering Thermodynamics, System Analysis and Technology Assessment, Stuttgart, Germany (Teske and Pregger 2015).

All the studies listed above share the same modelling horizon until 2050 and focus clearly on the fast and massive deployment of renewable energy resources (RES). Options with large uncertainties in terms of techno-economic, societal, and environmental risks, such as large hydro power, nuclear power, or unsustainable biomass use, carbon capture and storage (CCS), and geoengineering are excluded. However, each of these studies has a specific strength. On the one hand, the analyses from Stanford University and the University of Technology Lappeenranta include an hourly simulation of power demand and supply, in addition to the pathway modelling. On the other hand, the Energy [R]evolution study covers the complete energy sector, with detailed insights into the heat and transport sectors. However, all these studies cover only CO_2 emissions from the energy system, without further investigation of other GHG sources.

Therefore, our project combines these strengths into a single approach by combining a set of models. The approach is based on the scenario modelling used for the Energy [R]evolution scenario series developed by the authors between 2004 and 2015. It models scenarios of comprehensive pathways for power, heat, and fuel supply in 5-year steps, and includes specific insights from a transport model. The scenario building is also complemented by a simulation with hourly resolution to calculate the electricity storage demand and to increase the spatial resolution from 10 to 72 regions. Another significant improvement over existing studies is its combination with a climate model. The interaction between non-energy GHG pathways and a high-resolution integrated energy assessment model (IAM) provides additional information on how to achieve the goals of the Paris Agreement.

3.1 100% Renewable Energy—Modelling Approach

The complete decarbonisation of the global energy supply requires entirely new technical, economic, and policy frameworks for the electricity, heating, and cooling sectors, and the transport system. Such new framework conditions and the political and regulative interventions necessary for their implementation are widely discussed in the literature. However, assessing their feasibility and effectiveness requires an in-depth analysis of specific regional and national conditions and mechanisms. Therefore, societal frameworks, measures, and policy interventions are not explicitly discussed in this scenario analysis, but they are implicit elements in the definition of the narratives and assumptions as core step of scenario development (see Chap. 5).

Modelling Approach
To develop a global plan, the authors combined various established computer models:

- Global GHG Model: The non-energy GHG emissions scenarios are calculated with the following models:
  - Generalized Equal Quantile Walk (GQW): This statistical method is used to complement the CO_2 pathways with the non-CO_2 regional emissions for the relevant GHGs and aerosols, based on a statistical analysis of the large number (~700) of multi-gas emission pathways underlying the recent IPCC Fifth Assessment Report and the recently published IPCC Special Report on 1.5 °C. The GQW method calculates the median non-CO_2 gas emission levels every 5 years, conditional on the energy-related CO_2 emission level percentile of the 'source' pathway. This method is further developed in this project—building on an earlier 'Equal Quantile Walk' method—and is now better able to capture the emission dynamics of low-mitigation pathways.
  - Land-based sequestration design: A Monte Carlo analysis across temperate, boreal, subtropical, and tropical regions has been performed based on various literature-based estimates of sequestration rates, sequestration periods, and the areas available for a number of sequestration options. This approach can be seen as a quantified literature synthesis of the potential for land-based CO_2 sequestration, which is not reliant on bioenergy with sequestration and storage (BECCS)
  - Carbon cycle and climate modelling (MAGICC): This study used the MAGICC climate model, which also underlies the classification of both the IPCC Fifth Assessment Report and the IPCC Special Report on 1.5 °C in terms of the ability of various scenarios to limit the temperature increase to below 2.0 °C or 1.5 °C. MAGICC is constantly evolving, but its core goes back to the 1980s, and it represents one of the most established reduced-complexity climate models in the international community.
- Renewable Resource Assessment [R]E-SPACE: This is based on a Geographic Information Systems (GIS) approach and provides maps of the solar and wind potentials in space-constrained environments. GIS attempts to emulate processes in the real world, at a single point in time or over an extended period (Goodchild 2005). The primary purpose of GIS mapping is to ascertain whether renewable energy resources (primarily solar and wind) are sufficiently available in each region. It also provides an overview of the existing electricity infrastructures for fossil fuel and renewable sources.
- Transport model (TRAEM): The transport scenario model allows the representation of long-term transport developments in a consistent and transparent way. The model disaggregates transport into a set of different modes and calculates the final energy demand by multiplying the specific transport demand of each transport mode with the powertrain-specific energy demands, using passenger–km and tonne–km activity-based bottom-up approaches. The model applied is an accounting system, without system or ownership cost-optimization.
- Energy system model (EM): The scenario model is a mathematical accounting system for the energy sector that applies different methodologies. It aims to

model the development of energy demand and supply according to the energy potentials, future costs, emissions, specific fuel consumptions, and physical flows between processes. The data available and the objectives of the analysis significantly influence the model architecture and approach. It is very important to differentiate between an energy model and a scenario. An energy model is the technical basis for a scenario. Scenarios are the results of the energy model, which have been calculated with different input data and assumptions. The energy model is used in this study to develop long-term scenarios for the energy systems across all sectors (power, heat, transport, and industry) without the application of cost-optimization based on uncertain cost assumptions. However, an ex-post analysis of costs and investments shows the main economic effects of the pathways.

- Power system model [R]E 24/7: This simulates the electricity system on an hourly basis and at geographic resolution to assess the requirements for infrastructure, such as grid connections, between different regions and electricity storages, depending on the demand profiles and power-generation characteristics (Teske 2015). High-penetration or renewable-energy-only scenarios will contain significant proportions of variable solar photovoltaic (PV) and wind power because they are inexpensive. Therefore, a power system model is required to assess the demand and supply patterns, the efficiency of power generation, and the resulting infrastructural needs. On the generation side, meteorological data, typically in 1 h steps, are required and historical solar and wind data are used to calculate the possible renewable power generation. On the demand side, either historical demand curves are used, or—if unavailable—demand curves are calculated based on assumptions of consumer behaviour, the electrical equipment and common electrical appliances.

Figure 3.1 provides an overview of the interactions between the energy- and GIS-based models. The climate model is not directly connected but provided the probabilistic temperatures for the 2.0 °C and 1.5 °C Scenarios. The land-use and non-CO_2 emissions modules provide information on additional gases based on the energy-related CO_2 emissions (output of the energy model). Besides the climate and energy models, the effects on employment and the requirements for selected metal resources have been calculated (see Sects. 3.6 and 3.7).

3.2 Global Mapping—Renewable Energy Potential in Space-Constrained Environments: [R]E-SPACE

The primary purpose of GIS mapping is to ascertain the renewable energy resources (primarily solar and wind) available in each region. It also provides an overview of the existing electricity infrastructures for fossil fuel and renewable sources.

In this project, mapping was undertaken with the computer software QGIS. QGIS is a free, cross-platform, open-source desktop GIS application that supports the viewing, editing, and analysis of geo-spatial data. It analyses and edits spatial information and composes and exports graphical maps, and was used to allocate solar

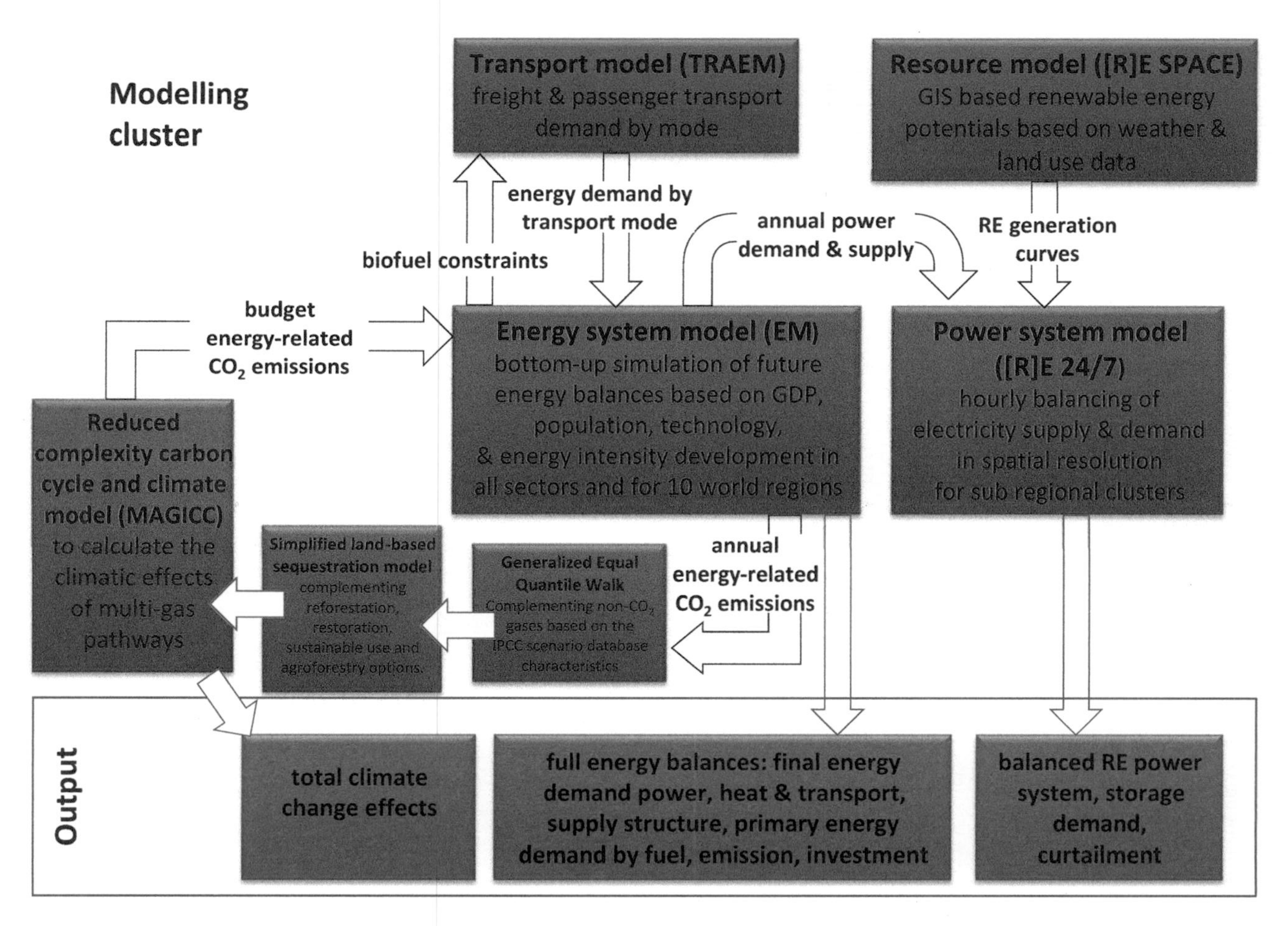

Fig. 3.1 Interaction of models in this study

Table 3.1 Overview of regions and sub-regions used in the analysis

Regions	Cluster/Sub-regions	Regions	Cluster / Sub-regions
North America	USA-Alaska West Canada East Canada North-West USA North-East USA South-West USA South East USA Mexico Mexico	Eurasia	Central Asia Eastern Europe East Caspian West Caspian Kazakhstan Mongolia Russia
Latin America	Argentina Brazil Caribbean Central America Central—South America Chile North Latin America Uruguay	Non-OECD Asia	Asia West: Pakistan, Afghanistan, Nepal, Bhutan Sri Lanka Asia Central North: Viet Nam, Laos and Cambodia Asia North West: Bangladesh, Myanmar, Thailand Asia South-West: Malaysia, Brunei Pacific Island States Indonesia Philippines
Europe	Balkans & Greece Baltic Central Europe Nordic Iberian Peninsula Turkey UK & Ireland	India	East India North India Northeast India South India including Islands West India
Africa	Central Africa East Africa North Africa South Africa Southern Africa West Africa	China	Central China East China North China Northeast China Northwest China South China Taiwan Tibet
Middle East	East ME North ME Iran Iraq Israel Saud Arabia UAE	OECD Pacific	South Korea North Japan South Japan North New Zealand South New Zealand Australia—NEM Australia—SWIS

and wind resources and for demand projection for each region analysed. Open-source data and maps from various sources were used to visualize each country and its regions and districts. The regions and districts are divided into clusters. The regions are divided along geographic boundaries, using the IEA regions as a guide. Some of the larger countries, such as China and India, have been extracted to create individual scenarios. The clusters are also divided on geographic and political bases. A list of regions and their respective clusters is given in Table 3.1.

Wind speed data at different levels, in metres per second (m/s), were obtained from Vaisala 2017. For this analysis, wind speed at a height of 80 m was used to determine the electricity-generation potential. Wind speeds are categorized and mapped within the range of 5–12 m/s to gain an understanding of the potential generation across the regions. Speeds under 5 m/s are ignored when plotting optimal sites. Land-cover types were constrained to bare soil and grasslands. The model only accounts for the onshore wind-generation potential.

Land-cover data were obtained from the Global Land Cover 2000 project (Global Land Cover 2015), hosted by the European Commission's Joint Research Centre. The classification was based on the FAO Land Cover Classification System.

Similarly, solar resource data were obtained from the Global Solar Atlas (Global Solar Atlas 2016), owned by the World Bank Group and provided by SolarGis. Data categorized by direct normal irradiation were mapped to estimate the potential PVs in the different regions. To avoid conflict with competing uses of land, only the land-cover types 'bare soil' and 'grasslands' were included in the analysis.

The area of land available for potential solar and wind power generation was calculated at the cluster level using the Geometry tool in the QGIS-processing toolbox. Intersects (overlapping areas between different layers) were created between the transmission-level layers and the solar/wind utility vector layers to break down the total land area available into clusters. A correction was put in place manually for sites that intersected the cluster boundaries and were part of two clusters.

For some maps (India, China, the Middle East, and OECD Pacific) with large data files, the analysis was performed using raster files for land use and renewable potentials. The raster tools 'clipper' (used to cut a raster file to the size of the cluster) and 'merge' (used to extract common areas between two layers) were used. This input was fed into the calculations for the [R]E 24/7 Model.

The regional maps illustrate the different clusters that were identified for scenario modelling. The existing infrastructure maps highlight the power plants and transmission networks in the regions. The wind and solar potential maps indicate the land available for new power generation given the current land-use patterns. These maps show utility-scale installations. There are much larger expanses of land available for small-scale distributed energy generation.

The following types of maps were created for 10 world regions:

<u>Regional breakdown into a maximum of eight clusters:</u>

The example given in Fig. 3.2 shows OECD North America—one of the 10 world regions—broken down into eight sub-regions (clusters). The [R]E 24/7 power system analysis (see Sect. 3.5) calculates an electricity demand and supply scenario for each of those eight clusters. The clusters can exchange electricity with each other (see Sect. 3.8).

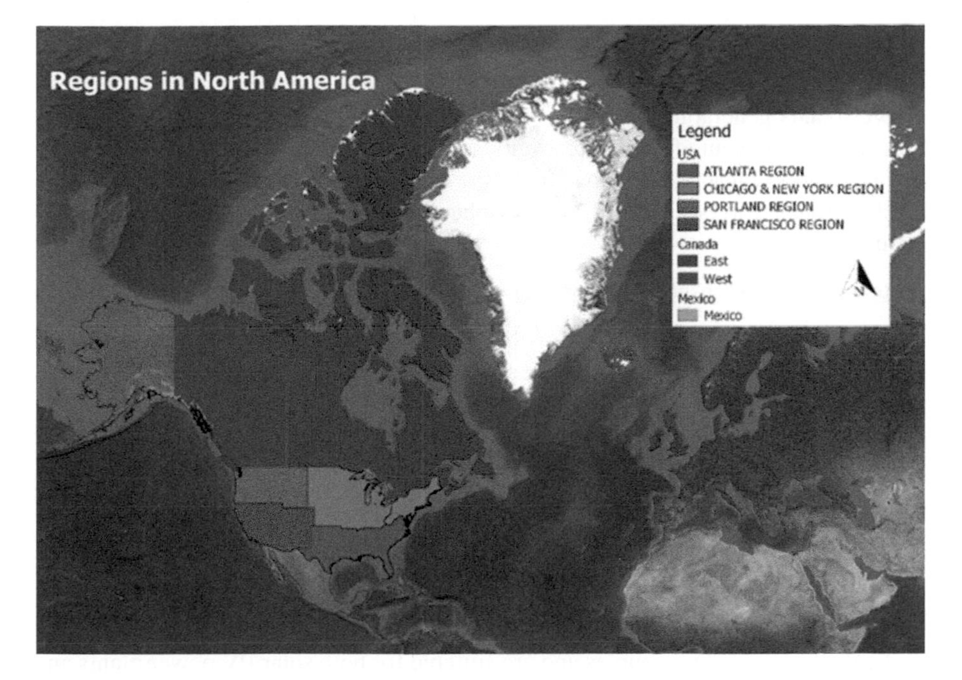

Fig. 3.2 OECD North America broken down into eight sub-regions

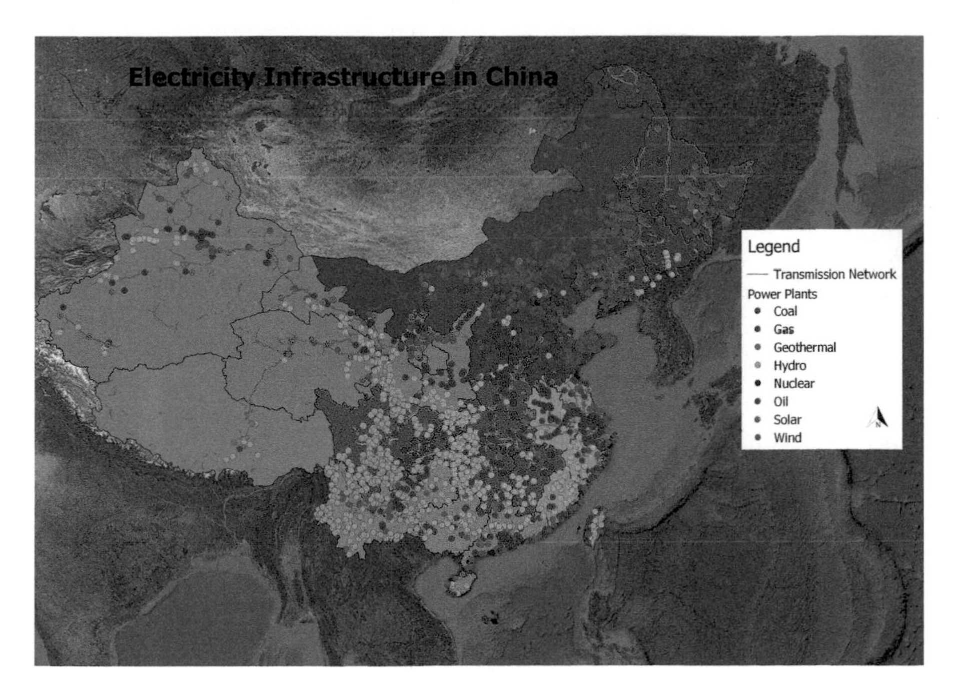

Fig. 3.3 Current electricity infrastructure in China

Current electricity infrastructure

The example given in Fig. 3.3 shows the current electricity infrastructure—power plants generating > 1 MW—and high-voltage transmission lines in China. For the development of future electricity scenarios, it is important to know whether the generation capacity for dispatch and the transmission capacity to transport electricity from utility-scale wind and/or solar power plants to demand centres are available.

Potential sites for onshore wind power

Figure 3.4 gives an overview of the potential onshore wind-power-generation sites in Africa. Only the blue areas are available for new wind development, whereas the remaining regions are used for nature conservation, agriculture, settlement, or other forms of land use that do not allow the installation of wind farms. The darker the blue area, the better the wind potential.

Potential sites for utility-scale solar power plants

Figure 3.5 shows the suitable sites for utility-scale solar power sites in Central and South America. The scale from yellow to orange to red indicates increasing available solar radiation. Red areas—in this example, the Atacama Desert in Chile—indicate the best solar resources and are suitable for both solar PV power plants and concentrated solar power plants.

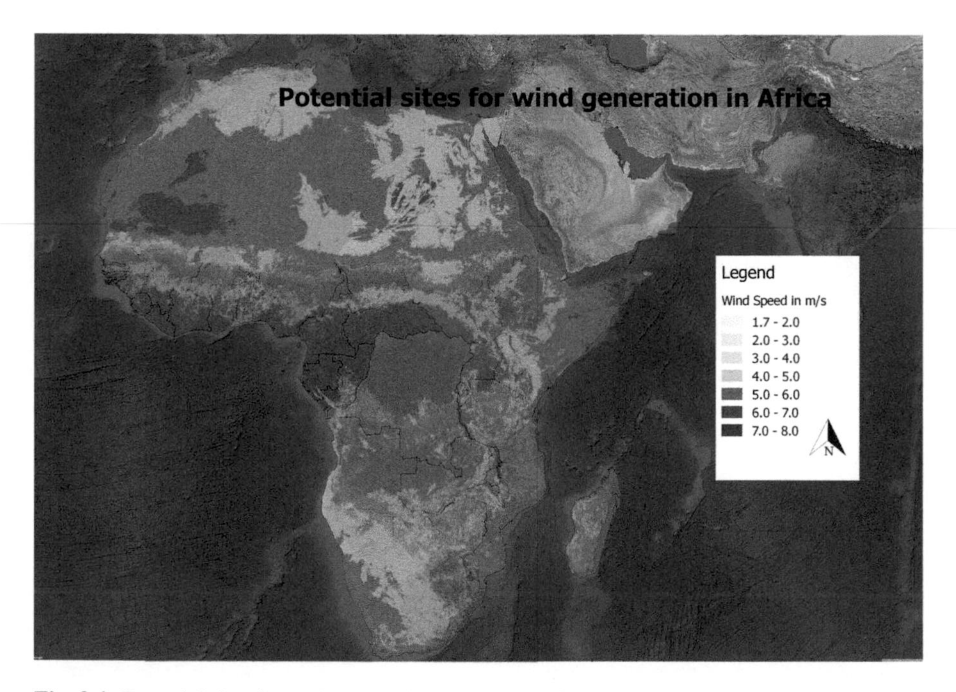

Fig. 3.4 Potential sites for onshore wind generation in Africa

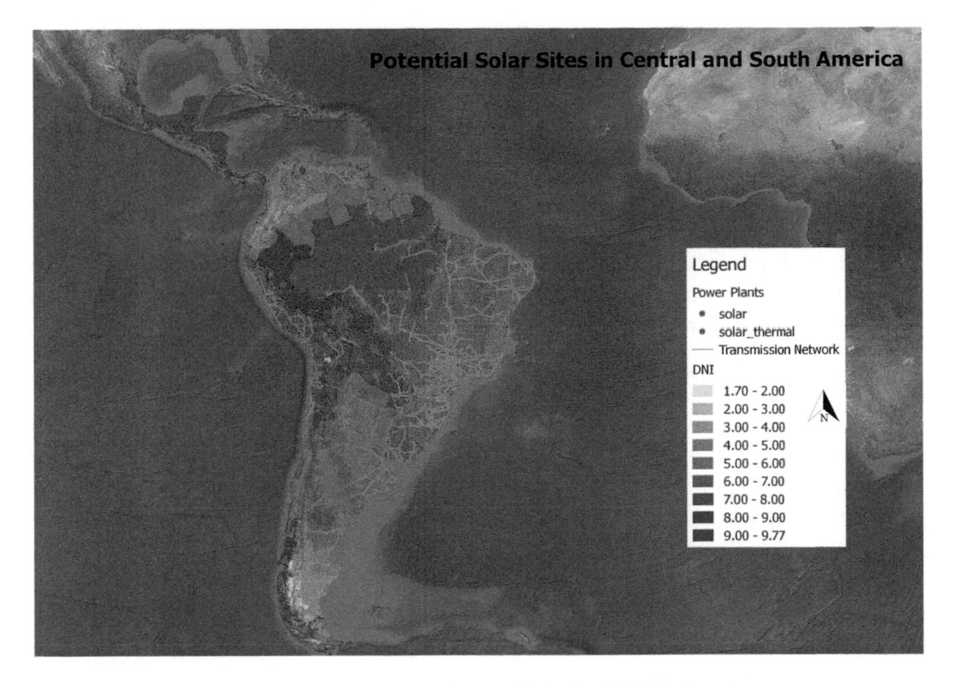

Fig. 3.5 Existing and potential solar power sites in Central and South America

3.3 Transport Energy Model-TRAEM

3.3.1 Transport Model Structure

The transport scenario model TRAEM (TRAnsport Energy Model) allows the modelling of long-term transport developments for the 10 world regions. It is divided into several sub-models according to the transport modes, which are discussed below. All 10 world regions are aggregated in the world model using a passenger–km (pkm) and tonne–km (tkm) activity-based bottom-up approach. The model calculates the final energy demand by multiplying the specific transport demand of each transport mode with the powertrain-specific energy demands. This gives the annual energy demand for electricity, fossil fuels (diesel, petrol), natural gas, bio-based fuels, synthetically produced fuels (also called 'synfuels'), and hydrogen for each of the 10 world regions. The calculation is performed in 5-year steps, from 2015 to 2050.

For all scenarios (5.0 °C, 2.0 °C, and 1.5 °C), the 2015 energy demand by region was adjusted to the IEA World Energy Balances 2017 and is therefore identical in all scenarios. The projected total energy demands for the reference scenario (5.0 °C) from 2020 until 2040 follow the IEA World Energy Outlook 2017 Current Policies Scenario (IEA 2017b). The total energy demands by region for the years 2045 and 2050 were extrapolated linearly based on the 2035–2040 change rates. The 2.0 °C

Scenario was adjusted from 2020 onwards to 2050 in line with the carbon budget of the 2.0 °C pathway and the 1.5 °C pathway.

In the transport model, the CO_2 emissions from biofuels are given a GHG emission factor of zero, because the downstream emissions level out with the upstream emissions. The CO_2 emissions from synthetic fuels are also given a value of zero, because the CO_2 used for producing the fuels upstream level out the downstream emissions. The upstream emissions from electricity and hydrogen production and all other fuels are factored into the energy system model described in Section 5 with which the transport model has a data interface. The model distinguishes between road, rail, aviation, and maritime passenger and freight transport modes.

Road passenger transport modes include:

- Light duty vehicles (cars): automobiles, vans and sports utility vehicles with up to eight seats for private transport, which are further distinguished into small, medium and large cars;
- 2- and 3-wheel vehicles: includes rollers, motorbikes, and rickshaws;
- Busses: urban, suburban, and long-distance buses and minibuses serving public and private-company transport services.

Rail passenger:

- Urban metro/light rail vehicles;
- Regional/intercity trains;
- High-speed trains.

Aviation (passenger):

- Small and medium aircrafts for domestic flights;
- Medium and large aircrafts for international flights, distinguishing narrow-body, wide-body, and regional jets.

Road freight:

- Light-duty trucks (< 3.5 t gross vehicle weight [GVW]);
- Medium-duty trucks (3.5–15 t GVW);
- Heavy-duty trucks (> 15 t GVW).

Rail freight:

- Ordinary freight, intermodal, and low-density high-value freight trains are distinguished.

Navigation (freight):

- Inland navigation;
- Coastal ships for domestic navigation and maritime shipping are distinguished in the model.

We assume that energy efficiency improves over time. The changes in the powertrain shares over time are mainly driven by fleet electrification. Energy intensities per pkm and per tkm are region-dependent, based on the occupancy rates of the passenger transport modes and the loading factors for freight vehicles. The energy demands of all transport modes (passenger and freight) are summed to the total energy demand by region.

Backcasting transport scenarios are modelled iteratively by fitting the drivetrain shares, transport performance (pkm or tkm), and modal shares until the specific downstream CO_2 budgets of the world regions are met. The emission reductions are based on a combination of technical, operational, and behavioural measures during modelling—such as powertrain electrification, the use of biomass-based and synthetically produced fuels, efficiency increases within transport modes, and modal shifts towards more-efficient modes.

The replacement of internal combustion engines by electric powertrains is prioritized in our modelling. However, the rapid electrification of fleets is quantity-restricted over the immediately subsequent years until the capacities for battery production, battery recharging, hydrogen production, and refuelling stations have ramped up ubiquitously. Therefore, a shift towards more energy-efficient and electrified passenger and freight transport modes, such as railways, is required and is therefore one measure implemented in the model. Such modal shifts are especially required in the OECD countries, to reduce carbon emissions while maintaining transport performance at the current levels. Supply constraints on biomass and especially synfuel production will also limit rapid decarbonisation right from the start, and motivate modal shifts and general restrictions to overall transport activities by carbon-intensive transport modes. The 1.5 °C Scenario requires electrification, modal shifts, and alternative fuel uptake to start earlier than the 2.0 °C Scenario and particularly the 5.0 °C Scenario, and their more rapid implementation. However, because electrification will remain quantity-restricted until the 2020s in any case, widespread modal shifts and changes in mobility behaviour are modelled more stringently within the 1.5 °C Scenario. The detailed modelling results are discussed in Chap. 6.

3.3.2 Transport Data

We have derived historical and current data on transport activities (pkm, tkm) and total energy consumption levels according to transport mode from statistical agencies, governmental and intergovernmental organizations, etc., including:

- IEA Mobility Model;
- OECD statistics;
- World Bank Open Data;
- National and supranational statistical bodies;
- UIC IEA Railway Handbook;
- UIC Railway Synopsis;
- Railway operators data;

- HBEFA (Handbuch Emissionsfaktoren);
- EIA Open Data.

However, statistical data are often unavailable or lack consistency with other derived data (for example, on vehicle stock or occupancy rates in certain world regions). In these cases, we applied best guesses based on the scientific and grey literature. Data for energy intensity per transport mode were derived from the German Aerospace Centre (DLR) vehicle databases and the state-of-the-art literature.

3.3.3 Transport Model Output

Based on the TRAEM model, energy consumption and CO_2 emissions can be calculated for each transport sub-category.

The final energy demand (ED) of the passenger and freight transport modes is calculated for every world region and all powertrains in 5-year steps from 2015 to 2050 in the following way:

$$TTED(t) = \sum_{wr=1}^{WR}\sum_{m=1}^{M}\sum_{i=1}^{I} TPP_{m,i}^{wr}(t) \times SECP_{m,i}^{wr}(t) + TPF_{m,i}^{wr}(t) \times SECF_{m,i}^{wr}(t)$$

with:

- $\mathrm{SECF}_{m,i}^{wr}(t)$: specific freight mode energy consumption of powertrain *i* and mode *m* in world region *wr* at time step *t* [MJ/tkm]
- $\mathrm{SECP}_{m,i}^{wr}(t)$: specific passenger mode energy consumption of powertrain *i* and mode *m* in world region *wr* at time step *t* [MJ/pkm]
- $\mathrm{TPF}_{m,i}^{wr}(t)$: freight transport performance of powertrain *i* and mode *m* in world region *wr* at time step *t* [tkm/a]
- $\mathrm{TPP}_{m,i}^{wr}(t)$: passenger transport performance of powertrain *i* and mode *m* in world region *wr* at time step *t* [pkm/a]
- TTED(t): total transport (final) energy demand at time step *t* [PJ/year]

The estimated plug-in hybrid electric vehicles, battery electric vehicles, and fuel-cell-electric vehicles stocks are considered mode by mode, using their respective battery capacities, vehicle-specific life expectancies, total battery capacity by mode, world region, and year, to estimate the total transport battery demand (Chap. 11).

3.4 Energy System Model (EM)

The focus of this study is the development of normative, long-term scenarios. The scenarios are target-oriented. Starting from the identified desirable future in 2050, they use a backcasting process to deliver potential transformation pathways for the energy system. Technical bottom-up scenarios are developed to meet the climate targets in terms of cumulative CO_2 emissions and are then compared with a reference case. The scenarios are based on detailed input data sets that consider defined targets, renewable and fossil fuel energy potentials, and specific parameters for power, heat, and fuel generation in the energy systems. The scenarios are represented in the Energy System model (EM) developed by the DLR, which is implemented in the energy simulation platform Mesap/PlaNet (Seven2one 2012; Schlenzig 1998). Mesap/PlaNet is an accounting framework that allows the calculation of detailed and complete energy system balances, from demand to energy supply, in 5-year steps up to 2050. The model consists of two independent modules:

- a flow calculation module, which balances energy supply and demand annually, and
- a cost calculation module for the calculation of the corresponding investment, generation, and fuel costs.

The strength of the model framework is in its flexible and transparent modelling of different normative paths. The approach requires exogenously defined expansion rates and market shares. It explicitly renounces economic optimization because of the uncertainty of long-term cost assumptions. Therefore, scenario development using this modelling approach is mainly based on background knowledge and derived narratives, and the experience and knowledge of the scenario developer is essential to the success of the scenario-building process. The model acts as a framework for integrating a wide variety of aspects of the transformation of energy systems, and therefore differs fundamentally from optimization models. The standardized cost calculation for the power sector is used for the ex-post evaluation of the scenarios. The modelling framework combines a database with a graphical programming interface. The database allows the management of both the input parameters and the simulation output for the different scenarios calculated. The graphical interface allows the definition of the structure of the modelled system and the quantitative interdependences between the individual structural elements at different structural depths.

The scope of the scenario model allows the increasing electrification processes in the heating and transport sectors to be considered, such as electric vehicles, electric boilers, heat pumps, and hydrogen use. Co-generation in different sectors is also explicitly represented in the model. The EM is implemented in this framework and Figure 3.6 gives an overview of its structure.

Details of the structure and relevant model equations were also recently described by Simon et al. (2018). The model calculates the energy flows of a system on an

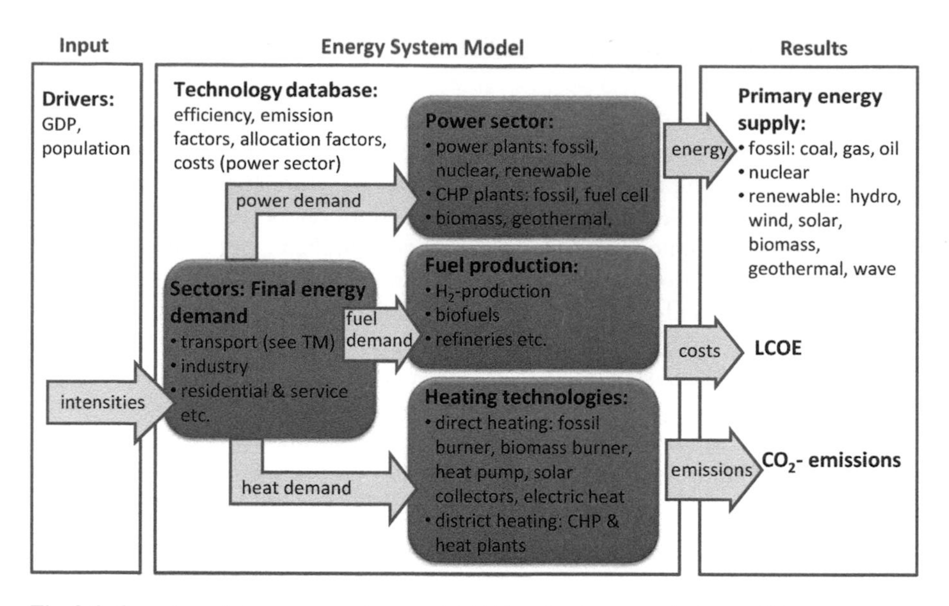

Fig. 3.6 Overview of the energy system model (EM) as implemented in Mesap/PlaNet

annual basis. These flows connect a set of technologies in each sector and for all relevant energy carriers, using linear equations. The equation system is solved sequentially and the model thus balances demand and supply. This approach is applied over the scenario period in 5-year steps until 2050. Ultimately, the overall final energy is calculated as described in the following equations:

$$FD_{ss}^{fe}(t) = \sum_{et} UED_{ss}(t) \cdot MS_{ss}^{et}(t) \cdot \eta_{fe}^{et}(t)$$

$$FD^{fe}(t) = \sum_{ss} FD_{ss}^{fe}(t)$$

$$TFD(t) = \sum_{fe} FD^{fe}(t) = \sum_{ss}\sum_{fe} FD_{ss}^{fe}(t) = \sum_{ss}\sum_{fe}\sum_{et} UED_{ss}(t) \cdot MS_{ss}^{et}(t) \cdot \eta_{fe}^{et}(t)$$

with:

- $FD_{ss,}^{fe}(t)$: demand of (final) energy carrier *fe* in sub-sector *ss*[1] at time step *t* [PJ/year]
- $FD_{ss,}^{fe}(t)$: total demand of (final) energy carrier *fe* at time step *t* [PJ/year]
- TFD(t): total final energy demand at time step *t* [PJ/year]

[1]The sub-sectors include 'heat' and 'non-heat electrical appliances' in the sectors 'Industry' and 'Residential and other', aviation, road transport, navigation, rail transport, non-energy consumption, the conversion sector, and storage and transmission losses for power and district heat. Conversion losses are taken into account in the calculation of the primary energy demand.

- $UED_{ss}(t)$: useful energy demand / transport services in sub-sector *ss* at time step *t* [PJ/year]
- $MS_{ss}^{et}(t)$: market share of end-sector technology *et* in sub-sector *ss* [dimensionless]
- $\eta_{fe}^{et}(t)$: efficiency of end-sector technology *et* using energy carrier *fe*[2] at time step *t* [dimensionless]
- *t*: time step

The indices denote:

ss: sub-sector
fe: (final) energy carrier
et: end-sector technology

The primary energy demand (without exports) is calculated as follows:

$$PD^{pe}(t) = \sum_{ct}\sum_{fe} FD^{fe}(t) \cdot MS_{fe}^{ct}(t) \cdot \eta_{fe}^{ct}(t)$$

$$TPD(t) = \sum_{pe} PD^{pe}(t)$$

with

- $PD^{pe}(t)$: total demand of (primary) energy carrier *pe* at time step *t* [PJ/year]
- $TPD(t)$: total primary energy demand at time step *t* [PJ/year]
- $MS_{fe}^{ct}(t)$: market share of conversion technology *ct* in the generation of final energy carrier *fe* [dimensionless]
- $\eta_{fe}^{ct}(t)$: efficiency of conversion technology[3] *ct* using the final energy carrier *fe* at time step *t* [dimensionless]

The indices denote:

- *pe*: (primary) energy carrier
- *ct*: conversion sector technology[4]

The drivers of energy consumption include forecasts of population growth, gross domestic product (GDP), and energy intensities. Specific energy intensities are assumed for:

- electricity and heat consumption per person and per GDP;
- the ratio of industrial heat demand to GDP;

[2] Note that some technologies (e.g., electric heat pumps) require two energy carriers as inputs (electricity and environmental heat), with a specific efficiency for each energy carrier.

[3] Some conversion technologies produce more than one output, e.g. CHP plants, leading to constraints on efficiencies or market shares.

[4] Power and district heat generation, biofuel, synfuel, and H_2 generation, and refineries.

- demand for energy services, such as useful heat;
- different transport modes based on the Transport Model (see Sect. 3.4).

The model consists of a broad technology database across the heat, fuel, and power sectors, including sector coupling via combined heat and power (CHP), power-to-heat, and power-to-fuels technologies, and electric mobility.

For both heat and electricity production, the model distinguishes between different technologies, characterized by their primary energy sources, efficiency, and costs. Examples include biomass or gas burners, heat pumps, solar thermal and geothermal technologies, and several power-generation technologies, such as PV, onshore and offshore wind, biomass, gas, coal, nuclear, and CHP. In the transport sector, the model is directly linked to the results of the transport model (Sect. 3.3). For each technology, the market share with respect to total heat or electricity production is specified according to a range of assumptions, including targets, potential costs, and societal, structural, and economic barriers. The model eventually calculates the annual energy flows for a set of energy carriers.

The main inputs of the Energy System Model are:

- IEA World Energy Balances 2017 (IEA 2016a) for the calibration of the model for each world region in the years 2005–2015;
- IEA World Energy Outlook 2016/2017 (IEA 2016b, 2017a) for the parameterization of the model for the reference case (5.0 °C Scenario);
- various studies and statistics used for the assumption of further specific values, such as the power-to-heat ratios of co-generation plants, coefficients of performance of heat pumps, and the efficiency of hydrogen electrolysers and synthetic fuel production plants;
- narratives and assumptions regarding the further development of demand and supply technologies in line with the climate targets and by taking into account RES potentials and costs, stable market developments, and the constraints imposed by production capacities and regional implementation. These assumptions and narratives are described in detail in Chap. 5, Sect. 4.

The main outputs of the model are:

- the final and primary energy demands, broken down by fuel, technology, and energy sectors, as defined by the International Energy Agency (IEA)—industry, power generation, transport, and other (buildings, forestry, and fisheries);
- the results broken down by the three main types of energy demand—electricity, heating, and mobility (transport); specifically, the energy required, technology deployment, and financial investment for each of these energy demand types;
- total energy budget, which is the total cost of energy for the whole power system;
- energy-related CO_2 emissions over the projected period.

3.5 [R]E 24/7 (UTS-ISF)

The long-term scenarios calculated with the EM for 2020, 2030, 2040, and 2050 (see Sect. 3.4) are used as the input data for the dispatch modelling described in this section. The [R]E 24/7 model transforms a long-term scenario for a specific year into hourly load and generation curves. The annual electricity demand is transformed into an hourly load curve (see Sect. 3.2) and the annual power generation is transformed into a generation time series for variable power generation from regional solar and wind data and dispatchable power-generation data via interchangeable dispatch orders (see Sect. 3.7). The [R]E24/7 model is an accounting framework used to calculate the complete power system balance at 1 h resolution, and consists of two modules:

1. a flow calculation module, which balances the energy supply and demand; and
2. a cost calculation module, which calculates the corresponding generation and fuel costs.

The [R]E 24/7 model examines the influence of the dispatch order of power-generation technologies, the storage technologies, and the interconnection of up to eight regions. It calculates the impact of these variables on the overall system costs. [R]E 24/7 also calculates load curves for the residential, industry, and transport sectors based on the sector-specific energy intensity factors and applications that are in use. The factors and applications used depend on the GDP and population (see Sect. 3.2).

3.5.1 [R]E 24/7—Model Structure

Teske (2015) has developed a three-level grid model called '[R]E 24/7' as a grid analysis tool that differentiates between four voltage levels. For this analysis, the model has been simplified to eight interconnected clusters to reduce the data volume and the calculation time. High resolution, with multiple voltage levels, is impractical for a global energy scenario, because the required input data would not be available for all regions and—if the data were available—the calculation time would be extremely long. Therefore, the simplified [R]E 24/7 model uses eight clusters that can exchange electricity on an hourly basis with a user-defined interconnection capacity (see Sect. 3.8). Different voltage levels are not calculated. Figure 3.7 provides an overview of the different modules of the [R]E 24/7 model. In the first step, a database provides the main input data for the base year, including socio-economic parameters, the currently available power generation, and the energy infrastructure. The data are partly with the GIS tool (see Sect. 3.2) and partly from other information resources, such as publicly available databases of populations (UN PD DB 2018), GDP (CIA 2018; ST 7-2018), and energy efficiency indicators (WEC 2018),

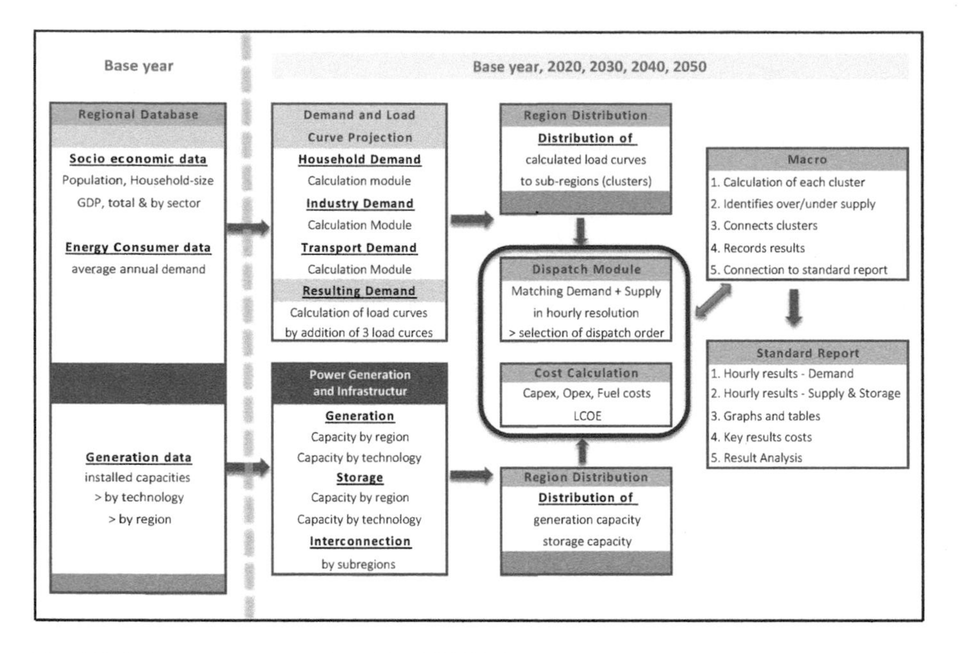

Fig. 3.7 Schematic representation of the [R] E24/7 model structure

and statistical data on renewable power generation from IRENA (REN21-GSR 2018) and the World Resources Institute (WRI 2018).

3.5.2 Development and Calculation of Load Curves

Energy demand projections and the calculation of load curves are important factors in calculating supply security and the dispatch and storage capacities required, especially for energy supply concepts with high proportions of variable renewable power generation. The [R]E 24/7 model calculates the development of the future power demand and the resulting possible load curves, because:

(a) Actual demand curves are not available for all countries and/or regions and are sometimes classified information.
(b) Future load curves with high penetration of storage, electric heating systems and electric mobility will have a very different shape than current load curves.
(c) For developing countries with low access to energy rates or little access to sufficient data, the curves must be calculated based on a set of assumptions because actual curves are neither available nor representative of future load curves.

The model generates load curves and the resulting annual power demands for three different consumer groups/sectors:

- households;
- industry and business; and
- transport (public and individual electric mobility).

Although each sector has its specific consumer groups and applications, the same set of parameters is used to calculate load curves:

- electrical applications in use;
- demand pattern (24 h);
- efficiency progress (base year 2015) for 2020 until 2050, and individual efficiency input for each year.

The calculations involve detailed bottom-up projections of the increased use of electricity for heating in buildings, for industrial process heat, for electric mobility, and for the production of synthetic fuels and hydrogen. They also include increased access to energy in developing countries based on the applications used, the demand patterns, and the household types. This allows detailed demand projections to be made.

Infrastructure needs, such as power grids combined with storage facilities, require in-depth knowledge of local loads and generation capacities. In this project, the annual electricity demand for each of the 10 world regions was calculated with the long-term EM. The [R]E 24/7 model breaks each region into up to eight subregions (or clusters) to calculate hourly load and generation curves.

3.5.3 Load Curve Calculation for Households

The model differentiates nine household groups, with various degrees of electrification and equipment:

- Rural—phase 1: Minimal electrification stage
- Rural—phase 2: White goods are introduced and increase the overall demand
- Rural—phase 3: Fully equipped western-standard household with electrical cooking and air conditioning and electric vehicle(s)
- Urban single: Single-person household with minimal equipment
- Urban shared flat: 3–5 persons share one apartment; fully equipped western household, but without electric vehicles
- Urban—Family 1: 2 adults and 2–3 children; middle income, middle western standard
- Urban—Family 2: 2 adults and > 3 children and/or higher income, full western standard
- Suburbia 1: average family, middle income, full equipment, high transport demand due to extensive commuting
- Suburbia 2: High-income household, fully equipped, extremely high transport demand due to high-end vehicles and extensive commuting

The following electrical equipment and applications can be selected:

- Lighting: 4 different light bulb types (LED, three efficiency classes of CFLs),
- Cooking: 10 different cooking stoves (2+4 burners, electricity, gas, firewood)
- Entertainment: 3 different efficiency levels of computers, TV, and radio types
- White goods: 2 different efficiency levels each for washing machine, dryer, fridge, freezer
- Climatization: 2 different efficiency levels each for fan, air-conditioning
- Water heating: A selection of direct electric, heat-pump, and solar heating systems

3.5.4 Load Curve Calculation for Business and Industry

The industrial sector is clustered into eight groups based on widely used statistical categories:

- Agriculture
- Manufacturing
- Mining
- Iron and steel
- Cement industry
- Construction industry
- Chemical industry
- Service and trade

Each sector has a definite energy intensity in energy per dollar GDP (MJ/$\$_{GDP}$), which is been converted to electrical units (kW/$\$_{GDP}$) based on an estimated fuel efficiency factor, electricity shares, and operational hours per year. The calculated electricity intensity per dollar GDP conversion can only show the required connected load and the specific consumption of an industrial sector to a first approximation because there is a variety of uncertainty factors, such as:

(a) significant regional differences;
(b) significant demand differences within one industry sector, such as *manufacturing* or *chemical industry*;
(c) lack of standardized data on industry energy demands, especially for the electricity sector.

Despite the high degree of uncertainty, we decided to apply this methodology because after an initial calibration, the current statistically recorded industrial electricity consumption in some well-documented countries (e.g., USA) and regions (e.g., Europe) can be recalculated with a tolerance of ± 10%. However, this methodology requires further research.

3.5.5 Load Distribution by Cluster

The spatial concept of the [R]E 24/7 is shown in Fig. 3.8. The model calculates the load distribution for one region, which can be broken down further to a maximum of eight sub-regions (or 'clusters'). Therefore, the 10 world regions modelled in this analysis are calculated separately. OECD North America, for example, includes Canada, USA, and Mexico. These three countries can be subdivided into up to eight clusters. A cluster can be a country (e.g., Mexico), a province/state of a country (e.g., Alaska), or a selection of several provinces/states (e.g., West Canada = British Columbia, Alberta, Yukon Territory, and North-West Territories). A cluster is defined to capture the existing interconnected power supply areas of a region, a country, or across several provinces. In Europe, for example, one cluster is the Iberian Peninsula (Spain and Portugal), a region within Europe that has only very limited interconnection with the central European grid system (UCT-E). However, data availability and the model limitations (maximum of eight clusters) force simplifications, and countries or state/provinces must be bundled together in one cluster even though they may have significant differences. Therefore, further research is required to obtain more detailed results for selected countries or provinces.

The distribution of the regional load, calculated in Sects. 3.3 and 3.4, is connected to the projected GDP, population, and power plant capacities for each cluster.

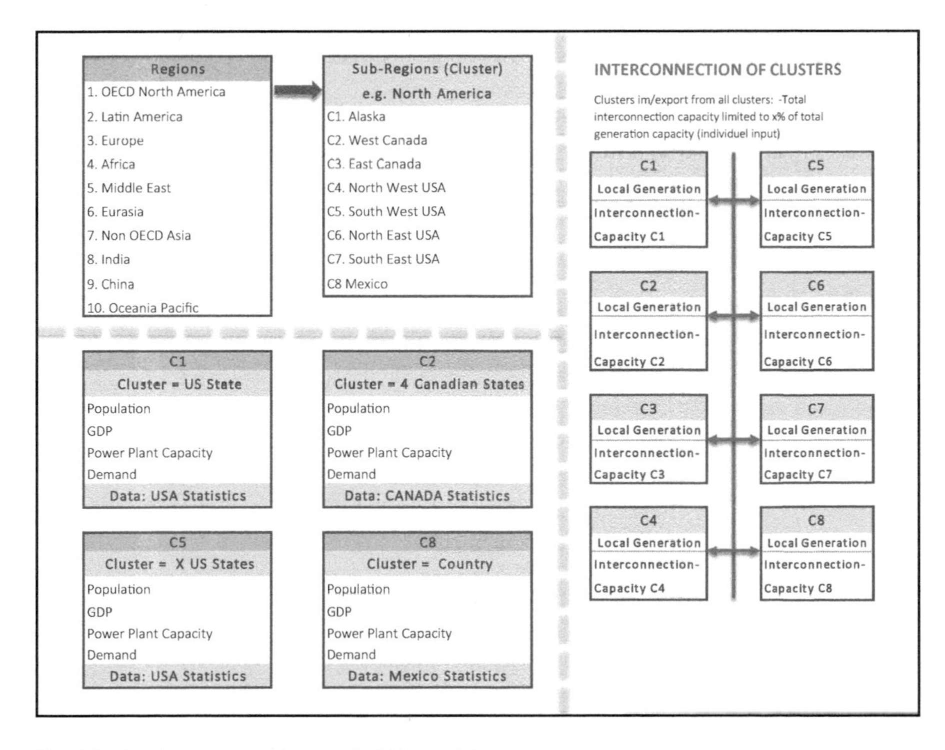

Fig. 3.8 Spatial concept of the [R]E 24/7 model

The cluster-specific data for the base year (2015) are taken from the model's database interface to calculate the demand and supply for the base year. When data are not available for each sub-region, the input data from the entire region will be broken down (in percentages) by cluster, according to the population—as a result of the GIS analysis. For the global analysis, the spatial distribution of the population, GDP, and power plant capacities remain unchanged over the modelled years (2020–2050) for all 10 regions and their respective sub-regions. In the next step, the resulting population and GDP values for each cluster are multiplied by the normalized load curves calculated as described in Sects. 3.3 and 3.4. Each cluster has an hourly load curve over one entire year (8760 h). Thus, one region (e.g., OCED North America) has eight different load curves.

3.5.6 The [R]E 24/7 Dispatch Module

Although the dispatch module for the [R]E 24/7 energy access model has been developed specifically for this study, integral parts have been taken from a model developed to analyse the generation and storage needs for a micro grid on Kangaroo Island (Dunstan and Fattal 2016), the Australian Storage Requirements (Rutovitz and James 2017), and a 100% Renewable Energy Analysis for Tanzania (Teske and Morris 2017). The key objective of this modelling is to calculate the theoretical generation and storage requirements for energy adequacy in each cluster and for the whole survey region (Tables 3.2 and 3.3).

Figure 3.9 provides an overview of the dispatch calculation process for one cluster. The key inputs include the generation capacities by type, the demand projections and load curves for each cluster, the interconnection with other clusters, and the meteorological data from which to calculate the solar and wind power genera-

Table 3.2 Input parameters for the dispatch model

Input parameter		
$L_{Cluster}$	Load Cluster	[MW]
$L_{Interconnection}$	Load Interconnection (Im- or Export)	[MW]
$L_{Initial}$	Initial Load (Cluster + Interconnection)	[MW]
$Cap_{Var.RE}$	Installed capacity *Variable Renewables*	[MW]
$Meteo_{Norm}$	Meteorological data for solar and wind	[MW/MW$_{INST}$]
$L_{Post\_Var.RE}$	Load after *Variable Renewable* Supply	[MW]
$Cap_{Storage}$	Capacity *Storage*	[MW]
$CapFact_{Max\_Storage}$	Max capacity factor storage technologies	[h/year]
$L_{Post\_Storage}$	Load after *Storage* Supply	[MW]
$Cap_{Dispatch}$	Capacity *Dispatch Power Plants*	[MW]
$CapFact_{Max\_Dispatch}$	Max capacity factor *Dispatch Power Plants*	[h/year]
$L_{Post\_Dispatch}$	Load after *Dispatch Power Plant* Supply	[MW]
$Cap_{Interconnection}$	Capacity *Interconnection*	[MW]

Table 3.3 Output parameters for the dispatch model

Output parameter		
$L_{Initial}$	Initial Load (Cluster + Interconnection)	[MW]
$L_{Post\_Var.RE}$	Load after *Variable Renewable* Supply	[MW]
$S_{EXECC\_VAR.RE}$	Access supply Renewables	[MW]
$L_{Post\_Storage}$	Load after *Storage* Supply	[MW]
$S_{Storage}$	Storage Requirement/Curtailment	[MW]
$CapFact_{Actual\_Storage}$	Utilization Factor Storage	[h/year]
$L_{Post\_Dispatch}$	Load after *Dispatch Power Plant* Supply	[MW]
$S_{Dispatch}$	Dispatch Requirement	[MW]
$CapFact_{Actual\_Dispatch}$	Utilization Factor *Dispatch Power Plants*	[h/year]
$L_{Post\_Interconnection}$	Load after *Interconnection* Supply	[MW]
$S_{Interconnection}$	Interconnection Requirement	[MW]
$CapFact_{Actual\_Interconnection}$:	Utilization Factor *Interconnection*	[h/year]

tion at hourly resolution. The calculation of one region with eight sub-regions will require eight calculation intervals. Table 3.4 shows the four different supply technology groups: variable renewables, dispatch power plants, storage technologies, and interconnections. The model allows the order in which the technology groups will be utilized to be changed to satisfy the demand. Storage and interconnection cannot be selected as the first supply technology. Within each technology group, the dispatch order can be changed. Tables 3.5, 3.6, and 3.7 provide an overview of all the available technologies and examples of different dispatch scenarios. While CSP plants with storage are dispatchable to some extent—depending on the storage size and the available solar radiation—they are part of the variable renewable group in the [R]E 24/7 model. Although the model allows the dispatch order to change, the 100% renewable energy analysis always follows the same dispatch logic. The model identifies excess renewable production, which is defined as potential wind and solar PV generation greater than the actual hourly demand in MW during a specific hour. To avoid curtailment, the surplus renewable electricity should be stored with some form of electric storage technology or exported to a different cluster. Within the model, the excess renewable production accumulates through the dispatch order. If storage is present, it will charge the storage within the limits of the input capacity. If no storage is present, this potential excess renewable production is reported as 'potential curtailment' (pre-storage).

Limitations: It is important to note that the calculation of possible interconnection capacities for transmission grids between sub-regions does not replace technical grid simulation. Grid services, such as inductive power supply, frequency control, and stability, should be analysed, although this is beyond the scope of this analysis. The results of [R]E 24/7 provide a first rough estimate of whether the increased use of storage or increased interconnection capacities or a mix of both will reduce systems costs.

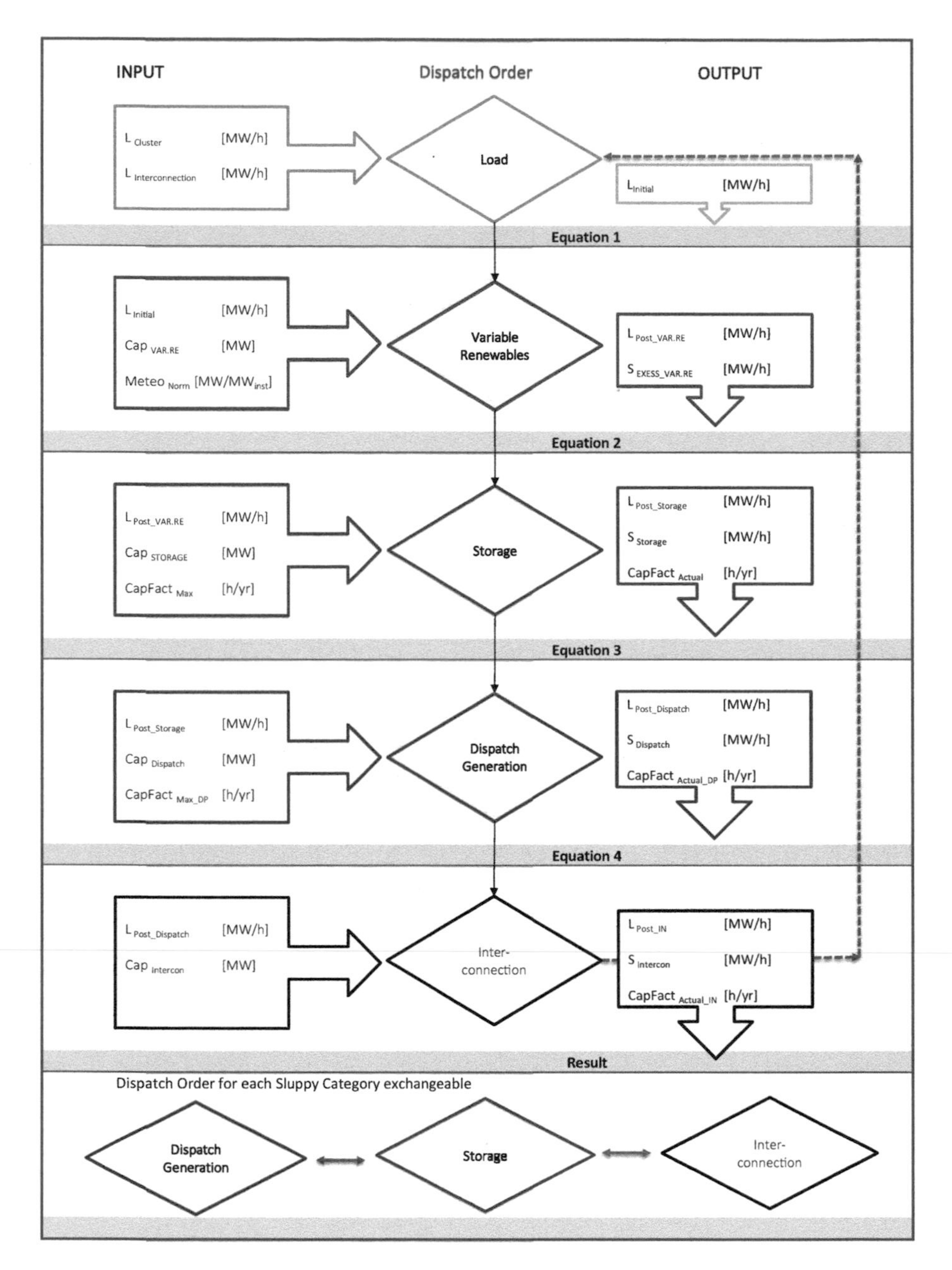

Fig. 3.9 Dispatch order module of the [R]E 24/7 model

Table 3.4 Technology groups for dispatch order selection

Technology options	Input
Variable renewables	Variable renewables
Storage	Storage
Dispatch generation	Dispatch generation
Interconnector	Interconnector

Table 3.5 Technology options—variable renewable energy

Variable renewable power technology options	Input
Photovoltaic—roof top -	Photovoltaic—roof top
Photovoltaic—utility scale -	Photovoltaic—utility scale
Wind—onshore -	Wind—onshore
Wind—offshore -	Wind—offshore
CSP (Dispatchable)	CSP

Table 3.6 Technology options—dispatch generation

Dispatch generation	
Technology options	Input
Bioenergy	Hydropower
Geothermal	Bioenergy
Hydropower	CoGen Bio
Ocean	Geothermal
Oil	CoGen Geothermal
Gas	Ocean
CoGen bio	Gas
CoGen geothermal	CoGen Gas
CoGen gas	Coal
CoGen coal	CoGen Coal
Coal	Brown Coal
Brown coal	Oil
Nuclear	Nuclear

3.5.7 Meteorological Data

Variable power-generation technologies are dependent on the local solar radiation and wind regimes. Therefore, all installed capacities of this technology group are connected to cluster-specific time series. The data were derived from the database *Renewable Ninja* (RE-N DB 2018), which allows the simulation of the hourly power output from wind and solar power plants at specific geographic positions throughout the world. Weather data, such as temperature, precipitation, and snowfall, for the year 2014 are also available.

Table 3.7 Technology options—storage technologies

Storage	
Technology option	Input
Battery	Battery
Hydro pump STORAGE	Hydro Pump storage
H2	H2

To utilize climatization technologies for buildings (air-conditioning, electric heating), the demand curves for households and services are connected to the cluster-specific temperature time series. The demand for lighting is connected to the solar time series to accommodate the variability in lighting demand across the year, especially in northern and southern regions, which have significantly longer daylight periods in summer and very short daylight periods in winter.

3.5.7.1 Solar and Wind Time Series

For every region included in the model, hourly output traces are utilized for onshore wind, offshore wind, utility solar, CSP, and roof-top solar energies. Given the number of clusters and the geographic extent of the study, and the uncertainty associated with the prediction of the spatial distribution of future generation systems, an representative site was selected for each of the five generation types. For utility solar and CSP, the indicative sites were situated in areas of high solar output, close to the transmission network or regional centre or city, and in areas without competing land uses (as described in the mapping methodology). A roof-top solar indicative site was chosen in the demographic centre of the region, usually the capital city.

The onshore wind indicative site selected for each region was situated in an area of non-competing land use with the highest average wind speed and close to the transmission network or regional centre or city. The offshore wind indicative site was an area within 100 km of the shore with the highest average wind speed, and close to the transmission network or regional centre or city. In some cases, no acceptable wind area within a region was available, in which case the wind potential was set to zero.

Once the indicative sites were chosen, the hourly output values for typical solar arrays and wind farms were selected using the database of Stefan Pfenninger at ETH Zurich and Iain Stafell (Renewables.ninja; see above). The model methodology used by the Renewables.ninja database is described by Pfenninger and Staffell (2016a, b), and is based on weather data from global reanalysis models and satellite observations (Rienecker and Suarez 2011; Müller and Pfeifroth, 2015; SARAH 2018). It was assumed that the utility solar sites were optimized, and as such, a tilt angle was selected within a couple of degrees of the latitude of the indicative site. For roof-top solar, this was left at the default 35° because it is likely that the panels matched the roof tilt.

The wind outputs for both onshore and offshore wind were calculated at an 80 m hub height because this reflects the wind data sets used in the mapping exercise. Although onshore wind and offshore wind are likely be higher than this, 80 m was considered a reasonable approximation and made our model consistent with the mapping-based predictions. A turbine model of Vestas V90 2000 was used.

Limitations: The solar and wind resources can differ within one cluster. In some cases, there are even different climate zones within one large cluster, e.g., in Australia and Russia. Therefore, the potential generation output can vary within a cluster and across the model period (2020–2050). Furthermore, some clusters extend significantly across several time zones, such as Russia. The model can only take into account the time variations in sunrise and sunset between different clusters, but not within a single cluster. The effect of time differences within clusters with a large east–west spread requires high-resolution modelling, which is possible with the [R]E 24/7 model but beyond the scope of this research project.

3.5.8 Interconnection Capacities

The interconnection capacities are set as a function of the total generation capacity within a cluster and a manually set percentage. Defining the relevant percentage of a country's overall (peak) capacity and/or total generation capacity is based on European energy policy. The European Union (EU) proposed in 2002 that all EU member states must establish a transmission capacity of at least 10% of the peak demand (in megawatts) by 2005 (EMP-BARCELONA 2002). The EU developed this regulation further, improved the calculation method, and increased the target to 15% (EU-EG 2017), whereas the [R]E 24/7 model implements a simplified approach by taking a percentage of the overall installed capacity. Clusters that are not connected at all to the real energy market (e.g., South Korea, Japan, Australia, and New Zealand in the OECD Pacific region) are assigned 0% interconnection capacity. Responsibly well-connected clusters (such as the south-western USA) are set to 15%, and highly interconnected countries (such as Denmark) are assigned up to 40%.

Several simplifications have been made to the [R]E 24/7 model for ease of computation and to accommodate the paucity of data and uncertainty about the future when designing the interconnector algorithms:

- Interconnections between the project-defined regions are the only ones considered, so all intra-regional interconnections or line constraints are excluded (*'copper plate'*);
- Optimal load flow is neglected because policy and price signals are considered to be the factors dominating the international and inter-regional load flow;
- Non-adjacent inter-regional interconnections are neglected for computational reasons, e.g., one region cannot buy power from a region with which it does not share a border.

The algorithm devised for the function of the interconnectors is based on three key pieces of information for each region in a cluster:

- Excess generation capacity;
- Unmet load;
- Interconnection capacity with each adjacent region, both in and out.

The excess generation capacity and unmet load were calculated by running the model without the interconnectors to determine the excess or shortfall in generation when the load within the region is met. These excesses and shortfalls were calculated at the point in the dispatch cascade at which the interconnectors provide or consume power, for example, after the variable renewables and dispatchable generators and before the storage elements.

The interconnection capacity between adjacent regions was defined based on a percentage of the maximum regional load. The capacity was defined in a matrix, both to and from every region to every other region. For non-adjacent regions, the capacities were set to zero. A priority order for each region to every other region was given, so that if the region had an unmet load, it would be served sequentially with the excess generation of the loads in their defined order.

For every hour and for every region in each cluster, the possible interconnection inflow or outflow for load balancing was calculated. Each region was considered in turn, and the algorithm attempted to meet the unmet load with excess generation by adjacent regions, keeping track of the residual excess load and interconnector capacities. Each region's internal load was met first, before its generation resources were considered for other interconnected regions.

Once the total inflow and outflow of the interconnectors were calculated, the hourly values were fed into the model once more at the position in the cascade to which they were assigned, and the model was run again to give the total system behaviour. For regions sending generation capacity to other regions, the interconnector element behaved as an increase in load, whereas for regions accepting power from neighbouring regions, the interconnector element behaved as an additional generator, from the model's perspective.

3.6 Employment Modelling (UTS-ISF)

Two of the key dimensions influencing the social and economic impacts of the transition from fossil-fuel to clean energy are the quantity and type of jobs that are lost and created. Currently, there are limited data on the volumes of jobs that will be lost and created within particular occupations and locations during the transition to clean energy. National statistical agencies classify and collect data on occupations within the fossil fuel sectors but not within the renewable energy sectors (ABS 2017). ISF has developed a model to estimate the volume of renewable energy jobs under different 100% global renewable energy scenarios (Rutovitz and Dominish 2015), and an increasing body of research is estimating the jobs created by renewable energy. The following section provides an overview of the basic methodology. Based on this, UTS/ISF has developed this methodology further, as presented in Sect. 3.2.

3.6.1 Quantitative Employment Calculation

In 2015, the Institute for Sustainable Futures (ISF) at the University of Technology Sydney (UTS) developed a quantitative employment model that calculates employment development in the electricity, heating, and fuel production sectors for the analysis of future energy pathways (Rutovitz and Dominish 2015). Figure 3.10 provides a simplified overview of how the calculations are performed, based on Rutovitz (2015b). The main inputs for the quantitative employment calculations are:

for each calculated scenario, e.g., the 5.0 °C (Sect. 5.1.1) and 2.0 °C Scenarios (Sect. 5.1.2),

- the electrical and heating capacity that will be installed each year for each technology;
- the primary energy demand for coal, gas, and biomass fuels in the electricity and heating sectors;
- the amount of electricity generated per year from nuclear power, oil, and diesel.

for each technology:

- 'employment factors', or the number of jobs per unit of capacity, separated into manufacturing, construction, operation, and maintenance, and per unit of primary energy for fuel supply;
- for the 2020, 2030, and 2050 calculations, a 'decline factor' for each technology, which reduces the employment factors by a certain percentage per year. This reflects the fact that employment per unit decreases as technology efficiencies improve.

for each region:

- the percentage of local manufacturing and domestic fuel production in each region, to calculate the proportions of jobs in manufacturing and fuel production that occur in the region;
- the percentage of world trade in coal and gas fuels, and traded renewable components that originates in each region.

A 'regional job multiplier', which indicates how labour-intensive the economic activity is in that region compared with the OECD, is used to adjust the OECD employment factors when local data are not available. The figures for the increase in electrical capacity and energy use from each scenario are multiplied by the employment factors for each of the technologies, and then adjusted for regional

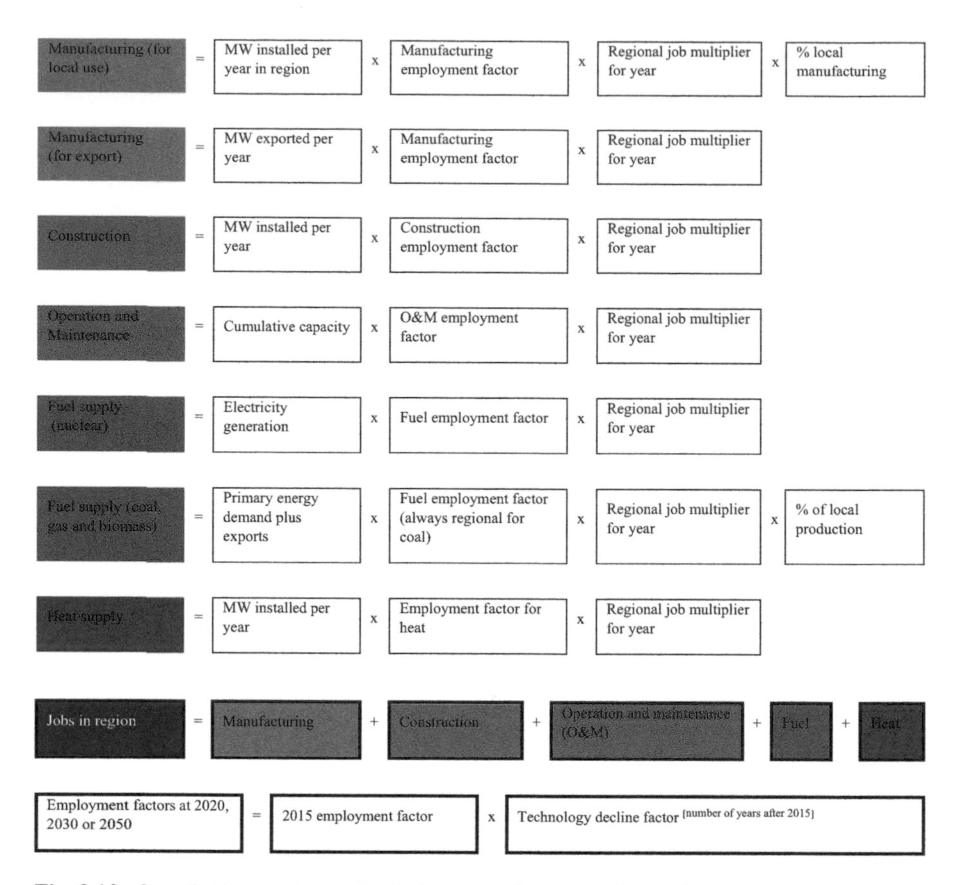

Fig. 3.10 Quantitative employment calculation: methodological overview

labour intensity and the proportion of fuel or manufacturing that occurs locally. The calculation is summarized in Fig. 3.10.

A range of data sources were used for the model inputs, including the International Energy Agency, US Energy Information Administration, BP Statistical Review of World Energy, US National Renewable Energy Laboratory, International Labour Organization, World Bank, industry associations, national statistics, company reports, academic literature, and the ISF's own research.

These calculations only take into account direct employment; for example, the construction team required to build a new wind farm. They do not include indirect employment; for example, the extra services provided in a town to accommodate the construction team. The calculations do not include jobs in energy efficiency because this is beyond the scope of this project. The large number of assumptions required to make these calculations means that employment numbers are only estimates, especially for regions where few data exist. However, within the limits of data availability, the figures presented are representative of employment levels under the 5.0 °C and 2.0 °C Scenarios.

3.6.2 Occupational Employment Modelling

The quantitative employment model documented in Sect. 3.6.1 were further developed to analyse the qualitative occupational composition of employment in the fossil fuel and renewable energy industries. UTS-ISF has developed a framework for modelling disaggregated occupational change, and this framework is described in this section.

Quantitative employment studies at the level of technology and project phases (manufacturing, construction, and O&M) are useful when providing estimates of aggregate job creation. However, more disaggregated, granular data on the locations and types of occupations are required to plan a just transition to renewable energy. For example, it is necessary to know how many electricians are currently employed in fossil fuel industries and how many will be employed in the renewable energy sectors. Although our forecasts will almost inevitably be wrong, key trends can be established. For example, we can direct our focus to areas of the workforce in which an increase in the supply of labour will probably be required, and to areas where the effects of dislocation will be greatest.

Using a variety of data sources, ISF has developed a framework for classifying and measuring job changes at different levels of occupational disaggregation, to provide a richer picture of the composition of this employment change. The methodology and key figures are detailed below.

Three primary studies that classify and measure the occupational compositions of renewable energy industries have been conducted by the International Renewable Energy Agency (IRENA). Using surveys of the participants in around 45 industries across a range of developed and developing nations, IRENA has estimated the percentages of person-days for the various occupations across the solar PV and onshore and offshore wind farm supply chains (IRENA 2017). Figure 3.11 is an example (in this case, for solar PV manufacturing).

IRENA's studies are the most detailed estimates available of the occupational compositions of the solar PV and onshore wind sectors. ISF has extended the application of IRENA's work. Chapter 10 provides more details about the methodology and the specific factors used in this analysis.

3.7 Material and Metal Resources Analysis (UTS-ISF)

3.7.1 Methodology—Material and Metal Resources Analysis

The future demands for metals have been modelled to better understand the resource requirements of the shift to renewable energy and transport systems. The future demands for metals have been modelled for the projection of 100% renewable energy and the full electrification of the transport system by 2050, as described in in Chap. 6.

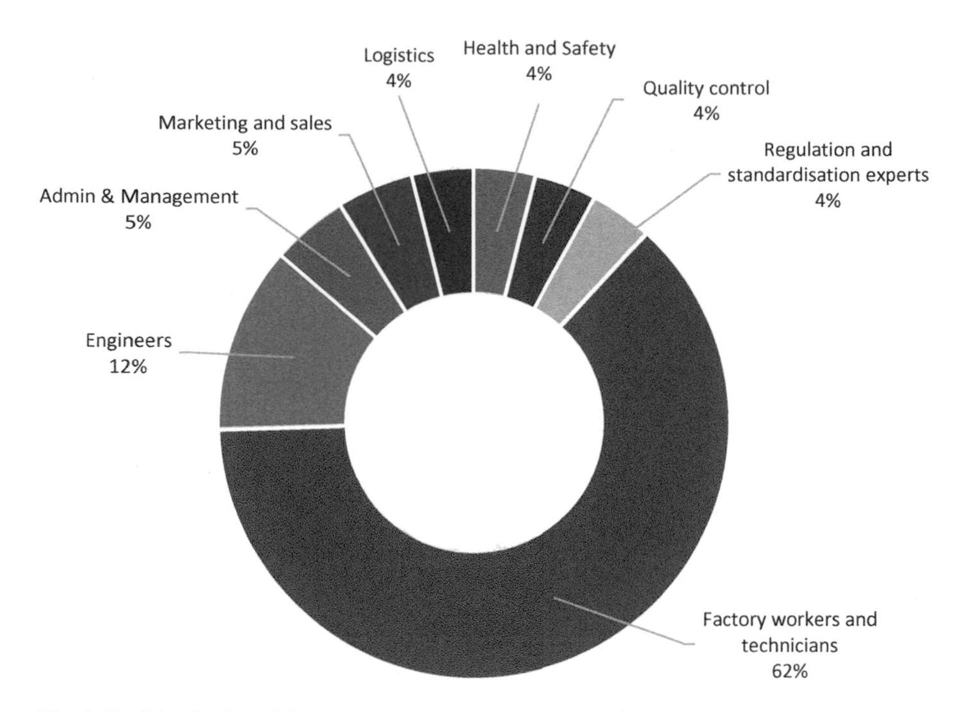

Fig. 3.11 Distribution of human resources required to manufacture the main components of a 50 MW solar photovoltaic power plant. (IRENA 2017)

The predicted demand for the metals required to produce clean energy each year is estimated based on the increase in capacity plus an additional amount required to replace the capacity or vehicles that reach the ends of their lives in each year (based on a lifetime distribution curve for the average lifetime). From this, the GW of capacity or number of vehicles introduced in each year is estimated (also accounting for the replacement stock for end-of-life technologies).

When assuming that the introduced amount of specific technologies in year t is I_t, the accumulated stock amount in year t (generation capacity or in-use stock) is S_t, and the discarded amount in year t is O_t, can be expressed by:

$$S_t = S_{t-1} + I_t - O_t \tag{3.1}$$

Where O_t depends on the number of use years of each product. This use year varies from product to product, and even within the same product group introduced into a society in the same year. The discarded year is not constant and has a lifetime distribution. Therefore, if the number of use years of the product is assumed as a, lifetime distribution can be defined as $g_{(a)}$. Hence, is given by following:

$$O_t = \sum_{a=0}^{a_{\max}} I_{t-a} g(a) \tag{3.2}$$

Where a_{max} is the maximum value of the product life. Therefore, I_t can be calculated with equation (3.3).

$$I_t = S_t - S_{t-1} + \sum_{a=0}^{a_{max}} I_{t-a} g(a) \tag{3.3}$$

In this book, the Weibull distribution is used to consider the life characteristics of the products described above, with the key assumptions shown in Table 3.8.

Based on the annual introduced amount of clean energy technologies given by equation (3.3), the metal demand for technology p in year t is estimated as:

$$Demand_{p,t} = I_{p,t} \cdot Metal\ intensity_{p,t} \tag{3.4}$$

Where $Metal\ intensity_{p,t}$ is the amount of required metal in technology p in year t. Because this value can change over time with technological developments, we assume that the various scenarios incorporating the material efficiency improvement.

The demand estimated with equation (3.4) indicates the total metal requirements for the introduction of clean energy technologies. This demand arises from primary production (mined from natural deposits) and secondary production (recovered from end-of-life products). Secondary production could play an important role in the future by increasing metal availability and reducing the environmental impact. Therefore, we evaluated the effects of recycling by estimating the potential reduction in primary production entailed. When the recycling of end-of-life products is considered, primary production is given by equation (3.5).

$$Primary\ production_{p,t} = Demand_{p,t} - Discard_{p,t} \cdot Recycling\ rate_p \tag{3.5}$$

Where $Discard_{p,t}$ is end-of-life technology in year , and is estimated from the Weibull distribution, and the $Recycling\ rate_p$ indicates the proportion of metals recovered from end-of-life technology. Since this value can be increased by technological improvements, the metal price, and the amount of end-of-life product available, we assumed both the current recycling rate and an improved recycling rate.

This recycling rate is based on the rate of recycling of the metal within the technology (e.g., silver discarded from solar panels can be recycled into new solar panels), rather than as an average across the use of the metal, as has been done in previous studies. This has been chosen as the most appropriate recycling rate to use because we assume that by using recycling rates specific to the technology, it is more likely to offset demand for new materials for that technology.

Table 3.8 Key assumptions

Technology	Lifetime (years)	Shape parameter
Solar PV	30	5.38
Battery	8	3.5

Ultimately, the mineral requirements estimated with equations (3.4) and (3.5) under the various assumptions were compared with metal reserves and annual production (in 2017). A 'reserve' is regarded as the amount economically extractable with the current technologies and at the current metal price, and can change significantly over time. However, comparing reserves with estimated future requirements can provide insight into how the introduction of clean energy technologies will affect the physical availability of metals in the future. We also compared current production with the estimated future requirements to estimate the likelihood of a rapid increase in requirements. The key results are presented in Chap. 11.

3.8 Climate Model

3.8.1 Deriving Non-CO_2 GHG Pathways

This section provides an overview of the methodology that has been used to complement the energy-related CO_2 emission pathways for non-energy-related CO_2 emissions, other GHG emissions, and aerosols.

The energy-related CO_2 emissions were derived using energy-system modelling frameworks, but two different approaches have been used to derive the land-use CO_2 emissions and other GHG emissions. First, we will describe the approach that was used to determine other GHG emissions. This approach can be summarized as a statistical analysis of currently published scenarios. To derive non-CO_2 pathways that are consistent with the relevant emission mitigation levels, the non-CO_2 emissions were regressed against the fossil fuel and industrial CO_2 emissions. These regression characteristics were then used to derive the non-CO_2 emissions. This method has been newly developed in the context of this study and can be regarded as a further development of the Equal Quantile Walk method introduced by Meinshausen et al. (2006).

One challenge in applying the collective knowledge that is enshrined in multi-gas-emission scenarios in the literature is that regional and sectoral definitions differ slightly between the various modelling groups. Because most IPCC scenarios work are based on the emission categories used by the IAM community, their emission categories and regions have been adopted in this analysis of non-CO_2 emission pathways. The steps in the analysis are described in the following sub-sections.

3.8.1.1 Regional Definitions

First, the regional energy-related CO_2 emissions developed in the previous sections must be transformed to match the five Renewable Communities Program (RCP) regions used by the IAM community, into the regions OECD90 (OECD countries,

Table 3.9 Regional definitions according to the Integrated Assessment Modelling community

Regions	Developing Asia	Europe	Africa	Middle East	Central & South Amer	Eurasia	China	India	North America	OECD Asia Oceania	Subtotal
RCP5_Asia	1,561	-	-	-	19	21	8,826	1,985	-	674	13,086
RCP5_REF	-	720	-	-	-	2,153	-	-	-	-	2,873
RCP5_MAF	-	-	1,174	2,021	-	-	-	-	-	-	3,195
RCP5_OECD90	2	3,031	-	-	-	-	-	-	5,766	1,577	10,376
RCP5_LAM	-	-	-	-	1,216	-	-	-	461	-	1,676
Subtotal	1,563	3,751	1,174	2,021	1,235	2,174	8,826	1,985	6,226	2,251	31,206

membership status as of 1990), ASIA (Asian countries), REF (economies in transition), LAM (Latin America), and MAF (Middle East and Africa). Table 3.9 (above) indicates the overlap and differences between the RCP regions with the other regions described in this report. As an indicator of how different the regional definitions are, we used the fossil fuel and industrial emissions for the year 2015 according to the 2017 update of the PRIMAP database (Gütschow et al. 2016).

Table 3.9 provides an overview to the regional definitions used in this study. The top row indicates the regions for the CO_2 fossil and industrial emissions, and the various rows refer to the five regions used in IAMs. To derive the non-CO_2 emissions, we used the IAM's five RCP regions. The numbers indicate the fossil fuel and industrial emissions in the year 2015 in $MtCO_2$, aggregated from country-level data. The colour shading of the cells indicates where most of the 2015 emissions occurred.

Table 3.9 Regional definitions according to the Integrated Assessment Modelling community (so-called 'RCP5' regions) compared with the other regions used in this study The overlap and differences between the two sets of regional definitions are shown with the 2015 fossil and industrial CO_2 emissions. For example, the first row indicates that the largest sub-region in the RCP5_Asia group is China, with 8,826 $MtCO_2$ of emissions.

The transfer of the energy-related CO_2 emission results to fit the IAM's regional categorization (which is consistent with IEA WEO reports) was performed by first disaggregating all the results to country-level data. A simple proportional scale was applied to the 2015 energy-related country-level CO_2 emissions from the PRIMAP database. The disaggregated country-level data were then re-aggregated at the RCP5 regional level.

3.8.1.2 Harmonization: Emission Category Adjustments

Before proceeding with the application of CO_2 versus non-CO_2 statistical relationships, a harmonization step is necessary. Various IAM use slightly different categories, emission factors, and activity data to estimate emissions. This can result in some spread in the current emission estimates for the same regions and categories. To address this issue, the standard practice in the IAM community is to work with harmonized emissions scenarios, meaning that the original emissions scenarios have either been scaled or shifted towards a common reference point. A recent historical emission level is normally chosen as this reference point. Here, we chose the 2015 emissions across the five RCP regions.

Harmonization was performed in two steps. First, emissions were added that were related to the CO_2 fossil and industrial emission categories (such as waste-related emissions) and that were outside the scope of emissions in the energy-related CO_2 emission chapters. The scenarios from which these 'other' energy-related CO_2 emissions were taken were: the SSP2_Ref_SPA0_V25_upscaled_MESSAGE_GLOBIOM (for the 5.0 °C reference scenario); SSP1_26_SPA1_V25_IMAGE (for the 2.0 °C Scenario), and SSP1_19_SPA1_V25_IMAGE scenario (for the 1.5°C Scenario). In the second harmonization step, the overall sum of the complemented 2.0 °C and 1.5 °C scenario CO_2 emissions were compared with the overall fossil and industrial sum of CO_2 emissions in the year 2015, which were used for scenario harmonization under the CMIP6 ScenarioMIP process (Meinshausen et al., in preparation). This comparison revealed that there were still differences between the complemented energy scenarios (see Chapter 8) and the harmonization emission levels for the various regions. These differences could again have resulted from different emission factors or activity assumptions, or they could simply reflect genuine uncertainty in the overall global and regional anthropogenic emissions. Consistent with the processing steps used in the CMIP6 process, we up- and down-scaled the raw and regionally disaggregated energy scenario emissions towards the harmonization emission levels. Figure 3.12 shows the differences between the raw emission scenario data, the data re-aggregated into the RCP regions, and the CMIP6 emission harmonization fossil and industrial CO_2 emission levels for 2015 (in GtC). The differences were bridged by applying a time-constant scaling factor.

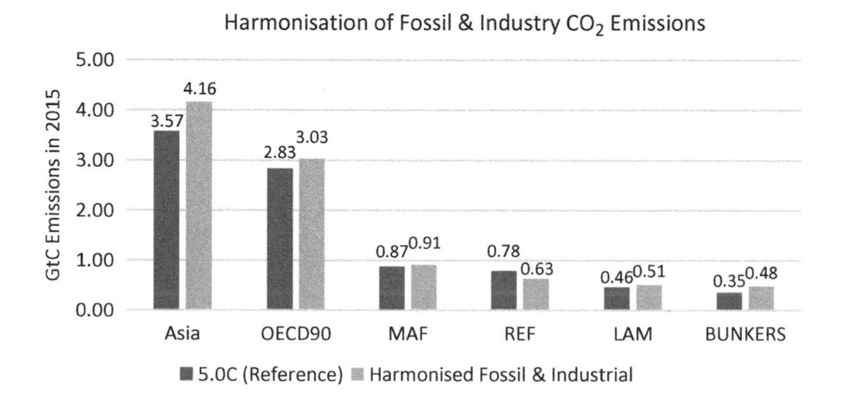

Fig. 3.12 Differences between the raw LDF emission scenario data

3.8.1.3 A New Quantile Regression Method for Non-CO_2 Gases

The completed fossil and industrial CO_2 emission time series can now be compared with the set of scenarios in the literature. In this study, we used 811 scenarios from CMIP6 databases or the databases underlying the IPCC SR1.5 report. These literature-reported studies are either reference scenarios or mitigation scenarios with a specific forcing target or climate target. Some of the scenarios aim for 1.5 °C levels of change, others for 450 ppm CO_2-equivalence concentrations, and yet others assume fragmented worlds, with regional rivalries and no consistent policy approach. In summary, the input assumptions of all these literature-reported scenarios vary widely, yet all have some formal energy-system modelling framework behind them that provides first-level assurance that the envisaged CO_2, methane, nitrous oxide, and other gas emission levels are not set below the limits considered technologically feasible under a certain set of boundary conditions, such as the requirement to continuing feeding the human population. The technological and economic feasibilities of emission pathways are fluid concepts, subject to change in response to technological advances and changes in policy settings.

This study and the approach it uses are not dependent on absolute levels of mitigation costs or precise definitions of technological feasibility. Instead, the method used assumes that that non-CO2 gases are reduced with a similar effort as that required to reduce CO_2 emissions. Therefore, using the emission characteristics from a large set of scenarios reported in the literature, we assume similar levels of technological feasibility, economic mitigation costs, and implementation opportunities will be required to reduce emissions of CO_2 and various other gases.

More specifically, we derived the non-CO_2 emissions in a particular year by ranking all the scenarios against the indicator of fossil and industrial CO_2 emissions in that year (see Fig. 3.11 below). By comparing them to the 'crowd' of other literature-reported scenarios, the LDF pathways could also be ranked. Specifically, the LDF reference scenario turned out to be around the 75th percentile of the distribution of the fossil and industrial CO_2 emissions across all 811 scenarios considered. By contrast, the lower 1.5 °C Scenario and 2.0 °C Scenario were not at the absolute lower boundary of the 811 scenario distribution, but were close to it. The 1.5 °C Scenario ranked between the zero and first percentile—that is, among the 1% most stringent scenarios in the literature for the years 2025–2045. In the period until 2050, the 2.0 °C Scenario was situated between the 5th and 10th percentiles of the scenario distribution (see Fig. 3.13).

Figure 3.13 shows the 1.5 °C and 2.0 °C Scenarios, their absolute fossil and industry CO_2 emissions until 2050 (upper panel), and their respective locations in the set of 811 literature-reported scenarios considered (lower panel). The post-2050 scenario extensions were extrapolated differently for fossil and industrial CO_2 and the non-CO_2 gases. To derive the non-CO_2 gases, the 2050 percentile location was assumed constant for the remainder of the twenty-first century. For the fossil and industrial CO_2 emissions in the 2.0 °C and 1.5 °C Scenarios, which do not assume

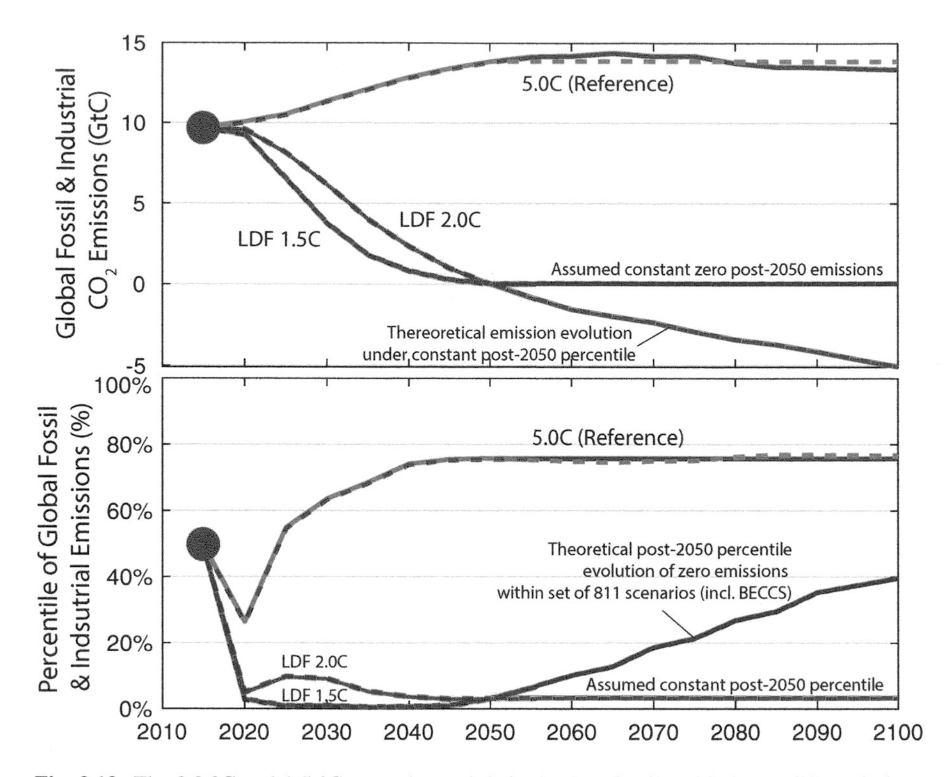

Fig. 3.13 The 2.0 °C and 1.5 °C scenarios and their absolute fossil and industry CO_2 emissions until 2050. The energy-related CO_2 emissions pathways from the other chapters are used until 2050, and then extended beyond 2050 by either keeping the CO_2 emissions constant (in the case of the 1.5 °C and 2.0 °C Scenarios, i.e., red and purple dashed lines beyond 2050 in the upper panel) or by keeping the percentile level within the literature-reported scenarios constant (in the case of the reference scenario, i.e., green solid line in the upper panel). The percentile rank within the other literature-reported scenarios is shown in the lower panel. The constant absolute emission level after 2050 in the case of the 1.5 °C and 2.0 °C Scenarios can be seen to result in an increasing percentile rank among all the literature-reported scenarios (increasing purple–red line in the lower panel)

BECCS to achieve net negative emission levels, a continuation of the constant zero emission level was assumed (straight constant emission level in the upper panel corresponding to the increasing percentile level of the red–blue dashed line in the lower panel).

3.8.1.4 'Pseudo' Fossil and Industrial CO_2 Extensions Beyond 2050

By the end of the century, almost 40% of all of 811 scenarios will feature net negative fossil and industrial CO_2 emissions, largely because there is some level of biomass and CCS deployment. Given that the energy scenarios developed for this study do not assume, by design, any BECCS-related emission uptake, the extended post-2050 energy scenarios are assumed to be consistent with other scenarios in which emissions will be around zero by the end of the century. However, those other scenarios with zero emissions by the end of the century tend to reflect a much lower level of mitigation effort. Therefore, to derive non-CO_2 emissions that involve a level of effort that is comparable to the mitigation effort involved in reducing energy-related CO_2 emissions, we assumed that the energy scenarios developed for this study were comparable to other scenarios that share zero emissions around 2050. This percentile 'stringency' level was then held constant for the remainder of the twenty-first century. Therefore, whereas the actual fossil and industrial CO_2 emissions in the LDF scenarios are assumed to remain constant at zero, the non-CO_2 gas emissions are derived from data in the existing literature, as if the scenarios remained at a stringency level of ~3% in the second half of the twenty-first century (see the lower panel in Fig. 3.13).

We now have the fossil and industrial CO_2 emission levels throughout the twenty-first century for each of the three scenarios, and have complemented these with the 'pseudo' CO_2 emission levels for the second half of the twenty-first century. Therefore, we can derive the corresponding non-CO_2 emissions.

In the first step, we derived the total non-CO_2 emissions for a specific year and for the world as a whole. In the second step, we determined the shares of global fossil and industrial emissions versus the land-use-related emissions—again regressed against the overall fossil CO_2 emission level as an indicator of the 'stringency' of the scenario. In the third step, we disaggregated these fossil and land-use-specific emission time series into regional time series. Again, the shares of the regional emissions were derived with the same quantile regressions shown in Fig. 3.13 above. With these three quantile regression steps, we inferred either the lower (if lower quantile ranges are chosen), medium (for a median 50% quantile regression), or higher emission levels of the other gases. In this study, we do not intend to provide probabilistic emission scenarios and therefore limited our quantile regression choice to the median 50% setting for all regions, sectoral divisions, and other global total gases.

The major advantage of this newly developed method compared with the EQW method developed earlier (Meinshausen et al. 2006) is that the negative correlations between CO_2 and other gases can also be taken into account. By performing all quantile regressions in the space defined by the global fossil and industrial CO_2 emissions in a particular year, any kind of non-linear, positive or negative relationship with other non-CO_2 gas emission levels, sectoral divisions, or regional divisions are automatically incorporated into the final result—reflecting the overall characteristics of the chosen set of emission scenarios. Not all the 811 emission scenarios contained details of all the sectoral and regional divisions, but the stepwise approach of this method can incorporate the characteristics from all the scenarios in whatever detail is available.

Figure 3.14 shows sample distributions of the emission scenario characteristics for the year 2040 and a subset of 21 GHGs. The x-axis of each plot shows the global fossil and industrial CO_2 emissions, and the y-axis shows the global emission levels of another GHG, with one marker (blue dot) for each literature-reported scenario considered. The five red lines are quantile regressions at the levels of 20%, 33%, 50% (median), 66%, and 80% of the scenario distribution. It can be clearly seen that some total gas emissions correlate strongly with the fossil and industrial CO_2 emissions

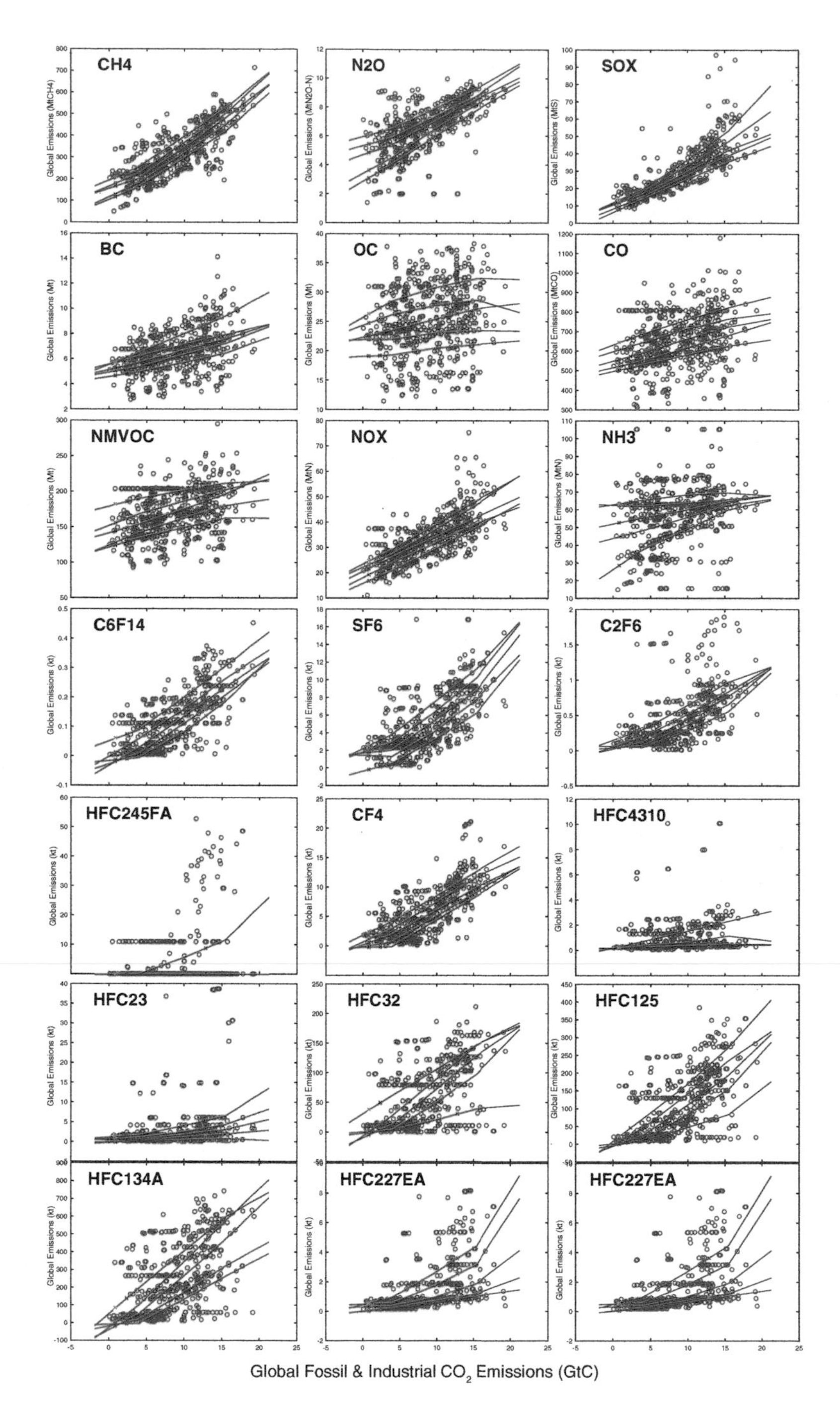

Fig. 3.14 Example distributions of emissions scenario characteristics

(such as the SO_X aerosols in the top-right panel), whereas others correlate less strongly. This method reflects the level of correlation in the finally derived multi-gas scenarios.

3.8.1.5 Land-Use Assumptions

In principle, the same methodological approach can be used for land-use emissions. In the IAM scenarios, the emission sequestration with the BECCS technology is reported as a negative emission in the fossil and industrial CO_2 emission categories. Therefore, the quantile regression approach can also be applied to the land-use CO_2 sector. However, to more explicitly define the land-use choices that are implied by various land-use scenarios, we developed a new (probabilistic) scenario in conjunction with another land-use emission project (Dooley et al. 2018)

This method is based on various literature-reported studies and we provide an overall synthesis of four different land-use-based sequestration pathways: 'forest ecosystem restoration', 'reforestation', 'sustainable use of forests', and 'agroforestry'. These land-use sequestration pathways are based on the premise that the better management of terrestrial ecosystems should allow previously degraded carbon stocks to be restored, entailing the removal of significant atmospheric CO_2 (DeCicco and Schlesinger 2018; Law et al. 2018; Mackey et al. 2013; Mackey 2014; Nabuurs et al. 2017). We derived the overall pathways separately for the temperate and boreal regions on the one side and the subtropical and tropical regions on the other. This distinction is largely consistent with the dominant distinction of different climate domain characteristics in the literature (Grace et al. 2014; Houghton and Nassikas 2018; Pan et al. 2011), although the temperate and boreal biomes are as different in terms of land-use and forest ecosystem characteristics as the tropical biomes are from each of them. However, we derived only two climate domains because several of the RCP regions cross both temperate and boreal biomes. A narrative for each of these pathways is available in Table 3.10 below.

Based on literature studies and Food and Agriculture Organization (FAO) statistics, we then defined the available areas (and their uncertainties) for each of the four sequestration pathways. Similarly, we sourced average estimates of the maximal annual sequestration rates for the biomes (and their levels of uncertainty) for those four sequestration pathways—again distinguished in the large temperate/boreal and subtropical/tropical climate domains. We assume that after a certain 'phase-in period', this maximal annual sequestration rate can be reached and sustained for a number of years. We assume that after some decades to centuries, the capacities of these terrestrial ecosystems as carbon sinks will slowly decline until they reach equilibrium, termed the 'saturation' period. At the equilibrium point, these ecosystems have a net zero effect on atmospheric CO_2 over the time scales of interest here (Houghton and Nassikas 2018). The period over which the maximal sequestration rate is assumed, is reduced by the half-length of the corresponding phase-in and phase-out periods to account for the cumulative carbon uptake in those periods. As the last element in this framework, we assume a cap on the median carbon density change that is achieved over the full period. The difference between a degraded for-

Table 3.10 Narrative for each sequestration pathway per climatic biome

Pathway	Climatic domain	Narrative
Forest ecosystem restoration (set aside areas of degraded natural forest to restore to primary forest—25% of total)	Te,B	Assume 25% of degraded natural forest put aside for full ecosystem and carbon stock recovery. Saturation times in temperate and boreal forests can be well over 100, or even 200 years (Luyssaert et al. 2008; Roxburgh et al. 2006). 25% set-aside is slightly higher than assumptions made in recent literature (Böttcher et al. 2018; Nabuurs et al. 2017), but in line with calls from conservation and indigenous movements.
	S,Tr	Assume 25% of degraded natural forests across the tropics set aside for full ecosystem and carbon stock recovery. Stopping all deforestation, wood harvest and temporary use, while traditional and customary uses continue. Net Primary Productivity (NPP) is higher across the tropics than in temperate and boreal biomes (Anav et al. 2015), hence sequestration rates are higher, but saturation times are shorter. We assume 60 years to ecosystem maturity (Arneth et al. 2017; Poorter et al. 2016). Sequestration rates across all biomes for forest ecosystem restoration are lower than post-logging recovery rates, as here we assume mixed age-class forests which have not been recently logged than (>20 years ago), which then saturates when forest reaches maturity.
Reforestation (forest expansion through natural regeneration)	Te	Forest expansion on recently deforested land via natural regeneration of forests (passive regeneration) or reforestation of mixed native species (assisted regeneration). Extent of forest expansion is assumed to occur in line with current political targets: 350 Mha of reforestation by 2030 under the Bonn Challenge. Further, this 350 Mha is assumed to be reforestation for conservation purposes, which creates an ongoing sink from 2030 to 2100, with saturation assumed at 100 years. Boreal areas excluded due to albedo effect (Houghton and Nassikas 2018).
	S,Tr	Natural forest expansion on recently deforested land as described above. We assume 80% of forest expansion occurs in the tropics, given 80% of Bonn Challenge pledges are in tropical regions (Wheeler et al. in press). All regeneration is assumed to be with natural forests rather than plantations, as this delivers the highest mitigation and biodiversity values (Grace et al., 2014; Wheeler et al. in press). Saturation of tropical regrowth forests is assumed at 60 years, although large trees can take well over a century to mature (Poorter et al. 2016).

(continued)

Table 3.10 (continued)

Pathway	Climatic domain	Narrative
Sustainable use of forests (secondary forests under continued (but reduced) forest harvest)	Te,B	Sustainable use of natural production forests (i.e. excluding plantations) under continued wood harvest. Multiple studies show the potential to double forest carbon stocks in production forests through reducing harvest intensity and extending rotation lengths (Law et al. 2018; Nabuurs et al. 2017; Pingoud et al. 2018). Wood harvest is slightly reduced, requiring reduced demand for wood products, more efficient wood use and compensation to land-owners (Law et al. 2018; Pingoud et al. 2018). Including harvested wood products (HWP) in calculations could increase mitigation values (Houghton and Nassikas 2018; Nabuurs et al. 2017), but the life-time of HWP is generally too short to realise mitigation value compared to residence times in forest biomass (Law et al. 2018; Keith et al. 2015).
	S,Tr	Reduced harvest intensity and management practices such as reduced impact logging have not been shown to increase carbon stock in tropical forests (Martin et al. 2015). Carbon stocks are concentrated in commercially-valuable hardwood trees taking >100 years to reach maturity; hence selective logging as practiced across the tropics significantly decreases standing carbon stocks (Lutz et al. 2018; Zimmerman and Kormos 2012). Our scenario assumes no commercial logging of tropical forests, and the extent of shifting cultivation is halved, allowing traditional practices to continue with lengthened fallow times and/or improved swidden practices (Mackey et al. 2018; Ziegler et al. 2012).
Agroforestry (Trees in croplands)	T,B S,Tr	We calculate biome-average sequestration rates from the literature for above-ground carbon uptake due to a broad range of agroforestry practices (Watson et al. 2000; Nabuurs et al. 2017; Ramachandran Nair et al. 2009), and subtract from this the baseline increase observed by Zomer et al. (2016). We apply this uptake to 20% of permanent cropland, and assume the resulting sequestration rate could be sustained for 50 years (Watson et al. 2000).

These domains are defined as temperate (Te), boreal (B), subtropical (S), or tropical (Tr). Note, this narrative overlaps with another land-use-related study (Dooley et al. 2018)

est ecosystem and its natural carbon-carrying capacity is the maximum potential for additional sequestration. Therefore, we used biome-averaged values for the per hectare carbon density of undisturbed forest ecosystems (Keith et al. 2009), rather than average global biome values (Liu et al. 2015), to define the maximum carbon density. Although the LDF scenarios only extend to 2050 or to 2100 for all the other GHGs, we modelled the land-use sequestration pathway assumptions until 2300 to be able to apply the overall 'added carbon density' cap.

Our climate-domain- and sequestration-specific assumptions regarding the median values, their uncertainty ranges, and confidence intervals are given in

Table 3.11. Our method of combining these input assumptions is a basic Monte Carlo ensemble. Given the symmetry or asymmetry and respective confidence intervals of the factors provided, we then created normal, lognormal, or skewed normal distributions (the latter is a linear combination of the normal and lognormal distributions to achieve the desired skewedness). We then made 500 independent draws from all four factors considered (area, maximum sequestration rate, phase-in time, and phase-out time). We repeated that process independently for each sequestration pathway and for each country within each climate domain. The areas for each country within the climate domains were assumed to be proportionally distributed by the relevant 'FAO Scaling Area' (see third column in Table 3.11), so that the climate domain aggregate areas matched our input assumptions for the respective sequestration pathway.

After combining all the country-specific and sequestration-pathway-specific time series for carbon uptake per sequestration pathway, we then checked whether the resulting cumulative sequestration over time (specifically its median) was at or below the specified maximum for the median carbon density change per hectare. If it was not, we scaled all the country-specific results proportionally, so that the median matched the cap on the carbon density gain.

3.8.2 Model for the Assessment of GHG-Induced Climate Change

To compute GHG concentrations and the implications for radiative forcing, global mean temperatures, and global mean sea-level rise, we used the reduced-complexity 'Model for the Assessment of Greenhouse-gas-induced Climate Change' (MAGICC), as described by Meinshausen et al. (2011). The model has recently been extended by the addition of a newly designed sea-level rise module, as described in Nauels et al. (2016).

The MAGICC model has at its core an upwelling-diffusion ocean with 50 layers, in both the northern and southern hemispheres. Some simpler model approaches, with only a diffusive one-box ocean, for example, tend to overestimate the medium-term warming compared with the longer-term warming, (i.e., they tend to reach equilibrium too quickly). In the short term, the MAGICC modelling structure provides faster warming, but a lower approach to equilibrium, due to the effective cooling cycle that mimics the sinking polar ocean waters.

Although simple in its general structure, the MAGICC model uses a broad coverage of GHGs and aerosols. This is much broader than for earth system models, because it would be too computationally expensive to carry around tracers for every minute GHG concentrations of, say, HFC227EA. Because of the breadth of the GHGs that MAGICC can model, its calibrated carbon cycle, and its calibrated climate system with feedbacks and heat exchange parameterizations, it is frequently used as a climate model in IAMs. For example, the IMAGE and MESSAGE teams both have MAGICC inbuilt.

IPCC Assessment Reports also frequently use MAGICC as the modelling framework to determine the exceedance probabilities of various emission pathways. For

Table 3.11 Assumptions regarding the four land-use sequestration pathways for two climate domain categories

Pathway	Climatic domain (Te = Temperature; B = Boreal; S = Subtropical; Tr = Tropical)	FAO scaling area	Assumed available area (Mha)	Related sources	Assumed median added carbon maximum (MgC/h)	Assumed maximum sequestration rate (MgC/ha/year)	Related sources	Saturation period (Years)	Related sources	Phase in period (Years)	Phase out period (Years)
Forest ecosystem restoration (set aside areas of secondary forest to restore to primary forest—25% of total)	Te,B	Other Natural Forest Area 2015	276 (80%: 248–303)	FAO (2016)	185 (Keith et al. 2009)	0.5 (90%: 0.25–1)	Pan et al. (2011)	100 (80%: 70–130)	Luyssaert et al. (2008) and Roxburgh et al. (2006).	20 (90%: 7–20)	30 (90%: 10–100)
	S,Tr		335 (80%: 302–369)	FAO (2016)	172	1.1 (90%: 0.55–2.2)	Pan et al. (2011)	60 (80%: 42–78)	Pan et al. (2011), Grace et al. (2014), and Asner et al. (2018)	15 (90%: 7–20)	20 (90%: 10–100)
Forest expansion via reforestation (land-use change from non-forest to forest through natural regeneration)	Te	Change in total forest area from 1990 to 2015	50 (80%: 45–55)	FAO (2016)	185	2.62 (80%: 0.56–7.05)	IPCC (2006)	100 (80%: 70–130)	Roxburgh et al. (2006) and Luyssaert et al. (2008)	25 (90%: 7–20)	30 (90%: 10 to 100)
	S,Tr		300 (80%: 270–330)	FAO (2016)	172	3.1 (80%: 0.42–8.46)	IPCC (2006)	60 (80%: 42–78)	Pan et al. (2011), Grace et al. (2014), and Poorter et al. (2016)	20 (90%: 7–20)	20 (90%: 10 to 100)

(continued)

Table 3.11 (continued)

Pathway	Climatic domain (Te = Temperature; B = Boreal; S = Subtropical; Tr = Tropical)	FAO scaling area	Assumed available area (Mha)	Related sources	Assumed median added carbon maximum (MgC/h)	Assumed maximum sequestration rate (MgC/ha/year)	Related sources	Saturation period (Years)	Related sources	Phase in period (Years)	Phase out period (Years)
Sustainable use of forests (secondary forests under continued but reduced forest harvest)	Te,B	Production Forest 2015	743 (80%: 669–817)	FAO (2016)	185 (an upper end estimate, possibly too high, cf. Liu et al. 2015)	0.4 (80%: 0.36–0.44)	Nabuurs et al. (2017)	100 (80%: 70–130)	Roxburgh et al. (2006) and Luyssaert et al. (2008)	20 (90%: 7–20)	30 (90%: 10–100)
	S,Tr		419 (80%: 377–461)	FAO (2016)	172	1.19 (80%: 1.07–1.31)	Houughton and Nassikas (2018)	60 (80%: 42 to 78)	Pan et al. (2011) and Grace et al. (2014)	15 (90%: 7–20)	20 (90%: 10–100)
Agroforestry (trees in croplands)	T,B	Permanent crop area 2015	100 (80%: 90–110)	Zomer et al. (2016) and Watson et al. (2000);	10 (Zomer et al. 2016)	0.65 (80%: 0.59–0.72)	Nabuurs et al. (2017) and Zomer et al. (2016)	50 (80%: 35 to 65)	Watson et al. (2000)	20 (90%: 7–20)	20 (90%: 10–100)
	S,Tr		300 (66%: 270–330)		30	1.09 (80%: 0.98–1.2)	Ramachdradan Nair et al (2009) and Zomer et al. (2016)	50 (80%: 35–65)	Watson et al. (2000)	15 (90%: 7–20)	20 (90%: 10–100)

example, see Chapter 6 of the Working Group 3 contribution to the Fifth Assessment Report and Chapter 2 of the IPCC Special Report on 1.5 °C. MAGICC has also been used in the preparation of the forthcoming IPCC Sixth Assessment Report to design the GHG concentration scenarios (Meinshausen et al., in preparation).

References

ABS (2017), Australian Bureau of Statistics (2017), Employment in Renewable Energy Activities—Explanatory Notes, for a summary. http://www.abs.gov.au/ausstats/abs@.nsf/Lookup/4631.0Explanatory+Notes12015-16. Accessed September 16, 2018.

Anav, A., et al. (2015), Spatiotemporal patterns of terrestrial gross primary production: A review, Rev. Geophys., 53, 785–818, doi:https://doi.org/10.1002/2015RG000483

Arneth, A., Sitch, S., Pongratz, J., Stocker, B.D., Ciais, P., Poulter, B., Bayer, A.D., Bondeau, A., Calle, L., Chini, L.P., Gasser, T., Fader, M., Friedlingstein, P., Kato, E., Li, W., Lindeskog, M., Nabel, J.E.M.S., Pugh, T.A.M., Robertson, E., Viovy, N., Yue, C., Zaehle, S., 2017. Historical carbon dioxide emissions caused by land-use changes are possibly larger than assumed. Nature Geoscience 10, 79–84. https://doi.org/10.1038/ngeo2882

Asner, G.P., Brodrick, P.G., Philipson, C., Vaughn, N.R., Martin, R.E., Knapp, D.E., Heckler, J., Evans, L.J., Jucker, T., Goossens, B., Stark, D.J., Reynolds, G., Ong, R., Renneboog, N., Kugan, F., Coomes, D.A., 2018. Mapped aboveground carbon stocks to advance forest conservation and recovery in Malaysian Borneo. Biological Conservation 217, 289–310. https://doi.org/10.1016/j.biocon.2017.10.020

Böttcher, H., Herrmann, L. M., Herold, M., Romijn, E., Román-Cuesta, R. M., Avitabile, V., ... & Schepaschenko, D. (2018). Independent Monitoring: Building trust and consensus around GHG data for increased accountability of mitigation in the land use sector.

Breyer (2016), Christian Breyer, Internet of Energy—A 100% renewable electricity system, Neo Carbon Energy, Lappeenranta University of Technology, Finland, 2016, https://www.lut.fi/web/en/news/-/asset_publisher/lGh4SAywhcPu/content/simulation-brings-global-100-renewable-electricity-system-alive-for-the-first-time

Breyer, C, Bogdanov, D (2018), Breyer Ch., Bogdanov D., Aghahosseini A., Gulagi A., Child M., Oyewo A.S., Farfan J., Sadovskaia K., Vainikka P., 2018. Solar Photovoltaics Demand for the Global Energy Transition in the Power Sector, Progress in Photovoltaics: Research and Applications, 26, 505-523, DOI: https://doi.org/10.1002/pip.2950

CIA (2018), Central Intelligence Agency, Library, World Factbook, Online database, viewed in July 2018 https://www.cia.gov/library/publications/the-world-factbook/fields/2195.html

DeCicco, J.M., Schlesinger, W.H., 2018. Reconsidering bioenergy given the urgency of climate protection. Proceedings of the National Academy of Sciences 115, 9642–9645. https://doi.org/10.1073/pnas.1814120115

Dooley, K., Stabinsky, D., Stone, K., Sharma, S., Anderson, T., Gurian-Sherman, D., Riggs, P., 2018. Missing Pathways to 1.5 °C: The role of the land sector in ambitious climate action. Climate, Land, Ambition and Rights Alliance. https://www.climatelandambitionrightsalliance.org/report/

Dunstan C, Fattal, A (2016), Dunstan C., Fattal A., James G., Teske S., 2016, Towards 100% Renewable Energy for Kangaroo Island. Prepared by the Institute for Sustainable Futures, University of Technology Sydney (with assistance from AECOM) for ARENA, Renewables SA and Kangaroo Island Council

Elliston, B, MacGill, Ian (2014), Ben Elliston, Iain MacGill, Mark Diesendorf, Comparing least cost scenarios for 100% renewable electricity with low emission fossil fuel scenarios in the Australian National Electricity, Market, Elsevier, Renewable Energy, Volume 66, June 2014,

Pages 196-204, https://www.sciencedirect.com/science/article/pii/S0960148113006745?via%3Dihub,
EMP-BARCELONA (2002)—Barcelona European Council, 15 and 16 March 2002, Presidency Conclusions. "In the field of Energy the European Council: (...) agrees the target for Member States of a level of electricity interconnections equivalent to at least 10% of their installed production capacity by 2005"; https://cordis.europa.eu/programme/rcn/805_en.html
EU-EG (2017), Towards a sustainable and integrated Europe, Report of the Commission Expert Group on electricity interconnection targets, November 2017, https://ec.europa.eu/info/news/moving-10-15-final-report-commission-expert-group-2030-electricity-interconnection-targets-2017-nov-09_en
FAO, 2016. Global forest resources assessment 2015: how are the world's forests changing? Food and Agriculture Organization of the United Nations
Fischedick, M., R. Schaeffer, A (2011), Fischedick, M., R. Schaeffer, A. Adedoyin, M. Akai, T. Bruckner, L. Clarke, V. Krey, I. Savolainen, S. Teske, D. Ürge-Vorsatz, R. Wright, 2011: Mitigation Potential and Costs. In IPCC Special Report on Renewable Energy, Sources and Climate Change Mitigation [O. Edenhofer, R. Pichs-Madruga, Y. Sokona, K. Seyboth, P. Matschoss, S. Kadner, T. Zwickel, P. Eickemeier, G. Hansen, S. Schlömer, C. von Stechow (eds)], Cambridge University Press, Cambridge, United Kingdom and New York, NY, USA, Chapter 10, http://www.ipcc.ch/pdf/special-reports/srren/Chapter%2010%20Mitigation%20Potential%20and%20Costs.pdf
Flato, G., J. Marotzke, B (2013), Flato, G., J. Marotzke, B. Abiodun, P. Braconnot, S.C. Chou, W. Collins, P. Cox, F. Driouech, S. Emori, V. Eyring, C. Forest, P. Gleckler, E. Guilyardi, C. Jakob, V. Kattsov, C. Reason and M. Rummukainen, 2013: Evaluation of Climate Models. In: Climate Change 2013: The Physical Science Basis. Contribution of Working Group I to the Fifth Assessment Report of the Intergovernmental Panel on Climate Change [Stocker, T.F., D. Qin, G.-K. Plattner, M. Tignor, S.K. Allen, J. Boschung, A. Nauels, Y. Xia, V. Bex and P.M. Midgley (eds.)]. Cambridge University Press, Cambridge, United Kingdom and New York, NY, USA, Chapter 9, http://www.ipcc.ch/pdf/assessment-report/ar5/wg1/WG1AR5_Chapter09_FINAL.pdf
(Global Land Cover 2015); Global Land cover 2000; Joint Research Centre The European Commission's science and knowledge service. 2015. Global Land Cover 2000—Products; Accessed 25 October 2018; ONLINE, http://forobs.jrc.ec.europa.eu/products/glc2000/products.php
(Global Solar Atlas 2016), Global Solar Atlas. 2016; download for varies countries and regions, Accessed 25 October 2018 ONLINE; https://globalsolaratlas.info/downloads
Goodchild, M. F. (2005). GIS and modeling overview. GIS, spatial analysis, and modeling. ESRI Press, Redlands, 1-18. http://www.geog.ucsb.edu/~good/papers/414.pdf
Grace, J., Mitchard, E., Gloor, E., 2014. Perturbations in the carbon budget of the tropics. Global Change Biology 20, 3238–3255. https://doi.org/10.1111/gcb.12600
Gütschow, J., M. L. Jeffery, R. Gieseke, R. Gebel, D. Stevens, M. Krapp and M. Rocha (2016). "The PRIMAP-hist national historical emissions time series." Earth Syst. Sci. Data Discuss. 2016: 1-44.
Hare, B, Roming, N (2016), Hare, B. , Roming, N. , Schaeffer, M. , Schleussner, C., (August 2016), Implications of the 1.5 °C limit in the Paris Agreement for climate policy and decarbonisation, Climate Analytics, Berlin, Germany, Perth, Australia; http://climateanalytics.org/files/1p5_australia_report_ci.pdf
Houghton, R.A., Nassikas, A.A., 2018. Negative emissions from stopping deforestation and forest degradation, globally. Global Change Biology 24, 350–359. https://doi.org/10.1111/gcb.13876
IEA (2016a) World Energy Balances (2016 edition). IEA energy statistics (Beyond 20/20). International Energy Agency, Paris
IEA (2016b) World Energy Outlook 2016. International Energy Agency, Organization for Economic Co-operation and Development, Paris
IEA (2017a) Energy Technology Perspectives. Modelling of the transport sector in the Mobility Model. https://www.iea.org/etp/etpmodel/transport/. Accessed 10 Oct 2018
IEA (2017b) World Energy Outlook 2017. OECD Publishing, Paris/IEA, Paris

IPCC, 2006. Chapter 4: Forest Land, in: 2006 IPCC Guidelines for National Greenhouse Gas Inventories.
IRENA (2017) Renewable Energy Benefits: Leveraging Local Capacity for Onshore Wind, IRENA, Abu Dhabi; http://www.irena.org/-/media/Files/IRENA/Agency/Publication/2017/Jun/IRENA_Leveraging_for_Onshore_Wind_2017.pdf
(Jacobson, M, Choi, C, 2017), Mark Jacobson, Charles Q. Choi, A Road Map to 100 Percent Renewable Energy in 139 Countries by 2050, Stanford Engineering, Stanford University, USA, September 2017, https://cee.stanford.edu/news/road-map-100-percent-renewable-energy-139-countries-2050
Keith, H., Mackey, B.G., Lindenmayer, D.B., 2009. Re-evaluation of forest biomass carbon stocks and lessons from the world's most carbon-dense forests. Proceedings of the National Academy of Sciences 106, 11635–11640. https://doi.org/10.1073/pnas.0901970106
Keith, H., Lindenmayer, D., Macintosh, A., Mackey, B., 2015. Under What Circumstances Do Wood Products from Native Forests Benefit Climate Change Mitigation? PLOS ONE 10, e0139640. https://doi.org/10.1371/journal.pone.0139640
Klaus, T, Vollmer, Clara, (2010), Thomas Klaus, Carla Vollmer, Kathrin Werner, Harry Lehmann, Klaus Müschen, Energy target 2050:100% renewable electricity supply, German Federal Environment Agency, Fraunhofer-Institut für Windenergie und Energiesystemtechnik (IWES), Kassel, Contractor of the research project „Modellierung einer 100-Prozent erneuerbaren Stromerzeugung in 2050",FKZ 363 01 277 Dessau-Roßlau, July 2010, https://www.umweltbundesamt.de/sites/default/files/medien/378/publikationen/energieziel_2050_kurz.pdf
Law, B.E., Hudiburg, T.W., Berner, L.T., Kent, J.J., Buotte, P.C., Harmon, M.E., 2018. Land use strategies to mitigate climate change in carbon dense temperate forests. Proceedings of the National Academy of Sciences 115, 3663–3668. https://doi.org/10.1073/pnas.1720064115
Liu, Y.Y., van Dijk, A.I.J.M., de Jeu, R.A.M., Canadell, J.G., McCabe, M.F., Evans, J.P., Wang, G., 2015. Recent reversal in loss of global terrestrial biomass. Nature Climate Change 5, 470–474. https://doi.org/10.1038/nclimate2581
Lutz, J.A., Furniss, T.J., Johnson, D.J., Davies, S.J., Allen, D., Alonso, A., Anderson-Teixeira, K.J., Andrade, A., Baltzer, J., Becker, K.M.L., Blomdahl, E.M., Bourg, N.A., Bunyavejchewin, S., Burslem, D.F.R.P., Cansler, C.A., Cao, K., Cao, M., Cárdenas, D., Chang, L.-W., Chao, K.-J., Chao, W.-C., Chiang, J.-M., Chu, C., Chuyong, G.B., Clay, K., Condit, R., Cordell, S., Dattaraja, H.S., Duque, A., Ewango, C.E.N., Fischer, G.A., Fletcher, C., Freund, J.A., Giardina, C., Germain, S.J., Gilbert, G.S., Hao, Z., Hart, T., Hau, B.C.H., He, F., Hector, A., Howe, R.W., Hsieh, C.-F., Hu, Y.-H., Hubbell, S.P., Inman-Narahari, F.M., Itoh, A., Janík, D., Kassim, A.R., Kenfack, D., Korte, L., Král, K., Larson, A.J., Li, Y., Lin, Y., Liu, S., Lum, S., Ma, K., Makana, J.-R., Malhi, Y., McMahon, S.M., McShea, W.J., Memiaghe, H.R., Mi, X., Morecroft, M., Musili, P.M., Myers, J.A., Novotny, V., de Oliveira, A., Ong, P., Orwig, D.A., Ostertag, R., Parker, G.G., Patankar, R., Phillips, R.P., Reynolds, G., Sack, L., Song, G.-Z.M., Su, S.-H., Sukumar, R., Sun, I.-F., Suresh, H.S., Swanson, M.E., Tan, S., Thomas, D.W., Thompson, J., Uriarte, M., Valencia, R., Vicentini, A., Vrška, T., Wang, X., Weiblen, G.D., Wolf, A., Wu, S.-H., Xu, H., Yamakura, T., Yap, S., Zimmerman, J.K., 2018. Global importance of large-diameter trees. Global Ecology and Biogeography 27, 849–864. https://doi.org/10.1111/geb.12747
Luyssaert, S., Schulze, E.-D., Börner, A., Knohl, A., Hessenmöller, D., Law, B.E., Ciais, P., Grace, J., 2008. Old-growth forests as global carbon sinks. Nature 455, 213–215. https://doi.org/10.1038/nature07276
Mackey, B., 2014. Counting trees, carbon and climate change. The Royal Statistical Society—Significance 19–23.
Mackey, B., Prentice, I.C., Steffen, W., House, J.I., Lindenmayer, D., Keith, H., Berry, S., 2013. Untangling the confusion around land carbon science and climate change mitigation policy. Nature Climate Change 3, 552–557. https://doi.org/10.1038/nclimate1804
Mackey, B., Ware, D., Buckwell, A., Nalau, J., Sahin, O., Flemming, C. M., Smart, J.C., Connollet, R., Hallgren, W., 2018. Options and Implementation for Ecosystem-based Adaptation, Tanna

Island, Vanuatu (No. Report No. ESRAM-3), Griffith Climate Change Response Program. Griffith University, Gold Coast.

Martin, P.A., Newton, A.C., Pfeifer, M., Khoo, M., Bullock, J.M., 2015. Impacts of tropical selective logging on carbon storage and tree species richness: A meta-analysis. Forest Ecology and Management 356, 224–233. https://doi.org/10.1016/j.foreco.2015.07.010

Meinshausen, M., B. Hare, T. M. L. Wigley, D. van Vuuren, M. G. J. den Elzen and R. Swart (2006). "Multi-gas emission pathways to meet climate targets." Climatic Change 75(1): 151–194.

Meinshausen, M., S. C. B. Raper and T. M. L. Wigley (2011). "Emulating coupled atmosphere-ocean and carbon cycle models with a simpler model, MAGICC6: Part I—Model Description and Calibration." Atmospheric Chemistry and Physics 11: 1417-1456.

Müller, R., Pfeifroth, U (2015), Müller, R., Pfeifroth, U., Träger-Chatterjee, C., Trentmann, J., Cremer, R. (2015). Digging the METEOSAT Treasure—3 Decades of Solar Surface Radiation. Remote Sensing 7, 8067–8101. doi: https://doi.org/10.3390/rs70608067

Nabuurs, G.-J., Delacote, P., Ellison, D., Hanewinkel, M., Hetemäki, L., Lindner, M., 2017. By 2050 the Mitigation Effects of EU Forests Could Nearly Double through Climate Smart Forestry. Forests 8, 484. https://doi.org/10.3390/f8120484

Nauels, A., M. Meinshausen, M. Mengel, K. Lorbacher and T. M. L. Wigley (2016). "Synthesizing long-term sea level rise projections—the MAGICC sea level model." Geosci. Model Dev. Discuss. 2016: 1–40.

Pan, Y., Birdsey, R.A., Fang, J., Houghton, R., Kauppi, P.E., Kurz, W.A., Phillips, O.L., Shvidenko, A., Lewis, S.L., Canadell, J.G., Ciais, P., Jackson, R.B., Pacala, S.W., McGuire, A.D., Piao, S., Rautiainen, A., Sitch, S., Hayes, D., 2011. A Large and Persistent Carbon Sink in the World's Forests. Science 333, 988–993. https://doi.org/10.1126/science.1201609

Pfenninger, S, Staffell, I. (2016a), Pfenninger, Stefan and Staffell, Iain (2016). Long-term patterns of European PV output using 30 years of validated hourly reanalysis and satellite data. Energy 114, pp. 1251-1265. doi: https://doi.org/10.1016/j.energy.2016.08.060

Pfenninger, S, Staffell, I. (2016b), Staffell, Iain and Pfenninger, Stefan (2016). Using Bias-Corrected Reanalysis to Simulate Current and Future Wind Power Output. Energy 114, pp. 1224-1239. doi: https://doi.org/10.1016/j.energy.2016.08.068

Pingoud, K., Ekholm, T., Sievänen, R., Huuskonen, S., Hynynen, J., 2018. Trade-offs between forest carbon stocks and harvests in a steady state—A multi-criteria analysis. Journal of Environmental Management 210, 96–103. https://doi.org/10.1016/j.jenvman.2017.12.076

Poorter, L., Bongers, F., Aide, T.M., Almeyda Zambrano, A.M., Balvanera, P., Becknell, J.M., Boukili, V., Brancalion, P.H.S., Broadbent, E.N., Chazdon, R.L., Craven, D., de Almeida-Cortez, J.S., Cabral, G.A.L., de Jong, B.H.J., Denslow, J.S., Dent, D.H., DeWalt, S.J., Dupuy, J.M., Durán, S.M., Espírito-Santo, M.M., Fandino, M.C., César, R.G., Hall, J.S., Hernandez-Stefanoni, J.L., Jakovac, C.C., Junqueira, A.B., Kennard, D., Letcher, S.G., Licona, J.-C., Lohbeck, M., Marín-Spiotta, E., Martínez-Ramos, M., Massoca, P., Meave, J.A., Mesquita, R., Mora, F., Muñoz, R., Muscarella, R., Nunes, Y.R.F., Ochoa-Gaona, S., de Oliveira, A.A., Orihuela-Belmonte, E., Peña-Claros, M., Pérez-García, E.A., Piotto, D., Powers, J.S., Rodríguez-Velázquez, J., Romero-Pérez, I.E., Ruíz, J., Saldarriaga, J.G., Sanchez-Azofeifa, A., Schwartz, N.B., Steininger, M.K., Swenson, N.G., Toledo, M., Uriarte, M., van Breugel, M., van der Wal, H., Veloso, M.D.M., Vester, H.F.M., Vicentini, A., Vieira, I.C.G., Bentos, T.V., Williamson, G.B., Rozendaal, D.M.A., 2016. Biomass resilience of Neotropical secondary forests. Nature 530, 211–214. https://doi.org/10.1038/nature16512

Ramachandran Nair, P.K., Mohan Kumar, B., Nair, V.D., 2009. Agroforestry as a strategy for carbon sequestration. Journal of Plant Nutrition and Soil Science 172, 10–23. https://doi.org/10.1002/jpln.200800030

RE-N DB (2018) Renewables.ninja, onione database for hourly time series for solar and wind data for a specific geographical position, viewed and data down load took place between May and July 2018, https://www.renewables.ninja/

REN21-GSR (2018), 2018; Renewables 2018 Global Status Report, (Paris: REN21 Secretariat), ISBN 978-3-9818911-3-3

Rienecker, M, Suarez MJ, (2011) Rienecker MM, Suarez MJ, Gelaro R, Todling R, et al. (2011). MERRA: NASA's Modern-Era Retrospective Analysis for Research and Applications. Journal of Climate, 24(14): 3624-3648. doi: https://doi.org/10.1175/JCLI-D-11-00015.1

Rogelj, J, den Elzen, M (2016), Joeri Rogelj, Michel den Elzen, Niklas Höhne, Taryn Fransen, Hanna Fekete, Harald Winkler, Roberto Schaeffer, Fu Sha, Keywan Riahi & Malte Meinshausen, (30 June 2016), Paris Agreement climate proposals need a boost to keep warming well below 2 °C, Nature volume 534, pages 631–639 (30 June 2016), https://www.nature.com/articles/nature18307

Roxburgh, S.H., Wood, S.W., Mackey, B.G., Woldendorp, G., Gibbons, P., 2006. Assessing the carbon sequestration potential of managed forests: a case study from temperate Australia: Carbon sequestration potential. Journal of Applied Ecology 43, 1149–1159. https://doi.org/10.1111/j.1365-2664.2006.01221.x

Rutovitz (2015) chapter 7 of: Teske, S, Pregger, T., Naegler, T., Simon, S., Energy [R]evolution—A sustainable

Rutovitz, J., Dominish, E (2015), Rutovitz, J., Dominish, E & Downes, J. (2015) Calculating Global Energy Sector Jobs: 2015 Methodology. Prepared for Greenpeace International by the Institute for Sustainable Futures, University of Technology Sydney.

Rutovitz, J, James, G (2017), Rutovitz, J., James, G., Teske S., Mpofu, S., Usher, J, Morris, T., and Alexander, D. 2017. Storage Requirements for Reliable Electricity in Australia. Report prepared by the Institute for Sustainable Futures for the Australian Council of Learned Academies.

Sarah (2018) – Online database of "renewable.ninja," https://dx.doi.org/10.5676/EUM_SAF_CM/SARAH/V001

Schlenzig C (1998) Energy planning and environmental management with the information and decision support system MESAP. International Journal of Global Energy Issues, 12(1–6):81–91.

Seven2one (2012) Mesap/PlaNet software framework: Seven2one Modelling, Mesap4, Release 4.14.1.9, Seven2one Informationssysteme GmbH, Karlsruhe, Germany

Simon S, Naegler T, Gils H (2018) Transformation towards a Renewable Energy System in Brazil and Mexico—Technological and Structural Options for Latin America. Energies 11 (4):907

ST 7 (2018), Statistic Times viewed July 2018, http://statisticstimes.com/economy/countries-by-gdp-sector-composition.php

Teske, S (2015), Thesis, Bridging the Gap between Energy- and Grid Models, Developing an integrated infrastructural planning model for 100% renewable energy systems in order to optimize the interaction of flexible power generation, smart grids and storage technologies, chapter 2, University Flensburg, Germany

Teske, S, Brown, T, (2012), Teske, S, Dr. Tom Brown, Dr. Eckehard Tröster, Peter-Philipp Schierhorn, Dr. Thomas

Teske, S, Dominish, E (2016), Teske, S., Dominish, E., Ison, N. and Maras, K. (2016) 100% Renewable Energy for Australia—Decarbonising Australia's Energy Sector within one Generation; Report prepared by ISF for GetUp! and Solar Citizens, March 2016, https://www.uts.edu.au/sites/default/files/article/downloads/ISF_100%25_Australian_Renewable_Energy_Report.pdf

Teske, S, Morris, T (2017), Teske, S., Morris, T., Nagrath, Kriti (2017) 100% Renewable Energy for Tanzania—Access to renewable and affordable energy for all within one generation. Report prepared by ISF for Bread for the World, October 2017.

Teske, S, Pregger, T (2015), Teske, S, Pregger, T., Naegler, T., Simon, S., Energy [R]evolution—A sustainable World Energy Outlook 2015, Greenpeace International with the German Aerospace Centre (DLR), Institute of Engineering Thermodynamics, System Analysis and Technology Assessment, Stuttgart, Germany. https://www.scribd.com/document/333565532/Energy-Revolution-2015-Full

UN PD DB (2018) United Nations, Population Division, online database, viewed June 2018—https://population.un.org/Household/index.html#/countries

UNFCCC (2015), The Paris Agreement builds upon the Convention and for the first time brings all nations into a common cause to undertake ambitious efforts to combat climate change and adapt to its effects, with enhanced support to assist developing countries to do so. As such, it charts a new course in the global climate effort. https://unfccc.int/process-and-meetings/the-paris-agreement/the-paris-agreement

(Vaisala 2017) Free Wind and Solar Resource Maps; Accessed 25 October 2018; ONLINE; https://www.vaisala.com/en/lp/free-wind-and-solar-resource-maps.

Watson, B., Noble, L., Bolin, B., et. al., 2000. Summary for policymakers: land use, land-use change, and forestry: a special report of the Intergovernmental Panel on Climate Change. Intergovernmental Panel on Climate Change.

WEC (2018) World Energy Council—Energy Efficiency Indicators, Online database, viewed June 2018), https://wec-indicators.enerdata.net/specific-electricity-use.html

Wheeler, C., Mitchard, E., Koch, A., Lewis, S.L., in press. The mitigation potential of large-scale tropical forest restoration: assessing the promise of the Bonn Challenge. in review.

WRI (2018), World Resources Institute, A Global Database of Power Plants, April 2018, viewed in July 2018, http://www.wri.org/publication/global-power-plant-database

Ziegler, A.D., Phelps, J., Yuen, J.Q., Webb, E.L., Lawrence, D., Fox, J.M., Bruun, T.B., Leisz, S.J., Ryan, C.M., Dressler, W., Mertz, O., Pascual, U., Padoch, C., Koh, L.P., 2012. Carbon outcomes of major land-cover transitions in SE Asia: great uncertainties and REDD+ policy implications. Global Change Biology 18, 3087–3099. https://doi.org/10.1111/j.1365-2486.2012.02747.x

Zimmerman, B., Kormos, C., 2012. Prospects for Sustainable Logging in Tropical Forests. BioScience 62, 479–487. https://doi.org/10.1525/bio.2012.62.5.9

Zomer, R.J., Neufeldt, H., Xu, J., Ahrends, A., Bossio, D., Trabucco, A., van Noordwijk, M., Wang, M., 2016. Global Tree Cover and Biomass Carbon on Agricultural Land: The contribution of agroforestry to global and national carbon budgets. Scientific Reports 6. https://doi.org/10.1038/srep29987

Chapter 4
Mitigation Scenarios for Non-energy GHG

Malte Meinshausen and Kate Dooley

Abstract Presentation of non-energy emission pathways in line with the new UNFCCC Shared Socio-Economic Pathways (SSP) scenario characteristics and the evaluation of the multi-gas pathways against various temperature thresholds and carbon budgets (1.5 °C and 2.0 °C) over time, and additionally against a 1.5 °C carbon budget in 2100, followed by a discussion of the results in the context of the most recent scientific literature in this field. Presentation of the non-energy GHG mitigation scenarios calculated to complement the energy-related CO_2 emissions derived in Chap. 8.

In this section, we present the results for the land-use CO_2 and non-CO_2 emissions pathways that complement the 2.0 °C and 1.5 °C energy-related CO_2 scenarios.

4.1 Land-Use CO_2 emissions

This section presents the aggregate results for the land-use sequestration pathways designed for this study. Figure 4.1 below shows the annual sequestration in the sequestration pathways over time, differentiated into climate domains. The pathways shown are the results of the Monte Carlo analysis described in Table 3.11 in Sect. 3.8.1.5 and the text. We focus on the median values (thick lines in Fig. 4.1). Note that the area under the curve for a given pathway is an indication of the cumulative CO_2 uptake. By far the most important sequestration may result from large-scale reforestation measures, particularly in the subtropics and tropics (see yellow pathways in Fig. 4.1 below). The second most important pathway in terms of the amount of CO_2 sequestered is the sustainable use of existing forests, which basically means reducing logging within those forests. Although effective mitigation is not

M. Meinshausen (✉) · K. Dooley
Australian-German Climate and Energy College, University of Melbourne, Parkville, Victoria, Australia
e-mail: malte.meinshausen@unimelb.edu.au; kate.dooley@unimelb.edu.au

S. Teske (ed.), *Achieving the Paris Climate Agreement Goals*,
https://doi.org/10.1007/978-3-030-05843-2_4

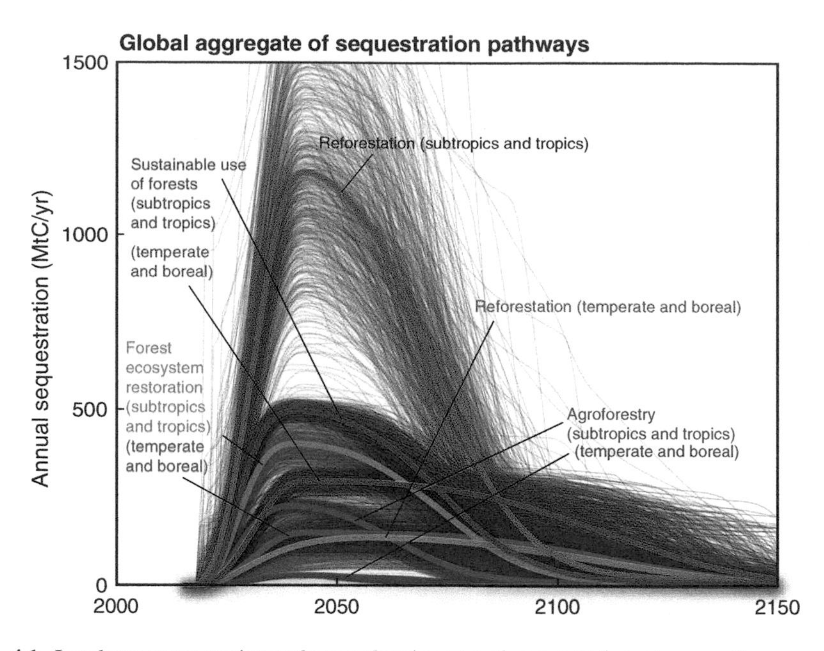

Fig. 4.1 Land-use sequestration pathways showing annual sequestration rates over time

achieved in the tropics (Martin et al. 2015), in the temperate and boreal regions, improved forest management could provide substantial additional carbon uptake over time. The time horizon for this sequestration option is assumed to be relatively long in temperate and boreal regions, consistent with the longer time it takes for these forest ecosystems to reach equilibrium (Roxburgh et al. 2006; Luyssaert et al. 2008). The 'forest ecosystem restoration' pathway is also important, which basically assumes a reduction to zero in logging rates in a fraction of the forest, allowing these forests to be restored to full ecosystem function, including their carbon stocks and resilience due to biodiversity (Mackey 2014).

Overall, the median of all the assumed sequestration pathways, shown in Fig. 4.1, would result in the sequestration of 151.9 GtC by 2150. This is approximately equivalent to all historical land-use-related CO_2 emissions to date (Houghton and Nassikas 2017; Mackey et al. 2013). The magnitude of these figures indicates the substantial challenges that go hand in hand with these sequestration pathways. Given the competing forms of land use that exist today, the challenge of converting overall terrestrial carbon stocks back to pre-industrial levels cannot be underestimated. There would be significant benefits, but also risks, if this sequestration option were to be used instead of mitigation. The benefits are clearly manifold, ranging from biodiversity protection, reduced erosion, improved local climates, wind protection, and potentially a reduction in air pollution (Mackey 2014). Despite this, terrestrial carbon sequestration is inherently impermanent. However, a future warming climate with an increased fire risk also brings with it the risk of large reversals in sequestered carbon. Similarly, prolonged droughts in some areas could reverse the gains in terrestrial carbon stocks. Although the increased resilience of natural and biodiverse ecosystems compared with that of monoculture plantations can guard against this risk (DellaSala, 2019; Lindenmayer and Sato 2018), a future mitigation pathway that relies on sequestration *instead of mitigation action* is ultimately always more susceptible to higher long-term climate change, given the risk of 'non-permanence'. However, in this study, the land-use CO_2 sequestration pathways complement some of the most ambitious mitigation pathways, and should therefore be regarded, not as 'offsetting' mitigation action, but as complementary measures to help reduce the CO_2 concentrations that have arisen from the overly high emissions in the past.

The thin lines in Fig. 4.1 indicate individual draws in the Monte Carlo analysis. The thick lines are the median values from the ensemble of draws for each sequestration pathway and domain.

We aggregated the four sequestration pathways from our country-level data to the five RCP regions (Fig. 4.2). The country-level data were subject to substantial uncertainties and simplifications because we used climate-domain average uptakes, carbon density caps, and saturation periods. The re-aggregated sequestration rates over the five RCP regions can be considered approximate illustrations of the biome-average sequestration rates if those sequestration pathways were pursued with a range of institutional and policy measures.

For the 1.5 °C Scenario, we assumed the full extent of sequestration shown in Fig. 4.1, whereas for the 2.0 °C pathway, we assumed that only a third of that sequestration will occur. The reference scenario is assumed to follow the SSP2 'middle of the road' reference scenario created by the MESSAGE-GLOBIOM modelling team. As illustrated in Fig. 4.2, the reference scenario does not assume a complete phasing-out of global land-use-related net emissions over the next 20 or 30 years. Instead, it assumes that they are not phased-out until approximately 2080.

The 2.0 °C pathway (brighter blue in Fig. 4.2) aligns relatively well with the SSP1 1.9 and SSP1 2.6 scenarios from the forthcoming CMIP6 model inter-comparison project. The 1.5 °C pathway, with three times the sequestration rates, is consistent with the lower land-use CO_2 scenarios analysed here—with mitigation rates of up to −2 GtC per annum from 2040 to 2050.

Figure 4.2 shows the land-use-related CO_2 emission and sequestration rates of the 2.0°C and 1.5 °C pathways in this study compared with those in the CMIP6 CEDS scenarios (turquoise) and the scenarios from the IPCC SR1.5 database (thin green lines). The global total pathway is the sum of the five regional pathways shown in the lower row of the panels.

4.1.1 *Other GHG and Aerosol Emissions*

This section examines the other main GHGs (methane and N_2O) and gives examples of some fluorinated gases. The full results, with the species-by-species time series, are provided in a data appendix.

Methane (CH_4) emissions are the second-largest contributor to anthropogenically induced climate change. Our approach, described in the Methods section earlier, derives pathways for the 1.5 °C and 2.0 °C Scenarios that are close to the lower end of the overall scenario distribution. This is mainly because the methane distribution at

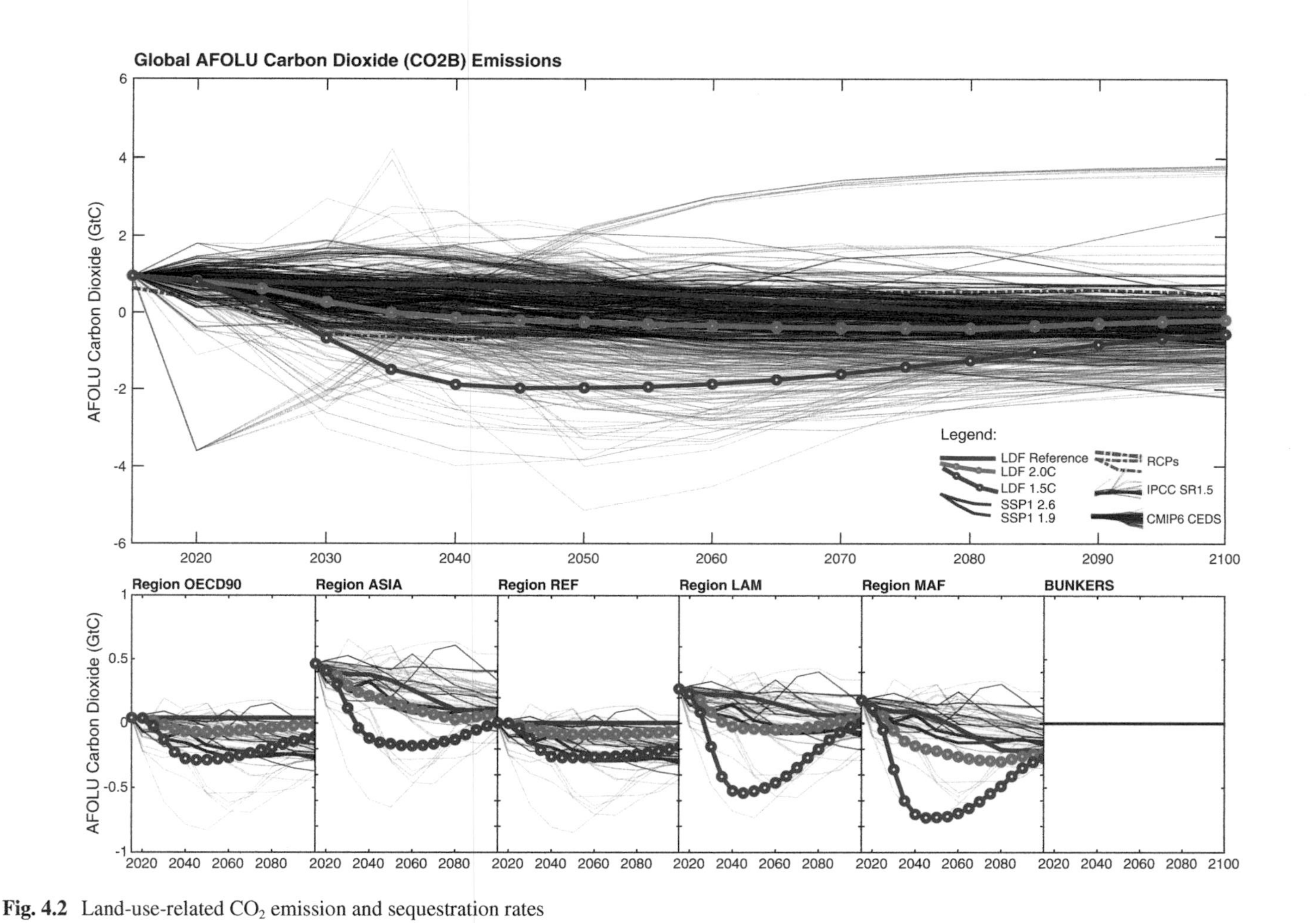

Fig. 4.2 Land-use-related CO_2 emission and sequestration rates

the lower end of the fossil fuel CO_2 emissions is relatively narrow, and there is a strong correlation between the fossil CO_2 and total CH_4 emissions in the scenarios in any given year (see top-left methane panel in Fig. 3.14). As with almost all literature-reported scenarios, a lower plateau of methane emissions is associated with agricultural activities required to feed the world's population. Our quantile regression method resulted in long-term methane emission levels that are quite similar to those in the two lower SSP scenarios, SSP1 1.9 and SSP1 2.6 (Fig. 4.3). The derived CH_4 pathways for 1.5 °C and 2.0 °C track towards the lower of the scenario distributions.

Nitrous oxide (N_2O) is one of the longer-lived GHGs, although the overall amounts in the atmosphere are much smaller than those of methane or CO_2. The relatively high plateau of global emissions, around 5 MtN_2O-N for N_2O, are reflected in the SSP1. 1.9 and SSP1 2.6 scenarios (dark green lines in Fig. 4.4) and in the quantile regression results for the 2.0 °C and 1.5 °C scenarios in this study. This plateau of emissions is related to agricultural activities, mainly the use of fertilizers, and combined with the long lifetime of N_2O, it means that the N_2O concentrations are projected to increase further over the course of the century, even for the lower 1.5 °C and 2.0 °C pathways.

The derived methane pathways for 1.5 °C and 2.0 °C track towards the lower of the scenario distributions.

Halocarbons and fluorinated gases are another group of important GHGs. Recently, some of these gases, such as HFCs, were also subjected to control under the Montreal Protocol, with clear phase-out schedules. Some of the halocarbons and fluorinated gases (such as tetrafluoromethane, CF_4) are only produced and emitted in relatively small quantities—largely for industrial purposes in the semi-conductor industry. Some also have applications in the agricultural sector, including methyl bromide, which is used for soil fumigation. SF_6 is one of the strongest GHGs on a per mass basis. It is controlled under the Kyoto Protocol and included in the nationally determined contributions (NDCs) by many countries under the Paris Agreement. Our applied quantile regression method practically phases-out many of the halogenated species over the course of the next 10–20 years, although some small background emissions remain. The full results for 40 halocarbons, HFCs, PFCs, and SF_6 are provided in a data appendix (Figs. 4.5 and 4.6).

Aerosols have an important temporary masking effect on GHG-induced warming. The most important anthropogenically emitted aerosol coolant in the climate system is sulfur dioxide or SO_X. With higher fuel standards and concerns about local air pollution, future SO_X emissions are projected to be substantially lower than current levels. In fact, most emission inventories assume that SO_X emissions peaked in the 1990s. Therefore, even in the most high-fossil-fuel-emitting reference scenarios, SO_X emissions are projected to decrease. Asia produces by far the most SO_X emissions of any continent because of the coal-fuelled power plants in China and India. In the 2.0 °C Scenario, our quantile regression method sets sulfate aerosol emissions at levels in between those in the SSP1 2.6 and SSP1 1.9 scenarios, whereas in the 1.5 °C Scenario, the level is even lower.

Similarly, the projected emissions of NO_X, which is largely a by-product of fossil fuel burning, are highest in Asia. In the derived 1.5 °C and 2.0 °C Scenarios, NO_X

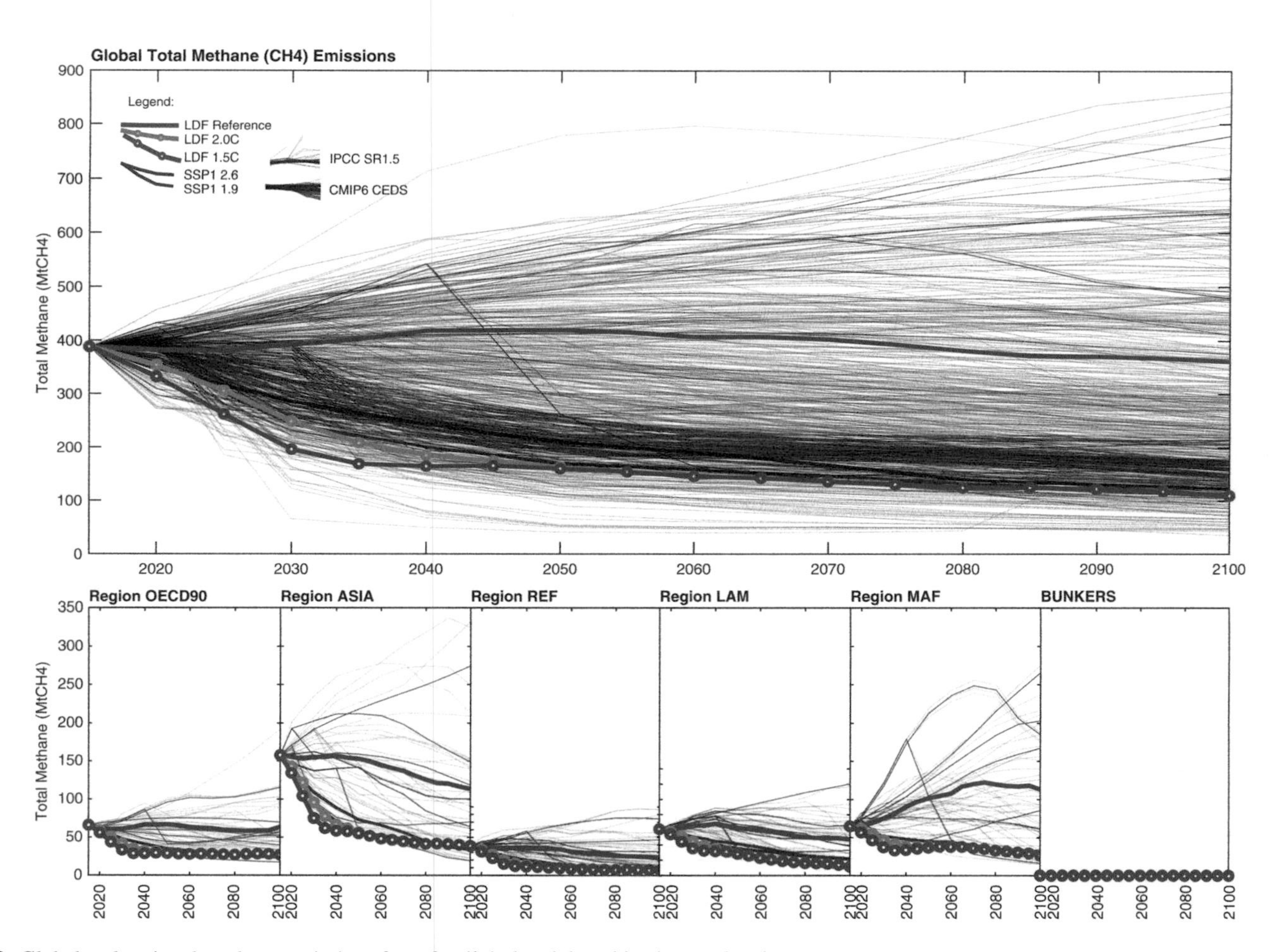

Fig. 4.3 Global and regional methane emissions from fossil, industrial, and land-use-related sources

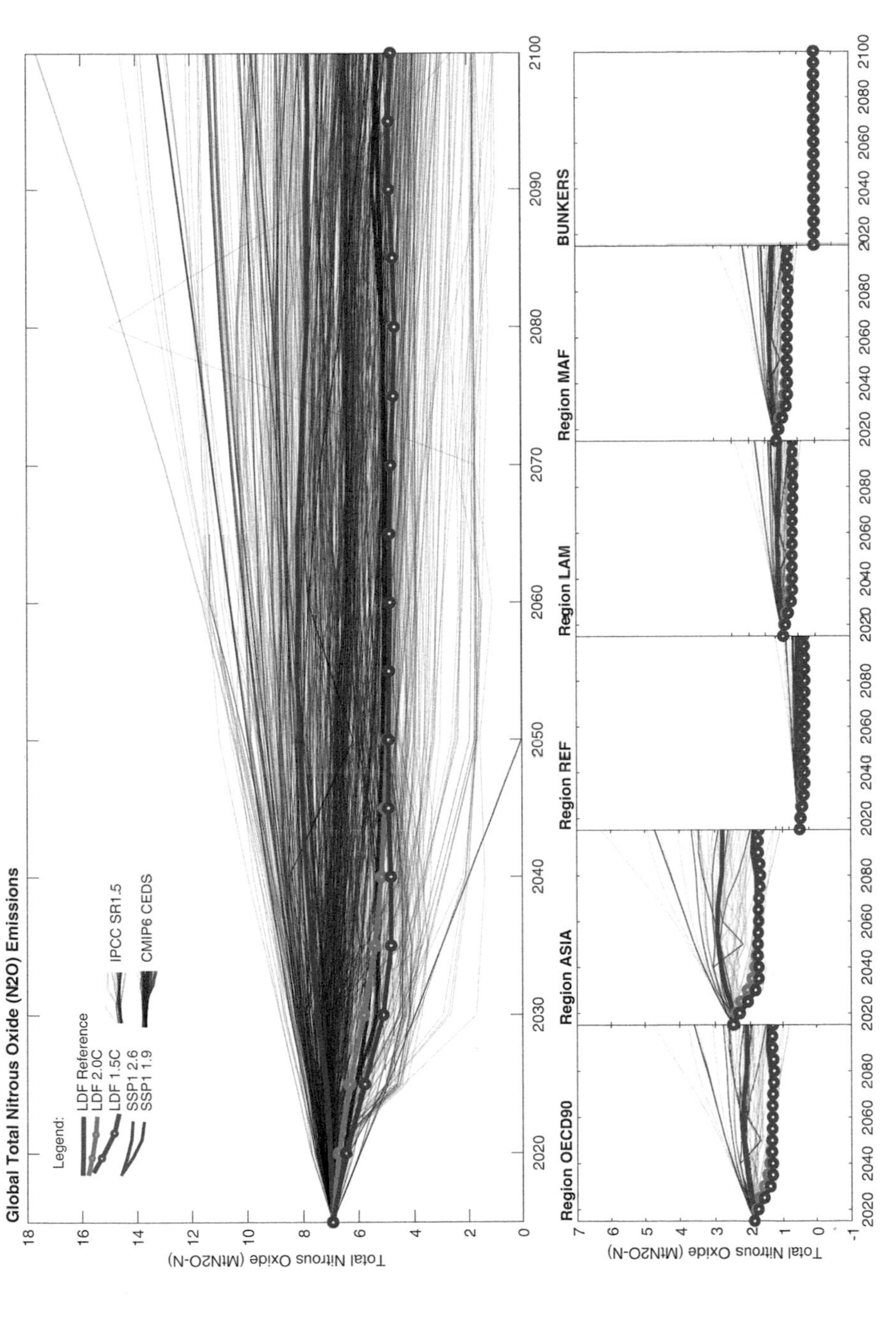

Fig. 4.4 Global and regional methane emissions from fossil, industrial, and land-use-related sources

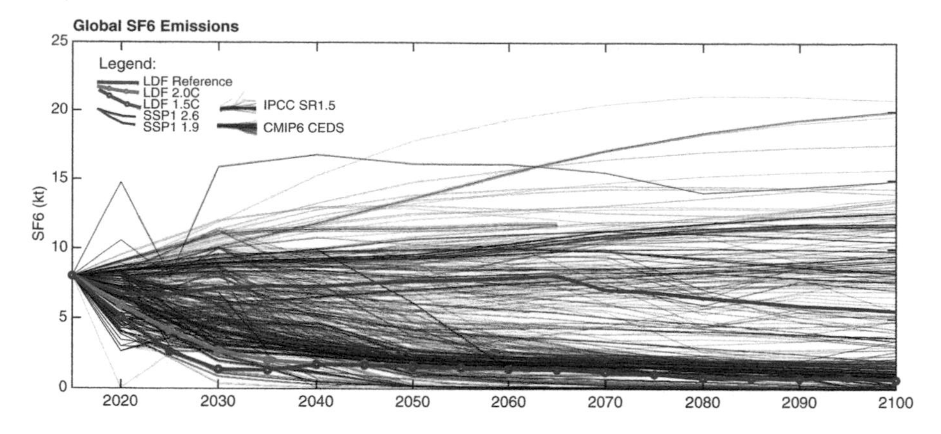

Fig. 4.5 Global SF_6 emission levels from literature-reported scenarios and the LDF pathways derived in this study

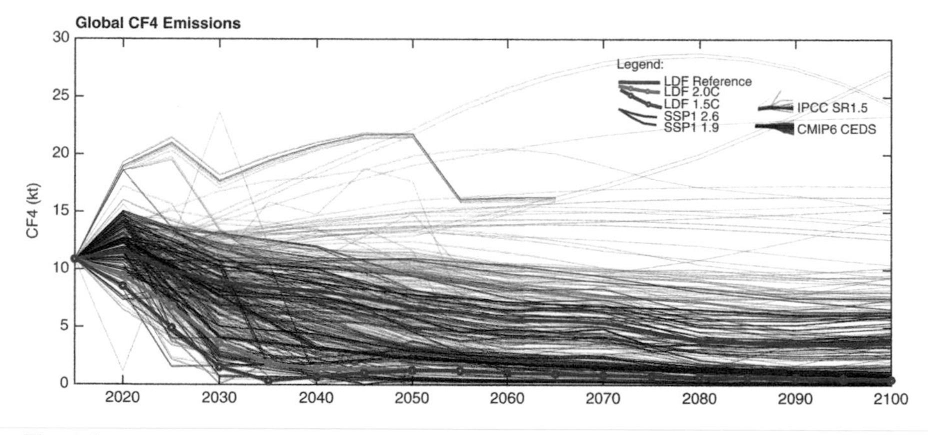

Fig. 4.6 Global tetrafluoromethane (CF_4) emissions from the collection of assessed literature-reported scenarios and the LDF pathways derived in this study

levels are between the levels in the SSP1 2.6 and SSP1 1.9 scenarios for most of the twenty-first century (Figs. 4.7 and 4.8).

Black and organic carbon emissions are also accruing substantially in the Middle East and Africa, largely from biomass burning (Figs. 4.9 and 4.10). Similar to other aerosol emissions, black and organic carbon emissions are projected to decrease. Although black carbon is a substantial warming agent, organic carbon is a net coolant. Because both species are often co-emitted, the net effect of policies to reduce black carbon do not have as large a mitigation benefit as might be initially assumed. This is because a reduction in the processes and activities that produce black carbon emissions will also lead to lower organic carbon emissions, partially offsetting both the warming and cooling effects. The emissions projected as part of the IMAGE model SSP1 2.6 and SSP1 1.9 scenarios are very low compared with those in other studies. Furthermore, the correlation between fossil CO_2 emissions and black or

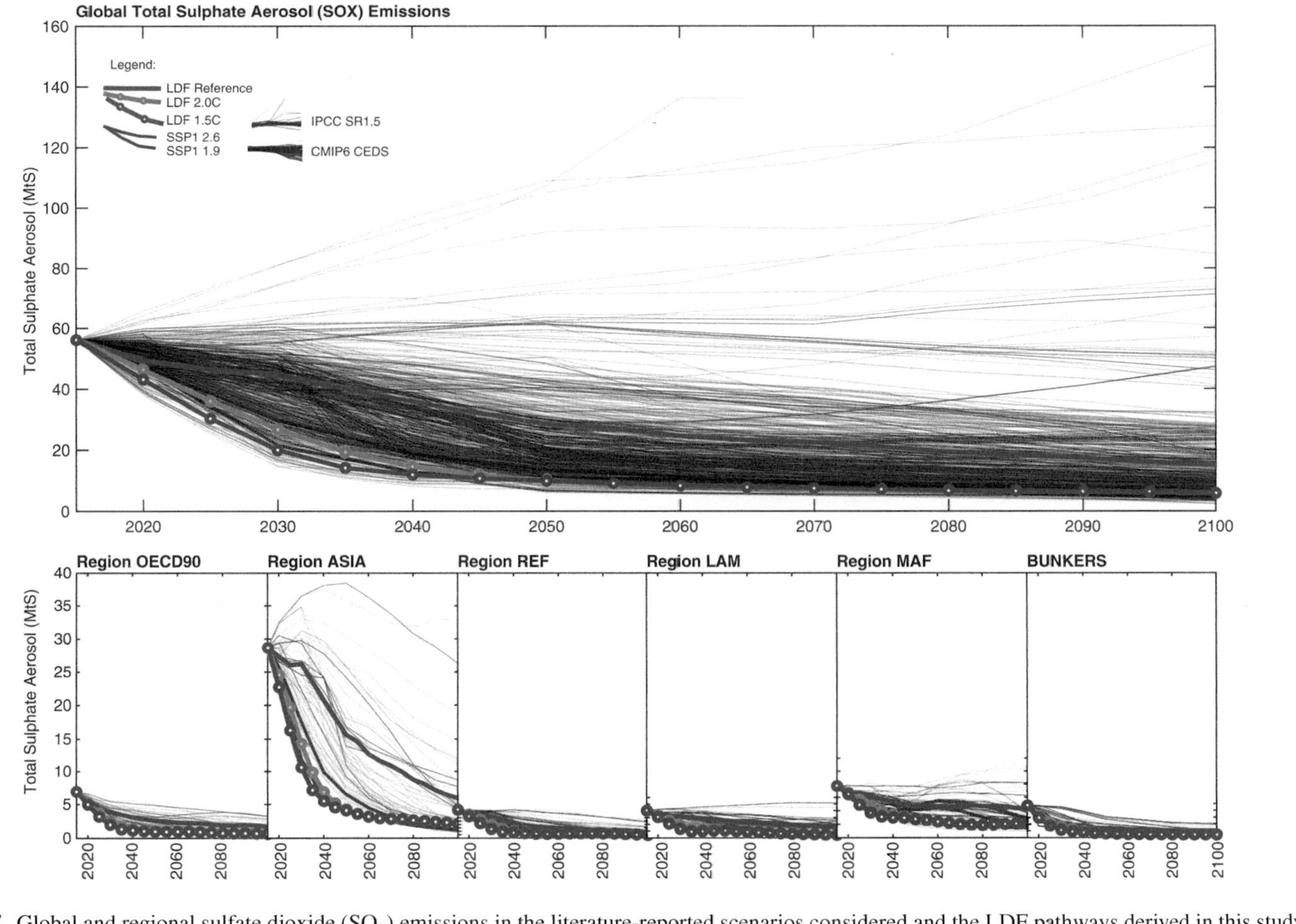

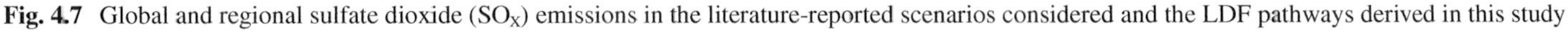
Fig. 4.7 Global and regional sulfate dioxide (SO_X) emissions in the literature-reported scenarios considered and the LDF pathways derived in this study

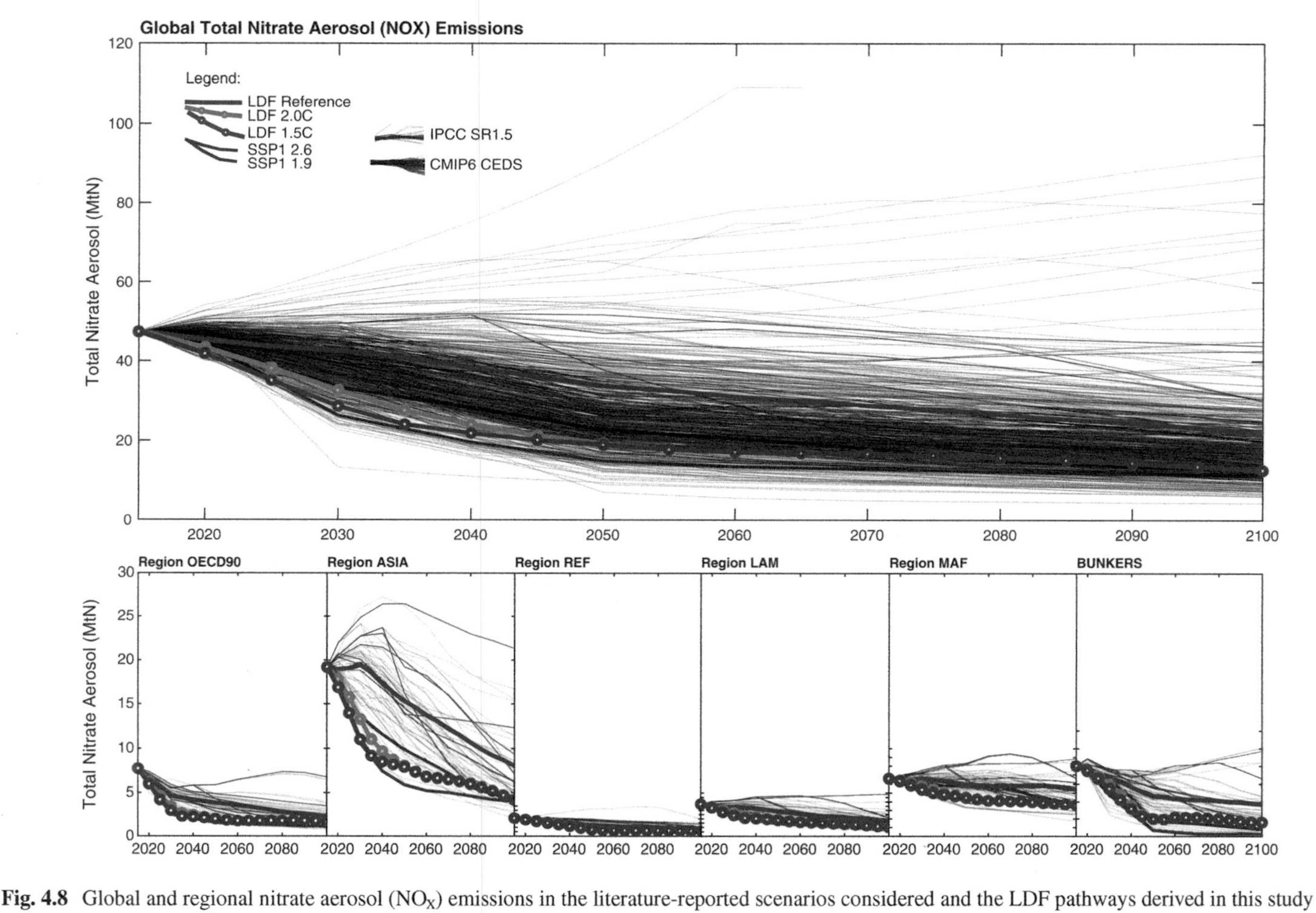

Fig. 4.8 Global and regional nitrate aerosol (NO_X) emissions in the literature-reported scenarios considered and the LDF pathways derived in this study

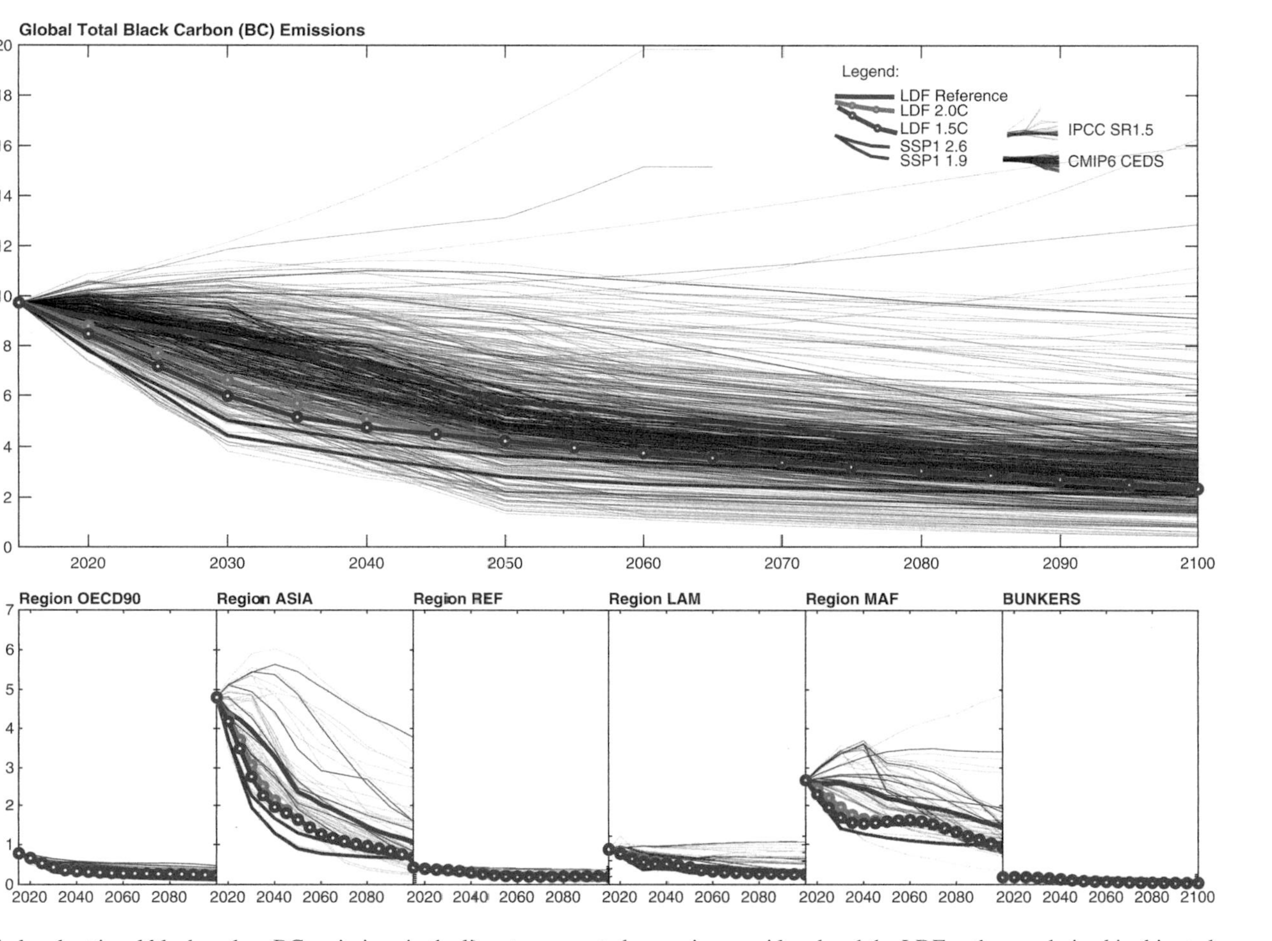

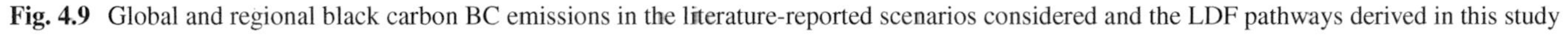
Fig. 4.9 Global and regional black carbon BC emissions in the literature-reported scenarios considered and the LDF pathways derived in this study

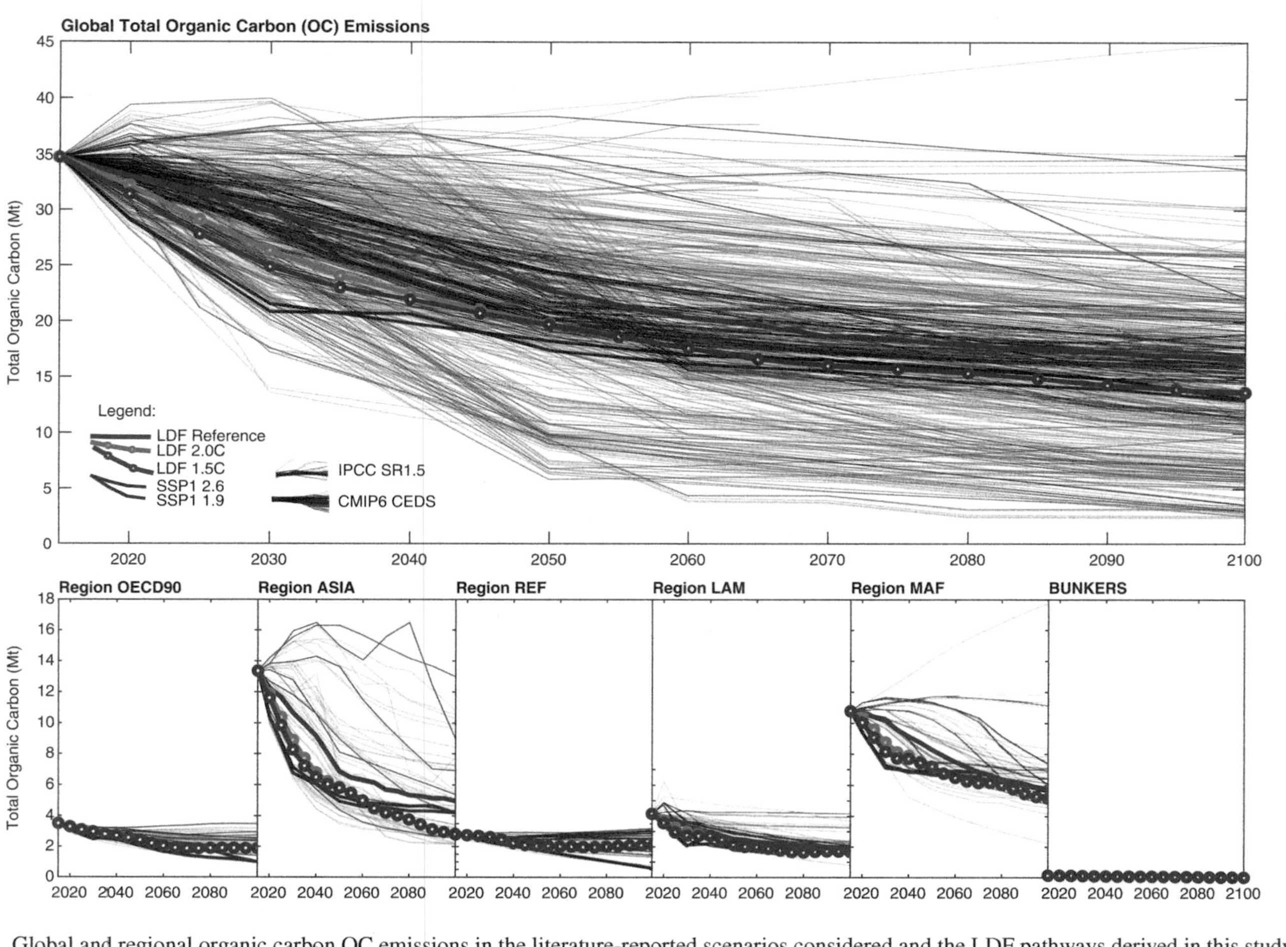

Fig. 4.10 Global and regional organic carbon OC emissions in the literature-reported scenarios considered and the LDF pathways derived in this study

organic carbon is less pronounced than the correlations of fossil CO_2 with other aerosols, such as NO_X and SO_X, partly because it results from biomass burning, which is not related to the burning of fossil fuels. Therefore, with our quantile regression method, the black carbon and organic carbon emission pathways are not as low as those found in the lower SSP scenarios (see Figs. 4.9 and 4.10).

For Tabular overview of three scenarios see Annex

References

DellaSala, D.L., 2019. "Real" vs. "Fake" Forests: Why Tree Plantations are Not Forests, in: Encyclopedia of the World's Biomes. Elservier, UK.

Houghton, R.A., Nassikas, A.A., 2017. Global and regional fluxes of carbon from land use and land cover change 1850-2015: Carbon Emissions From Land Use. Global Biogeochemical Cycles 31, 456–472. https://doi.org/10.1002/2016GB005546

Lindenmayer, D.B., Sato, C., 2018. Hidden collapse is driven by fire and logging in a socioecological forest ecosystem. Proceedings of the National Academy of Sciences 115, 5181–5186. https://doi.org/10.1073/pnas.1721738115

Luyssaert, S., Schulze, E.-D., Börner, A., Knohl, A., Hessenmöller, D., Law, B.E., Ciais, P., Grace, J., 2008. Old-growth forests as global carbon sinks. Nature 455, 213–215. https://doi.org/10.1038/nature07276

Mackey, B., 2014. Counting trees, carbon and climate change. The Royal Statistical Society - Significance 19–23.

Mackey, B., Prentice, I.C., Steffen, W., House, J.I., Lindenmayer, D., Keith, H., Berry, S., 2013. Untangling the confusion around land carbon science and climate change mitigation policy. Nature Climate Change 3, 552–557. https://doi.org/10.1038/nclimate1804

Martin, P.A., Newton, A.C., Pfeifer, M., Khoo, M., Bullock, J.M., 2015. Impacts of tropical selective logging on carbon storage and tree species richness: A meta-analysis. Forest Ecology and Management 356, 224–233. https://doi.org/10.1016/j.foreco.2015.07.010

Roxburgh, S.H., Wood, S.W., Mackey, B.G., Woldendorp, G., Gibbons, P., 2006. Assessing the carbon sequestration potential of managed forests: a case study from temperate Australia: Carbon sequestration potential. Journal of Applied Ecology 43, 1149–1159. https://doi.org/10.1111/j.1365-2664.2006.01221.x

Chapter 5
Main Assumptions for Energy Pathways

Thomas Pregger, Sonja Simon, Tobias Naegler, and Sven Teske

Abstract The aim of this chapter is to make the scenario calculations fully transparent and comprehensible to the scientific community. It provides the scenario narratives for the reference case (5.0 °C) as well as for the 2.0 °C and 1.5 °C on a global and regional basis. Cost projections for all fossil fuels and renewable energy technologies until 2050 are provided. Explanations are given for all relevant base year data for the modelling and the main input parameters such as GDP, population, renewable energy potentials and technology parameters.

Scenario studies cannot predict the future, but they can describe what is needed for a successful pathway in terms of technology implementation and investments. Scenarios also help us to explore the possible effects of transition processes, such as supply costs and emissions. The energy demand and supply scenarios in this study are based on information about current energy structures and today's knowledge of energy resources and the costs involved in deploying them. As far as possible, we also take into account potential constraints and preferences in each world region. However, this remains difficult due to large sub-regional variations. Our energy modelling primarily aims to achieve a transparent and consistent scenario, an ambitious but still plausible storyline from several possible techno-economic pathways. Knowledge integration is the core of this approach because we must consider different technical, economic, environmental, and societal factors. The scenario modelling follows a hybrid bottom-up/top-down approach, with no cost optimising objective functions. The analysis considers the key technologies required for a successful energy transition, and focuses on the roles and potential of renewable energies. Wind and solar energies have the highest economic potential and dominate the pathways on the

T. Pregger (✉) · S. Simon · T. Naegler
Department of Energy Systems Analysis, German Aerospace Center (DLR),
Institute for Engineering Thermodynamics (TT), Pfaffenwaldring, Germany
e-mail: thomas.pregger@dlr.de; sonja.simon@dlr.de; tobias.naegler@dlr.de

S. Teske
Institute for Sustainable Futures, University of Technology Sydney, Sydney, NSW, Australia
e-mail: sven.teske@uts.edu.au

S. Teske (ed.), *Achieving the Paris Climate Agreement Goals*,
https://doi.org/10.1007/978-3-030-05843-2_5

supply side. However, variable renewable power from wind and photovoltaics (PV) remains limited by the need for sufficient secured capacity in energy systems. Therefore, we also consider concentrated solar power (CSP) with high-temperature heat storage as a solar option that promises large-scale dispatchable and secured power generation.

5.1 Scenario Definition

Scenario modelling was performed for three main scenarios that can be related to different overall carbon budgets between 2015 and 2050 and derived mean global temperature increases. The (around) 5.0 °C Scenario was calculated based on the Current Policies scenario published by the International Energy Agency (IEA) in World Energy Outlook 2017 (IEA 2017), and the emission budget for this scenario simply uses and extrapolates from the corresponding narratives. The 2.0 °C and 1.5 °C Scenarios were calculated in a normative way to achieve defined emission budgets.

5.1.1 The 5.0 °C Scenario (Reference Scenario)

The reference case only takes into account existing international energy and environmental policies. Its assumptions include, for example, continuing progress in electricity and gas market reforms, the liberalization of cross-border energy trade, and recent policies designed to combat environmental pollution. The scenario does not include additional policies to reduce greenhouse gas (GHG) emissions. Because the IEA's projections only extend to 2040, we have extrapolated their key macroeconomic and energy indicators forward to 2050. This provides a baseline for comparison with the 2.0 °C and 1.5 °C Scenarios.

5.1.2 The 2.0 °C Scenario

The first alternative scenario aims to achieve an ambitious emissions reduction to zero by 2050 and a global energy-related CO_2 emissions budget between 2015 and 2050 of around 590 Gt. The scenario is close to the assumptions and results of the Advanced E[R] scenario published in 2015 by Greenpeace (Teske et al. 2015). However, the scenario includes an updated base year, more coherent regional developments of energy intensities, and reconsidered trajectories and shares of renewable energy resource (RES) deployment. The 2.0 °C Scenario represents a far more likely pathway than the 1.5 °C Scenario, because the 2.0 °C case takes into account unavoidable delays due to political, economic, and societal processes and stakeholders.

5.1.3 The 1.5 °C Scenario

The second alternative scenario aims to achieve a global energy-related CO_2 emission budget of around 450 Gt, accumulated between 2015 and 2050. The 1.5 °C Scenario requires immediate action to realize all available options. It is a technical pathway, not a political prognosis. It refers to technically possible measures and options without taking into account societal risks or barriers. Efficiency and renewable potentials must be deployed even more quickly than in the 2.0 °C Scenario. Furthermore, avoiding inefficient technologies and behaviours are essential strategies for developing regions in this time period.

5.2 Scenario World Regions and Clusters

The regional implementation of the long-term energy scenarios is defined according to the breakdown of the ten world regions of the IEA WEO 2016 (IEA 2016a, b). This approach has been chosen because the IEA also provides the most comprehensive global energy statistics and, in contrast to the regional breakdown of the IEA WEO 2017, it is also consistent with the Energy [R]evolution study series. Table 5.1 provides a country breakdown of the ten world regions considered in the scenarios.

Regional conditions play an important role in the layout of the scenario pathways. Therefore, scenario building tries to take into account important factors, such as current demand and supply structures, RES potentials, urbanization rates, and as far as possible, societal and behavioural factors. The following sections provide some regional information. Statistical data for the energy systems in the regions can be found in Sect. 5.3.

5.2.1 OECD North America

The energy system in OECD North America (USA, Canada, and Mexico) is dominated by developments in the USA, where more than 80% of the region's demand occurs. In the highly developed countries of the USA and Canada, reducing the demand for energy by increasing efficiency will play a crucial role in decarbonisation. However, the high energy intensity (i.e., the high demand per capita or per gross domestic product [GDP]) requires even more ambitious measures than in other regions to reduce the energy demand as quickly as possible. In Mexico, in contrast, increasing living standards and the increasing population will increase the difficulties associated with reducing the energy demand, despite ambitious increases in efficiency. Wind and solar power generation will be the backbone of the power supply in OECD North America. They will be supplemented by hydro power (mainly in Canada) and also concentrated solar power (CSP). The high potential for

Table 5.1 World regions used in the scenarios

World region	Countries
OECD Europe	Austria, Belgium, Czech Republic, Denmark, Estonia, Finland, France, Germany, Greece, Hungary, Iceland, Ireland, Italy, Israel, Luxembourg, the Netherlands, Norway, Poland, Portugal, Slovak Republic, Slovenia, Spain, Sweden, Switzerland, Turkey, United Kingdom
OECD North America	Canada, Mexico, United States of America
OECD Pacific	Australia, Japan, Korea (South), New Zealand
Eastern Europe/ Eurasia	Albania, Armenia, Azerbaijan, Belarus, Bosnia-Herzegovina, Bulgaria, Croatia, former Yugoslav Republic of Macedonia, Georgia, Kazakhstan, Kosovo, Kyrgyz Republic, Latvia, Lithuania, Montenegro, Romania, Russia, Serbia, Tajikistan, Turkmenistan, Ukraine, Uzbekistan, Cyprus, Gibraltar and Malta
China	People's Republic of China, including Hong Kong
India	India
Non-OECD Asia (without China and India)	Afghanistan, Bangladesh, Bhutan, Brunei Darussalam, Cambodia, Chinese Taipei, Cook Islands, East Timor, Fiji, French Polynesia, Indonesia, Kiribati, Democratic People's Republic of Korea, Laos, Macao, Malaysia, Maldives, Mongolia, Myanmar, Nepal, New Caledonia, Pakistan, Papua New Guinea, Philippines, Samoa, Singapore, Solomon Islands, Sri Lanka, Thailand, Tonga, Vanuatu, Vietnam,
Latin America	Antigua and Barbuda, Argentina, Aruba, Bahamas, Barbados, Belize, Bermuda, Bolivia, Brazil, British Virgin Islands, Cayman Islands, Chile, Colombia, Costa Rica, Cuba, Dominica, Dominican Republic, Ecuador, El Salvador, Falkland Islands, French Guyana, Grenada, Guadeloupe, Guatemala, Guyana, Haiti, Honduras, Jamaica, Martinique, Montserrat, Netherlands Antilles, Nicaragua, Panama, Paraguay, Peru, St. Kitts and Nevis, Saint Lucia, St. Pierre et Miquelon, St. Vincent and Grenadines, Suriname, Trinidad and Tobago, Turks and Caicos Islands, Uruguay, Venezuela
Africa	Algeria, Angola, Benin, Botswana, Burkina Faso, Burundi, Cameroon, Cape Verde, Central African Republic, Chad, Comoros, Congo, Democratic Republic of Congo, Cote d'Ivoire, Djibouti, Egypt, Equatorial Guinea, Eritrea, Ethiopia, Gabon, Gambia, Ghana, Guinea, Guinea-Bissau, Kenya, Lesotho, Liberia, Libya, Madagascar, Malawi, Mali, Mauritania, Mauritius, Morocco, Mozambique, Namibia, Niger, Nigeria, , Rwanda, Sao Tome and Principe, Senegal, Seychelles, Sierra Leone, Somalia, South Africa, South Sudan, Sudan, Swaziland, United Republic of Tanzania, Togo, Tunisia, Uganda, Western Sahara, Zambia, Zimbabwe
Middle East	Bahrain, Iran, Iraq, Jordan, Kuwait, Lebanon, Oman, Qatar, Saudi Arabia, Syria, United Arab Emirates, Yemen

CSP in Mexico and the southern parts of the USA will allow the large-scale use of CSP plants for grid balancing and grid stabilization. This will reduce the need for power storage, demand-side management, and other balancing strategies. In the large metropolitan areas in North America, electromobility and hydrogen cars will enter the market earlier and at a faster rate than in many other world regions. Large biomass potentials (residues) mean that biofuels could play important roles as climate-neutral fuels to bridge the gap until new powertrain technologies dominate the vehicle market. In the heating sector, particularly for process heat, solid biomass and biogas will be required as alternative fuels until the (direct or indirect) electrification of the heat sector is accomplished.

5.2.2 Latin America

Latin America's energy system is dominated by Brazil, which accounts for around half the region's energy demand. In the reference (5.0 °C) scenario, this region has a particularly high demand for electrification and a strong increase in CO_2 emissions per capita. Latin America has the highest urbanization rate of all non-OECD regions. This provides opportunities for efficiency measures and the large-scale electrification of the heat and transport sectors based on renewable resources. Latin America has a high overall potential for the use of renewable energies (Herreras Martínez et al. 2015) and the largest biomass potential of all regions. It already meets more than 60% of its power demand from renewable sources, and higher shares are the focus of research (Nascimento et al. 2017; Barbosa et al. 2017; Gils et al. 2017). However, in many studies, heat and transport demands are not integrated into the assessments, even though the region has a large potential for renewable heat and decarbonised transport. Given the abundance of biomass, there is potential for generating more than 12 EJ from residues (Seidenberger et al. 2008). Biomass will also play a significant role in the industry sector. Because the region has a long experience of biofuels, they will play a major role in the 2.0 °C and 1.5 °C Scenarios, especially in Brazil, where bioethanol for transport is already competitive (Lora and Andrade 2009; La Rovere et al. 2011; Nass et al. 2007). However, the high urbanization rate in Latin America means there is also an opportunity to develop electromobility early. In the power sector, the use of biomass from residues will help to balance the increasing share of variable renewable energy from the excellent solar and wind resources. Grid extensions will contribute to inter-regional stability (Nascimento et al. 2017).

5.2.3 OECD Europe

The OECD Europe region includes countries with quite different energy supply systems, different potentials for renewable energy sources, and different power and heat demand patterns. High solar potentials and low heat demand for buildings are characteristic of the south. The northern and western parts of Europe have high wind potentials, especially offshore wind. In northern and central Europe, there are high potentials for hydropower and a high energy demand for space heating (such as in Eastern Europe). Biomass potentials exist predominantly in the north and east, but are only limited in the southern regions. The industrial demands for electricity and process heat are quite different in highly industrialized countries, such as the Scandinavian countries, Germany, and France compared with some eastern and southern countries. Most European countries, particularly European Union (EU) member countries, already have policies and market mechanisms for the implementation of renewable energy. The European Network of Transmission System Operators (ENTSO-E) can be used as a well-established basis for the further development of an interconnected European grid, which would be able to implement the large-scale and long-range transmission of renewable power to demand centres. This may also lead to important interconnections to the Middle East/North Africa (MENA) region and Eastern Europe/Eurasia. The possible large-scale importation of solar thermal electricity from MENA countries via high-voltage direct-current lines has been described in many studies and still represents a promising option in the long term, despite the currently difficult political conditions.

5.2.4 Eastern Europe/Eurasia

The Eastern Europe/Eurasia region includes some eastern EU member countries that are not part of the OECD, some other countries of the former Yugoslav Republic, and several countries of the former Soviet Union. However, the region is dominated by the economy and energy system of Russia. The main energy carrier today is natural gas, followed by oil. The region has large energy resources in biomass and wind power, but also geothermal energy and PV. Eastern Europe/Eurasia is the only world region that may face a significant population decline with expected demographic developments, particularly in Russia. Today, the region has by far the highest final and primary energy demand per $GDP. This indicates the existence of energy-intensive industries, but also large efficiency potentials in all sectors. The high heat demand, large rural areas, enormous oil and natural gas potentials, and the uneven distribution of economic wealth are some of the major challenges in this region. So far, only low expansion rates for renewable energies have been achieved in the region.

5.2.5 The Middle East

The Middle East consists of a series of oil-dependent countries, all of which have tremendous solar potential. The transport demand in the Middle East is very high, as is the electrification rate in urban areas, where currently almost 70% of the fast-growing population lives. Therefore, the electrification of transport systems is a major target in our scenarios. For many Middle East countries, water scarcity is a problem, and there are opportunities to combine large CSP plants with water desalination, to reduce the pressure on water supply systems. Biomass is very scarce, so its use must be limited to high-temperature process heat, especially in industry, where other renewable sources cannot be used. This will lead to a high demand for hydrogen or synthetic fuels. Naturally, this also limits the potential for combined heat and power generation (CHP), which is primarily seen as a transition technology to provide the most efficient use of the remaining fossil fuels and low-value biomass wastes. However, because the Middle East has extraordinary solar and wind potentials (Nematollahi et al. 2016; Hess 2018), the solar market is taking off. Projects with a capacity of 11 GW are planned for 2018 (MESIA 2018). With the extraordinarily high number of full-load hours, there is also the potential to use high-temperature solar heat. These resources also provide excellent conditions for hydrogen production, which are extensively exploited in the 2.0 °C and the 1.5 °C Scenarios. Therefore, the Middle East is a model solar and hydrogen region.

5.2.6 Africa

Africa is a very heterogeneous region, both economically and geographically. One of the few things African nations have in common is their very fast population growth. Africa includes the arid regions of North Africa, the undeveloped sub-Saharan region, and the emerging market of South Africa. North Africa features a high electrification rate and a strong dependence on oil. The water and biomass potentials for energy are very low, because water and biomass are prioritized for nutrition (or at least nutrition competes strongly with energy use). The region has outstanding solar irradiation, an excellent renewable energy source. Sub-Saharan Africa is characterized by low urbanization and a lack of access to electricity for two-thirds of its people (IEA 2014). Its energy supply is characterized by a high share of low-efficiency forms of generation, such as traditional biomass use. There is a general lack of energy services. Modernizing traditional biomass use could lead to significant reductions in energy demand, while maintaining or improving energy services (van der Zwaan et al. 2018). A broad variety of renewable energy sources, including biomass, hydro, geothermal, solar, and wind, have great potential. However, it will be a major challenge to find the investment required to tap these power sources under the present economic conditions (van der Zwaan et al. 2018). The picture is somewhat different in South Africa, which has a coal-based energy system and a comparatively stable and well-connected electricity grid, with access to electricity for more than 85% of its population (IEA 2014). The dependence on traditional biomass is extensive in the household and commerce sectors. Over 700 million people rely on fuel wood or charcoal for cooking on inefficient cooking stoves or open fires, with an efficiency of 10–20%. Modern biomass technologies provide multiple advantages. The introduction of more-efficient technologies, even those as simple as improved cooking stoves (with an average efficiency of 25%) or biogas stoves (with an average efficiency of 65%) (IEA 2014), will reduce the biomass input and thus the primary energy demand. This will also alleviate the heavy pressure on the ecosystem from the unsustainable exploitation of natural forests. The introduction of modern technologies will improve the supply of useful energy, lower indoor pollution, and improve living standards. Therefore, we assume in our scenarios that the overall biomass efficiency will improve from 35% in 2015 to 65% in 2050, while biomass's share will decrease and be partially replaced by electric power and solar heat.

5.2.7 *Non-OECD Asia*

The Non-OECD Asia region includes all the developing countries of Asia, except China and India. This group covers a large spectrum in terms of size, economy, stability, and developmental status. The region is spread over a large area from the Arabian Sea to the Pacific. Electricity access varies widely in these countries, according to WEO 2014. In Southeast Asia, the average access is 77%, with only 30% access in Myanmar and Cambodia, and nearly 100% access in Singapore, Thailand, and Vietnam. In Indonesia, there is 76% access (92% in urban areas), in Bangladesh 60% (90%), and in Pakistan 69% (88%). In Southeast Asia, 46% of the population still relies on traditional biomass, with the highest use in Myanmar (93%) and Cambodia (89%). In Indonesia, 42% of the population used biomass for cooking in 2012; in Bangladesh the figure was 89%; and in Pakistan, it was 62%. The lowest values are in Singapore (0%), Malaysia (0%), and Thailand (24%). The scenarios thus cover the whole band-width of renewable resources and technological development, even though the outlooks for individual countries deviate widely from the average.

5.2.8 *India*

India has a fast-growing population of over 1.2 billion people and is the world's seventh largest country by area. However, the population density is already 2.7 times higher than that in China. Due to its climate, India has a rather limited CSP potential but a large potential for PV power generation. Its wind power potential is expected to be limited by land-use constraints, but the technical potential estimated from available meteorological data is large. According to the WEO 2014 database, electricity access is on average 75%, with 94% in urban areas and 67% in rural areas. In India, about 815 million people still relied on the traditional use of biomass for cooking in 2012. Due to population and GDP growth, increasing living standards, and increasing mobility, it is expected that the energy demand in India will increase significantly, although large potentials for efficiency savings exist. Electrification is a core strategy for decarbonisation in India, which, combined with the rising demand for energy services, will lead to strong growth in the per capita and overall electricity demand. It is also expected that the need for mobility in India will increase rapidly and more strongly than in other regions of the world.

5.2.9 *China*

China has great potential renewable energy resources, especially for the generation of solar thermal power in the west, onshore wind in the north, and offshore wind in the east and southeast. Photovoltaic power generation could play an important role throughout all parts of China. The expansion of hydropower generation is currently also seen as a major strategy, but the potential for small hydro systems is rather low. China will face further large increases in energy demand in all sectors of the energy system. Chinese economic prosperity has mainly been underpinned by coal, which provides over two-thirds of China's primary energy supply today (IEA WEO 2014). The increase in electricity use due to higher electrification rates will be a major factor in the successful expansion of renewable energy in the industry, building, and transport sectors. In China, nearly all households are connected to the electricity grid. However, according to WEO 2014, about 450 million Chinese still relied on the traditional use of biomass for cooking in 2012. China has pledged to reduce CO_2 emissions before 2030, and already has some ambitious political targets for renewable energy deployment.

5.2.10 OECD Pacific

OECD Pacific consists of Japan, New Zealand, the peninsula of South Korea, and the continent of Australia. The region is dominated by the high energy demand in Japan, which has rather limited renewable energy resources. The lack of physical grid connections prevents power transmission between these countries. Therefore, it is a huge effort to supply the large Japanese nation with renewable energy and to stabilize the variable wind power without tapping the large solar potential in other countries, such as Australia. Here, hydrogen and synfuels will not only be used for the long-term storage of renewable power, but also as an option for balancing the renewable energy supply across borders. The early market introduction of fuel-cell cars in Japan may support such a strategy. Following the accident at Fukushima and the implementation of feed-in tariffs for renewable electricity, the expansion rates for renewable energies, in particular PV, have risen sharply.

5.3 Key Assumptions for Scenarios

5.3.1 Population Growth

Population growth is an important driver of energy demand, directly and through its impact on economic growth and development. The assumptions made in this study up to 2050 are based on United Nations Development Programme (UNDP) projections for population growth (UNDP 2017 (medium variant)). Table 5.2 shows that according to the UNDP, the world's population is expected to grow by 0.8% per year on average over the period 2015–2050. The global population will increase from 7.4 billion people in 2015 to nearly 9.8 billion by 2050. The rate of population growth will slow over this period, from 1.1% per year during 2015–2020 to 0.6% per year during 2040–2050. From a regional perspective, Africa's population growth

Table 5.2 Population growth projections (in millions)

Million capita	2015	2020	2025	2030	2035	2040	2045	2050	Change 2015–2050
OECD North America	482	503	524	543	560	575	588	599	24%
OECD Pacific	207	208	208	208	206	204	201	198	−4%
OECD Europe	570	582	587	592	596	598	599	598	5%
Eastern Europe/ Eurasia	343	346	347	346	345	343	341	339	−1%
Middle East	234	254	276	295	314	331	348	363	56%
Latin America	506	531	552	571	587	599	609	616	22%
China	1405	1433	1447	1450	1442	1426	1403	1374	−2%
Africa	1194	1353	1522	1704	1897	2100	2312	2528	112%
India	1309	1383	1452	1513	1565	1605	1636	1659	27%
Non-OECD Asia	1132	1203	1269	1329	1382	1428	1467	1499	32%
Global	7383	7795	8185	8551	8893	9210	9504	9772	32%

Source: UN World Population Prospects—2017 revision, medium variant

will continue to be the most rapid (on average 2.2%/year), followed by the Middle East (1.3%/year). In contrast, in China and OECD Pacific, a population decline of about 0.1%/year. is expected. The populations in OECD Europe and OECD North America are expected to increase slowly through to 2050. The proportion of the population living in today's non-OECD countries will increase from its current 81% to 85% in 2050. China's contribution to the world population will drop from 19% today to 15% in 2050. Africa will remain the region with the highest population growth, leading to a share of 26% of world population in 2050. Satisfying the energy needs of a growing population in the developing regions of the world in an environmentally friendly manner is the fundamental challenge in achieving a sustainable global energy supply.

5.3.2 GDP Development

Economic growth is a key driver of energy demand. Since 1971, each 1% increase in the global GDP has been accompanied by a 0.6% increase in primary energy consumption. Therefore, the decoupling of energy demand and GDP growth is a prerequisite for the rapid decarbonisation of the global energy industry. In this study, the economic growth in the model regions is measured in GDP, expressed in terms of purchasing power parity (PPP) exchange rates. Purchasing power parities compare the costs in different currencies of fixed baskets of traded and non-traded goods and services. GDP PPP is a widely used measure of living standards and is independent of currency exchange rates, which might not reflect a currency's true value (purchasing power) within a country. Therefore, GDP PPP is an important basis of comparison when analysing the main drivers of energy demand or when comparing the energy intensities of countries.

Although PPP assessments are still relatively imprecise compared with statistics based on national incomes, trade, and national price indices, it is argued that they provide a better basis for global scenario development. Therefore, all the data on economic development in the IEA World Energy Outlook 2016 (WEO 2016a, b) refer to purchasing-power-adjusted GDP in international US$ (2015). However, because WEO 2016 only covers the time period up to 2040, projections for 2040–2050 in the 5.0 °C, 2.0 °C, and 1.5 °C Scenarios are based on German Aerospace Center (DLR) estimates, which are mainly used to extrapolate the GDP trends in the world regions used in our modelling.

GDP growth in all regions is expected to slow gradually over the coming decades (Table 5.3). It is assumed that world GDP will grow on average by 3.2% per year over the period 2015–2050. China, India, and Africa are expected to grow faster than other regions, followed by the Middle East, Africa, other non-OECD Asia, and Latin America. The growth of the Chinese economy will slow as it becomes more mature, but it will nonetheless become the economically strongest region in the world in PPP terms by 2020. The GDP in OECD Europe and OECD Pacific is assumed to grow by 1.3–1.5% per year over the projection period, while economic growth in OECD North America is expected to be slightly higher (2.1%). The OECD's share of global PPP-adjusted GDP will decrease from 45% in 2015 to 28% in 2050.

Table 5.3 GDP development projections based on average annual growth rates for 2015–2040 from IEA (WEO 2016a, b) and on our own extrapolations

Billion $ (2015) PPP	2015	2020	2025	2030	2035	2040	2045	2050	change 2050/2015
OECD North America	22,123	24,787	27,650	30,513	34,038	37,562	41,675	45,788	107%
OECD Pacific	8284	8880	9644	10,407	11,125	11,842	12,462	13,081	58%
OECD Europe	21,632	23,883	26,076	28,269	30,538	32,807	34,885	36,963	71%
Eurasia	6397	6757	7919	9081	10,467	11,853	13,439	15,025	135%
Middle East	5380	6236	7646	9055	10,853	12,650	14,909	17,167	219%
Latin America	7181	7473	8807	10,141	11,951	13,761	16,218	18,675	160%
China	20,179	28,567	37,997	47,427	56,207	64,986	74,906	84,825	320%
Africa	5851	7118	9247	11,376	14,437	17,498	21,950	26,403	351%
India	8021	11,515	17,084	22,652	30,309	37,966	46,020	54,074	574%
Other Asia	10,061	11,361	14,577	17,794	21,835	25,876	30,055	34,234	240%
Global	**115,108**	**136,578**	**166,646**	**196,715**	**231,758**	**266,801**	**306,519**	**346,236**	201%

5.3.3 Technology Cost Projections

The parameterization of the models requires many assumptions about the development of the particular characteristics of technologies, such as specific investment and fuel costs. Therefore, because long-term projections are highly uncertain, we must define plausible and transparent assumptions based on background information and up-to-date statistical and technical information.

The speed of an energy system transition also depends on overcoming economic barriers. These largely relate to the relationships between the costs of renewable technologies and their fossil and nuclear counterparts. For our scenarios, the projection of these costs is vital in making valid comparisons of energy systems. However, there have been significant limitations to these projections in the past in relation to investment and fuel costs.

In addition, efficiency measures also generate costs which are usually difficult to determine depending on technical, structural and economic boundary conditions. In the context of this study, we have therefore assumed uniform average costs of 3 ct per avoided kWh of electricity consumption in our cost accounting.

During the last decade, fossil fuel prices have seen huge fluctuations. Figure 5.1 shows oil prices since 1997. After extremely high oil prices in 2012, we are currently in a low-price phase. Gas prices saw similar development (IEA 2017). Therefore, fossil fuel price projections have also seen considerable variations (IEA 2013, 2017) and have had a considerable influence on scenario results ever since.

Although in the past, oil-exporting countries provided the best oil price projections, institutional price projections have become increasingly accurate, with the International Energy Agency (IEA) leading the way in 2018 (Roland Berger 2018).

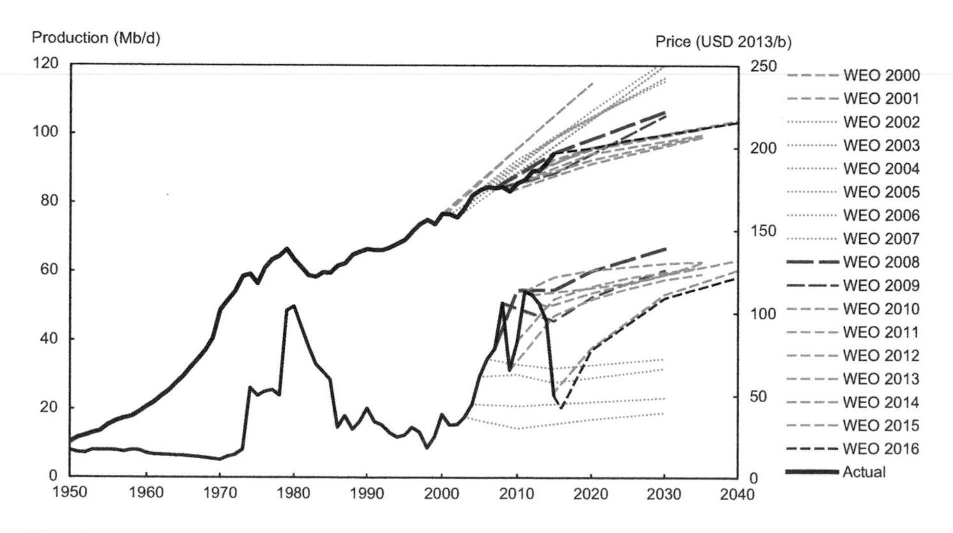

Fig. 5.1 Historic development and projections of oil prices (bottom lines) and historical world oil production and projections (top lines) by the IEA according to Wachtmeister et al. (2018)

An evaluation of the oil price projections of the IEA since 2000 by Wachtmeister et al. (2018) showed that price projections have varied significantly over time. Whereas the IEA's oil production projections seem comparatively accurate, oil price projections showed errors of 40–60%, even when made for only 10 years ahead. Between 2007 and 2017, the IEA price projections for 2030 varied from $70 to $140 per barrel, providing significant uncertainty regarding future costs in the scenarios. Despite this limitation, the IEA provides a comprehensive set of price projections. Therefore, we based our scenario assumptions on these projections, as described below.

However, because most renewable energy technologies provide energy without fuel costs, the projections of investment costs become more important than fuel cost projections, and this limits the impact of errors in the fuel price projections. It is only for biomass that the cost of feedstock remains a crucial economic factor for renewables. Today, these costs range from negative costs for waste wood (based on credit for the waste disposal costs avoided), through inexpensive residual materials, to comparatively expensive energy crops. Because bioenergy holds significant market shares in all sectors in many regions, a detailed assessment of future price projections is provided below.

Investment cost projections also pose challenges for scenario development. Available short-term projections of investment costs depend largely on the data available for existing and planned projects. Learning curves are most commonly used to assess the future development of investment costs as a function of their future installations and markets (McDonald and Schrattenholzer 2001; Rubin et al. 2015). Therefore, the reliability of cost projections largely depends on the uncertainty of future markets and the availability of historical data.

Fossil technologies provide a large cost data set featuring well-established markets and large annual installations. They are also mature technologies, where many cost reduction potentials have already been exploited.

For renewable technologies, the picture is more mixed. For example, hydro power is, like fossil fuels, well established and provides reliable data on investment costs. Other technologies, such as PV and wind, are currently experiencing tremendous developments in installation and cost reduction. Solar PV and wind are the focus of cost monitoring, and considerable data are already available on existing projects. However, their future markets are not easily predicted, as can be seen from the evolution of IEA market projections over recent years in the World Energy Outlook series (compare for example IEA 2007, 2014, 2017). For PV and wind, small differences in cost assumptions will lead to large deviations in the overall costs, and cost assumptions must be made with special care. Furthermore, many technologies feature only comparably small markets, such as geothermal, modern bioenergy applications, and CSP, for which costs are still high and for which future markets are insecure. The cost reduction potential is correspondingly high for these technologies. This is also true for technologies that might become important in a transformed energy system but are not yet widely available. Hydrogen production, ocean power, and synthetic fuels might deliver important technology options in the long term after 2040, but their cost reduction potential cannot be assessed with any certainty today.

Thus, cost assumptions are a crucial factor in evaluating scenarios. Because costs are an external input into the model and are not internally calculated, we assume the same progressive cost developments for all scenarios. In the next sections, we present a detailed overview of our assumptions for power and renewable heat technologies, including the investment and fuel costs, and the potential CO_2 costs in the scenarios.

5.3.3.1 Power and CHP Technologies

The focus of cost calculations in our scenario modelling is the power sector. We compared the specific investment costs estimated in previous studies (Teske et al. 2012, 2015), which were based on a variety of studies, including the European Commission-funded NEEDS project (NEEDS 2009), projections from the European Renewable Energy Council (Zervos et al. 2010), investment cost projections by the IEA (2014), and current cost assumptions by IRENA and IEA (IEA 2016b). We found that investment costs generally converged, except for PV. Therefore, for consistency reasons, the investment costs and operation and maintenance costs for the power sector are based primarily on the investment costs within WEO 2016 (IEA 2016b) up to 2040, including their regional disaggregation. We extended the projections until 2050 based on the trends in the preceding decade.

For renewable power production, we used investment costs from the 450 ppm scenario from IEA 2016b. For technologies not distinguished in the IEA report (such as geothermal CHP), we used cost assumptions based on our own research, from the Energy [R]evolution Scenario 2015 (Teske et al. 2015). As the cost assumptions for PV systems by the IEA do not reflect recent cost degressions, we based our assumptions on a more recent analysis by Steurer et al. (2018), which projects lower investment costs for PV in 2050 than does the IEA. The costs for onshore and offshore wind in Europe were adapted from the same source, in order to reflect more recent data. The cost assumptions for hydrogen production come from our own analysis in the PlanDelyKaD project (Michalski et al. 2017). Table 5.4 summarizes the cost trends for power technologies derived from the assumptions discussed above for OECD Europe. It is important to note that the cost reductions are, in reality, not a function of time, but of cumulative capacity (production of units), so dynamic market development is required to achieve a significant reduction in specific investment costs. Therefore, we might underestimate the costs of renewables in the reference (5.0 °C) scenario compared with the 2.0 °C and 1.5 °C Scenarios. However, our approach is conservative when we compare the reference scenario with the 2.0 °C or 1.5 °C Scenarios. The cost assumptions for the other nine regions are in the same range, but differ slightly for different renewable energy technologies. Fossil fuel power plants have a limited potential for cost reductions because they are at an advanced stage of technology and market development. Gas and oil plants are relatively cheap, at around $670/kW and $822/kW, respectively. CHP applications and coal plants are more expensive, ranging between $2000/kW and $2500/kW. The IEA sees some cost reduction potential for expensive nuclear plants, tending towards $4500/kW by 2050, whereas gas might even increase in cost.

In contrast, several renewable technologies have seen considerable cost reductions over the last decade. This is expected to continue if renewables are deployed extensively. Fuel cells are expected to outpace other CHP technologies, with a cost reduction potential of more than 75% (from currently high costs). Hydro power and biomass remain stable in terms of costs. Tremendous cost reductions are still expected for solar energy and offshore wind, even though they have experienced significant reductions already. Although CSP might deliver dispatchable power at half its current cost in 2050, variable PV costs could drop to 35% of today's costs. Offshore wind could see cost reductions of over 30%, whereas the cost reduction potential for onshore wind seems to have been exploited already to a large extent (Table 5.4).

Table 5.4 Investment cost assumptions for power generation plants (in $2015/kW) in the scenarios until 2050

Investment costs power generation plants in Europe						
		2015	2020	2030	2040	2050
CHP Coal	$/kW	2500	2500	2500	2500	2500
CHP Gas	$/kW	1000	1000	1000	1000	1000
CHP Lignite	$/kW	2500	2500	2500	2500	2500
CHP Oil	$/kW	1310	1290	1240	1180	1130
Coal power plant	$/kW	2000	2000	2000	2000	2000
Diesel generator	$/kW	900	900	900	900	900
Gas power plant	$/kW	670	500	500	500	670
Lignite power plant	$/kW	2200	2200	2200	2200	2200
Nuclear power plant	$/kW	6600	6000	5100	4500	4500
Oil power plant	$/kW	950	930	890	860	820
CHP Biomass	$/kW	2550	2500	2450	2350	2250
CHP Fuel cell	$/kW	5000	5000	2500	2500	1120
CHP Geothermal	$/kW	13,200	11,190	8890	7460	6460
Biomass power plant	$/kW	2400	2350	2300	2200	2110
Geothermal power plant	$/kW	12,340	2800	2650	2500	2400
Hydro power plant[a]	$/kW	2650	2650	2650	2650	2650
Ocean energy power plant	$/kW	6950	6650	4400	3100	2110
PV power plant	$/kW	1300	980	730	560	470
CSP power plant[b]	$/kW	5700	5000	3700	3050	2740
Wind turbine offshore	$/kW	4000	3690	3190	2830	2610
Wind turbine onshore	$/kW	1640	1580	1510	1450	1400
Hydrogen production	$/kW	1380	1220	920	700	570

[a]Costs for a system with solar multiple of two and thermal storage for 8 h of turbine operation
[b]Values apply to both run-of-the-river and reservoir hydro power

In the 2.0 °C and 1.5 °C Scenarios, hydrogen is introduced as a substitute for natural gas, with a significant share after 2030. Hydrogen is assumed to be produced by electrolysis. With electrolysers just emerging on larger scale on the markets, they have considerable cost reduction potential. Based on the Plan-DelyKaD studies (Michalski et al. 2017), we assume that costs could decrease to $570/kW in the long term.

5.3.3.2 Heating Technologies

Assessing the costs in the heating sector is even more ambitious than in the power sector. Costs for new installations differ significantly between regions and are interlinked with construction costs and industry processes, which are not addressed in this study. Moreover, no data are available to allow the comprehensive calculation of the costs for existing heating appliances in all regions. Therefore, we concentrate on the additional costs resulting from new renewable applications in the heating sector.

Our cost assumptions are based on a previous survey of renewable heating technologies in Europe, which focused on solar collectors, geothermal, heat pumps, and biomass applications. Biomass and simple heating systems in the residential sector are already mature. However, more-sophisticated technologies, which can provide higher shares of heat demand from renewable sources, are still under development and rather expensive. Market barriers will slow the further implementation and cost reduction of renewable heating systems, especially for heating networks. Nevertheless, significant learning rates can be expected if renewable heating is increasingly implemented, as projected the 2.0 °C and 1.5 °C Scenarios.

Table 5.5 presents the investment cost assumptions for heating technologies for OECD Europe, disaggregated by sector. Geothermal heating displays the same high costs in all sectors. In Europe, deep geothermal applications are being developed for

Table 5.5 Specific investment cost assumptions (in $2015) for heating technologies in the scenarios until 2050

Investment costs heat generation plants in OECD Europe							
			2015	2020	2030	2040	2050
Geothermal		$/kW	2390	2270	2030	1800	1590
Heat pumps		$/kW	1790	1740	1640	1540	1450
Biomass heat plants		$/kW	600	580	550	510	480
Residential biomass stoves	Industrialized countries	$/kW	840	810	760	720	680
Residential biomass stoves	Developing countries	$/kW	110	110	110	110	110
Solar collectors	Industry	$/kW	850	820	730	650	550
	In heat grids	$/kW	970	970	970	970	970
	Residential	$/kW	1060	1010	910	800	680

heating purposes at investment costs ranging from €500/kW_{th} (shallow) to €3000/kW_{th} (deep), with the costs strongly dependent on the drilling depth. The cost reduction potential is assumed to be around 30% by 2050.

Heat pumps typically provide hot water or space heat for heating systems with relatively low supply temperatures, or they supplement other heating technologies. Therefore, they are currently mainly used for small-scale residential applications. Costs currently cover a large band-width and are expected to decrease by only 20% to $1450/kW by 2050.

For biomass and solar collectors, we assume significant differences between the sectors. There is a broad portfolio of modern technologies for heat production from biomass, ranging from small-scale single-room stoves to heating or CHP plants on an MW scale. Investment costs show similar variations: simple log-wood stoves can be obtained from $100/kW, but more sophisticated automated heating systems that cover the whole heat demand of a building are significantly more expensive. Log-wood or pellet boilers range from $500 to 1300/kW. Large biomass heating systems are assumed to reach their cheapest costs in 2050 at around $480/kW for industry. For all sectors, we assume a cost reduction of 20% by 2050. In contrast, solar collectors for households are comparatively simple and will become cheap at $680/kW by 2050. The costs of simple solar collectors for swimming pools might have been optimized already, whereas their integration in large systems is neither technologically nor economically mature. For larger applications, especially in heat grid systems, the collectors are large and more sophisticated. Because there is not yet a mass market for such grid-connected solar systems, we assume there will be a cost reduction potential until 2050.

5.3.4 *Fuel Cost Projections*

5.3.4.1 Fossil Fuels

Although fossil fuel price projections have seen considerable variations, as described above, we based our fuel price assumptions up to 2040 on the WEO 2017 (IEA 2017). Beyond 2040, we extrapolated from the price developments between 2035 and 2040. Even though these price projections are highly speculative, they provide a set of prices consistent to our investment assumptions. Fuel prices for nuclear energy are based on the values in the Energy [R]evolution report 2015 (Teske et al. 2015), corrected by the cumulative inflation rate for the Eurozone of 1.82% between 2012 and 2015 (Table 5.6).

Table 5.6 Development projections for fossil fuel prices in $2015 (IEA 2017)

Development projections for fossil fuel prices							
Reference scenario			2015	2020	2030	2040	2050
Oil	All	$/GJ	8.5	12.3	21.5	24.2	35.1
Gas	OECD North America	$/GJ	2.5	3.3	5.5	6.2	8.9
	OECD Europe	$/GJ	6.6	7.2	9.2	10.0	12.9
	China	$/GJ	9.2	9.5	10.3	10.5	11.4
	OECD Pacific	$/GJ	9.8	10.0	10.7	10.9	11.8
	Others	$/GJ	2.5	3.3	5.5	6.2	8.9
Coal	OECD North America	$/GJ	2.3	2.5	2.9	3.0	5.3
	OECD Europe	$/GJ	2.6	3.1	4.1	4.3	5.3
	China	$/GJ	3.2	3.5	4.3	4.5	5.3
	OECD Pacific	$/GJ	2.6	3.3	4.4	4.5	5.3
	Others	$/GJ	2.9	3.3	4.2	4.4	5.3
Nuclear	All	$/GJ	1.1	1.2	1.5	1.8	2.1
2.0 °C and 1.5 °C scenarios							
Oil	all	$/GJ	8.5	10.2	12.6	13.0	14.3
Gas	OECD North America	$/GJ	2.5	2.8	4.5	5.1	7.6
	OECD Europe	$/GJ	6.6	6.6	8.5	9.4	13.0
	China	$/GJ	9.2	8.5	9.2	10.0	12.9
	OECD Pacific	$/GJ	9.8	9.0	9.6	10.3	13.2
	Others	$/GJ	2.5	2.8	4.5	5.1	7.6
Coal	OECD Europe	$/GJ	2.9	2.7	2.5	2.5	2.4
	OECD North America	$/GJ	2.3	2.6	2.9	2.9	2.7
	China	$/GJ	2.6	2.9	3.1	3.1	2.9
	OECD Pacific	$/GJ	3.2	3.4	3.5	3.5	3.4
	Others	$/GJ	2.9	2.8	3.0	3.0	2.8
Nuclear	All	$/GJ	1.1	1.2	1.5	1.8	2.1

5.3.4.2 Biomass Prices

Biomass prices depend on the quality of the biomass (residues or energy crops) and the regional supply and demand. The variability is large. Lamers et al. (2015) found a price range of €4–4.8/GJ for forest residues in Europe in 2020, whereas agricultural products might cost €8.5–12/GJ. Lamers et al. modelled a range for wood pellets from €6/GJ in Malaysia to 8.8€/GJ in Brazil. IRENA modelled a cost supply curve on a global level for 2030 (see Fig. 5.2), ranging from $3/GJ for a potential of 35 EJ/year. up to $8–10/GJ for a potential up to 90–100 EJ/year (IRENA 2014) (and up to $17/GJ for an potential extending to 147 EJ).

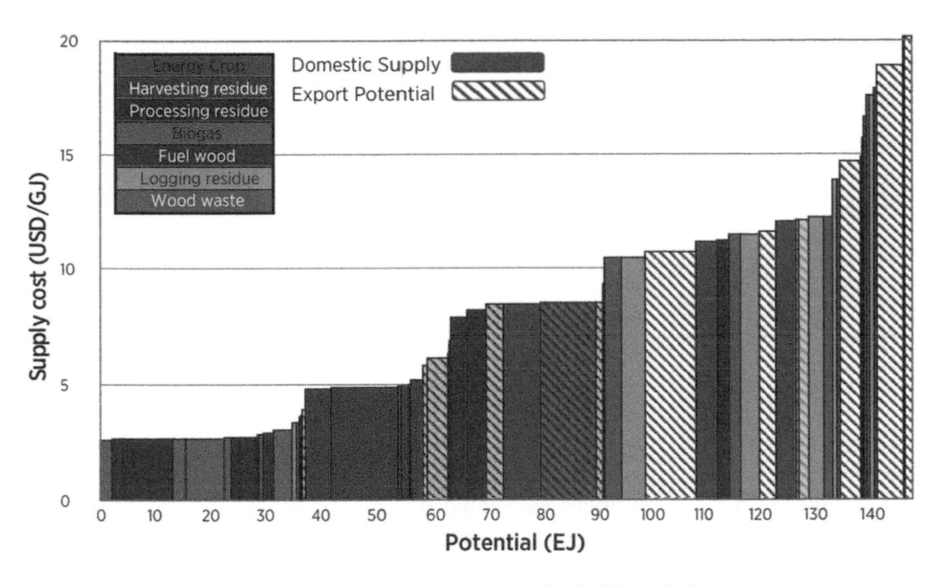

Fig. 5.2 Global supply curve for primary biomass in 2030 (IRENA 2014)

IRENA projected regional supply costs for liquid and other biomass sources in 2030 based on a global biomass use of around 108 EJ, using current primary biomass prices as a proxy (see Table 5.7). Liquid biofuels demand higher prices because of their production and transformation processes; 'other biomass' includes primary biomass, such as fuel wood, energy crops, and residues.

The prices cited above hold true for modern biomass applications. Traditional biomass use is still often based on firewood or other biomass, which is acquired without a price (and with the labour cost not considered). No price data are yet available for a considerable range of residues. Therefore, the average primary biomass costs across the complete energy system in many regions are lower than the available market prices for biomass commodities. Consequently, today's market prices represent the upper limit of today's biomass costs.

Therefore, for our scenarios, we assumed a lower average biomass price in all regions, starting from the lower end of the cost supply curve at around $7.50/GJ for OECD regions, with predominantly modern applications. For Africa, Latin America, and Asia, including Russia, which have abundant biomass residue potential, current prices were assumed to be $3/GJ. For the remaining regions (the Middle East, and Eastern Europe), we assumed $5/GJ.

The prices for primary biomass will increase proportionately to the IRENA reference price for 'other biomass' by 2030, following the increasing uptake of modern

Table 5.7 Biomass price projections for 2030 at 108 EJ of the biomass demand (IRENA 2014)

	Liquid biofuel reference price	Other biomass reference price
	US $/GJ	US $/GJ
Africa	36	10
Asia	40	7
Europe	58	18
North America	34	15
OECD Pacific	61	15
Latin America	59	12
World	42	11

biomass technologies and increasing trade, representing a further biomass potential uptake along the supply curve. For the period until 2050, we consider that biomass prices will be stable. The prices calculated by IRENA are valid for a demand of 108 EJ/year. The biomass demand considered in this study never exceeds a total of 100 EJ/year. However, the international trade in biomass may heavily influence biomass prices in the future, representing a significant source of uncertainty in our assumptions.

5.3.5 *CO_2 Costs*

The WEO 2017 (IEA 2017) considers the future price of CO_2 in the power and industry sectors. There is considerable variation between the current policy scenario, the new policy scenario, and the 450 ppm scenario, not only in value, but also in regional range. Various studies have indicated a close relationship between decarbonisation and the implicit or explicit CO_2 price (regardless of the most efficient implementation measure). On the one hand, the carbon price is a precondition for a decarbonisation of the energy sector (Lucena et al. 2016), but on the other hand, decarbonisation may limit the costs of CO_2 emissions if an efficient pricing measure is in place (Jacobson et al. 2017). Because the scenarios in this study rely heavily on effective reductions in CO_2 emissions, we used the CO_2 prices of the 450 ppm scenario in the 2.0 °C and 1.5 °C Scenarios. In the reference case, we deviated from the WEO 2017, which applies rather low CO_2 emission costs. Instead, we applied CO_2 costs equivalent to the cost of the resulting climate damage. Based on existing studies of fossil-energy-induced damage (Anthoff and Tol 2013; Stern et al. 2006), we assumed that $78/t of CO_2 is a plausible cost estimate in the wide range of estimates of the social costs of CO_2 emissions (Table 5.8).

Table 5.8 CO_2 cost assumptions in the scenarios

CO_2 costs							
			2020	2025	2030	2040	2050
Reference	All regions	\$/t CO_2	0	42	69	78	78
2.0 °C and 1.5 °C	OECD Economies	\$/t CO_2	0	62	87.6	138	189.0
	Other regions	\$/t CO_2	0	42	69.5	124	177.5

5.4 Energy Scenario Narratives and Assumptions for World Regions

The scenario-building process involves many assumptions and explicit, but also implicit, narratives about how future economies and societies, and ultimately energy systems, may develop under the overall objective of 'deep and rapid decarbonisation' by 2050. These narratives depend on three main strategic pillars:

- Efficiency improvement and demand reduction: stringent implementation of technical and structural efficiency improvements in energy demand and supply. These will lead to a continuous reduction in both final and primary energy consumption. In the 1.5 °C Scenario, these measures must be supplemented with responsible energy consumption behaviour by the consumer.
- Deployment of renewable energies: massive implementation of new technologies for the generation of power and heat in all sectors. These will include variable renewable energies from solar and wind, which have experienced considerable cost reductions in recent years, but also more expensive technologies, such as large-scale geothermal and ocean energy, small hydro power, and CSP.
- Sector coupling: stringent direct electrification of heating and transport technologies in order to integrate renewable energy in the most efficient way. Because this strategy has its limitations, it will be complemented by the massive use of hydrogen (generated by electrolysis) or other synthetic energy carriers.

Some alternative or probably complementary future technical options are explicitly excluded from the scenarios. In particular, those options with large uncertainties with respect to technical, economic, societal, and environmental risks, such as large hydro and nuclear power plants, unsustainable biomass use, carbon capture and storage (CCS), and geoengineering, are not considered on the supply side as mitigation measures or—in the case of hydro—not expanded in the future. The sustainable use of biomass will partly substitute for fossil fuels in all energy sectors. However, this use will be limited to an annual global energy potential of less than 100 EJ per year for sustainability reasons, according to the calculations of Seidenberger et al. (2008), Thrän et al. (2011), and Schueler et al. (2013).

The transformations described in the two alternative scenarios are constrained, to a certain degree, by current short- to medium-term investment planning, as described in the reference case, because most technical and structural options to change the demand or supply side require years of planning and construction. This means that both alternative scenarios start deviating significantly from the reference case only after 2025. However, some short-term developments shown in the IEA WEO Current Policies scenario have been corrected and are not adopted in the alternative scenarios because there is newer statistical information (that renders the reference development implausible). This is the case for the IEA estimates of demand development in some regions and sectors, and it is partly true for investments in fossil-fuel-based heat and power generation.

5.4.1 Efficiency and Energy Intensities

It is obvious that a major increase in energy efficiency is the backbone of each ambitious transition scenario, because energy efficiency significantly reduces the need for energy conversion and infrastructure investment. The development of the future global energy demand is determined by three key factors:

- Population growth, which affects the number of people consuming energy or using energy services. Associated with this, increasing access to energy services in developing countries and emerging economies is an additional influencing factor, bearing in mind that this could mean power grid access or the implementation of isolated, usually small-scale, local power systems.
- Economic development, which is commonly measured as GDP. In general, GDP growth triggers an increase in energy demand, directly via additional industrial activities and indirectly via an increase in private consumption arising from the higher incomes associated with a prospering economy.
- Energy intensity, which is a measure of how much final energy is required in the industrial sector to produce a unit of GDP. Efficiency measures help to reduce energy intensity and can result in a decoupling of economic growth and final energy consumption. In the 'Residential and other' sector, energy intensity refers to the per capita demand for final energy (for electrical appliances and heat generation). Efficiency improvement is also a result of reduced conversion losses, in particular those achieved by replacing thermal power generation with renewable technologies, which leads to a further reduction in the primary energy intensity.

The reference scenario and both target scenarios are based on the same projections of population and economic growth. Therefore, the scenarios represent the specific, although widely accepted, development of future societies. However, the future development of energy intensities differs between the reference and alternative scenarios, taking into account the different efficiency pathways and therefore the successful implementation of measures to intensify required investments in efficient technologies or to change consumer behaviour.

The assumptions made about the potential to further increase the economic and technical efficiency in all sectors are based on various external studies. However, the lower benchmarks for the assumptions on efficiency potentials are derived from the Current Policies scenario of the IEA WEO 2017 (IEA 2017). The upper benchmarks for efficiency potentials per world region are taken from Graus et al. (2011), Kermeli et al. (2014), and recently published low-energy-demand scenarios developed by Grubler et al. (2018).

5.4.1.1 Industrial Electricity Demand

'Industrial electricity demand' refers to many appliances of different sizes and purposes. Large potentials for saving electricity have been identified in various studies in most branches of industry. This particularly applies to electric drives for compressed air, pumps, and fans. The scenario model approach distinguishes between electric appliances and power-to-heat devices for space and process heating. The consumption of electricity per GDP varies widely between regions, depending on their industrial structures and efficiency standards. The trajectories for industrial electricity demands are constrained by the abovementioned lower and upper benchmarks and aim for similar electricity uses per $GDP in the industrial sectors in all regions by 2050. The resulting trajectories for OECD and non-OECD countries are shown in Table 5.9. The average global electricity demand for electric appliances in 'industry' (without power-to-heat) will decrease from 55 kWh/$1000 GDP in 2015 to 36 kWh/$1000 in 2050 in the reference case, but to 24 kWh/$1000 in the 2.0 °C Scenario and 23 kWh/$1000 in the 1.5 °C Scenario. However, the increased electrification of industrial heat in both alternative scenarios almost cancels out the greater efficiency increases in those two scenarios when compared with the reference case. The average power-to-heat share in industry will increase in this period from 6% to 34% in 2050 in the 2.0 °C Scenario and to 37% in the 1.5 °C Scenario. In the 1.5 °C Scenario, the annual electricity demand for industrial electrical appliances will be around 5% lower than in the 2.0 °C Scenario between 2020 and 2025, and up to 10% lower between 2025 and 2035. However, between 2035 and 2050, the electricity demand for electric appliances in the industry sector converges under the two scenarios.

Table 5.9 Assumed average development of specific (per $GDP) electricity use for electrical appliances in the 'Industry' sector

kWh/$1000	2015	2020	2025	2030	2035	2040	2045	2050	Change 2050/2015
Reference case									
OECD regions	42.8	40.8	38.2	35.8	33.5	31.5	29.7	28.2	−34%
Non-OECD regions	65.8	62.5	55.7	52.9	48.7	45.7	41.9	39.1	−41%
2.0 °C Scenario									
OECD regions	42.8	40.4	35.6	32.2	28.7	25.7	23.5	21.7	−49%
Non-OECD regions	65.8	60.9	49.2	42.9	37.2	32.9	28.3	24.8	−62%
1.5 °C Scenario									
OECD regions	42.8	40.1	33.4	27.4	24.3	22.4	20.7	20.0	−53%
Non-OECD regions	65.8	58.6	46.4	39.8	34.4	30.8	27.4	24.7	−62%

5.4.1.2 Demand for Fuel to Produce Heat in the Industry Sector

Industrial heat is required for different purposes and at different temperatures. Currently, industrial (process) heat is mainly produced by burning fossil fuels. Biomass plays a minor role, except in heat use from the combustion of residues and biogenic waste. Some low- and medium-temperature heat is produced by co-generation plants with combined heat and power provisions. Power-to-heat makes up only a small percentage of the industrial energy demand for heat. Regional differences in the nature of industry, especially in terms of the presence of energy-intensive heavy or manufacturing industries, strongly influence the amounts of low-, medium-, and high-temperature heat that must be produced today and in the future (because we assume that the regional industry structure will remain the same). Various technological improvements, process substitutions, and innovations are technically possible and have been implemented to some extent already. An important example is highly efficient waste heat recovery. Shorter investment cycles and incentives to replace old technologies will help to reduce energy consumption as quickly as possible. It is obvious that incentives are essential to trigger rapid innovation and the implementation of new technologies. Any political strategy for introducing such a pathway requires strong support from various industry stakeholders and regional or even global governance to overcome the economic and technical obstacles and conflicting interests. Therefore, both alternative scenarios assume that the conditions exist to allow rapid technological change. Table 5.10 provides the resulting final energy demands for heating per $GDP for OECD and non-OECD countries. The average global values will decrease from 680 MJ/$1000 GDP in 2015 to 366 MJ/$1000 in 2050 in the reference case, and to 185 MJ/$1000 in the 2.0 °C Scenario and 172 MJ/$1000 in the 1.5 °C Scenario. Compared with the 2.0 °C Scenario, the 1.5 °C Scenario assumes a significantly more rapid reduction in the industrial heat demand. Between 2020 and 2025, the annual energy demand for heat will be up to 8% lower under the 1.5 °C Scenario than under the 2.0 °C Scenario, and up to 17% lower between 2025 and 2035. After 2035, the difference will decrease again, and by 2050, it will be around 7%.

Table 5.10 Assumed average development in final energy use for heating in the industry sector (including power-to-heat) (per $GDP)

MJ/$1000	2015	2020	2025	2030	2035	2040	2045	2050	Change 2050/2015
Reference case									
OECD regions	406	417	384	352	320	293	270	249	−39%
Non-OECD regions	911	823	687	620	553	506	453	410	−55%
2.0 °C Scenario									
OECD regions	406	383	330	284	242	207	177	157	−61%
Non-OECD regions	911	791	608	491	381	302	238	196	−79%
1.5 °C Scenario									
OECD regions	406	377	306	239	199	177	158	143	−65%
Non-OECD regions	911	762	558	428	318	259	212	182	−80%

5.4.1.3 Electricity Demand in the 'Residential and Other' Sector

The electricity demand in the 'Residential and other' sector includes electricity use in households, for commercial purposes, and in the service and trade sectors, fishery, and agriculture. Besides lighting, information, and communication, a large amount of electricity is used for cooking, cooling, and hot water. It has been estimated that in 2015, electricity use for heating had a global average share of 5% of the final energy use for heating. It is assumed that this share will increase significantly to 30% in 2050 in the 2.0 °C Scenario and to 37% in the 1.5 °C Scenario. These increases are attributed to sector coupling, the provision of storage for variable renewable energy in the heat sector, and the provision of high-temperature heat without fuel combustion. The average global electricity use for appliances in the 'Residential and other' sector will decrease in the reference case from 78 kWh/$1000 GDP in 2015 to 60 kWh/$1000 in 2050, whereas it will decrease to 38 kWh/$1000 in the 2.0 °C Scenario and to 37 kWh/$1000 in the 1.5 °C Scenario, a reduction of more than 50% relative to today's energy consumption. The average global electricity use for appliances in the 'Residential and other' sector, which is related to per capita consumption, will increase in the reference scenario from 1350 kWh/capita in 2015 to 2370 kWh/capita in 2050, whereas it will increase to only 1490 kWh/capita in the 2.0 °C Scenario and to 1460 kWh/capita in the 1.5 °C Scenario. Table 5.11 shows the changes in electricity use for appliances in OECD and non-OECD countries (without electricity for heating). Significant reduction potentials are assumed for all world regions. Similar to the development in the industry sector, between 2020 and 2025, the annual power demand for electrical appliances in the 'Residential and other' sector will be around 5% lower in the 1.5 °C Scenario than in the 2.0 °C Scenario, and more than 10% lower between 2025 and 2035. After 2035, the two scenarios will converge again, so that in 2050, the global demand will be only 2% higher in the 2.0 °C Scenario than in the 1.5 °C Scenario.

Table 5.11 Assumed average developments of per capita electricity use in the 'Residential and other' sector for electrical appliances (without power-to-heat)

kWh/capita	2015	2020	2025	2030	2035	2040	2045	2050	Change 2050/2015
Reference case									
OECD regions	4457	4585	4753	4972	5189	5419	5626	5837	31%
Non-OECD regions	712	851	977	1191	1366	1532	1661	1788	151%
2.0 °C Scenario									
OECD regions	4457	4526	4366	4137	3837	3590	3304	3023	−32%
Non-OECD regions	712	834	894	1004	1083	1143	1193	1238	74%
1.5 °C Scenario									
OECD regions	4457	4502	4078	3346	2987	2896	2889	2872	−36%
Non-OECD regions	712	806	842	928	1001	1086	1164	1224	72%

5.4.1.4 Fuel Demand for Heat in the 'Residential and Other' Sector

The fuel demand for heat in households has quite different characteristics depending on the consumption structures in each world region and their climatic conditions. In regions with harsh winters, the heat demand is dominated by the building sector (space heat and hot water in private and commercial buildings), but in regions with a comparatively warm climate, the demand for space heat is generally low and heat is predominantly used for cooking and as low-temperature heat for hot water. In the commercial sector, the energy mix for heat is more diverse. The medium- to high-temperature process heat demand arises in this sector. Reducing the final energy use for heating will involve reducing the demand (e.g., by improving the thermal insulation of building envelopes) and replacing inefficient procedures and technologies, such as the traditional use of biomass, which is still widely used for cooking and heating in some regions. In contrast to traditional biomass, the efficiency of electrical appliances can improve significantly, and they produce zero direct emissions and no air pollution. Table 5.12 shows the assumed average final energy demand for heating in OECD and non-OECD countries. The average global consumption will decrease from 560 MJ/$1000 GDP in 2015 to 280 MJ/$1000 in 2050 in the reference scenario, but to 173 MJ/$1000 in the 2.0 °C Scenario and to 160 MJ/$1000 in the 1.5 °C Scenario. The average global per capita energy demand for fuels in 'Residential and other sectors' will decrease from around 12,600 MJ/capita per year in 2015 to 11,700 MJ/capita in the reference case. This will mainly be due to a shift in the global population shares towards the developing regions. Compared with the reference scenario, the energy intensity will decrease to 7300 MJ/capita in the 2.0 °C Scenario and to 6700 MJ/capita in the 1.5 °C Scenario.

Table 5.12 Assumed average development of specific final energy use for heating in the 'Residential and other' sector (including power-to-heat)

MJ/capita	2015	2020	2025	2030	2035	2040	2045	2050	Change 2050/2015
Reference case									
OECD regions	24,932	24,421	24,163	23,980	23,794	23,696	23,821	24,121	−3%
Non-OECD regions	10,047	9933	9786	9749	9678	9628	9623	9666	−4%
2.0 °C Scenario									
OECD regions	24,932	24,282	22,300	20,578	19,064	17,677	16,599	15,800	−37%
Non-OECD regions	10,047	9650	8961	8209	7498	6807	6217	5868	−42%
1.5 °C Scenario									
OECD regions	24,932	24,047	20,413	16,222	15,143	14,538	14,172	13,901	−44%
Non-OECD regions	10,047	9549	8593	7515	6737	6234	5808	5510	−45%

The 1.5 °C Scenario assumes a significantly stronger reduction in demand than the 2.0 °C Scenario. In the 1.5 °C Scenario, additional efficiency measures will reduce the final energy demand until 2025 by around 5%, and by up to 13% between 2025 and 2035 (compared with the 2.0 °C Scenario). Thereafter, the differences will become smaller, finally reaching around 8% by 2050.

5.4.1.5 Resulting Energy Intensities by Region

Figure 5.3 shows the final energy intensities related to $GDP for each of the ten world regions between 2015 and 2050 and for both alternative scenarios. The final energy use per GDP will decrease significantly in all regions, but the decreases will be larger (and faster) in OECD countries. This will result in smaller regional differences in the final energy demand compared with the current situation. Compared with the very ambitious assumptions of Grubler et al. (2018) for the specific final energy demands in northern and southern world regions, the assumptions made in this study are conservative. In Grubler et al. (2018), the annual global final energy use, including non-energy consumption, will decrease from 363 EJ in 2015 to 245 EJ by 2050, whereas in our study, the annual global value will decrease to 310 EJ in the 2.0 °C Scenario and to 284 EJ in the 1.5 °C Scenario compared with 586 EJ in the reference case (see Chap. 8). Because the 1.5 °C target requires a significant reduction in emissions before 2030, the 1.5 °C Scenario necessarily reduces the energy demand more rapidly than the 2.0 °C Scenario, but only a slightly lower annual consumption is assumed in 2050.

5.4.2 RES Deployment for Electricity Generation

The power demand will increase significantly in all scenarios. In the 2.0 °C and 1.5 °C Scenarios, this will result from the continuous electrification of the heating and transport sectors, and the increasing production of synthetic fuels for indirect electrification and sector coupling. The available energy sources for renewable

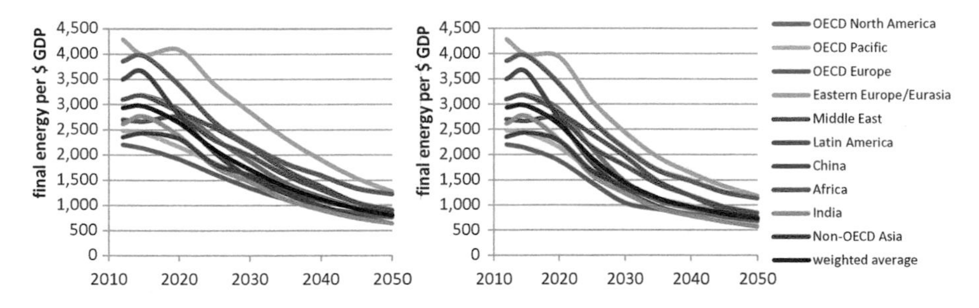

Fig. 5.3 Development of the specific final energy use (per $GDP) in all stationary sectors (i.e., without transport) per world region under the 2.0 °C Scenario (left) and 1.5 °C Scenario (right)

power generation and their costs vary from region to region. Therefore, the scenarios follow regionally different strategies and storylines, taking the different regional conditions into account on the supply side. The core strategy is the replacement of conventional thermal power and heat generators with solar, wind, geothermal, and other renewable options for the highly efficient generation of electricity for final energy consumption and the generation of synthetic fuels.

Our estimates of the potential for renewable power generation are based on the results of the REMix EnDat tool developed (Scholz 2012; Stetter 2014; Pietzcker et al. 2014). The technical potentials for solar and wind power in each world region were estimated while taking into account the different exclusion criteria and constraints documented by Stetter (2014). The analysis was used to estimate the 'economic' potential for each world region, which is the upper limit of the technological expansion under the different scenarios. 'Economic' potentials were derived by assuming the minimum annual full-load hours for each technology. In the case of PV, the assumed global economic potential was estimated to be in the order of 5.4 million TWh per year. The potential of CSP was even larger, at around 5.6 million TWh per year. The annual wind power potentials were estimated to be in the order of 500,000 TWh for onshore wind and around 100,000 TWh for offshore wind.

The harvesting of global solar radiation by PVs has enormous economic potential worldwide. In the last few years, economies of scale have led to a significant cost degression for PV modules, and large PV production capacities have been created. The PV technology also plays a major role in our scenarios because of its decentralized characteristics, which make it easy to build cost-efficient renewable power supplies in rural and isolated areas. However, its restriction to sunny hours causes high daily and seasonal variability. This results in rather low annual full-load hours. Therefore, large quantities of storage capacity must also be installed for short-term storage to support the major expansion of PV. The storage options considered are pumped hydro storage (e.g., by the enhancement of existing hydro sites) and a massive expansion of battery storage. This leads to uncertainties in the total infrastructure costs for the integration of high shares of PV into the power system and the mineral resources required for this. For this reason, the share of PV globally remains in the range of about 30% of total power generation, with the highest shares in the Middle East (40%), followed by OECD North America and Other Asia (35% each). The lowest shares, in the range of 20–25%, are in Eurasia, OECD Europe, Latin America, and China.

Wind power on land will achieve an average global generation share of 25% in 2050 in both alternative scenarios (compared with 8% in the reference case). The highest generation shares for onshore wind are assumed in India and Eastern Europe/Eurasia, at about 30% each. The lowest shares will be in China, the Middle East, and Non-OECD Asia, at 18–23%. For offshore wind, the global generation share will rise to 8% by 2050 under both alternative scenarios, compared with only 1% in the reference scenario. The highest offshore wind shares of 10–15% will be achieved under the 2.0 °C Scenario in the OECD regions and Eastern Europe/ Eurasia. The lowest shares are predicted in the Middle East (2%), India (6%), and China (6.5%), where the potential is rather limited. The offshore shares under the

1.5 °C Scenario will tend to be slightly lower because of the stronger focus on PV and onshore wind as the best options for a very rapid expansion of RES.

Compared with PV and wind power, CSP plants promise highly flexible power and heat generation, with high capacity factors due to high-temperature heat storage. The use of heat for desalination can also contribute to secure water supplies in the sunbelt of the world. We assume that its multi-purpose uses and dispatchable generation capability can lead to a significant role for CSP in the medium- to long-term future, although levelized costs of CSP are today still much higher than they are for PV or wind, and investment costs for batteries for short-term electricity storage might also decrease in the future. Therefore, it is assumed that CSP will achieve an average global electricity generation share of 15% by 2050 in the 2.0 °C Scenario and 13% in the 1.5 °C Scenario, compared with 0.5% in the reference case. A high share, close to 30%, will be achieved, especially in the Middle East. This will include electricity generation for export to OECD Europe of up to 120 TWh per year by 2050. High CSP potentials are also assumed for Africa (16%), which will also export up to 280 TWh/year in 2050 from North Africa to OECD Europe—and for China (18–20%) and Non-OECD Asia (15–17%).

Hydro power generation will increase only moderately under the alternative scenarios compared with the reference case. This source has already been tapped and large hydro plants usually have significant ecological and societal consequences. Therefore, the average global power generation share will decrease in the alternative scenarios from today's 16% to around 8%, whereas the reference scenario assumes a generation share of 14% in 2050. Nevertheless, a 30% increase in the global hydro power generation is assumed between 2015 and 2050. The highest power generation shares in 2050 are assumed to be in Latin America (24%), followed by OECD Europe (11%) and China (10%). The generation of hydro power plays only a minor role in the long term in the Middle East (1%), Africa, and India (each 4%).

Even smaller contributions are assumed for geothermal energy, ocean energy, and biomass. All three options have comparably high power-generation costs, but offer complementary characteristics and availabilities that may stabilize future electricity supply systems. The global average share of geothermal power generation is assumed to be about 5% in 2050, ocean energy use will contribute another 2%, and biomass, including co-generation, will achieve a maximum of 6–7% around 2030 and 5% in 2050. The highest share for geothermal power generation is assumed to occur in Eastern Europe/Eurasia (11%), for ocean energy in OECD Pacific (4%), and for biomass use in Latin America, Eastern Europe/Eurasia, and OECD Europe (8–9%). It is predicted that hydrogen will take an increasing share of the remaining thermal power generation, as a substitute for natural gas in gas turbines and combined cycle gas turbine plants and with increasing contributions from hydrogen fuel cells to co-generation. Figure 5.4 shows the basic storyline in global power generation under the 2.0 °C Scenario.

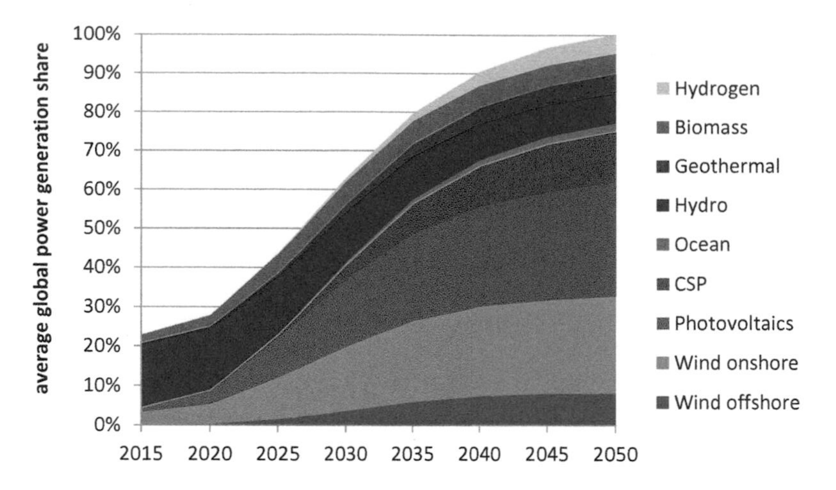

Fig. 5.4 Development of the average global RES shares in total power generation in the 2.0 °C Scenario

5.4.3 RES Deployment for Heat Generation

Heat generation covers a broad range of processes, including district heat (either from co-generation or public heating plants), direct heating in buildings, and process heat in industry, commerce, and other sectors. Different technologies are considered for each sector, and strong RES expansion is assumed. The increasing heat extraction from co-generation will trigger increasing district heat use, which will stabilizes in the long term or decrease by 2050 with the declining heat demand attributable to ambitious efficiency measures. However, CHP will only contribute significantly in regions with a tradition of district heating and/or a high heat demand in the building sector. No strong expansion of heat grids is assumed in regions without existing heat grids and a low demand for space heat.

Electricity for heating is assumed to play a significant role in future energy systems. In contrast to today's technology, flexible use is assumed, adjusted to variable feed-ins from variable renewable generation. This implies the availability of heat storages, smart operation controls, and flexible electricity tariffs. Electricity can be used with relatively low investment for space heating and is therefore easily combined with other heating technologies. For example, simple electric heaters can easily be integrated into heat storage or heating grids. However, more-efficient electric heat pumps generally require higher investment in both the heat pump itself and in the heat distribution within the building. Electricity can also be used to provide process heat in industry at high-temperature levels. The alternative scenarios globally assume a significant increase in the average electricity share of the final energy for heating in 'Industry' from 6% in 2015 to around 34% in the 2.0 °C Scenario and 37% in the 1.5 °C Scenario by 2050. All regions achieve shares of between 23% and 43%. Electricity shares in the 'Residential and other' sector are assumed to grow from 5% in 2015 to 30% in the 2.0 °C Scenario and 35% in the 1.5 °C Scenario.

However, the need for process heat at high temperature levels, problems associated with process integration, and specific process requirements call for additional process-specific strategies for replacing fossil fuels in the industry sector. As mentioned above, electrification is a comparatively efficient strategy. Other strategies are the use of biomass and hydrogen or—with some limitations—concentrated solar energy. While hydrogen is used to provide high temperature process heat in this study, it could—at least partially—be replaced by other synthetic energy carriers, such as synthetic methane, which can be generated from hydrogen and a (renewable) carbon source. This power-to-gas option has the advantage that it can be fed in into the gas grid, and act in storage and transport. However, energy losses are around 20% higher (compared with hydrogen). As a consequence, synthetic methane is not taken into account in the scenarios.

Overall, biomass use for heating in the 'Residential and other' sector is decreasing as the traditional and currently inefficient use of biomass is replaced by advanced efficient technologies, and biomass is used in a more efficient way in the energy system. Thus, the average global share of biomass as a final energy source for heating will decrease from today's 34% to 22% by 2050 in the 2.0 °C Scenario and to 17% in the 1.5 °C Scenario, when direct electrification and indirect electrification (via synthetic gases and fuels) play a stronger role. In contrast, biomass use in the 'Industry' sector will increase continuously as the combustion of biomass can be used to generate high-temperature heat for many industrial processes for which renewable low-temperature heat sources are unsuitable. The average global share of biomass for the final energy for heating in the 'Industry' sector will increase from 9% in 2015 to 19% by 2050 in the 2.0 °C Scenario and to 14% in the 1.5 °C Scenario. The largest shares are assumed for Latin America (47%) and Africa (35%) in the 2.0 °C Scenario, where biomass residues are still rather abundant, with much lower values for the 1.5 °C Scenario (18–21%). In that scenario, biomass as a transition technology will be avoided due to the earlier development of alternative renewable technologies and electrification. The lowest shares are assumed for the Middle East (4%) and China (7%) in both alternative scenarios because of their limited sustainable potentials.

Solar collectors are suitable for hot water preparation and for supporting heating systems using heat storage. In heat grids, large heat storage systems can also be used to balance the seasonal heat demand and solar generation, and the integration of solar heat at low costs in the long term. The contributions of solar collectors to the heat supply are limited by the temperatures that collectors can provide (below 120 °C for traditional collectors and up to 300 °C for concentrated collectors) and the seasonal variations in regions with significant space heat demand. In the alternative scenarios, the global average share of solar thermal final energy for heating will rise to 19% in the 'Residential and other' sector. The largest shares are assumed to be in the Middle East (30–25%), where the abundance of solar radiation can be exploited by concentrated solar heat applications. All other regions have shares between 15% and 22%, considering the limited applicability of concentrated systems. The global average solar share in the 'Industry' sector will increase to 16% by 2050. The largest shares are achieved in the Middle East (25%) and Africa (20%)

and the lowest shares are assumed in Eastern Europe/Eurasia (9%), followed by the OECD regions (10–13%).

Heat pumps allow very efficient heat supply. System-wide CO_2 emissions depend on the CO_2 emissions in the power mix. Because of their generation of low-temperature heat, heat pumps play a role in space heating in regions with moderate or cold climates, but large industrial heat pumps can also generate low-temperature heat for industrial processes and tap the enormous potential of waste heat. A continuous improvement in the coefficient of performance of heat pumps, which describes the ratio of useful energy in the form of heat to the required compressor energy in the form of electricity, in all existing plants from an average of 3 today to a value of 4 in 2050 is assumed. The application potential of heat pumps (in terms of MW_{th}) is assumed to be limited by the low temperature of the heat provided, the increasing share of the grid-connected heat supply, the assumed increase in extensive building insulation, and the low space heating demand in some world regions. In addition to heat pumps, an increase in deep geothermal energy use is assumed for the 'Industry' sector, with an increase of up to 11% in the global average share of final energy by 2050 under both alternative scenarios.

Hydrogen use for direct heating is linked to the increasing substitution of natural gas with hydrogen in all sectors, except transport. This will lead to a hydrogen share of 12–14% in final energy for heating in the 'Industry' sector under the two alternative scenarios and 3% in the 'Residential and other' sector under the 2.0 °C Scenario and 4% under the 1.5 °C Scenario. The highest hydrogen shares in 'Industry' are assumed for OECD North America (22–23%), followed by other OECD regions, Eastern Europe/Eurasia, China, and the Middle East with shares of 13%–18%. The lowest shares, below 5%, are assumed for Latin America and Africa, where biomass from residues provides a flexible alternative. The highest hydrogen shares in the 'Residential and other' sector will be in the Middle East (7–10%), followed by OECD Pacific (6–8%) and OECD Europe (up to 9%) (Figs. 5.5 and 5.6).

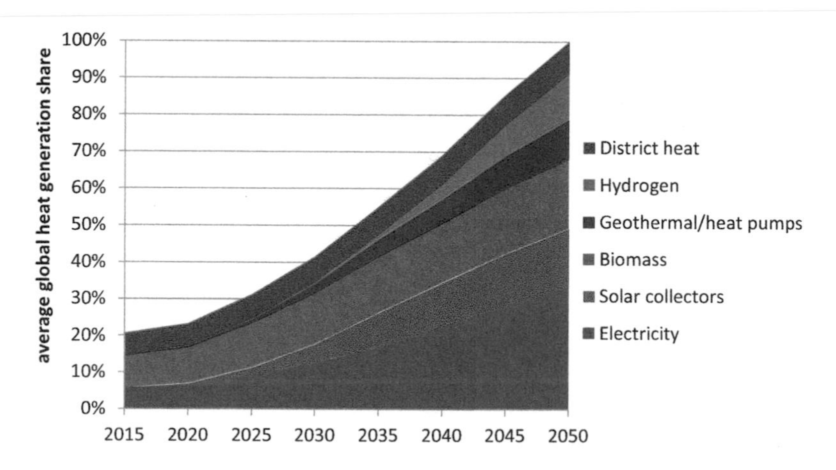

Fig. 5.5 Development of the average global RES shares of future heat generation options in 'Industry' in the 2.0 °C scenario

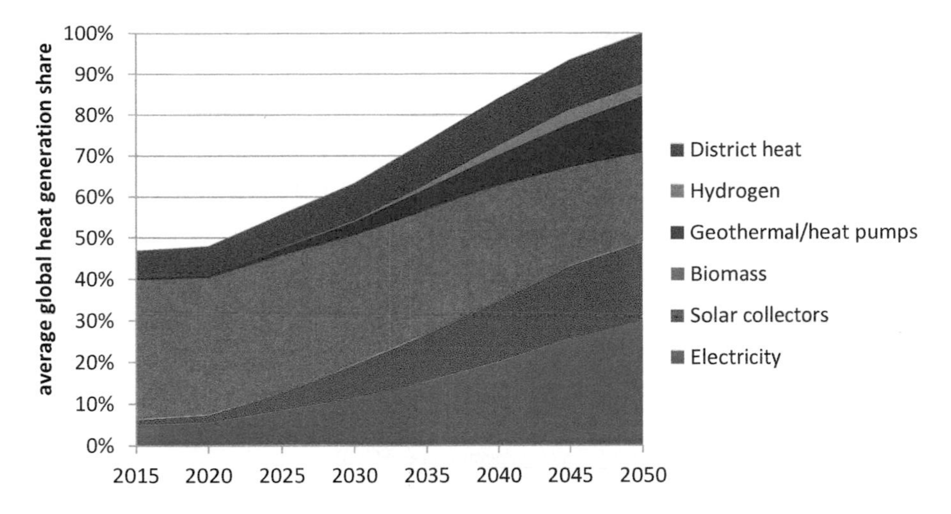

Fig. 5.6 Development of the average global shares of future heat-generation options in the 'Residential and other' sector under the 2.0 °C scenario

5.4.4 Co-generation of Heat and Power and District Heating

Compared with condensing power plants with high efficiency losses due to waste heat, the co-generation of heat and power (combined heat and power, CHP) allows the highly efficient use of fossil and renewable fuels. If this approach is made more flexible with the help of heat storage, and if it is powered by renewable fuels, such as biomass or hydrogen from renewable electricity, CHP promises to generate not only renewable power in a flexible way but also to integrate efficiently large shares of renewable heat into energy systems via large and small district heating systems. Therefore, co-generation is a particularly good option in regions with high low- to medium-temperature heat demands (e.g., industrial consumers and space heating). Our modelling distinguishes between public generation in large CHP plants and CHP autoproduction. The latter comprises industrial CHP generators but also smaller plants in the 'Residential and other' sector. Power-to-heat ratios, efficiencies, and assumed costs reflect these different structural options. The IEA Energy Balances provides statistical values with which to calibrate the co-generation for each world region. The scenarios then assume the similar development of these parameters according to a defined advanced state of technology, with overall efficiencies of 85–90%.

Although absolute electricity generation from CHP will increase in all regions, all scenarios assume a decreasing power generation share for CHP in the long term

Table 5.13 Development of power from co-generation per $GDP

kWh/$1000	2015	2020	2025	2030	2035	2040	2045	2050	Change 2050/2015
Reference case									
OECD regions	18.8	17.4	15.9	14.3	13.0	12.1	11.3	10.7	−43%
Non-OECD regions	34.0	28.7	23.1	19.3	16.3	14.1	12.1	10.7	−69%
2.0 °C Scenario									
OECD regions	18.8	18.0	18.8	18.8	18.1	16.7	15.1	13.6	−28%
Non-OECD regions	34.0	29.4	25.9	23.5	20.9	18.6	16.3	14.3	−58%
1.5 °C Scenario									
OECD regions	18.8	18.0	18.1	17.7	16.6	15.3	13.8	12.4	−34%
Non-OECD regions	34.0	29.4	25.9	23.5	20.8	18.7	16.3	14.2	−58%

Table 5.14 Development of heat from co-generation per $GDP

MJ/$1000	2015	2020	2025	2030	2035	2040	2045	2050	Change 2050/2015
Reference case									
OECD regions	39.5	39.7	38.8	39.4	40.2	41.8	43.6	47.2	19%
Non-OECD regions	108.8	94.8	78.3	67.9	60.0	54.8	50.1	47.9	−56%
2.0 °C Scenario									
OECD regions	39.5	40.5	49.3	58.1	65.2	70.1	75.2	78.4	98%
Non-OECD regions	108.8	101.2	92.6	89.8	86.3	83.4	77.6	70.4	−35%
1.5 °C Scenario									
OECD regions	39.5	41.0	51.9	58.4	60.4	61.6	63.8	66.0	67%
Non-OECD regions	108.8	101.1	93.8	93.3	89.8	84.5	77.9	69.9	−36%

due to the decommissioning of fossil-fuel generators, the limited availability of biomass, and the assumption that heat losses in CHP technologies will be reduced, leading to higher overall efficiency. Tables 5.13 and 5.14 show the resulting developments for power and heat as intensities summed across the 'Industry' and 'Residential and other' sectors. Whereas the absolute power supply from CHP will increase in all regions, the power-related intensities will decrease. Under the 2.0 °C Scenario, higher CHP power production is assumed to be higher than in the reference case, as a balancing option for variable renewable sources. CHP plants will play a smaller role in the future, particularly in non-OECD countries.

For heat, the situations are more diverse between the scenarios and regions. While the intensity of heat use from CHP will increase in the OECD regions, it will decrease in the non-OECD regions. Even though the intensity converges between OECD and non-OECD regions under all scenarios, the 2.0 °C and 1.5 °C Scenarios will achieve higher levels of intensity. However, the early reduction in demand in the 1.5 °C Scenario will preclude any additional demand for CHP.

5.4.5 *Other Assumptions for Stationary Processes*

Although the energy losses in the production of synthetic fuels are significant, these fuels are expected to be mandatory in the deep decarbonisation scenarios for sectors and processes in which the direct use of other renewable sources, including renewable power, is not technically feasible. We assume an optimistic increase in the efficiency[1] of electrolytic hydrogen generation, from 66% today to 77% by 2050 (ratio of energy output [H_2] to energy input [electricity]). The generation of synthetic fuels (such as Fischer-Tropsch fuels) from hydrogen, using CO_2 as the carbon source, is assumed to be a complementary option that will allow the decarbonisation of long-range transport, particularly aviation and international bunkers, without exceeding the defined maximum sustainable biomass use. Therefore, the assumed shares of power-to-liquid synfuels in the aggregated biofuel/synfuel fraction of all transport modes (according to Sect. 6.3.3.) is a result of the sectoral allocation of the limited biomass potentials in each world region. The assumed efficiency[2] of synfuel generation will increase from 35% in 2020 to 42% in 2050. As well as in fuel-cell vehicles, hydrogen can be used to replace natural gas in stationary processes by 2030 in both scenarios, mostly for use in industry and co-generation plants. The scenarios do not assume a hydrogen economy in the long term. The development of hydrogen infrastructure is expected to be inefficient, especially for residential use. However, because hydrogen is assumed to be fed into the gas grid with a share of up to 100% by 2050, it will also be partly used in the 'Residential and other' sector, which includes various commercial applications.

The allocation of the limited biomass to the power, heat, and transport sectors differs significantly between the available scenarios. In this study, the biomass in the alternative scenarios is mainly for use in transport, co-generation, and industry. Traditional biomass use is strongly reduced, but biomass remains an important pillar of heat supply in the 'Residential and other' sector under the assumption that the most-efficient technologies are implemented. Biofuel production for transportation remains limited. An increase in overall efficiency of up to 75% is assumed, implying the use of residues for heat and power generation.

The efficiency of fossil technologies will also increase, especially for gas power plants. This goes along with the decreasing utilization rates that result from the variable feed-in from renewable energies. Therefore, the scenarios implicitly assume that the future innovations in all combustion technologies will focus on maximum efficiency at lower utilization rates. Gas power plants will be used for backup, requiring low investment but providing high flexibility to the power system. Therefore, a part of the energy transition is the rapid replacement of inflexible medium- to base-load power plant capacities with flexible gas power plants.

---

[1] Ratio energy output (H_2) to energy input (electricity)

[2] Ratio of energy output (synfuels) to energy input (electricity).

References

Anthoff D, Tol RS (2013) The uncertainty about the social cost of carbon: A decomposition analysis using fund. Climatic Change 117 (3):515–530

Barbosa LdSNS, Bogdanov D, Vainikka P, Breyer C (2017) Hydro, wind and solar power as a base for a 100% renewable energy supply for South and Central America. PLOS ONE 12 (3):e0173820. doi:https://doi.org/10.1371/journal.pone.0173820

Gils H, Simon S, Soria R (2017) 100% Renewable Energy Supply for Brazil—The Role of Sector Coupling and Regional Development. Energies 10 (11):1859

Graus WJ, Blomen E, Worrell E (2011) Global energy efficiency improvement in the long term: A demand- and supply-side perspective. Energy Efficiency, 4, (3), August 2011, pp. 435–463

Grubler A, Wilson C, Bento N, Boza-Kiss B, Krey V, McCollum DL, Rao ND, Riahi K, Rogelj J, De Stercke S, Cullen J, Frank S, Fricko O, Guo F, Gidden M, Havlík P, Huppmann D, Kiesewetter G, Rafaj P, Schoepp W, Valin H (2018) A low energy demand scenario for meeting the 1.5 °C target and sustainable development goals without negative emission technologies. Nature Energy 3 (6):515–527. doi:https://doi.org/10.1038/s41560-018-0172-6

Herreras Martínez S, Koberle A, Rochedo P, Schaeffer R, Lucena A, Szklo A, Ashina S, van Vuuren DP (2015) Possible energy futures for Brazil and Latin America in conservative and stringent mitigation pathways up to 2050. Technological Forecasting and Social Change 98:186–210. doi:https://doi.org/10.1016/j.techfore.2015.05.006

Hess D (2018) The empirical probability of integrating CSP and its cost optimal configuration in a low carbon energy system of EUMENA. Solar Energy

IEA (2007) World Energy Outlook 2007. International Energy Agency, Organization for Economic Co-operation and Development, Paris

IEA (2013) World Energy Outlook 2013. International Energy Agency, Organization for Economic Co-operation and Development, Paris

IEA (2014) World Energy Outlook 2014. International Energy Agency, Organization for Economic Co-operation and Development, Paris

IEA (2016a) World Energy Outlook 2016. International Energy Agency, Organization for Economic Co-operation and Development, Paris

IEA (2016b) World Energy Outlook 2016—power generation assumptions. International Energy Agency, Organization for Economic Co-operation and Development, Paris

IEA (2017) World Energy Outlook 2017. International Energy Agency, Organization for Economic Co-operation and Development, Paris

IRENA (2014) Global Bioenergy Supply and Demand Projections: A working paper for REmap 2030. International Renewable Energy Agency Abu Dhabi, United Arab Emirates

Jacobson MZ, Delucchi MA, Bauer ZAF, Goodman SC, Chapman WE, Cameron MA, Bozonnat C, Chobadi L, Clonts HA, Enevoldsen P, Erwin JR, Fobi SN, Goldstrom OK, Hennessy EM, Liu J, Lo J, Meyer CB, Morris SB, Moy KR, O'Neill PL, Petkov I, Redfern S, Schucker R, Sontag MA, Wang J, Weiner E, Yachanin AS (2017) 100% Clean and Renewable Wind, Water, and Sunlight All-Sector Energy Roadmaps for 139 Countries of the World. Joule 1 (1):108–121. doi:https://doi.org/10.1016/j.joule.2017.07.005

Kermeli K, Graus WJ, Worrell E (2014) Energy efficiency improvement potentials and a low energy demand scenario for the global industrial sector. Energy Efficiency 7 (6):987–1011. doi:https://doi.org/10.1007/s12053-014-9267-5

La Rovere EL, Pereira AS, Simões AF (2011) Biofuels and Sustainable Energy Development in Brazil. World Development 39 (6):1026–1036. doi:https://doi.org/10.1016/j.worlddev.2010.01.004

Lamers P, Hoefnagels R, Junginger M, Hamelinck C, Faaij A (2015) Global solid biomass trade for energy by 2020: an assessment of potential import streams and supply costs to North-West Europe under different sustainability constraints. GCB Bioenergy 7 (4):618–634. doi:https://doi.org/10.1111/gcbb.12162

Lora ES, Andrade RV (2009) Biomass as energy source in Brazil. Renewable and Sustainable Energy Reviews 13 (4):777–788. doi:https://doi.org/10.1016/j.rser.2007.12.004
Lucena AFP, Clarke L, Schaeffer R, Szklo A, Rochedo PRR, Nogueira LPP, Daenzer K, Gurgel A, Kitous A, Kober T (2016) Climate policy scenarios in Brazil: A multi-model comparison for energy. Energy Economics 56:564–574. doi:https://doi.org/10.1016/j.eneco.2015.02.005
McDonald A, Schrattenholzer L (2001) Learning rates for energy technologies. Energy Policy 29 (4):255–261. doi:https://doi.org/10.1016/S0301-4215(00)00122-1
MESIA (2018) Solar Outlook Report 2018. Middle East Solar Industry Association
Michalski J, Bünger U, Crotogino F, Donadei S, Schneider G-S, Pregger T, Cao K-K, Heide D (2017) Hydrogen generation by electrolysis and storage in salt caverns: Potentials, economics and systems aspects with regard to the German energy transition. International Journal of Hydrogen Energy 42 (19):13427–13443. doi:https://doi.org/10.1016/j.ijhydene.2017.02.102
Nascimento G, Huback V, Schaeffer R, Lima L, Ludovique C, Paredes JR, Lucena AF, Vasquez E, Viviescas C, Emenson F (2017) Contribution of Variable Renewable Energy to Increase Energy Security in Latin America. IDB Monograph (Infrastructure and Energy Sector Energy Division); IDB-MG-562
Nass LL, Pereira PAA, Ellis D (2007) Biofuels in Brazil: An Overview. Crop Science 47 (6):2228–2237. doi:https://doi.org/10.2135/cropsci2007.03.0166
NEEDS (2009) The NEEDS Life Cycle Inventory Database. http://www.needs-project.org/needswebdb/index.php. Accessed 15th April 2016 2016
Nematollahi O, Hoghooghi H, Rasti M, Sedaghat A (2016) Energy demands and renewable energy resources in the Middle East. Renewable and Sustainable Energy Reviews 54:1172–1181. doi:https://doi.org/10.1016/j.rser.2015.10.058
Pietzcker RC, Stetter D, Manger S, Luderer G (2014) Using the sun to decarbonize the power sector: The economic potential of photovoltaics and concentrating solar power. Applied Energy 135:704-720. doi:https://doi.org/10.1016/j.apenergy.2014.08.011
Roland Berger (2018) 2018 oil price forecast: who predicts best? Roland Berger study of oil price forecasts. https://www.rolandberger.com/en/Publications/pub_oil_price_forecast_2015.html. Accessed 10.9.2018 2018
Rubin ES, Azevedo IML, Jaramillo P, Yeh S (2015) A review of learning rates for electricity supply technologies. Energy Policy 86:198–218. doi:https://doi.org/10.1016/j.enpol.2015.06.011
Scholz Y (2012) Renewable energy based electricity supply at low costs : development of the REMix model and application for Europe. Hochschulschrift, Universität Stuttgart
Schueler V, Weddige U, Beringer T, Gamba L, Lamers P (2013) Global biomass potentials under sustainability restrictions defined by the European Renewable Energy Directive 2009/28/EC. GCB Bioenergy 5 (6):652–663. doi:https://doi.org/10.1111/gcbb.12036
Seidenberger T, Thrän D, Offermann R, Seyfert U, Buchhorn M, Zeddies J (2008) Global biomass potentials – investigation and assessment of data – remote sensing in biomass potential research – country specific energy crop potential. Deutsches BiomasseForschungsZentrum (DBFZ), Leipzig/Germany, June 2008
Stern N, Peters S, Bakhshi V, Bowen A, Cameron C, Catovsky S, Crane D, Cruickshank S, Dietz S, Edmonson N (2006) Stern Review: The economics of climate change, vol 30. HM treasury London
Stetter D (2014) Enhancement of the REMix energy system model : global renewable energy potentials, optimized power plant siting and scenario validation. Hochschulschrift, University of Stuttgart
Steurer M, Brand H, Blesl M, Borggrefe F, Fahl U, Fuchs A-L, Gils HC, Hufendiek K, Münkel A, Rosenberg M, Scheben H, Scheel O, Scheele R, Schick C, Schmidt M, Wetzel M, Wiesmeth M (2018) Energiesystemanalyse Baden-Württemberg: Datenanhang zu techoökonomischen Kenndaten. Ministerium für Umwelt Klima und Energiewirtschaft Baden-Württemberg, STrise: Universität Stuttgart, Deutsches Zentrum für Luft- und Raumfahrt, Zentrum für Sonnenenergie- und Wasserstoff-Forschung Baden-Württemberg, Stuttgart

Teske S, Muth J, Sawyer S, Pregger T, Simon S, Naegler T, O'Sullivan M, Schmid S, Graus W, Zittel W, Rutovitz J, Harris S, Ackermann T, Ruwahata R, Martensen N (2012) Energy [R]evolution— a sustainable world energy outlook. Greenpeace International, European Renewable Energy Council (EREC), Global Wind Energy Council (GWEC), Deutsches Zentrum für Luft- und Raumfahrt (DLR), Amsterdam

Teske S, Sawyer S, Schäfer O, Pregger T, Simon S, Naegler T, Schmid S, Özdemir ED, Pagenkopf J, Kleiner F, Rutovitz J, Dominish E, Downes J, Ackermann T, Brown T, Boxer S, Baitelo R, Rodrigues LA (2015) Energy [R]evolution—A sustainable world energy outlook 2015. Greenpeace International

Thrän D, Bunzel K, Seyfert U, Zeller V, Buchhorn M, Müller K, Matzdorf B, Gaasch N, Klöckner K, Möller I, Starick A, Brandes J, Günther K, Thum M, Zeddies J, Schönleber N, Gamer W, Schweinle J, Weimar H (2011) Global and Regional Spatial Distribution of Biomass Potentials-Status Quo and Options for Specification. DBFZ Report No. 7. Deutsches BiomasseForschungsZentrum (DBFZ)

UNPD (2017) World Population Prospects: The 2017 Revision. United Nations, Department of Economic and Social Affairs, Population Division. https://population.un.org/wpp/. Accessed 30.04.2018 2018

van der Zwaan B, Kober T, Longa FD, van der Laan A, Jan Kramer G (2018) An integrated assessment of pathways for low-carbon development in Africa. Energy Policy 117:387–395. doi:https://doi.org/10.1016/j.enpol.2018.03.017

Wachtmeister H, Henke P, Höök M (2018) Oil projections in retrospect: Revisions, accuracy and current uncertainty. Applied Energy 220:138–153. doi:https://doi.org/10.1016/j.apenergy.2018.03.013

Zervos A, Lins C, Muth J (2010) RE-thinking 2050: a 100% renewable energy vision for the European Union. European Renewable Energy Council (EREC)

Chapter 6
Transport Transition Concepts

Johannes Pagenkopf, Bent van den Adel, Özcan Deniz, and Stephan Schmid

Abstract Detailed background for all transport scenarios and development pathways including all key parameters, and story-lines for the 5.0 °C, 2.0 °C and 1.5 °C transport scenario pathways. Mode specific efficiency improvement over time for road-, rail- and aviation transport technologies. Explanations of all vehicle technologies are included in the scenarios, along with the rationale for their selection. Description of key technology parameters for all relevant transport modes such as energy demand per passenger, and per freight tonne. Detailed regional breakdown for developments in regard to transport energy demand for ten world regions and all transport modes are provided.

6.1 Introduction

Global transport accounted for 23% of total anthropogenic CO_2 emissions in 2010 and those emissions have increased at a rapid rate in recent decades, reaching 7 Gt in 2010 according to the IPCC Fifth Assessment Report (Sims et al. 2014). The reason for this steady increase in emissions is that passenger and freight transport activities are increasing in all world regions, and there is currently no sign that this growth will slow down in the near future. The increasing energy demand in the transport sector has mainly been met by greenhouse gas (GHG)-emitting fossil fuels. Although (battery) electric mobility has recently surged considerably, it has done so from a very low base, which is why, in terms of total numbers, electricity still plays a relatively minor role as an energy carrier in the transport sector.

Apart from their impacts on climate, increasing transport levels, especially of cars, trucks, and aeroplanes, also have unwanted side-effects, including accidents, traffic jams, the emission of noise and other pollutants, visual pollution, and the disruption of landscapes by the large-scale build-up of the transport infrastructure.

J. Pagenkopf (✉) · B. van den Adel · Ö. Deniz · S. Schmid
Department of Vehicle Systems and Technology Assessment, German Aerospace Center (DLR), Institute of Vehicle Concepts (FK), Pfaffenwaldring, Germany
e-mail: johannes.pagenkopf@dlr.de; Bent.vandenAdel@dlr.de; oezcan.deniz@dlr.de; stephan.schmid@dlr.de

S. Teske (ed.), *Achieving the Paris Climate Agreement Goals*,
https://doi.org/10.1007/978-3-030-05843-2_6

However, road, rail, sea, and air transport are also an integral part of our globalized and interconnected world, and guarantee prosperity and inter-cultural exchange. Therefore, if we are to cater to people's desire for mobility while keeping the economy running and meeting the Paris climate goals, fundamental technical, operational, and behavioural measures are required immediately.

In this transport chapter, we discuss potential transport activity pathways and technological developments by which the requirement that warming does not exceed pre-industrial levels by more than 2.0 °C or 1.5 °C can be met—while at the same time maintaining a reasonable standard of mobility.

For our transport scenario modelling, the global warming limits of 2.0 °C and 1.5 °C were translated into transport CO_2 budgets. We structured our scenario designs around the following key CO_2-reducing measures[1]:

- Powertrain electrification;
- Enhancement of energy efficiency through technological development;
- Use of bio-based and synthetically produced fuels;
- Modal shifts (from high- to low-energy intensity modes) and overall reductions in transport activity in energy-intensive transport modes.

These measures are outlined in more detail in the subsequent chapters.

6.2 Global Transport Picture in 2015

The world final energy demand in the transport sector totalled 94,812 PJ[2] in 2015, according to the IEA Energy Balances (IEA 2017a, b, c). Based on this estimate, we used TRAEM (Sect. 3.3 to model the freight and passenger transport performance in our transport model with statistical data and energy efficiency figures.

The following paragraphs outline the 2015 transport structure modelled in TRAEM, which is the starting point for the subsequent scenario building until 2050.

As can be seen from Fig. 6.1, road passenger transport had the biggest transport final energy share of 51% in 2015. Most of this comprised individual road passenger modes (mostly cars, but also two- and three-wheel vehicles), which accounted for 45% of all end energy in the transport sector. In total, road transport (passenger and freight) accounted for around 90% of total final energy demand for transport.

The majority of total passenger–km (pkm) in passenger transport (around 85% of total pkm) is contributed by road transport modes. Freight is much more rail-oriented, and has a 42% share of total tonne–km (tkm), as shown in Fig. 6.2. The tkm share is much larger than the energy share arising from the much higher energy efficiency of railways compared with trucks.

Figure 6.3 shows the powertrain split of all transport modes in 2015 (by pkm or tkm respectively). With a few exceptions, the majority of modes were still heavily dependent on conventional internal combustion engines (ICE). A small number of buses had electric powertrains, which were mainly trolleybuses and increasingly

[1] See also Teske et al. (2015).

[2] International aviation and navigation bunkers are not included in this figure.

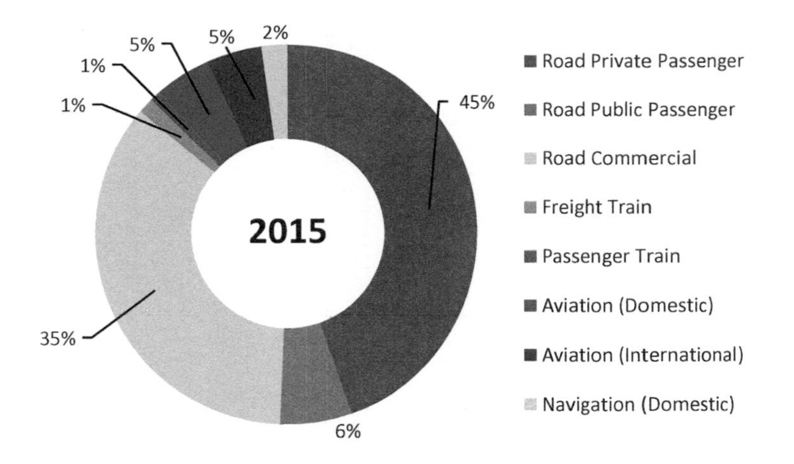

Fig. 6.1 World final energy use by transport mode in 2015

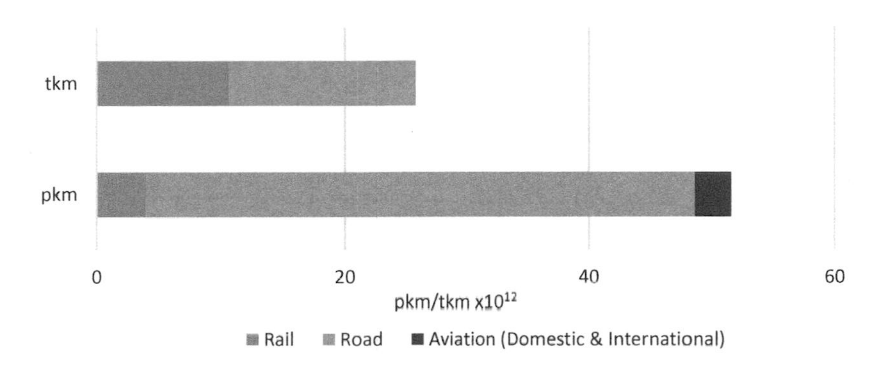

Fig. 6.2 Transport mode performances of road, rail, and aviation

also battery-powered electric buses, predominantly in China. China also has a particularly large number of electric two- and three-wheel vehicles. Almost all battery electric scooters worldwide were in China. Passenger rail was electrified to a large extent (e.g., metro and high-speed trains), whereas freight trains were predominantly not electrified.

OECD America and OECD Europe together make up nearly half the total energy demand (Fig. 6.4), and China is almost on the same level as OECD Europe, although it has about twice as many inhabitants as OECD Europe.

6.3 Measures to Reduce and Decarbonise Transport Energy Consumption

This section describes the measures required to reduce the final energy demand and decarbonise the transport sector. A variety of actions will be required so that the transport sector can conform to the <2.0 °C or 1.5 °C global warming pathways. The

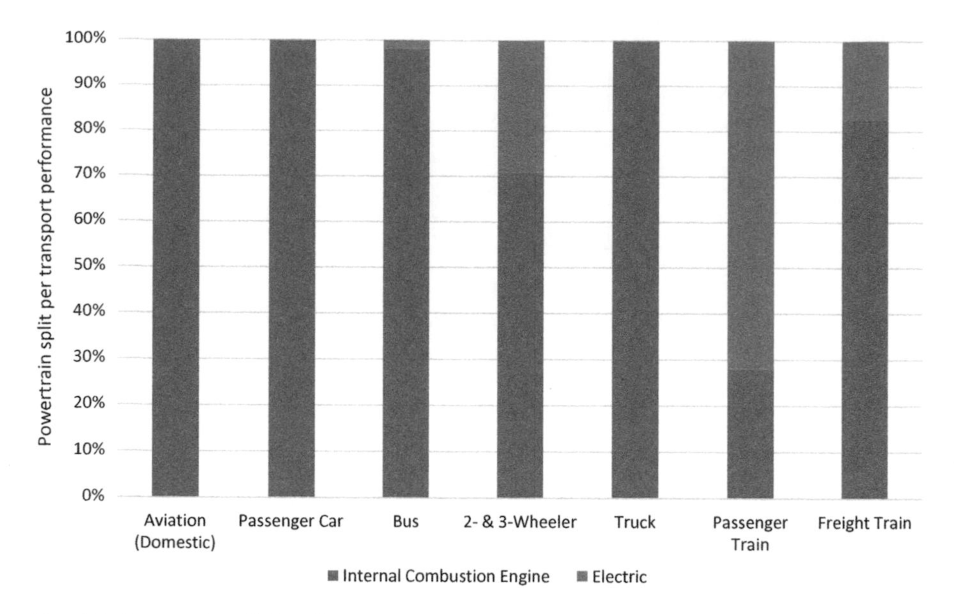

Fig. 6.3 Powertrain split for all transport modes in 2015 by transport performance (pkm or tkm)

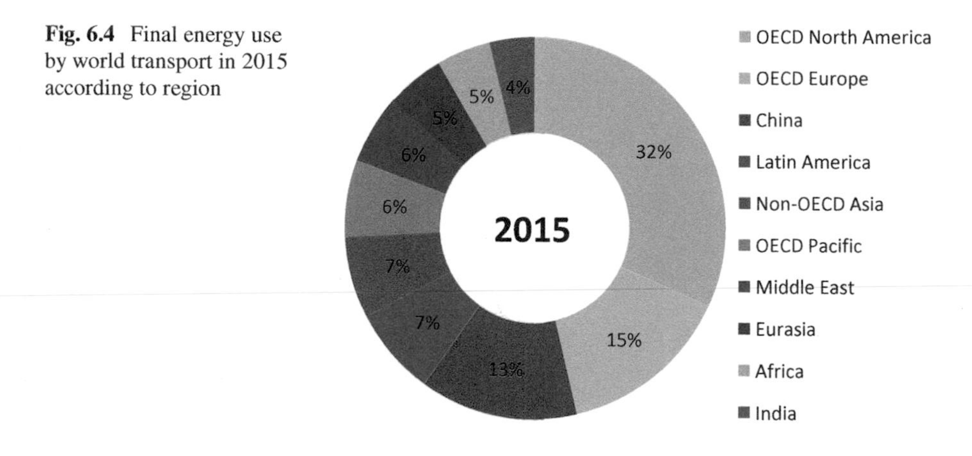

Fig. 6.4 Final energy use by world transport in 2015 according to region

set of actions described can be clustered into technical and operational measures (e.g., energy efficiency increases, drivetrain electrification); behavioural measures (e.g., shifts to less-carbon-intensive transport carriers and an overall reduction in transport activity); and accompanying policy measures (e.g., taxation, regulations, urban planning, and the promotion of less-harmful transport modes). This study focuses on the 2.0 °C and 1.5 °C Scenarios and sets out the differences between these scenarios and the business-as-usual 5.0 °C Scenario.

We found that urgent and profound measures must be taken because the emissions reduction window will soon close. Therefore, temporary reductions in fossil-fuel-related transport activities (in terms of pkm and tkm of passenger cars, trucks, and aviation) in OECD countries seem nearly unavoidable until the electrification (based on renewable energy production) of the transport sector undergoes a breakthrough.

6.3.1 *Powertrain Electrification*

Increasing the market penetration of highly efficient (battery and fuel-cell) electric vehicles, coupled with clean electricity generation, is a powerful lever and probably also the most effective means of moving toward a decarbonised transport system. All electric vehicles have the highest efficiency levels of all the drivetrain options. Today, only a few countries have significant proportions of electric vehicles in their fleets. The total numbers of electric vehicles, particularly in road transport, are insignificant, but because road transport is by far the largest CO_2 emitter in overall transport, it offers a very powerful lever for decarbonisation. In terms of drivetrain electrification, we cluster the world regions into three groups, according to the diffusion theory (Rogers 2003):

- *Innovators*: OECD North America (excluding Mexico), OECD Europe, OECD Pacific, and China
- *Moderate*: Mexico, Non-OECD Asia, India, Eurasia, and Latin America
- *Late adopters*: Africa and the Middle East.

Although this clustering is rough, it sufficiently mirrors the basic tendencies we modelled. The regions differ in the speed with which novel technologies, especially electric drivetrains, will penetrate the market.

6.3.1.1 The 5.0 °C Scenario

The 5.0 °C Scenario follows the IEA Current Policies Scenario (IEA 2017a) until 2040, with extrapolation to 2050. We model only minor electrification over all transport modes (see Fig. 6.5), with passenger cars and buses making relevant gains in *electric vehicle* (EV) shares. For example, we project a share of 30% for *battery electric vehicles* (BEV) in China by 2050 due to foreseeable legislation and technological advancements in that country (Cui and Xiao 2018), whereas for the world car fleet, the share is projected to increase to only around 10%. The growth in the share of the commercial road vehicle fleet and of the fleet of two- and three-wheel vehicles held of electric powertrains will be small, as will be the increase in further rail electrification. Aviation and navigation (shipping) will remain fully dependent on conventional kerosene and diesel, respectively.

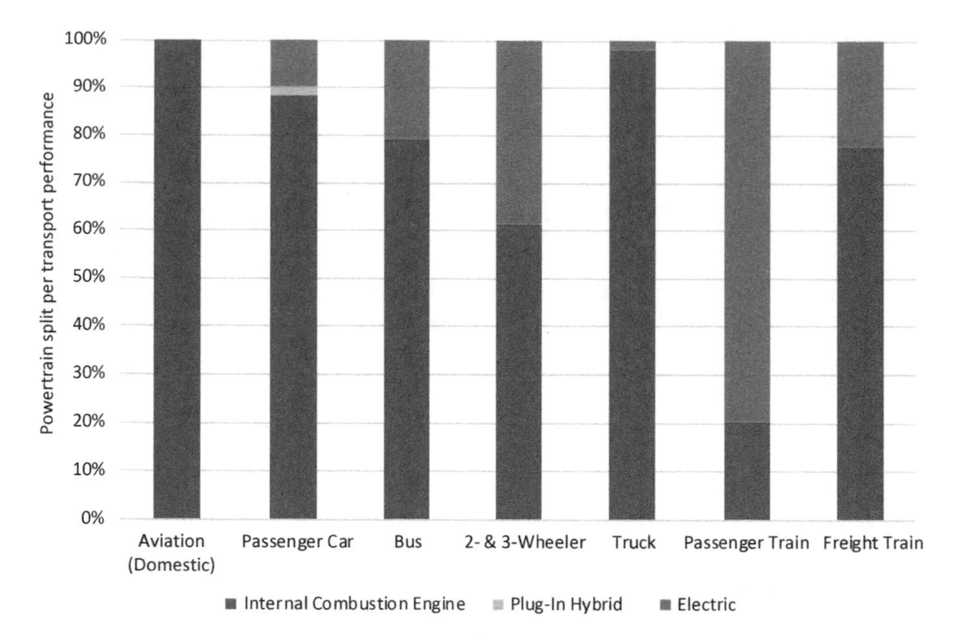

Fig. 6.5 Powertrain split for all transport modes in 2050 under the 5.0 °C Scenario in terms of transport performance

6.3.1.2 The 2.0 °C Scenario

Based on the low market share of BEV observed today (2018), minimal progress in electrification until 2020 is assumed in the 2.0 °C Scenario. Moving towards 2030, the *innovator* regions will experience strong electrification, encouraged by purchase incentives, EV credit systems, and tightened CO_2 fleet emission targets. Passenger cars and light commercial vehicles are projected to achieve shares of BEVs in the regional stocks between 21% and 30%, whereas heavy commercial vehicles and buses will attain even higher EV fleet shares of between 28% and 52% by 2030. This will require a massive build-up of battery production capacity in the coming years. Electric city buses and some trolley trucks will make a significant contribution to this development. Two- and three-wheel vehicles will be nearly completely electrified (batteries and fuel cells) in a couple of regions. OECD Pacific will head the *fuel-cell-electric vehicle* (FCEV) market introduction, which will account for up to 6% of passenger cars and light commercial vehicles by 2030, with Japan and South Korea the main market drivers. Higher shares of FCEV in *innovator* regions are more likely in the bus and heavy truck sector, which reach up to 10% in 2030. In 2050 in the innovative regions, only a minor proportion of vehicles will have ICE (up to 9%). Passenger cars and light commercial vehicles will predominantly be electrified, with a BEV share of around 80%. FCEV will also gain a significant share of 17% in OECD Pacific and OECD North America.

Looking ahead to 2050, 60–70% of buses and heavy trucks will probably be (battery) electric, whereas FCEV will increase their market share to around 37%. The proportion of buses and heavy trucks that will be BEV in 2050 will be around 60–70%. Two- and three-wheel vehicles will be nearly fully electrified in all regions (80–100%).

In the *moderate* regions, the BEV share of road transport vehicles is set in a range of 1–15%. Except for Latin America, the *moderate* regions will have an ICE share of 83% or less for passenger cars. For example, India is progressing with its current electrification strategy and up to 14% of its passenger cars will be battery-powered by 2030. It is likely that ICE will dominate buses and trucks in the *moderate* regions in 2030. Fuel-cell cars now have small shares of 1–2%. Non-OECD Asia is positively influenced by its *innovator* neighbour region, OECD Pacific. In the 2.0 °C Scenario, *moderate* regions will reach BEV shares of up to 67% for passenger cars and light commercial vehicles and between 54% and 65% for buses and heavy trucks by 2050. Compared with the *innovator* regions, the *moderate* regions will not experience a significant uptake of FCEVs. Africa and the Middle East will remain mostly dependent on ICE in 2050, with shares of around 90%. Only slow electrification will occur in Africa, with small and cheap BEVs (Fig. 6.6).

As an example of how different the electrification speeds will be across the world regions, Fig. 6.7 shows the uptake of electric and fuel-cell drivetrains in buses and two- and three-wheel vehicles, region by region, in terms of pkm. In 2015, a substantial proportion (15%) of China's buses were already electrified. OECD Pacific and OECD Europe will follow, with substantial electrification after 2020, as will India, Non-OECD Asia, and Eurasia after 2030. The remaining regions will electrify their fleets predominantly after 2035. Fuel-cell drivetrains will not begin to penetrate the market to a significant extent until 2025. We project a fleet that is 40–70% electric (battery and trolley) and 10–30% fuel-cell electric by 2050 in the 2.0 °C Scenario.

Figure 6.8 plots the projected electrification of passenger and freight rail in terms of final energy demand. Substantial electric passenger rail was present in OECD Europe, OECD Pacific, and China in 2015, and substantial electric freight transport in OECD Europe and Eurasia. In most other world regions, freight transport by rail predominantly relied on diesel locomotives.

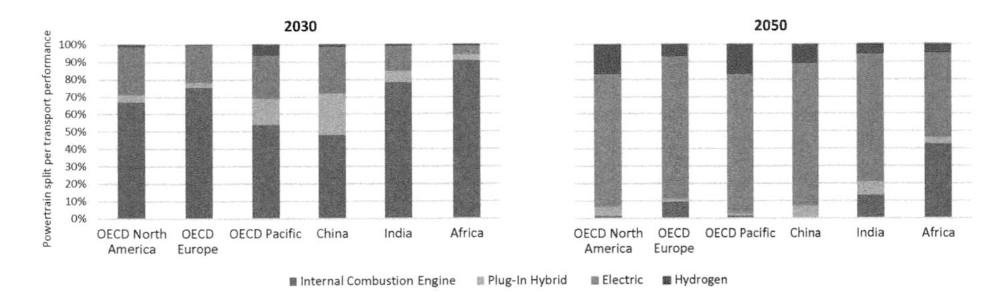

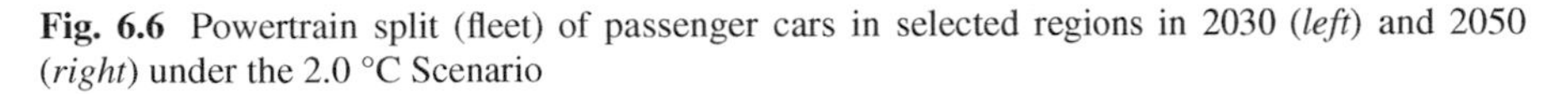

Fig. 6.6 Powertrain split (fleet) of passenger cars in selected regions in 2030 (*left*) and 2050 (*right*) under the 2.0 °C Scenario

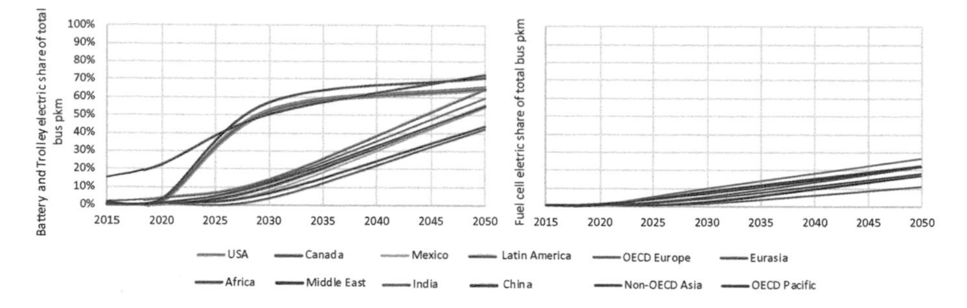

Fig. 6.7 Battery and trolley electric bus share of total bus pkm in the 2.0 °C Scenario (*left*) and fuel-cell electric bus share of total bus pkm in the 2.0 °C Scenario (*right*)

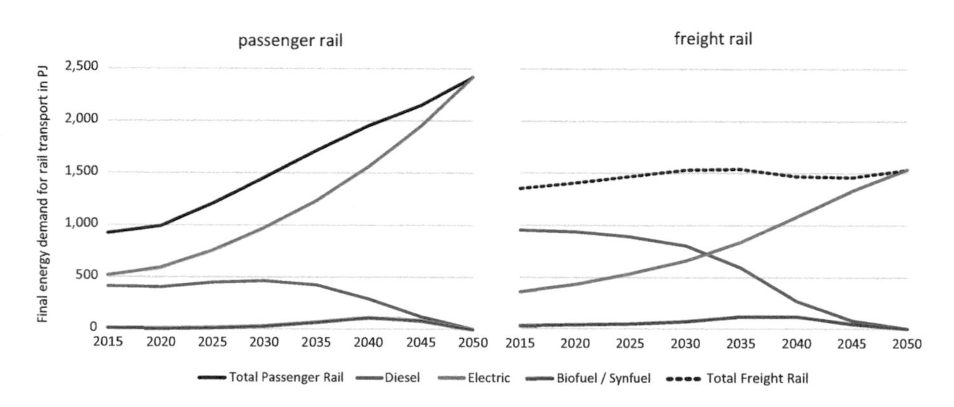

Fig. 6.8 Electrification of passenger rail (*left*) and freight rail (*right*) under the 2.0 °C Scenario (in PJ of final energy demand)

In most world regions, nearly all rail traffic is projected to be electric after 2040 in the 2.0 °C Scenario. Total diesel consumption in rail operations is projected to increase slowly until 2030, mainly because railway vehicles have long lifespans, and once diesel cars are put into operation, they are not replaced overnight. Furthermore, line electrification usually requires several years of planning and construction.

Aviation will probably remain predominantly powered by liquid fossil fuels (kerosene and bio- and synfuel derivatives) in the medium to long term because of limitations in electrical energy storage. We project a moderate increase in domestic pkm flown in electric aircrafts starting in 2030, with larger shares in OECD Europe because the flight distances are shorter than, for example, in the USA (Fig. 6.9). Norway has announced plans to perform all short-haul flights electrically by 2040 (Agence France-Presse 2018).

However, no real electrification breakthrough in aviation is foreseeable unless the attainable energy densities of batteries increase to 800–1000 Wh/kg, which would require fast-charging capable post-lithium battery chemistries.

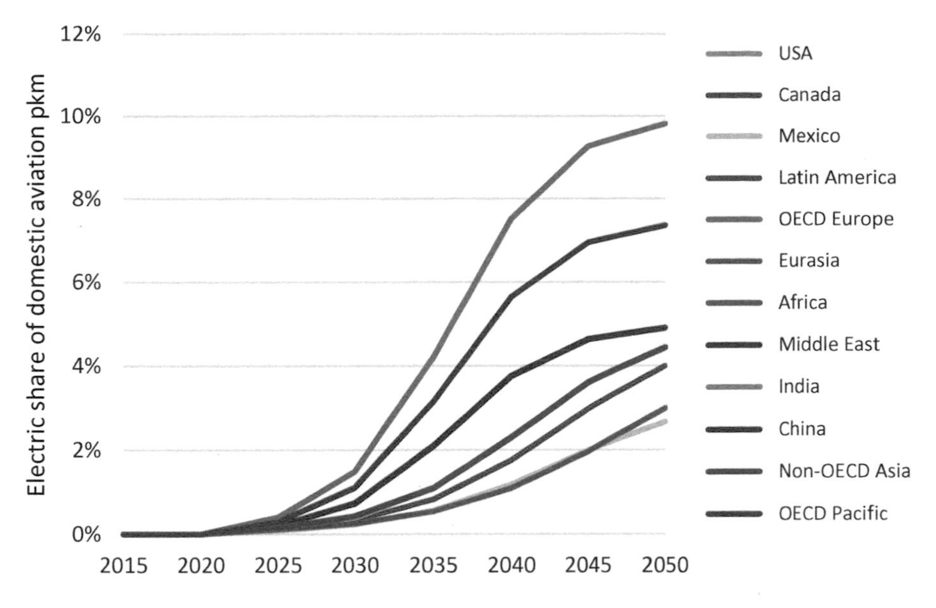

Fig. 6.9 Electricity-performed pkm in domestic aviation under the 2.0 °C Scenario

6.3.1.3 The 1.5 °C Scenario

In the 1.5 °C Scenario, an earlier and more rapid ramp-up of electric powertrain penetration is required than in the 2.0 °C Scenario and the *innovative* regions will be at the forefront. The *moderate* regions will also need to electrify more rapidly than in the 2.0 °C Scenario, but will end up with only a minimally higher share by 2050. In the passenger car sector in particular, plug-in hybrid electric vehicles (PHEVs) will ensure a sharp reduction in conventional combustion engine vehicles between 2030 and 2050. In the *late adopter* regions, there is no difference between the 2.0 °C and 1.5 °C Scenarios. The phasing out of internal combustion engine (ICE) vehicles will occur more quickly under the 1.5 °C Scenario than under the 2.0 °C Scenario (Fig. 6.10).

6.3.2 Mode-Specific Efficiency and Improvements Over Time

In passenger transport, trains and buses are much more energy efficient per pkm than passenger cars or aeroplanes. This situation does not change fundamentally if only electric drivetrains are compared (Fig. 6.11). It is apparent that railways and especially ships are clearly more energy efficient than trucks in transporting freight (Fig. 6.12). The 2015 figures are the starting point for a more detailed discussion, mode by mode, later in this chapter, and these figures are the basis for the rationale of our discussion in terms of modal shift (Chap. 6, Sect. 4). The efficiency data are

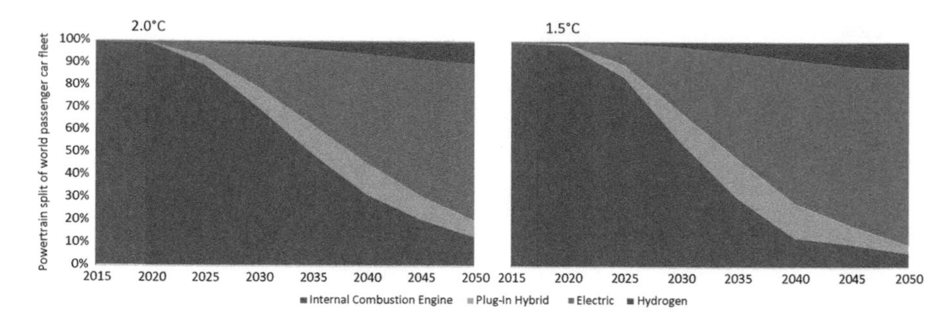

Fig. 6.10 Powertrain split of the world passenger car fleet in the 2.0 °C Scenario (*left*) and 1.5 °C Scenario (*right*)

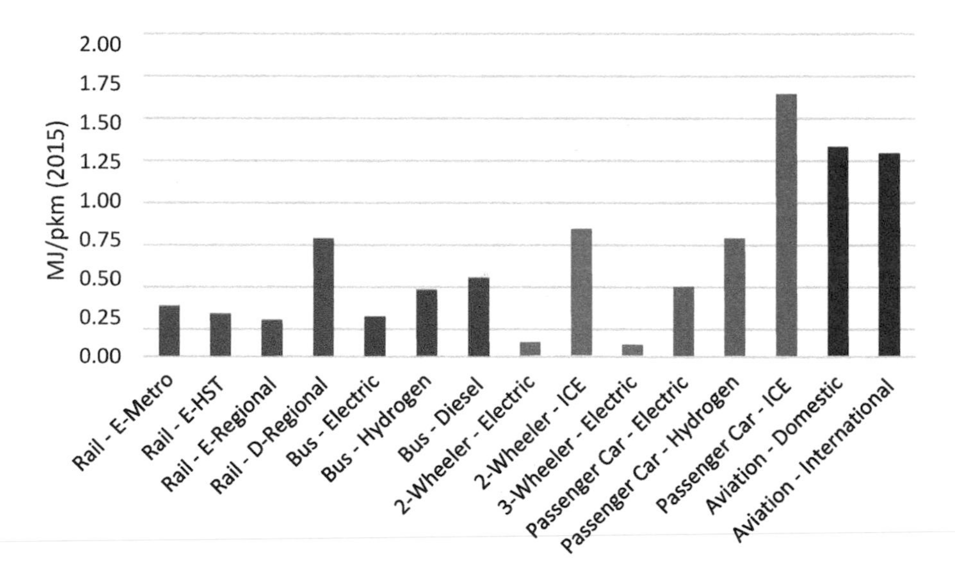

Fig. 6.11 Final energy demand in urban and inter-urban passenger transport modes in 2015 (world averages)

based on literature-reported and transport operator information. The efficiency levels in terms of pkm or tkm depend to a large extent on the underlying capacity utilization of the vehicles, which differs between world regions. The numbers are average values and differences are evaluated at the regional level.

In addition to powertrain electrification, there are other potential improvements in energy efficiency, and their implementation will steadily improve energy intensity over time. Regardless of the types of powertrains and fuels used, efficiency improvements on the MJ/pkm or MJ/tkm level will result from (for example):

- Reductions in powertrain losses through more-efficient motors, gears, power electronics, etc.;
- Reductions in aerodynamic drag;

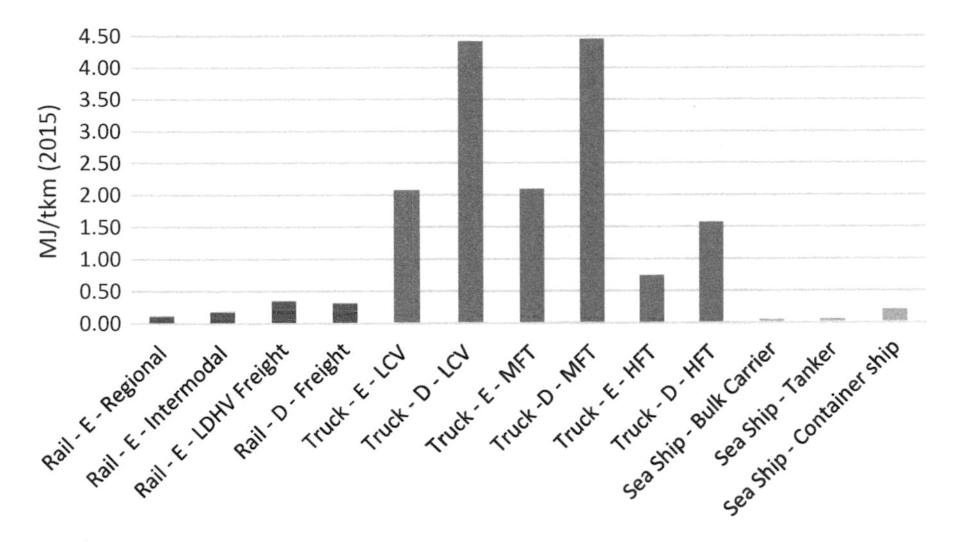

Fig. 6.12 Final energy demand in freight transport modes in 2015 (world averages)

- Reductions in vehicle mass through light-weighting;
- The use of smaller vehicles;
- Operational improvements (e.g., through automatic train operation, load factor improvements).

The measures are discussed in the following mode-specific sub-chapters.

6.3.3 *Road Transport*

6.3.3.1 Passenger Cars

As of 2017, 99% of the passenger cars produced worldwide were estimated to be equipped with an ICE: the majority of them gasoline or diesel (95%), 3.4% hybrid electric vehicles (HEV), and 0.7% PHEV (BCG 2017). Only about 0.9% of the cars sold were pure battery electric vehicles (BEV). There are several options for energy efficiency improvements. The fuel consumption reduction potential of petrol engines as a result of engine improvements and hybridization is around 25–30%, and it is around 15–20% for diesel engines (van Basshuysen and Schäfer 2015). Maximum efficiencies of 38–40% can be reached by ICE (Schäfer 2016), whereas electric drivetrains have efficiencies of 80–85% (including [re]charging losses). Besides the reduction in engine losses, both lightweight construction and reductions in rolling resistance will result in additional fuel savings.

Hybrid systems increase the complexity of powertrains, resulting in higher masses and higher costs, but they offer additional fuel-saving potentials. Energy can thus be recovered by recuperation during braking in hybridized and all-electric

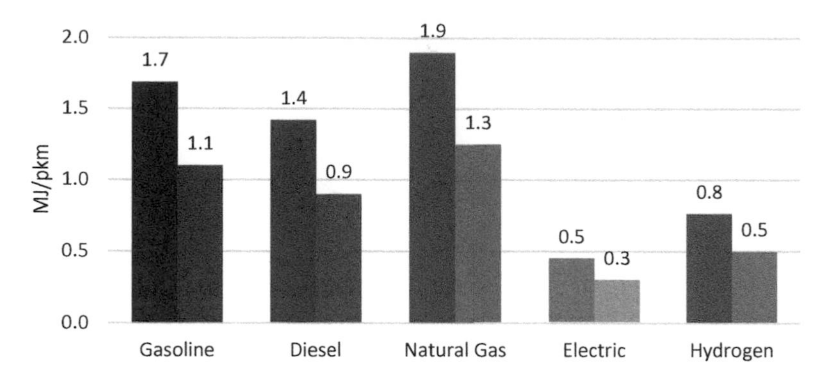

Fig. 6.13 World average energy consumption development for passenger cars per powertrain in 2015 (*left*) and 2050 (*right*)

vehicles. In BEV and FCEV, advanced battery technologies can reduce the overall vehicle mass. However, post-lithium technologies, such as lithium–sulfur and solid-state batteries with increased energy densities and lower systems masses compared with today's Li-ion battery technologies, will probably not enter the transport sector before 2030 (Schmuch et al. 2018).

In total, we project a 1% increase in annual efficiency on a per passenger/km basis (the same for all drivetrains) for the 5.0 °C and 2.0 °C Scenarios. The efficiency improvements for 2050 in the 2.0 °C Scenario will be achieved in 2040 under the 1.5 °C Scenario (Fig. 6.13).

6.3.3.2 Light and Heavy Freight Vehicles

Like passenger cars, trucks are currently driven almost exclusively by conventional ICE. With a market share of 84% of diesel-fuelled vehicles in 2015 (IEA 2017c), the global fleet of trucks was predominantly operating on diesel (IEA 2017b). In some regions, such as the Middle East, trucks are also powered by gasoline to a considerable degree. Electric drivetrains will enter the truck sector gradually in coming years because of changes in exhaust emissions legislation and because the higher efficiency of electric drivetrains compared with ICE will offer economic advantages to road carriers (Fig. 6.14). Compared with diesel-powered trucks, fuel-cell electric drivetrains in trucks can substantially reduce the energy intensity per tkm, and allow higher operating ranges than battery-powered trucks. To achieve rapid improvements in energy efficiency in the truck sector, the hybridization of diesel powertrains, especially those operating in stop-and-go intensive urban environments, is promising (Burke and Hengbing 2017; Lischke 2017). However, after 2030, the hybrid diesel powertrain will be seen merely as a transitional technology before the advent of fully electric powertrains (overhead catenary, battery electric, and fuel-cell electric). Therefore, the development of battery recharging and hydrogen refuelling infrastructures will require massive investments in coming years. In

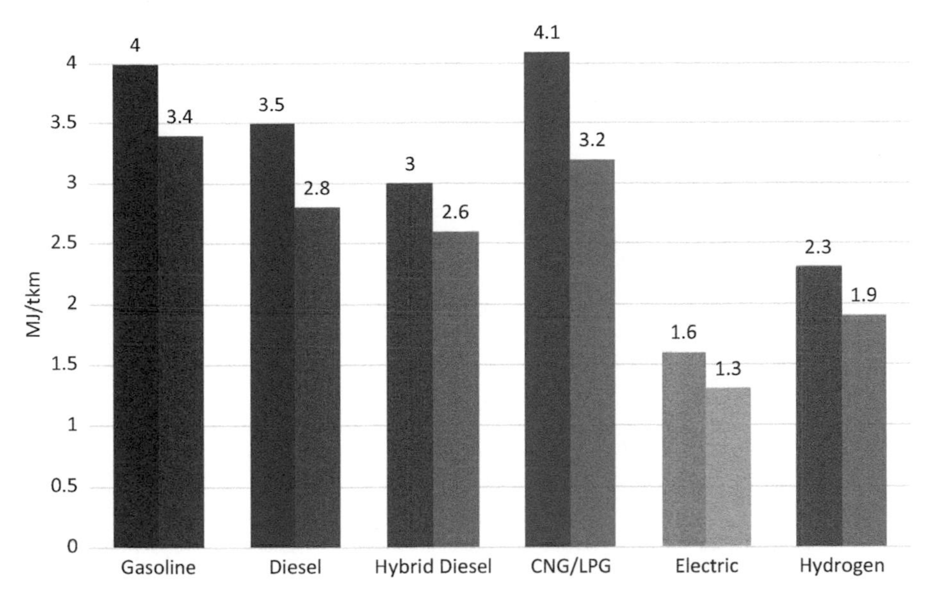

Fig. 6.14 Average global energy intensities of truck drivetrain technologies in 2015 and 2050

the commercial sector, it is likely that many electrical charging stations will be set up on the private grounds of logistics operators. The market growth of BEV in the truck sector will require considerable public spending on the installation of overhead catenary lines along major highways for trolley trucks (hybridized with diesel, fuel cells, or batteries). The first pilot lanes are being developed in Germany, Sweden, and California (USA) (Siemens 2017).

6.3.3.3 Buses

State-of-the-art city buses operate on ICE (diesel or gas) or are trolley buses (all-electric or hybridized with an additional battery or diesel motor). Diesel hybrids have entered the city bus market and constitute a large proportion of the bus fleets in a range of cities today. The short distances between stations, small operating radii, and moderate daily mileages make the use of batteries in city buses a viable option, complemented by fuel cells for routes with higher ranges or difficult terrains. Battery electric buses are between the prototype/experimental stage and a mature technology. In China, battery electric buses are already an integral part of public bus transport systems, and the city of Shenzen has had a 100% electric bus fleet since 2016 (over 16,000 battery electric buses) (Sisson 2018). Fuel-cell electric buses still lag behind battery electric buses in terms of numbers, but are increasing. In 2017, about 80 fuel-cell electric buses were in operation in Europe (Element Energy Ltd. 2018) and 26 in the USA (Eudy and Post 2017). Full battery operation is more difficult to achieve in regional and long-distance buses and coaches, so the

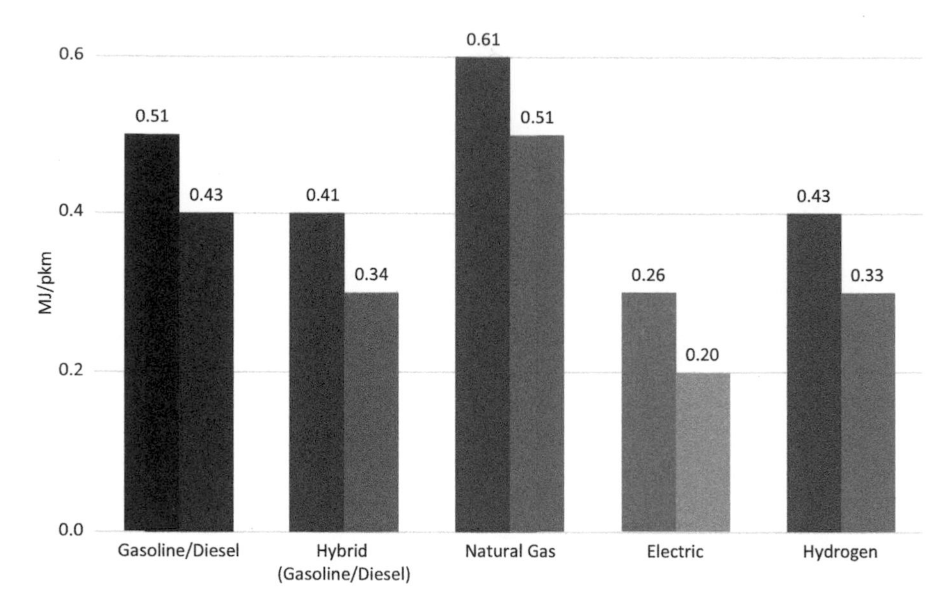

Fig. 6.15 Average global energy intensities of bus drivetrain technologies in 2015 (*left*) and 2050 (*right*)

uptake of full battery electric and also fuel-cell powertrains will be slower than in city buses. Diesel-powered buses will remain in the market longer, complemented by diesel hybrid powertrains. Ultimately, fuel-cell-powered inter-urban coaches will probably be more common than fuel-cell-powered city buses. In our model, we divided the regions into *innovators* (i.e., China and OECD countries) and all the other regions lagging behind the innovators. We also identified a clear trend towards electrification.

We project a 0.5% (diesel, diesel-hybrid and natural gas) to 0.8% (electric and fuel cell hydrogen electric) increase in efficiency on a per passenger per km basis in all three scenarios (Fig. 6.15).

6.3.3.4 Two- and Three-Wheel Vehicles

Two-wheel vehicles are probably the most important component of everyday traffic in large parts of Asia and Africa. The drivetrain efficiencies of e-scooters are reported to be 1.2–3 kWh/100 vehicle–km. China is by far the biggest market for electric scooters in the world, with more than 200 million electric scooters on the road by 2015, that translates to an electric share of about 70%.

Three-wheel vehicles (also country-specifically called 'rickshaws' or 'tuk-tuks') have at least two, and usually three or more seats, and are often overloaded in daily traffic. India alone is reported to have 2.5 million rickshaws on the road, each travelling 70–150 km/day (Abu Mallouh et al. 2010). Most three-wheel vehicles are fuelled by gasoline and some by liquid petroleum gas (LPG), although some com-

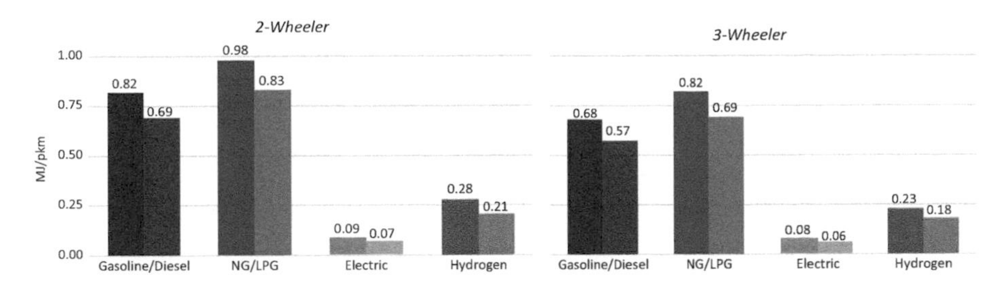

Fig. 6.16 Average global energy intensities of two-wheel vehicles (*left*) and three-wheel vehicles (*right*) by drivetrain technology in 2015 (*left bar*) and 2050 (*right bar*)

munities incentivize the conversion of two-stroke engines to battery electric three-wheel vehicles. Electric tuk-tuks are increasingly emerging in South-East Asia, together with battery swap stations and solar-powered rickshaws (Moran 2018; Reddy et al. 2017). Thailand has announced that it plans to convert all existing two-stroke-powered tuk-tuks to battery electric powertrains within 5 years (Coconuts Bangkok 2017). In India, too, plans are repeatedly announced to electrify all new two- and three-wheel vehicles within the next two decades (Ghoshal 2017).

The efficiency of battery electric drivetrains is much better than that of conventional two-stroke engines in two- and three-wheel vehicles (Fig. 6.16). We project a 0.5% annual increase in efficiency on a per pkm basis (the same for all drivetrains) in all three scenarios.

6.3.3.5 Rail Transport

No other transport mode is more suited to operate electrically than railways. Urban rail systems, for instance, are invariably electric. Electric trains consume about 60–70% less energy than diesel trains when their final energy use is compared (on catenary and tank levels, respectively). According to the International Railway Association, about 32% of the worldwide rail network was electrified in 2015 (UIC 2017). However, in terms of transport performance (pkm or tkm), the ratio of electric to diesel is higher because electrified rail lines usually experience more traffic than non-electrified lines. The electrification of rail lines with overhead catenaries has been the state-of-the-art technology for decades and new lines are almost exclusively equipped with overhead power lines right from the start.

However, line electrification, especially of existing lines, is often difficult to achieve due to unsettled and complex right-of-way issues, narrow tunnel diameters, or simply because line electrification is not economically viable due to low line utilization. The use of on-board batteries for mixed electrified/non-electrified lines and shunting operations and the use of fuel-cell hybrid powertrains for longer lines with little or no electrification whatsoever could be feasible fully electric alternatives to diesel power or full-line electrification. (Fuel-cell electric trains have not been modelled in this research).

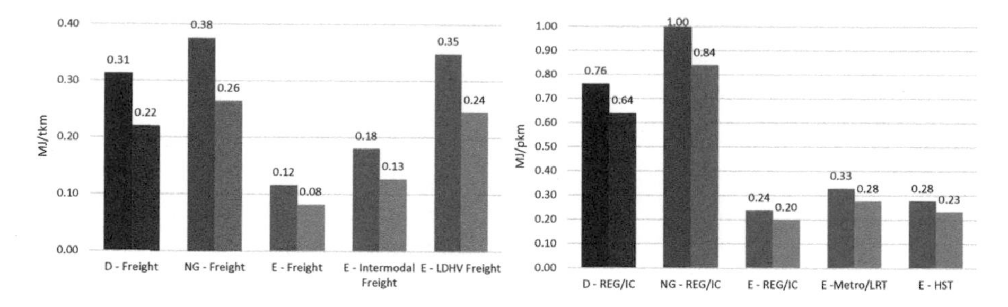

Fig. 6.17 MJ/tkm of freight rail trains (*left*) and MJ/pkm of passenger rail trains (*right*) for 2015 (*left*) and 2050 (*right*)

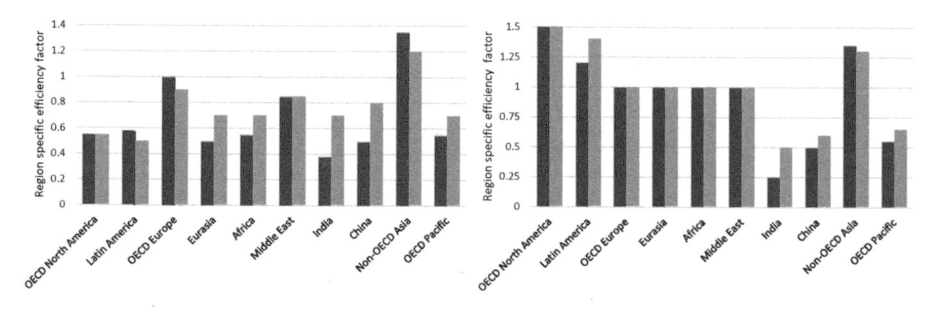

Fig. 6.18 Region-specific MJ/tkm and MJ/pkm in 2015 and 2050 for freight rail trains (*left*) and passenger rail trains (*right*)

We project a 0.5% annual increase in efficiency on a per passenger per km basis and a 0.8% annual increase in efficiency on a per tonne per km basis (the same for all drivetrains) in all three scenarios (Fig. 6.17).

The actual efficiency depends on the type of train and the world region. These differences are considered in every transport sub-model and are exemplified in Fig. 6.18 for freight and passenger trains.

6.3.3.6 Water and Air Transport

Inland navigation will probably remain predominantly powered by ICE in the next few decades. Therefore, we did not model the electrification of inland navigation vessels. However, pilot projects using diesel hybrids, batteries, and fuel cells are in preparation (DNV GL 2015). We assumed the same increase in the share of bio- and synthetic fuels over time as in the road and rail sectors.

In aviation, energy efficiency can be improved by measures such as winglets, advanced composite-based lightweight structures, powertrain hybridization, and enhanced air traffic management systems (Madavan 2016; Vyas et al. 2013). We project a 1% annual increase in efficiency on a per pkm basis.

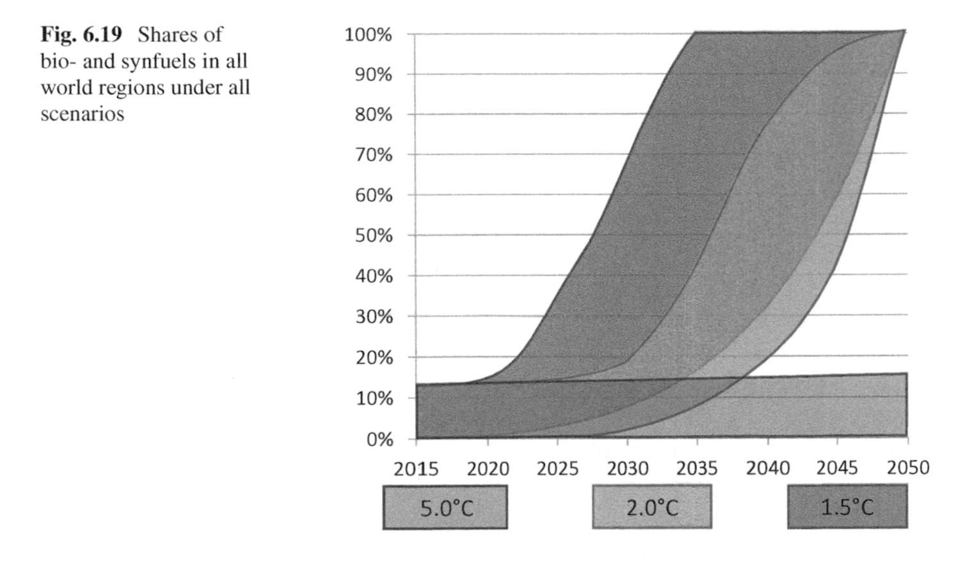

Fig. 6.19 Shares of bio- and synfuels in all world regions under all scenarios

6.3.4 *Replacement of Fossil Fuels by Biofuels and Synfuels*

The use of biofuels in the transport sector offers a potential lever to reduce the CO_2 emissions from fossil fuels. Biofuels can be used either as a direct drop-in or as admixtures to fossil fuels. Biofuels are widely used, especially in Latin America (e.g., E85 in Brazil, a blend of 85% ethanol and 15% gasoline). Biofuels will be replaced by synthetic fuels (synfuels) within the next few decades.

In the three scenarios, we use region-specific shares of bio- and synfuels to replace fossil fuels such as diesel, gasoline, and kerosene. Figure 6.19 shows the scenario-specific band-widths. In the 5.0 °C Scenario, Latin America will increase its proportion of bio- and synfuels to around 15%, whereas in the Middle East, Mexico, and Eurasia will even not reach a 0.5% share by 2050. Bio-fuel and especially synfuel shares will remain constant in the 5.0 °C Scenario, whereas they will increase in the 2.0 °C Scenario between 2030 and 2050 and in the 1.5 °C Scenario from 2020 onwards.

6.3.5 *Operational Improvements and Novel Service Concepts*

In addition to technical improvements, such as the energy efficiency savings described in previous chapters, behavioural changes can potentially reduce energy demand and help decarbonise the transport sector as a whole. These measures include, among other things, the efficient operation of vehicles, for example by increasing the occupancy rates of moving vehicles, reductions in mileages, and reductions in vehicle stocks.

6.3.5.1 Passenger Transport

The classic fixed boundary between private and public transport will blur in the future as novel platforms and sharing concepts emerge. This trend will potentially reduce the number and usage of privately owned vehicles. Private car ownership is generally becoming less a 'must' for many people in the car-oriented OECD regions. Driven by digitization, app-based cars and ride-sharing concepts are emerging. They include traditional car sharing (e.g., Zipcar), free-floating car sharing businesses (e.g., car2go), and ride-hailing mobility concepts (e.g., Uber, lyft). In Europe, private ride-sharing mobility platforms, such as *BlaBlaCar*, are seen as alternative passenger car usership schemes. All shared-use mobility concepts have in common that they reduce the vehicle stock and increase occupation rates on trips. To own a car is no longer considered a status symbol, but has become simply a mode of transport in the Western world.

Flexibility and passenger-friendly usability is becoming increasingly important to users of public transport. Various app-based on-demand mobility services are currently being tested in pilot projects—for example, in the German city of Schorndorf (Brost et al. 2018). Under these business models, innovative vehicle concepts are tested for suitability. Automated battery electric (mini) buses can make public transit more flexible. Their modularity also allows them to connect to other modes of transport, such as rail and air transport, and thus opens up fundamentally new transport chains that are more energy efficient than traditional passenger car usage. However, decision-makers and transport planners must consider the potentially detrimental rebound effects of highly personalized automated mobility (Pakusch et al. 2018). For example, an increase in total energy consumption may occur as a result of the higher transport demand arising from transport options that are seen as more attractive than more-energy-efficient mass transit systems or cycling. Therefore, novel transport concepts could cancel out the gains in energy efficiency on a per vehicle level.

6.3.5.2 Freight Transport

The automatization of road freight traffic can reduce fuel consumption. The technology for partially automated trucks is already available and being tested to make it widely available. In the Society of Automotive Engineers (SAE) automation level 3 (SAE 2018), (multi-brand) truck platooning has become possible, which is a driver-assistance technology (Janssen et al. 2015). With communication between the first truck and the following trucks, the control of the following vehicles is handed over to the leading vehicle. Thus, the following vehicles are driving in the slipstream and lower air drag is achieved, which can reduce energy consumption by up to 5% (Daimler 2018).

In urban freight transport, last-mile delivery concepts based on light commercial vehicles can help reduce fuel consumption. By fragmenting the transport routes into the main leg and urban traffic (last mile), different vehicle types and modes can be used more efficiently for specialized transport tasks. Like light commercial vehicles (electric), cargo bikes could shape the urban freight traffic of the future as they do already today in many countries in Asia.

Further novel logistic concepts, such as the use of drones for delivery and decentralized production using 3D printers, will help to reduce the overall transport demand and thus energy consumption through more-efficient operational solutions.

6.4 Transport Performance

We outline achievable 2.0 °C and 1.5 °C Scenario paths that will allow us to achieve very ambitious emission reduction targets. The levels of pkm and tkm for every world region in 2015 were calibrated against statistical data, the literature, and our own estimates when concrete data were not available.

To complement the anticipated technologically driven mode-inherent improvements in efficiency that we took as the input for our modelling in the previous sections of this chapter, we now describe the scenario assumptions in terms of the transport mode choices and transport demand. The main idea is to model low-emission pathways that shift from fossil-dependent, low-energy-efficiency modes toward more-energy-efficient and electrified modes. We also look at the general level of performance of transport modes and suggest region-specific development pathways. We first outline the scenario-specific pkm and tkm pathways to 2050 and then discuss the rationales for the mode choices that will result in the transport performance projections.

For the 5.0 °C Scenario, we extrapolate current trends in transport performance until 2050. In relative terms, the transport performance of all transport carriers will increase from current levels. Aviation, passenger car, and commercial road transport are particularly projected to grow strongly (Fig. 6.20). These modes consume more energy than trains, ships, and buses, as discussed in Sect. 6.3.2. Even if the full efficiency potential of these transport modes is realized, energy intensity per pkm or tkm will remain higher than that of trains, ships, or buses.

In the 2.0 °C Scenario and 1.5 °C Scenario, we project a strong increase in rail traffic (starting from a relatively low level) and a slower growth or even a reduction in the use of the other modes in all world regions (Fig. 6.21). The next two sections describe the specific changes and developments for each type of passenger and freight transport.

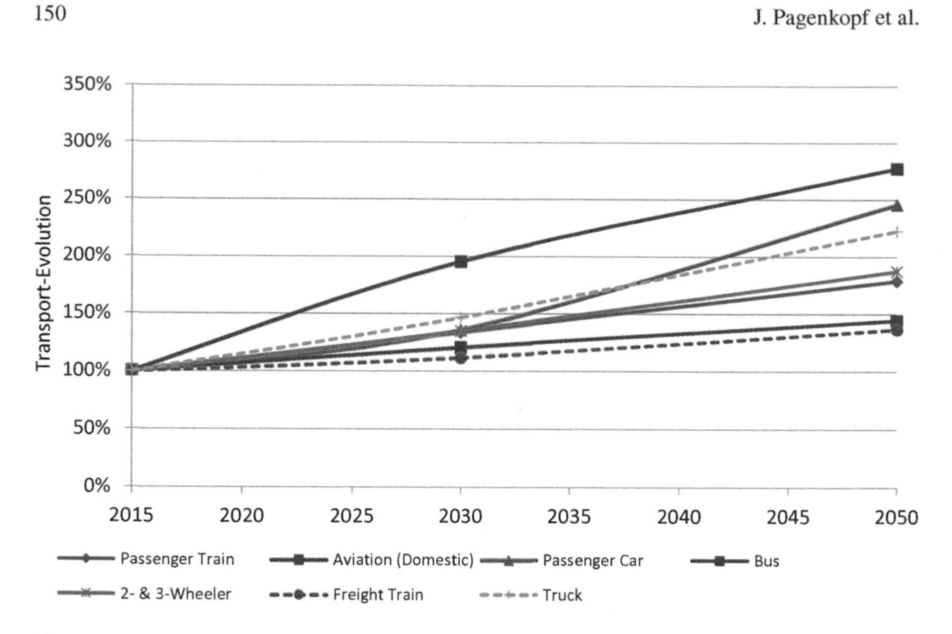

Fig. 6.20 Relative growth in world transport demand (2015 = 100% pkm/tkm) in the 5.0 °C scenario

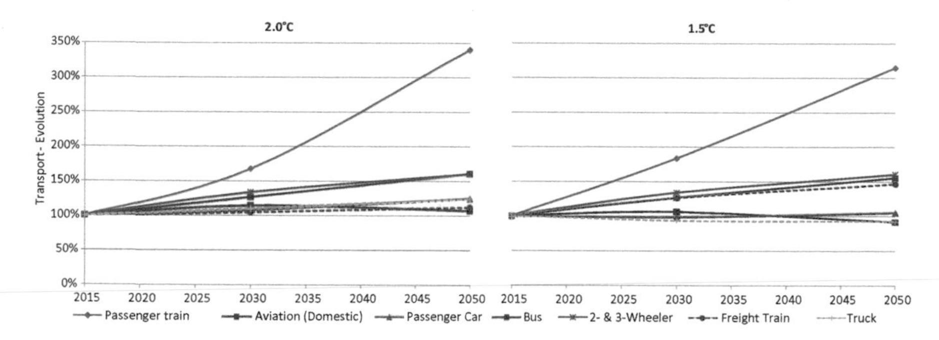

Fig. 6.21 Relative growth in world transport demand (2015 = 100% pkm/tkm) in the 2.0 °C Scenario (*left*) and 1.5 °C Scenario (*right*)

6.4.1 Passenger Transport Modes

To reduce transport-related CO_2 emissions, a shift towards low-energy-intensity modes of transport is required for both ambitious climate-protection scenarios. Travelling by rail is the most energy-efficient form of transport, and therefore we suggest a strong shift from domestic aviation to trains, especially high-speed trains and magnetic levitation trains. The assumed shifts from domestic aviation to trains are shown in Table 6.1. The mode shift potential differs, depending predominantly on the country-specific distance between origin–destination pairs. The shifts from international aviation to trains were not analysed because the potential is lower

Table 6.1 Pkm "per km" shift from domestic aviation to trains (in %)

	2015	2020	2025	2030	2035	2040	2045	2050
2.0 °C scenario	0	0	0.5–0.75	0.7–1.1	1–1.7	1.4–2.5	1.9–3.8	2.7–5.7
1.5 °C scenario	0	2	4	6	8	10	12	14

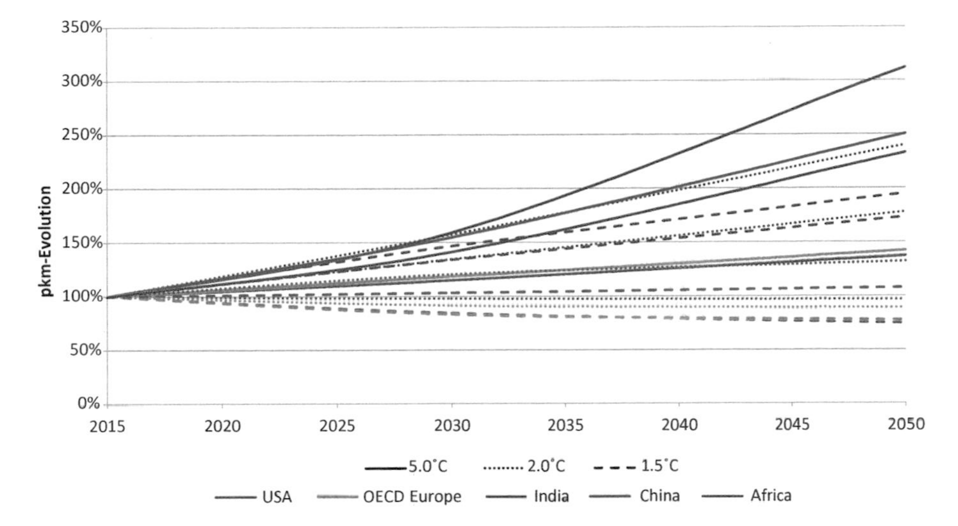

Fig. 6.22 Regional pkm development

(although not zero). The maximum global shift from domestic aviation to trains will be stronger in the 1.5 °C Scenario than in the 2.0 °C Scenario.

In the urban context, investments in public transit systems and limitations on the use of private cars are cornerstones of a more-energy-efficient transport system. The increased use and integration of homes and offices, and video conferencing can reduce traffic. With further urbanization, two- and three-wheel vehicles are suitable for travelling small distances quickly. A shift to small and medium cars and away from larger cars and SUVs will also allow lower energy intensity. Whereas occupancy rates remain steady for the passenger car sector in the 2.0 °C Scenario, in the 1.5 °C Scenario, we assume a significant increase in occupancy rates. Ultimately, pkm will stabilize, whereas vehicle–km (vkm) will decline in response to incentives such as high-occupancy vehicle lanes and novel ride-sharing services.

Figure 6.22 shows the development of pkm in all scenarios for sample world regions. In the 2.0 °C Scenario, pkm in the OECD countries will predominantly remain on the same level in 2050 as in 2015 due to saturation and sufficiency effects. In all the other world regions, we project a growth in pkm, reflecting economic catch-up processes in the developing world induced by increases in population and

GDP. As can be seen for the 1.5 °C Scenario, India, which represents a strongly growing economy, will double its pkm by 2050, whereas OECD areas, such as the USA and Europe, will experience a decline. Although China is expected to experience continuous economic growth over the next few decades, pkm will rise slowly compared with that in the 5.0 °C Scenario.

Pkm in absolute numbers in 2050 will be highest in the 5.0 °C Scenario. It will also increase in the 2.0 °C and 1.5 °C Scenarios, but at a slower rate (Figs. 6.23 and 6.24).

Looking more closely at the 2.0 °C Scenario, the transport modes will evolve differently in the world regions, both quantitatively and in relative terms (Fig. 6.25) due to the diverse mobility patterns. For example, Africa has a high bus share in total pkm today, whereas OECD Europe has a high passenger car share (LDV), but pkm must decrease by 2020 and in subsequent years to meet the CO_2 reduction targets because fleet electrification will not be able to keep up. In parallel, rail pkm

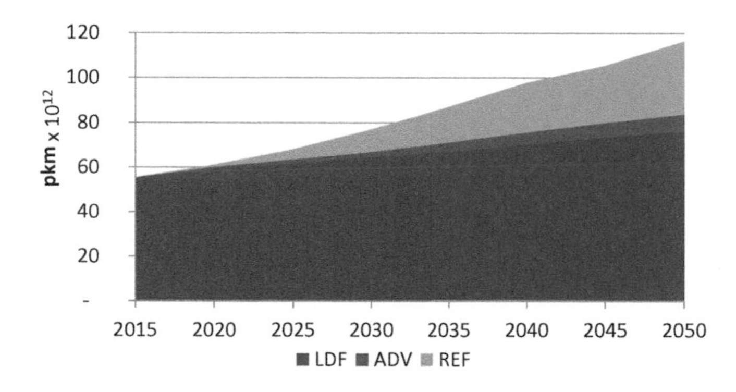

Fig. 6.23 World pkm development in all scenarios

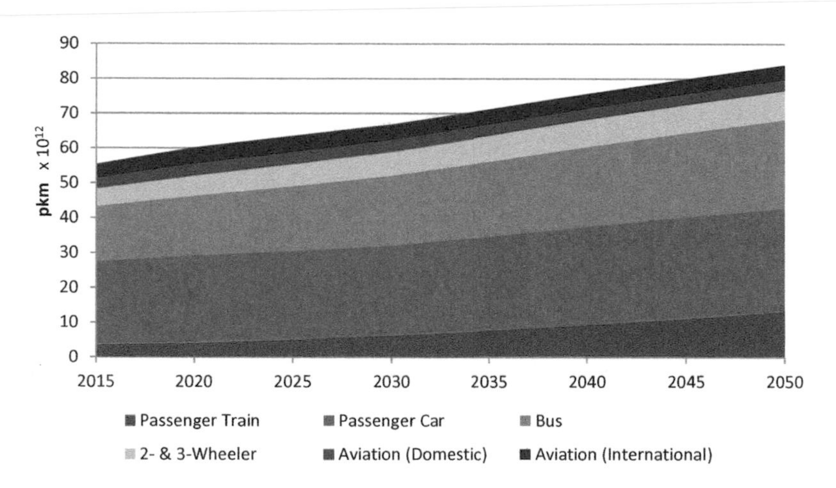

Fig. 6.24 World pkm development in the 2.0 °C Scenario

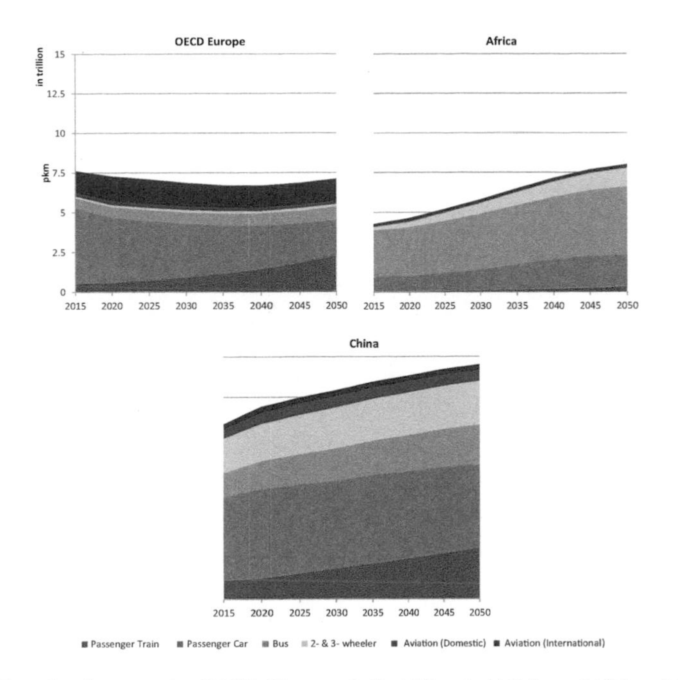

Fig. 6.25 Pkm development in OECD Europe (*left*) Africa (middle), and China (*right*) in the 2.0 °C Scenario

will increase strongly until 2050 and this will compensate, in part, the decline in passenger car pkm. Population and GDP are very likely to catch up in Africa and will result in a sharp rise in mobility demand. We project that most of this rise to be covered by informal and formal public transport systems, with buses and mini-buses. In China, the pkm split among modes is more balanced. All modes are projected to rise in pkm in the future.

6.4.2 *Freight Transport Modes*

Total freight activity is modelled to increase strongly in the 5.0 °C Scenario and more slowly in the 2.0 °C Scenario (Fig. 6.26). In the 1.5 °C Scenario, freight transport activity in 2050 is projected to remain at the 2015 level. Freight intensity will stagnate or decrease in the 1.5 °C Scenario in the OECD countries and increase in other regions, such as China and India (Figs. 6.26 and 6.27).

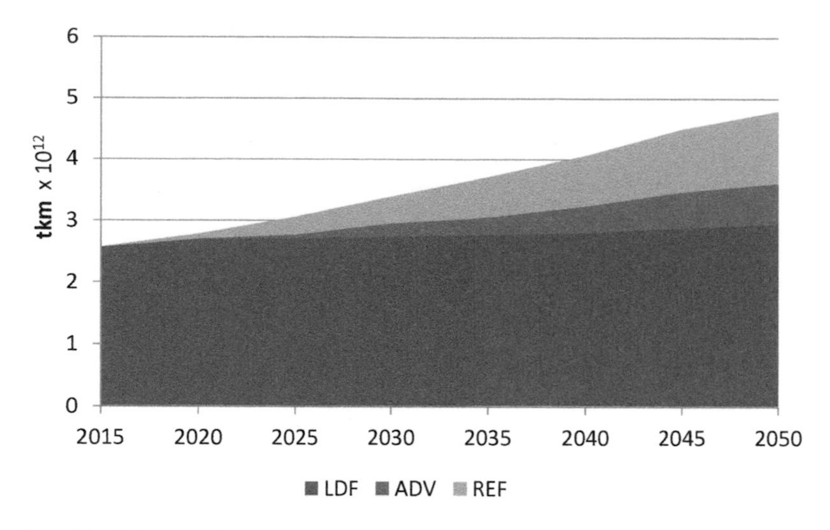

Fig. 6.26 World tkm development in all scenarios

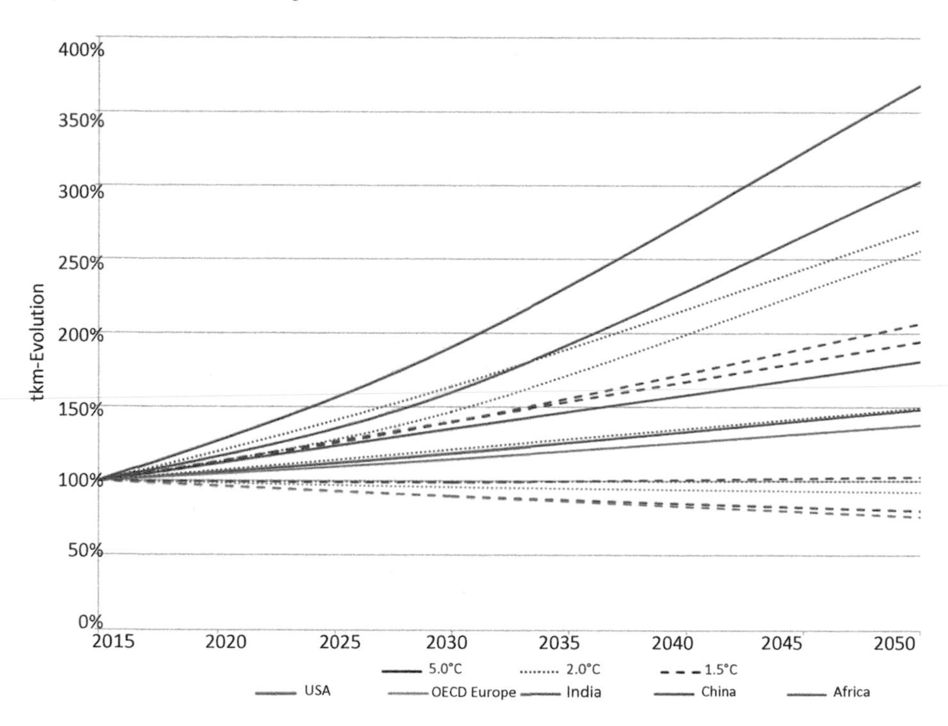

Fig. 6.27 Regional tkm development

Table 6.2 Global tkm shifts from truck to train in the 2.0 °C and 1.5 °C Scenarios (in %)

	2015	2020	2025	2030	2035	2040	2045	2050
LFT (2.0)	0	0	0	0	0	0	0	0
LFT (1.5)	0	1	4	6	7–8	9–10	11–12	13–14
MFT (2.0)	0	1	2	3	4	5	6	7
MFT (1.5)	0	2–3	5–10	9–14	13–18	13–19	13–19	13–18
HFT (2.0)	0	3	5	8	10	13	15	18
HFT (1.5)	0	2–3	8–18	12–22	16–27	16–27	16–27	18–27

We modelled the shift from high-energy-intensity modes to low-energy-intensity modes, especially from road to rail freight. This will require substantial investments in additional rail infrastructure. Table 6.2 shows the assumed global tkm shift from truck to train in the 2.0 °C and 1.5 °C Scenarios. Heavy freight trucks (HFT) are more likely to operate over long-haul distances, and are therefore more suitable for the shift to rail freight than light freight trucks (LFT) or medium freight trucks (MFT). In the 2.0 °C Scenario, we assume an average shift of 18% for HFT, whereas in the 1.5 °C Scenario, the shift is projected to reach up to 27%. The ramp-up of shift potential is slow, because the provision of rail and terminal infrastructures and rolling stock will require considerable time.

We modelled tkm for truck (road) and rail freight transport. In the 5.0 °C Scenario, transport activity is projected to increase clearly until 2050. In the 2.0 °C and 1.5 °C Scenarios, this increase will be slower and the tkm on rail will exceed the road tkm in numbers by 2050 (Fig. 6.28).

In the 2.0 °C Scenario, road tkm will decrease in the OECD countries and stagnate or increase slightly in China (Fig. 6.29). Rail tkm is projected to increase in all other world regions (Fig. 6.29). Rail tkm in China will temporarily decrease slightly because of an anticipated medium-term decline in coal transport (Fig. 6.30).

In 2015, rail's share of total tkm differed between regions, but will increase in the 2.0 °C Scenario in all regions except India, where road tkm will increase more than rail tkm (Fig. 6.31).

Energy Scenario Results

Our scenario modelling involves merging the transport performance for all modes and powertrains with specific energy demands, and yields the accumulated demands for electricity, hydrogen, gas, and liquid fuels across all world regions between 2015 and 2050. The transport energy scenario results are outlined in Chap. 8.

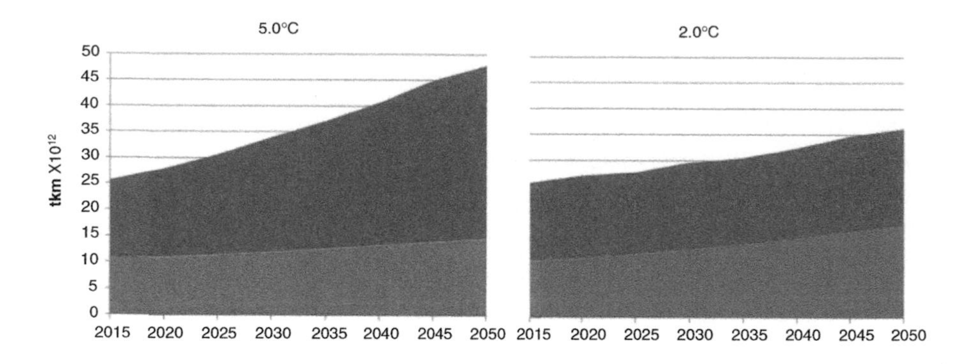

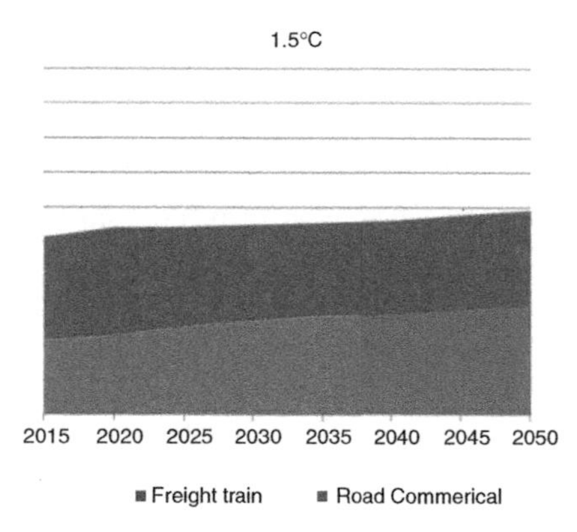

Fig. 6.28 World tkm development in the 5.0 °C, 2.0 °C, and 1.5 °C Scenarios

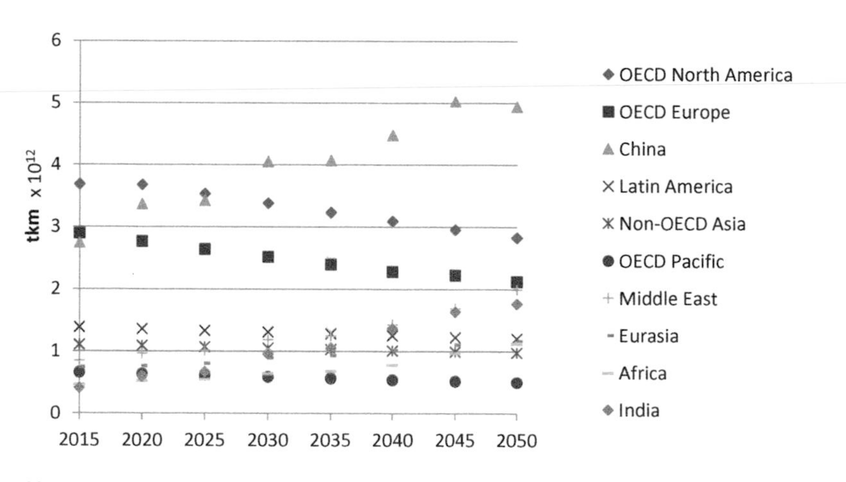

Fig. 6.29 Road tkm in the 2.0 °C Scenario

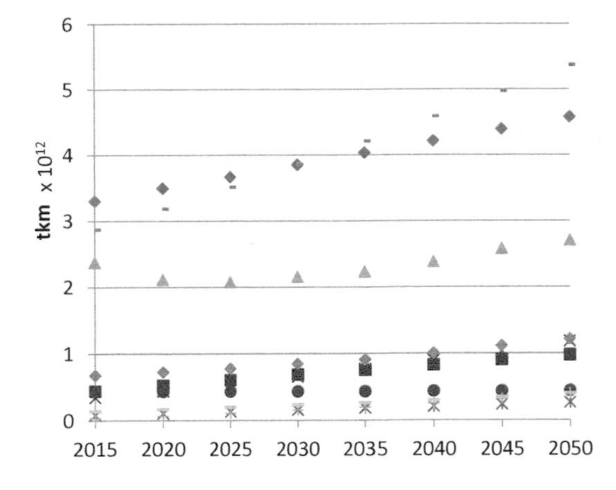

Fig. 6.30 Rail tkm in the 2.0 °C Scenario

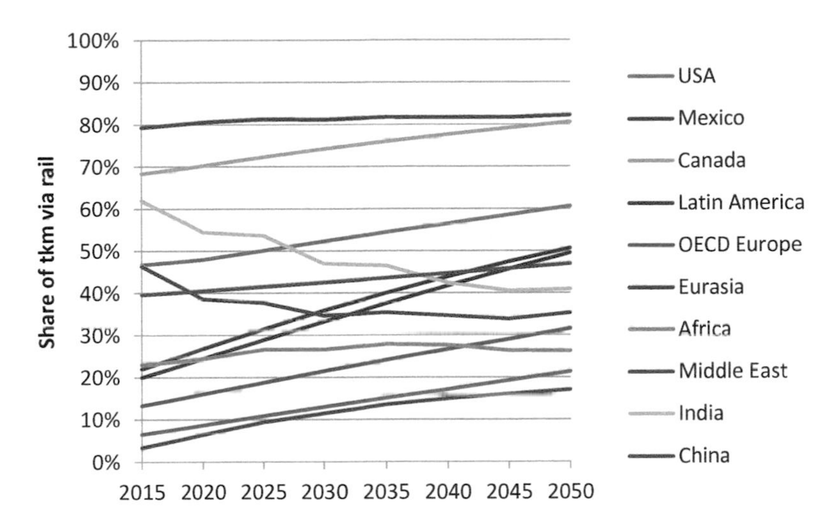

Fig. 6.31 Share of rail tkm in total rail + road tkm in the 2.0 °C Scenario

References

Abu Mallouh M, Denman B, Surgenor B, Peppley B (2010) A Study of Fuel Cell Hybrid Auto Rickshaws Using Realistic Urban Drive Cycles. Jordan Journal of Mechanical and Industrial Engineering 4:

Agence France-Presse (2018) Norway aims for all short-haul flights to be 100% electric by 2040. In: The Guardian Online Edition. https://www.theguardian.com/world/2018/jan/18/norway-aims-for-all-short-haul-flights-to-be-100-electric-by-2040

BCG (2017) The Electric Car Tipping Point. Research highlights. The Boston Consulting Group (BCG)

Brost M, Klötzke M, Kopp G, et al (2018) Development, Implementation (Pilot) and Evaluation of a Demand-Responsive Transport System. World Electric Vehicle Journal 9:. doi: https://doi.org/10.3390/wevj9010004
Burke A, Hengbing Z (2017) Fuel Economy Analysis of Medium/Heavy-duty Trucks - 2015-2050. EVS 30, Stuttgart, Germany
Coconuts Bangkok (2017) Thai ministry announces plan to convert gas-powered tuk-tuks to electric vehicles. https://coconuts.co/bangkok/news/thai-ministry-announces-plan-convert-gas-powered-tuk-tuks-electric-vehicles/
Cui H, Xiao G (2018) Fuel-efficiency technology trend assessment for LDVs in China: Hybrids and electrification. The International Council on Clean Transportation
Daimler (2018) Grünes Licht von ganz oben. https://blog.daimler.com/2015/10/05/highway-pilot-oder-gruenes-licht-von-ganz-oben-truck-actros-autonomes-fahren-a8/. Accessed 5 Sep 2018
DNV GL (2015) In Focus – The Future is Hybrid – a guide to use of batteries in shipping
Element Energy Ltd. (2018) Fuel cell buses in Europe: Latest developments and commercialisation pathway. CTE – US ZEB webinar. https://www.apta.com/resources/standards/quarterly-webinar-series/Documents/Lessons%20Learned_Zero-Emission_Bus_Expansion_Across_Europe.pdf. Accessed 4 Oct 2018
Eudy L, Post M (2017) Fuel Cell Buses in U.S. Transit Fleets: Current Status 2017. NREL, Technical Report NREL/TP-5400-70075
Ghoshal D (2017) EV, or not to be? India's electric vehicle revolution will begin with auto-rickshaws running on swappable batteries. https://qz.com/india/1001518/indias-electric-vehicle-revolution-will-begin-with-auto-rickshaws-running-on-swappable-batteries/. Accessed 1 Oct 2018
IEA (2017a) World Energy Outlook 2017. OECD Publishing, Paris/IEA, Paris
IEA (2017b) World Energy Balances 2017. IEA
IEA (2017c) The Future of Trucks. OECD/IEA 2017
Janssen R, Zwijnenberg H, Blankers I, Kruijff J de (2015) Truck Platooning. TNO
Lischke A (2017) Drive Trains, Fuels And Technologies For Heavy Duty Vehicles In 2030 And Beyond. In: 8th International Congress on Transportation Research (ICTR 2017), The Future of Transportation: A Vision for 2030. Thessaloniki, Greece
Madavan N (2016) A Nasa Perspective on Electric Propulsion Technologies for Commercial Aviation. 5 th UTIAS Workshop on Aviation and Climate Change, Toronto, Canada, 18.-20. May 2016
Moran G (2018) Battery swapping: The future of electric vehicle charging. In: Forbes India. http://www.forbesindia.com/blog/technology/battery-swapping-the-future-of-electric-vehicle-charging/. Accessed 8 Oct 2018
Pakusch C, Stevens G, Boden A, Bossauer P (2018) Unintended Effects of Autonomous Driving: A Study on Mobility Preferences in the Future. Sustainability 10:. doi: https://doi.org/10.3390/su10072404
Reddy KS, Aravindhan S, Mallick TK (2017) Techno-Economic Investigation of Solar Powered Electric Auto-Rickshaw for a Sustainable Transport System. Energies 10:. doi: https://doi.org/10.3390/en10060754
Rogers EM (2003) Diffusion of innovations, 5. The Free Press, New York
SAE (2018) SAE J3016 Taxonomy and Definitions for Terms Related to Driving Automation Systems for On-Road Motor Vehicles. SAE
Schäfer K (2016) Motoren: Mit Vollgas in die Sackgasse. https://www.heise.de/tr/artikel/Motoren-Mit-Vollgas-in-die-Sackgasse-3223514.html. Accessed 5 Sep 2018
Schmuch R, Wagner R, Hörpel G, et al (2018) Performance and cost of materials for lithium-based rechargeable automotive batteries. Nature Energy 3:267–278. doi: https://doi.org/10.1038/s41560-018-0107-2
Siemens AG (2017) eHighway – Solutions for electrified road freight transport. https://www.siemens.com/press/en/feature/2015/mobility/2015-06-ehighway.php. Accessed 5 Oct 2018

Sims R, Schaeffer R, Creutzig F, et al (2014) Transport. In: Climate Change 2014: Mitigation of Climate Change. Contribution of Working Group III to the Fifth Assessment Report of the Intergovernmental Panel on Climate Change [Edenhofer, O., R. Pichs-Madruga, Y. Sokona, E. Farahani, S. Kadner, K. Seyboth, A. Adler, I. Baum, S. Brunner, P. Eickemeier, B. Kriemann, J. Savolainen, S. Schlömer, C. von Stechow, T. Zwickel and J.C. Minx (eds.)]. Cambridge University Press, Cambridge, United Kingdom and New York, NY, USA.

Sisson P (2018) How a Chinese city turned all its 16,000 buses electric. https://www.curbed.com/2018/5/4/17320838/china-bus-shenzhen-electric-bus-transportation. Accessed 5 Oct 2018

Teske S, Sawyer S, Schäfer O, Pregger T, Simon S, Naegler T, Schmid S, Özdemir ED, Pagenkopf J, Kleiner F, Rutovitz J, Dominish E, Downes J, Ackermann T, Brown T, Boxer S, Baitelo R, Rodrigues LA (2015) Energy [R]evolution – A sustainable world energy outlook 2015. Greenpeace International

UIC (2017) 2016 Synopsis: Railway Statistics. ETF Publication

van Basshuysen R, Schäfer F (eds) (2015) Handbuch Verbrennungsmotor. Springer Fachmedien Wiesbaden, Wiesbaden

Vyas AD, Patel DM, Bertram KM (2013) Potential for Energy Efficiency Improvement Beyond the Light-Duty-Vehicle Sector. Transportation Energy Futures Series. Prepared for the U.S. Department of Energy by Argonne National Laboratory, Argonne, IL. DOE/GO-102013-3706. 82 pp.

Chapter 7
Renewable Energy Resource Assessment

Sven Teske, Kriti Nagrath, Tom Morris, and Kate Dooley

Abstract Literature overview of published global and regional renewable energy potential estimates. This section provides definitions for different types of RE potentials and introduces a new category, the economic renewable energy potential in space constrained environments. The potential for utility scale solar and onshore wind in square kilometre and maximum possible installed capacity (in GW) are provided for 75 different regions. The results set the upper limits for the deployment of solar- and wind technologies for the development of the 2.0 °C and 1.5 °C energy pathways.

There is a wide range of estimates of global and regional renewable energy potentials in the literature, and all conclude that the total global technical renewable energy potential is substantially higher than the current global energy demand (IPCC/SRREN 2011). Furthermore, the IPCC has concluded that the *global technical renewable energy potential will not limit continued renewable energy growth* (IPCC/SRREN 2011). However, the technical potential is also much higher than the sustainable potential, which is limited by factors such as land availability and other resource constraints.

This chapter provides an overview of various estimates of global renewable energy (RE) potential. It also provides definitions of different types of RE potential and presents mapping results for the spatial RE resource analysis (see Sect. 1.3 in Chap. 3)—[R]E-SPACE. The [R]E-SPACE results provide the upper limit for the deployment of all solar and wind technologies used in the 2.0 °C and 1.5 °C Scenarios.

S. Teske (✉) · K. Nagrath · T. Morris
Institute for Sustainable Futures, University of Technology Sydney, Sydney, NSW, Australia
e-mail: sven.teske@uts.edu.au; kriti.nagrath@uts.edu.au; tom.morris@uts.edu.au

K. Dooley
Australian-German Climate and Energy College, University of Melbourne, Parkville, Victoria, Australia
e-mail: kate.dooley@unimelb.edu.au

S. Teske (ed.), *Achieving the Paris Climate Agreement Goals*,
https://doi.org/10.1007/978-3-030-05843-2_7

7.1 Global Renewable Energy Potentials

The International Panel on Climate Change—Special Report on Renewable Energy Sources and Climate Change Mitigation (IPCC-SRRN 2011) defines renewable energy (RE) as:

> " (...) any form of energy from solar, geophysical or biological sources that is replenished by natural processes at a rate that equals or exceeds its rate of use. Renewable energy is obtained from the continuing or repetitive flows of energy occurring in the natural environment and includes low-carbon technologies such as solar energy, hydropower, wind, tide and waves and ocean thermal energy, as well as renewable fuels such as biomass."

Different types of renewable energy potentials have been identified by the German Advisory Council on Climate Change (WBGU)—World in Transition: Towards Sustainable Energy Systems, Chap. 3, page 44, published in 2003: (WBGU 2003) (Quote):

> "Theoretical potential: The theoretical potential identifies the physical upper limit of the energy available from a certain source. For solar energy, for example, this would be the total solar radiation falling on a particular surface. This potential does therefore not take account of any restrictions on utilization, nor is the efficiency of the conversion technologies considered.
>
> Conversion potential: The conversion potential is defined specifically for each technology and is derived from the theoretical potential and the annual efficiency of the respective conversion technology. The conversion potential is therefore not a strictly defined value, since the efficiency of a particular technology depends on technological progress.
>
> Technological potential: The technological potential is derived from the conversion potential, taking account of additional restrictions regarding the area that is realistically available for energy generation. [...] like the conversion potential, the technological potential of the different energy sources is therefore not a strictly defined value but depends on numerous boundary conditions and assumptions.
>
> Economic potential: This potential identifies the proportion of the technological potential that can be utilized economically, based on economic boundary conditions at a certain time [...].
>
> Sustainable potential: This potential of an energy source covers all aspects of sustainability, which usually requires careful consideration and evaluation of different ecological and socio-economic aspects [...]." (END Quote)

For the development of the 2.0 °C and 1.5 °C Scenarios, an additional renewable energy potential—the economic renewable energy potential in a space-constrained environment (Sect. 3.2 in Chap. 3)—has been analysed and utilized in this study.

The theoretical and technical potentials of renewable energy are significantly larger than the current global primary energy demand. The minimum technical potential of solar energy, shown in Table 7.1 (Turkenburg et al. 2012), could supply the global primary energy of 2015 112 times over.

However, the technical potential is only a first indication of the extent to which the resource is available. There are many other limitations, which must be considered. One of the main constraints on deploying renewable energy tech-

Table 7.1 Theoretical and technical renewable energy potentials versus utilization in 2015

Renewable energy resource	Theoretical potential (Annual energy flux) [EJ/year] IPCC 2011	Technical potential [EJ/year] Global energy assessment 2012, Chap. 11, p. 774	Utilization in 2015 [EJ/year] IEA-WEO 2017
Solar energy	3,900,000	62,000–280,000	1.3
Wind energy	6000	1250–2250	1.9
Bioenergy	1548	160–270	51.5
Geothermal energy	1400	810–1545	2.4
Hydropower	147	50–60	13.2
Ocean energy	7400	3240–10,500	0.0018
Total		76,000–294,500	(Total primary energy demand 2015) 555 EJ/year

nologies is the available space, especially in densely populated areas where there are competing claims on land use, such as agriculture and nature conservation, to name just two.

It is neither necessary nor desirable to exploit the entire technical potential. The implementation of renewable energy must respect sustainability criteria to achieve a sound future energy supply. Public acceptance is crucial, especially because the decentralised character of many renewable energy technologies will move systems closer to consumers. Without public acceptance, market expansion will be difficult or even impossible (Teske and Pregger 2015). The energy policy framework in a particular country or region will have a profound impact on the expansion of renewables, in terms of both the economic situation and the social acceptance of renewable energy projects.

7.1.1 *Bioenergy*

The discrepancy between the technical potential for bioenergy and the likely sustainable potential raises some issues that warrant further discussion. Recent analyses put the technical potential for primary bioenergy supply at 100–300 EJ/year. (GEA 2012; Smith et al. 2014). However, the dedicated use of land for bioenergy—whether through energy crops or the harvest of forest biomass—raises concerns over competition for land and the carbon neutrality of bioenergy (Field and Mach 2017; Searchinger et al. 2017). Research that focused on the trade-offs between bioenergy production, food security, and biodiversity found that less than 100 EJ / year. could be produced sustainably (Boysen et al. 2017; Heck et al. 2018), although such production levels would be dependent on strong global land governance systems (Creutzig 2017).

The carbon neutrality of bioenergy is based on the assumption that the CO_2 released when bioenergy is combusted is then recaptured when the biomass stock regrows (EASAC 2017). Most land is part of the terrestrial carbon sink or is used for food production, so that harvesting for bioenergy will either deplete the existing carbon stock or displace food production (Searchinger et al. 2015, 2017). The use of harvested forest products (e.g., wood pellets) for bioenergy is not carbon neutral in the majority of circumstances because an increased harvesting in forests leads to a permanent increase in the atmospheric CO_2 concentration (Sterman et al. 2018; Smyth et al. 2014; Ter Mikaelian et al. 2015). Leaving carbon stored in intact forests can represent a better climate mitigation strategy (DeCicco and Schlesinger 2018), because increased atmospheric concentrations of CO_2 from the burning of bioenergy may worsen the irreversible impacts of climate change before the forests can grow back to compensate the increase (EASAC 2017; Booth 2018; Schlesinger 2018).

Bioenergy sourced from wastes and residues rather than harvested from dedicated land can be considered carbon neutral, because of the 'carbon opportunity cost' per hectare of land (i.e., bioenergy production reduces the carbon-carrying capacity of land) (Searchinger et al. 2017). The supply of waste and residues as a bioenergy source is always inherently limited (Miyake et al. 2012). Although in some cases, burning residues can still release more emissions into the atmosphere in the mid-term (20–40 years) than allowing them to decay (Booth 2018), there is general agreement that specific and limited waste materials from the forest industry (for example, black liquor or sawdust) can be used with beneficial climate effects (EASAC 2017). The use of secondary residues (cascade utilization) may reduce the logistical costs and trade-offs associated with waste use (Smith et al. 2014).

7.2 Economic Renewable Energy Potential in Space-Constrained Environments

Land is a scarce resource. The use of land for nature conservation, agricultural production, residential areas, and industry, as well as for infrastructure, such as roads and all aspects of human settlements, limits the amount of land available land for utility-scale solar and wind projects. Furthermore, solar and wind generation require favourable climatic conditions, so not all available areas are suitable for renewable power generation. To assess the renewable energy potential of the available area, all ten world regions defined in Table 8 in Sect. 1 of Chap. 5 were analysed with the [R]E-SPACE methodology described in Sect. 3 of Chap. 3.

Given the issues involved in dedicated land-use for bioenergy outlined above, we assume that bioenergy is sourced primarily from cascading residue use and wastes, and do not analyse the availability of land for dedicated bioenergy crops.

This analysis quantifies the available land area (in square kilometres) in all regions and sub-regions with a defined set of constraints.

7.2.1 Constrains for Utility-Scale Solar and Wind Power Plants

The following land-use areas were excluded from the deployment of utility-scale solar photovoltaic (PV) and concentrated solar power plants:

- Residential and urban settlements;
- Infrastructure for transport (e.g., rail, roads);
- Industrial areas;
- Intensive agricultural production land;
- Nature conservation areas and national parks;
- Wetlands and swamps;
- Closed grasslands (a land-use type) (GLC 2000).

7.2.2 Mapping Solar and Wind Potential

After the spatial analysis, the remaining available land areas were analysed for their solar and wind resources. For concentrated solar power, a minimum solar radiation of 2000 kilowatt hours per square meter and year (kWh/m^2 year) is assumed as the minimum deployment criterion, and onshore wind potentials under an average annual wind speed of 5 m/s have been omitted.

In the next step, the existing electricity infrastructure of power lines and power plants was mapped for all regions with WRI (2018) data. Figure 7.1 provides an example of the electricity infrastructure in Africa. These maps provide important insights into the current situation in the power sector, especially the availability of transmission grids. This is of particular interest for developing countries because it allows a comparison of the available land areas that have favourable solar and wind conditions with the infrastructure available to transport electricity to the demand centres. This assessment is less important for OECD regions because the energy infrastructure is usually already fairly evenly distributed across the country—except in some parts of Canada, the United States, and Australia. For some countries, coverage is not 100% complete due to a lack of public data sources. This is particularly true for renewable energy generation assets such as solar, wind, biomass, geothermal energy, and hydropower resources.

Figure 7.2 shows the solar potential for utility-scale solar power plants—both solar PV and concentrated solar power—in Africa. The scale from light yellow to dark red shows the solar radiation intensity: the darker the area, the better the solar resource. The green lines show existing transmission lines. All areas that are not yellow or red are unsuitable for utility-scale solar because there is conflicting land use and/or there are no suitable solar resources. Africa provides a very extreme example of very good solar resources far from existing infrastructure. While roof-top

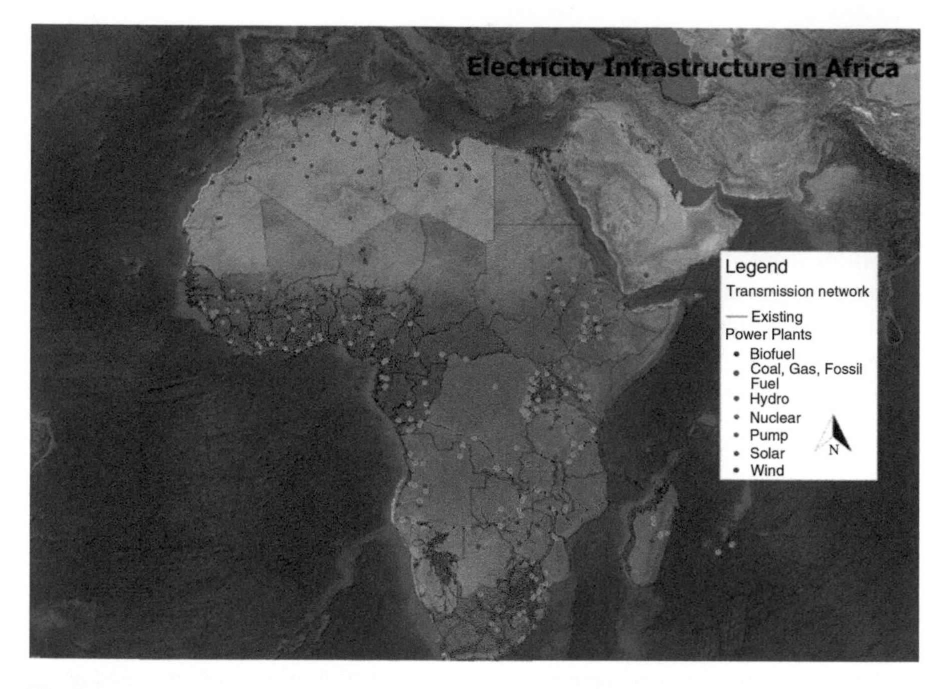

Fig. 7.1 Electricity infrastructure in Africa—power plants (over 1 MW) and high-voltage transmission lines

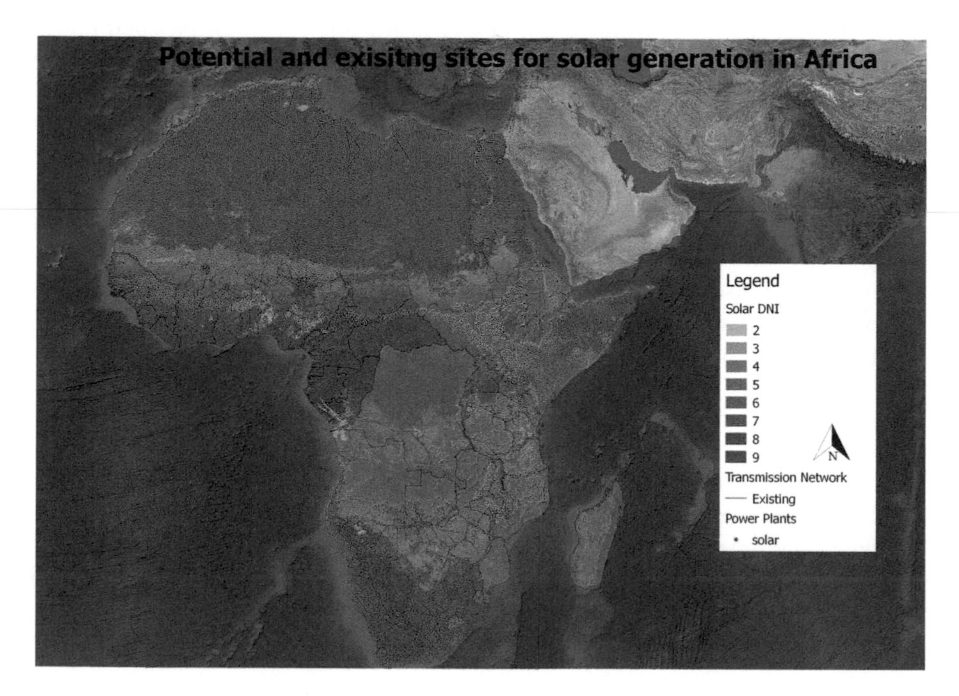

Fig. 7.2 Solar potential in Africa

solar PV can be deployed virtually anywhere and only needs roof space on any sort of building, bulk power supply via solar—to produce synthetic and hydrogen fuels—requires a certain minimum of utility-scale solar applications. The vast solar potential in the north of Africa—as well as in the Middle East—has been earmarked for the production of synthetic and hydrogen fuels and for the export of renewable electricity (via sub-sea cable) to Europe in the long-term energy scenarios in the 2.0 °C and 1.5 °C Scenarios.

Europe, in contrast, is densely populated and has fewer favourable utility-scale solar sites because of both its lower solar radiation and conflicting land-use patterns. Figure 7.3 shows Europe's potential for utility-scale solar power plants. Only the yellow and red dots across Europe, most visible in the south of Spain, south of the Alps, south-west Italy, and the Asian part of Turkey, mark regions suitable for utility-scale solar, whereas roof-top solar can be deployed economically across Europe, including Scandinavia.

However, Africa and Europe are in a good position, from a technical point of view, to form an economic partnership for solar energy exchange.

The situation for onshore wind power differs from that for solar energy. The best potential is in areas that are more than 30° north and south of the equator, whereas the actual equatorial zone is less suitable for wind installation. North America has significant wind resources and the resource is still largely untapped, even though there is already a mature wind industry in Canada and the USA. Figure 7.4 shows the existing and potential wind power sites. While significant wind power

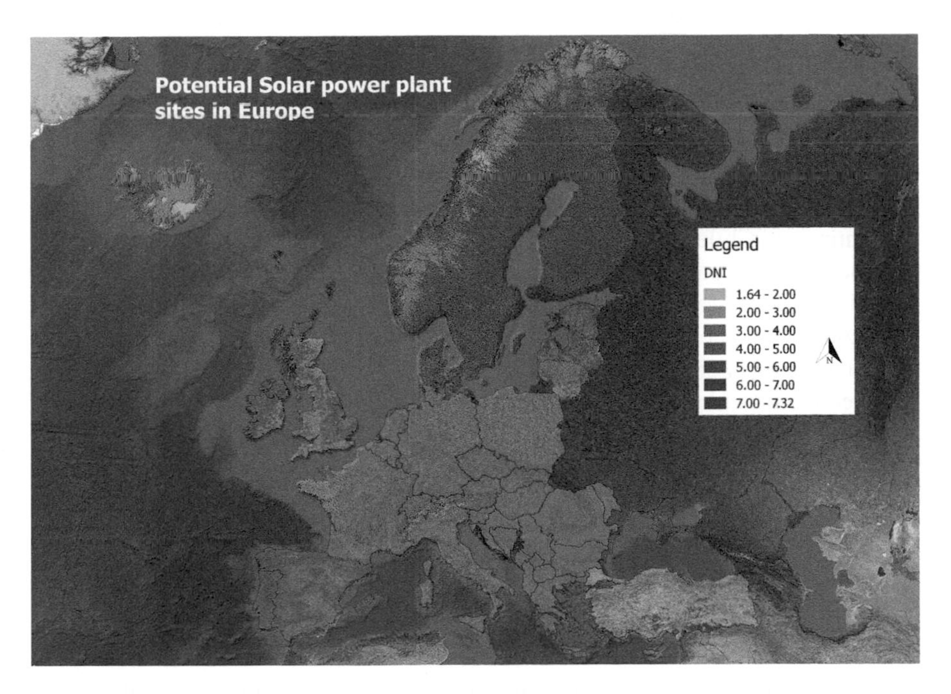

Fig. 7.3 Europe's potential for utility-scale solar power plants

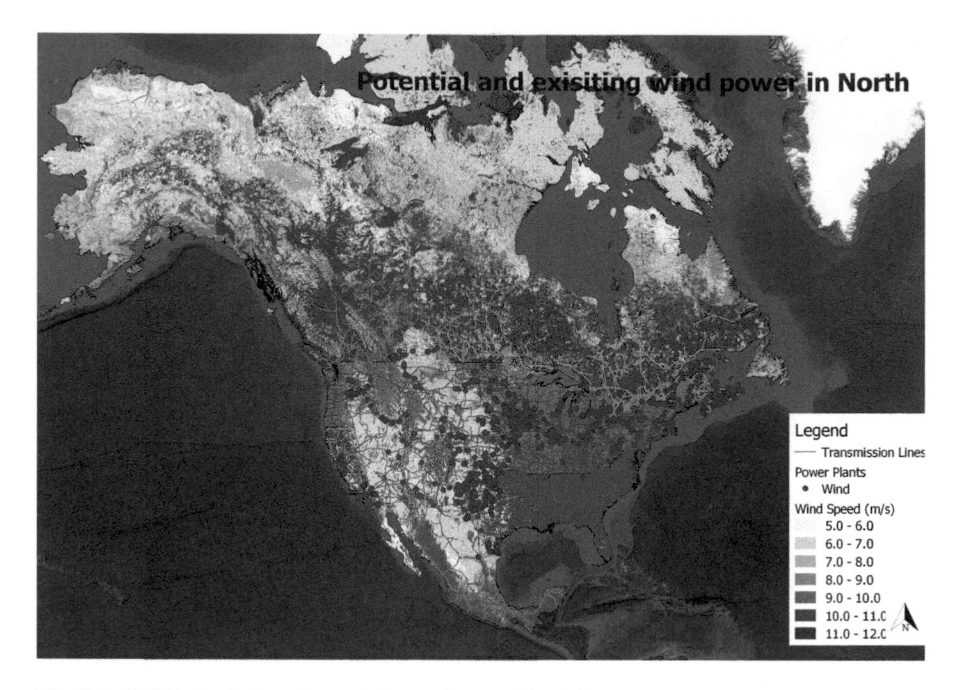

Fig. 7.4 OECD North America: existing and potential wind power sites

installations are already in operation, mainly in the USA, there are still very large untapped resources across the entire north American continent, in Canada and the mid-west of the USA.

Unlike the situation in the USA, wind power in Latin America is still in its initial stages and the industry, which has great potential, is still in its infancy. Figure 7.5 shows the existing wind farm locations—marked with blue dots—and the potential wind farm sites, especially in coastal regions and the entire southern parts of Argentina and Chile.

The available solar and wind potentials are distributed differently across all world regions. Whereas some regions have significantly more resources than others, all regions have enough potential to supply their demand with local solar and wind resources—together with other renewable energy resources, such as hydro-, bio-, and geothermal energies.

Table 7.2 provides an overview of the key results of the [R]E-SPACE analysis. The available areas (in square kilometres) are based on the space-constrained assumptions (see Sect. 2.1 in Chap. 7). The installed capacities are calculated based on the following space requirements (Table 7.2):

- Solar photovoltaic: 1 MW = 0.04 km^2
- Concentrated solar power: 1 MW = 0.04 km^2
- Onshore wind: 1 MW = 0.2 km^2

Note: Mapping Eurasia was not possible because the data files were incomplete.

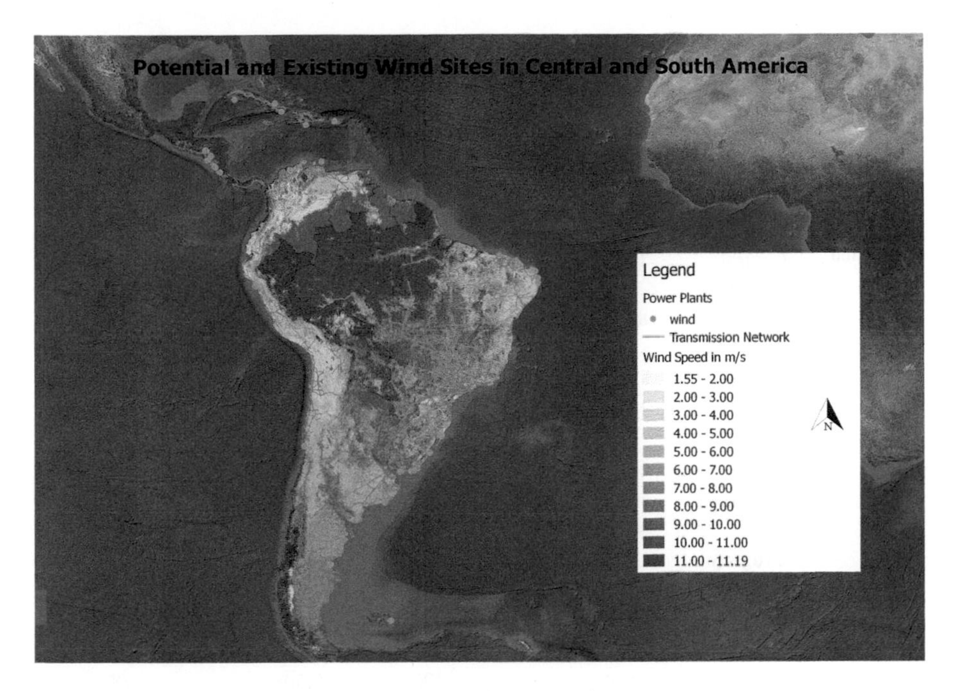

Fig. 7.5 Latin America: potential and existing wind power sites

Table 7.2 [R]E-SPACE: key results part 1

		Solar		Onshore wind	
		Potential availability for utility-scale installations	Space potential	Potential availability for utility-scale installations	Space potential
Region	Subregion	[km^2]	[GW]	[km^2]	[GW]
OECD North America	Canada East	2,742,668	68,567	2,530,232	12,651
	Canada West	2,242,715	56,068	2,180,271	10,901
	Mexico	3,365,974	84,149	3,341,940	16,710
	USA – South East	269,650	6741	254,976	1275
	USA – North East	1,043,033	26,076	1,043,026	5215
	USA – South West	1,847,162	46,179	1,840,980	9205
	USA – North West	431,277	10,782	427,709	2139
	USA – Alaska	1,152,288	28,807	1,091,698	5458

(continued)

Table 7.2 (continued)

Region	Subregion	Solar		Onshore wind	
		Potential availability for utility-scale installations	Space potential	Potential availability for utility-scale installations	Space potential
		[km²]	[GW]	[km²]	[GW]
Latin America	Caribbean	34,238	856	34,238	171
	Central America	17,529	438	17,603	88
	North Latin America	869,811	21,745	869,811	4349
	Brazil	1,623,625	40,591	1,623,625	8118
	Central South America	1,023,848	25,596	1,024,340	5122
	Chile	693,990	17,350	693,990	3470
	Argentina	1,651,168	41,279	1,651,168	8256
	CSA – Uruguay	32,360	809	32,360	162
Europe	EU – Central	146,797	3670	146,797	734
	EU – UK and Islands	22,406	560	22,406	112
	EU – Iberian Peninsula	15,608	390	15,608	78
	EU – Balkans + Greece	4825	121	4825	24
	EU – Baltic	32,090	802	32,090	160
	EU – Nordic	218,496	5462	218,496	1092
	Turkey	134,354	3359	134,354	672
Middle East	East – Middle East	165,302	4133	5738	29
	North – Middle East	91,970	2299	7123	36
	Iraq	119,967	2999	9104	46
	Iran	586,595	14,665	57,965	290
	United Arab Emirates	530	13	530	3
	Israel	386	10	217	1
	Saudi Arabia	13,284	332	13,284	66
Africa	North – Africa	9,726,388	243,160	9,784,694	48,923
	East – Africa	6,378,561	159,464	6,980,497	34,902
	West – Africa	8,336,960	208,424	8,669,628	43,348
	Central – Africa	7,229,129	180,728	7,509,351	37,547
	Southern – Africa	3,269,644	81,741	3,547,591	17,738
	Rep. South Africa	1,626,528	40,663	1,650,471	8252

Table 7.3 [R]E-SPACE: key results part 2

		Solar		Onshore wind	
		Potential availability for utility-scale installations	Space potential	Potential availability for utility-scale installations	Space potential
Region	Subregion	[km²]	[GW]	[km²]	[GW]
Non-OECD Asia	Asia-West (Himalaya)	1,315,395	32,885	801,044	4005
	South and South East Asia	9062	227	8184	41
	Asia North West	184,503	4613	43,710	219
	Asia Central North	138,861	3472	81,228	406
	Philippines	2634	66	941	5
	Indonesia	106,581	2665	12,162	61
	Pacific Island States	5510	138	673	3
India	North – India	229,314	5733	163,118	816
	East – India	32,511	813	5195	26
	West – West	224,355	5609	121,441	607
	South – Incl. Islands	129,346	3234	103,177	516
	Northeast – India	77,379	1934	1821	9
China	East – China	47,621	1191	39,648	198
	North – China	425,350	10,634	825,272	4126
	Northeast – China	193,006	4825	192,110	961
	Northwest – China	5,642,854	141,071	1,603,909	8020
	Central – China	256,272	6407	211,229	1056
	South – China	500,211	12,505	317,046	1585
	Taiwan	5862	147	2791	14
	Tibet	5460	137	377,610	1888
OECD Pacific	North Japan	8697	217	8213	41
	South Japan	3567	89	3036	15
	North Korea	10,724	268	9854	49
	South Korea	2411	60	1892	9
	North New Zealand	22,699	567	22,163	111
	South New Zealand	25,106	628	46,266	231
	Australia – South and East (NEM)	2,080,117	52,003	2,035,523	10,178
	Australia – West and North (NT)	2,813,791	70,345	2,762,499	13,812

References

Booth, M.S., 2018. Not carbon neutral: Assessing the net emissions impact of residues burned for bioenergy. Environmental Research Letters 13, 035001. https://doi.org/10.1088/1748-9326/aaac88

Boysen, L.R., Lucht, W., Gerten, D., Heck, V., Lenton, T.M., Schellnhuber, H.J., 2017. The limits to global-warming mitigation by terrestrial carbon removal: The limits of terrestrial carbon removal. Earth's Future 5, 463–474. https://doi.org/10.1002/2016EF000469

Creutzig, F., 2017. Govern land as a global commons. Nature 546, 28–29. https://doi.org/10.1038/546028a

DeCicco, J.M., Schlesinger, W.H., 2018. Reconsidering bioenergy given the urgency of climate protection. Proceedings of the National Academy of Sciences 115, 9642–9645. https://doi.org/10.1073/pnas.1814120115

EASAC. European Academies Science Advisory Council, Deutsche Akademie der Naturforscher Leopoldina (Eds.), 2017. Multi-functionality and sustainability in the European Union's forests, EASAC policy report. EASAC Secretariat, Deutsche Akademie der Naturforscher Leopoldina, Halle (Saale).

Field, C.B., Mach, K.J., 2017. Rightsizing carbon dioxide removal. Science 356, 706–707. https://doi.org/10.1126/science.aam9726

GEA (2012): Global Energy Assessment - Toward a Sustainable Future, Cambridge University Press, Cambridge, UK and New York, NY, USA and the International Institute for Applied Systems Analysis, Laxenburg, Austria. http://www.globalenergyassessment.org

GLC 2000, Joint Research Center, The European Commission's science and knowledge service, Global Land Cover 2000, webbased database, viewed October 2018, http://forobs.jrc.ec.europa.eu/products/glc2000/products.php

Heck, V., Gerten, D., Lucht, W., Popp, A., 2018. Biomass-based negative emissions difficult to reconcile with planetary boundaries. Nature Climate Change 8, 151–155. https://doi.org/10.1038/s41558-017-0064-y

Miyake, S., et al., 2012. Land-use and environmental pressures resulting from current and future bioenergy crop expansion: A review. *Journal of Rural Studies*, 28(4), pp. 650–658.

IPCC, 2011, IPCC: Summary for Policymakers. In: IPCC Special Report on Renewable Energy Sources and Climate Change Mitigation; [O. Edenhofer, R. Pichs-Madruga, Y. Sokona, K. Seyboth, P. Matschoss, S. Kadner, T. Zwickel, P. Eickemeier, G. Hansen,S. Schlömer, C. von Stechow (eds)], Cambridge University Press, Cambridge, United Kingdom and New York, NY, USA

IPCC-SRRN 2011, IPCC; ANNEX, Verbruggen, A., W. Moomaw, J. Nyboer, 2011: Annex I: Glossary, Acronyms, Chemical Symbols and Prefixes. In IPCC Special Report on Renewable Energy Sources and Climate Change Mitigation [O. Edenhofer, R. Pichs- Madruga, Y. Sokona, K. Seyboth, P. Matschoss, S. Kadner, T. Zwickel, P. Eickemeier, G. Hansen, S. Schlömer,C. von Stechow (eds)], Cambridge University Press, Cambridge, United Kingdom and New York, NY, USA.

Searchinger, T., Edwards, R., Mulligan, D., Heimlich, R., Plevin, R., 2015. Do biofuel policies seek to cut emissions by cutting food? Science 347, 1420–1422. https://doi.org/10.1126/science.1261221

Searchinger, T.D., Beringer, T., Strong, A., 2017. Does the world have low-carbon bioenergy potential from the dedicated use of land? Energy Policy 110, 434–446. https://doi.org/10.1016/j.enpol.2017.08.016

Schlesinger, W.H., 2018. Are wood pellets a green fuel? Science 359, 1328–1329. https://doi.org/10.1126/science.aat2305

Smith, P., Bustamante, M., Ahammad, H., et al., 2014. Chapter 11: Agriculture, Forestry and Other Land Use (AFOLU), in: Climate Change 2014 Mitigation of Climate Change: Working Group III Contribution to the Fifth Assessment Report of the Intergovernmental

Panel on Climate Change. Cambridge University Press, Cambridge. https://doi.org/10.1017/CBO9781107415416

Smyth, C.E., Stinson, G., Neilson, E., Lemprière, T.C., Hafer, M., Rampley, G.J., Kurz, W.A., 2014. Quantifying the biophysical climate change mitigation potential of Canada's forest sector. Biogeosciences 11, 3515–3529. https://doi.org/10.5194/bg-11-3515-2014

SRREN 2011, IPCC, 2011: IPCC Special Report on Renewable Energy Sources and Climate Change Mitigation, Chapter 1, Prepared by Working Group III of the Intergovernmental Panel on Climate Change [O. Edenhofer, R. Pichs-Madruga, Y. Sokona, K. Seyboth, P. Matschoss, S. Kadner, T. Zwickel, P. Eickemeier, G. Hansen, S. Schlömer, C. von Stechow (eds)]. Cambridge University Press, Cambridge, United Kingdom and New York, NY, USA, 1075 pp.

Sterman, J.D., Siegel, L., Rooney-Varga, J.N., 2018. Does replacing coal with wood lower CO_2 emissions? Dynamic lifecycle analysis of wood bioenergy. Environmental Research Letters 13, 015007. https://doi.org/10.1088/1748-9326/aaa512

Ter-Mikaelian, M.T., Colombo, S.J. & Chen, J., 2015. The Burning Question: Does Forest Bioenergy Reduce Carbon Emissions? A Review of Common Misconceptions about Forest Carbon Accounting. *Journal of Forestry*, 113(1), pp. 57–68.

Turkenburg, Arent et al 2012, Turkenburg, W. C., D. J. Arent, R. Bertani, A. Faaij, M. Hand, W. Krewitt, E. D. Larson, J. Lund, M. Mehos, T. Merrigan, C. Mitchell, J. R. Moreira, W. Sinke, V. Sonntag-O'Brien, B. Thresher, W. van Sark, E. Usher and E. Usher, 2012: Chapter 11 - Renewable Energy. In Global Energy Assessment - Toward a Sustainable Future, Cambridge University Press, Cambridge, UK and New York, NY, USA and the International Institute for Applied Systems Analysis, Laxenburg, Austria, pp. 761–900.

Teske, Pregger 2015, Teske, S, Pregger, T., Naegler, T., Simon, S., *Energy [R]evolution - A sustainable World Energy Outlook 2015*, Greenpeace International with the German Aerospace Centre (DLR), Institute of Engineering Thermodynamics, System Analysis and Technology Assessment, Stuttgart, Germany, https://www.scribd.com/document/333565532/Energy-Revolution-2015-Full

WBGU 2003, *World in Transition Towards Sustainable Energy Systems;* German Advisory Council on Global Change (WBGU), H. Graßl, J. Kokott, M. Kulessa, J. Luther, F. Nuscheler, R. Sauerborn, H.-J. Schellnhuber, R. Schubert, E.-D. Schulze

WRI 2018, World Resource Institute (WRI), Global Power Plant Database, web-based database, data download June 2018, http://datasets.wri.org/dataset/globalpowerplantdatabase

Chapter 8
Energy Scenario Results

Sven Teske, Thomas Pregger, Tobias Naegler, Sonja Simon, Johannes Pagenkopf, Bent van den Adel, and Özcan Deniz

Abstract Results for the 5.0 °C, 2.0 °C and 1.5 °C scenarios for ten world regions in regard to energy-related carbon-dioxide emissions, final-, primary-, transport- and heating demand and the deployment of various supply technologies to meet the demand. Furthermore, the electricity demand and generation scenarios are provided. The key results of a power sector analysis which simulates further electricity supply with high shares of solar- and wind power in one hour steps is provided. The ten world regions are divided into eight sub-regions and the expected development of loads, capacity-factors for various power plant types and storage demands are provided. This chapter contains more than 100 figures and tables.

This chapter provides a condensed description of the energy scenario results on a global scale, for each of the ten world regions. The descriptions include a presentation of the calculated energy demands for all sectors (power and heat/fuels for the following sectors: industry, residential and other, and transport) and of supply strategies for all the technologies considered, from 2015 to 2050. The results of the model-based analyses of hourly supply curves and required storage capacities are also discussed based on key indicators. Graphs, tables, and descriptions are provided in a standardized way to facilitate comparisons between scenarios and between regions.

S. Teske (✉)
Institute for Sustainable Futures, University of Technology Sydney, Sydney, NSW, Australia
e-mail: sven.teske@uts.edu.au

T. Pregger · T. Naegler · S. Simon
Department of Energy Systems Analysis, German Aerospace Center (DLR), Institute for Engineering Thermodynamics (TT), Pfaffenwaldring, Germany
e-mail: thomas.pregger@dlr.de; tobias.naegler@dlr.de; sonja.simon@dlr.de

J. Pagenkopf · B. van den Adel · Ö. Deniz
Department of Vehicle Systems and Technology Assessment, German Aerospace Center (DLR), Institute of Vehicle Concepts (FK), Pfaffenwaldring, Germany
e-mail: johannes.pagenkopf@dlr.de; Bent.vandenAdel@dlr.de; oezcan.deniz@dlr.de

S. Teske (ed.), *Achieving the Paris Climate Agreement Goals*,
https://doi.org/10.1007/978-3-030-05843-2_8

The following global summary of the regional results is presented in the same structure as that used for individual regions. Consistent with the regional results, these tables do not include the demand and supply details for the bunker fuels used in international aviation and navigation. Section 8.2 outlines a global demand and supply scenario for renewable bunker fuels in the long term, including estimates of additional CO_2 emissions from fossil bunker fuels between 2015 and 2050.

8.1 Global: Long-Term Energy Pathways

8.1.1 Global: Projection of Overall Energy Intensity

Combining the assumptions for the power, heat, and fuel demands for all sectors produced the overall final energy intensity (per $ GDP) development shown in Fig. 8.1. Compared with the 5.0 °C case based on the Current Policies Scenario of the IEA, the alternative scenarios follow more stringent efficiency levels. The 1.5 °C Scenario represents an even faster implementation of efficiency measures than the 2.0 °C Scenario. The 1.5 °C Scenario involves the decelerated growth of energy services in all regions, to avoid any further strong increase in fossil fuel use after 2020. The global average intensity drops from 2.4 MJ/$GDP in 2015 to 1.25 MJ/$GDP in 2050 in the 5.0 °C case compared with 0.65 MJ/$GDP in the 2.0 °C Scenario and 0.59 MJ/$GDP in the 1.5 °C Scenario. The average final energy consumption decreases from 46.3 GJ/capita in 2015 to 28.4 GJ/capita in 2050 in the 2.0 °C Scenario and to below 26 GJ/capita in the 1.5 °C Scenario. In the 5.0 °C case, it increases to 55 GJ/capita.

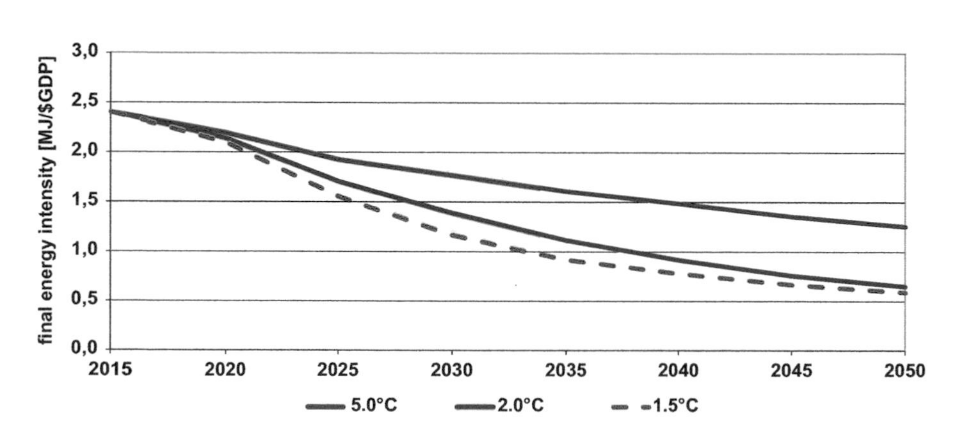

Fig. 8.1 Global: projection of final energy (per $ GDP) intensity by scenario

8.1.2 Global: Final Energy Demand by Sector (Excluding Bunkers)

Combining the assumptions for population growth, GDP growth, and energy intensity produced the future development pathways for the global final energy demand shown in Fig. 8.2 for the 5.0 °C, 2.0 °C, and 1.5 °C Scenarios. In the 5.0 °C Scenario, the total final energy demand will increase by 57% from 342 EJ/year in 2015 to 537 EJ/year in 2050. In the 2.0 °C Scenario, the final energy demand will decrease by 19% compared with the current consumption and reach 278 EJ/year by 2050, whereas the final energy demand in the 1.5 °C Scenario will reach 253 EJ, 26% below the 2015 demand. In the 1.5 °C Scenario, the final energy demand in 2050 is 9% lower than in the 2.0 °C Scenario. The electricity demand for 'classical' electrical devices (without power-to-heat or e-mobility) will increase from around 15,900 TWh/year in 2015 to 23,800 TWh/year (2.0 °C) or to 23,300 TWh/year (1.5 °C) by 2050. Compared with the 5.0 °C case (37,000 TWh/year in 2050), the efficiency measures in the 2.0 °C and 1.5 °C Scenarios will save 13,200 TWh/year and 13,700 TWh/year, respectively, by 2050.

Electrification will lead to a significant increase in the electricity demand by 2050. In the 2.0 °C Scenario, the electricity demand for heating will be about

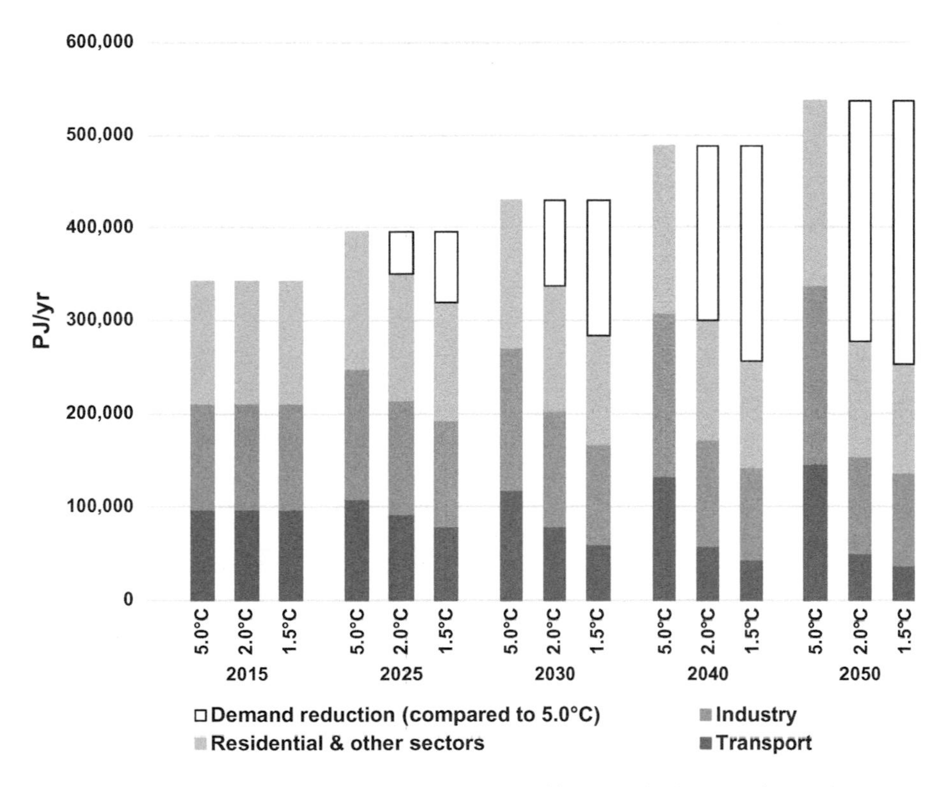

Fig. 8.2 Global: projection of total final energy demand by sector in the scenarios (without non-energy use or heat from combined heat and power [CHP] autoproducers)

12,600 TWh/year due to electric heaters and heat pumps, and in the transport sector there will be an increase of about 23,400 TWh/year due to increased electric mobility. The generation of hydrogen (for transport and high-temperature process heat) and the manufacture of synthetic fuels (mainly for transport) will add an additional power demand of 18,800 TWh/year The gross power demand will thus rise from 24,300 TWh/year in 2015 to 65,900 TWh/year in 2050 in the 2.0 °C Scenario, 34% higher than in the 5.0 °C case. In the 1.5 °C Scenario, the gross electricity demand will increase to a maximum of 65,300 TWh/year in 2050.

The efficiency gains in the heating sector could be even larger than in the electricity sector. In the 2.0 °C and 1.5 °C Scenarios, a final energy consumption equivalent to about 85.7 EJ/year and 95.4 EJ/year, respectively, is avoided through efficiency gains by 2050 compared with the 5.0 °C Scenario (Figs. 8.3, 8.4, 8.5, and 8.6).

8.1.3 Global: Electricity Generation

The development of the power system is characterized by a dynamically growing renewable energy market and an increasing proportion of total power coming from renewable sources. In the 2.0 °C Scenario, 100% of the electricity produced globally will come from renewable energy sources by 2050. 'New' renewables—mainly wind, solar, and geothermal energy—will contribute 83% of the total electricity generation. Renewable electricity's share of the total production will be 62% by 2030 and 88% by 2040. The installed capacity of renewables will reach about 9500

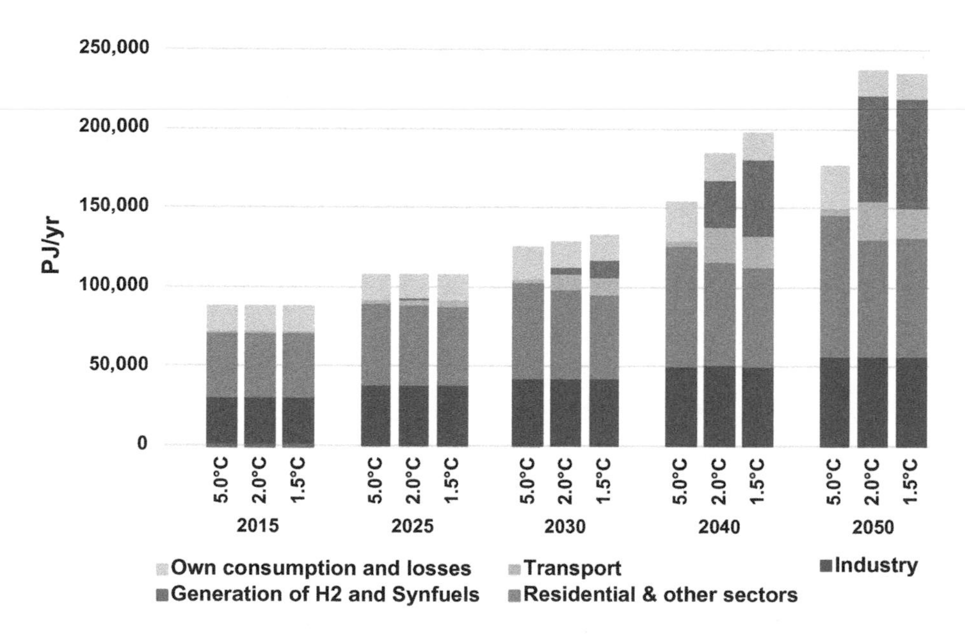

Fig. 8.3 Global: development of gross electricity demand by sector in the scenarios

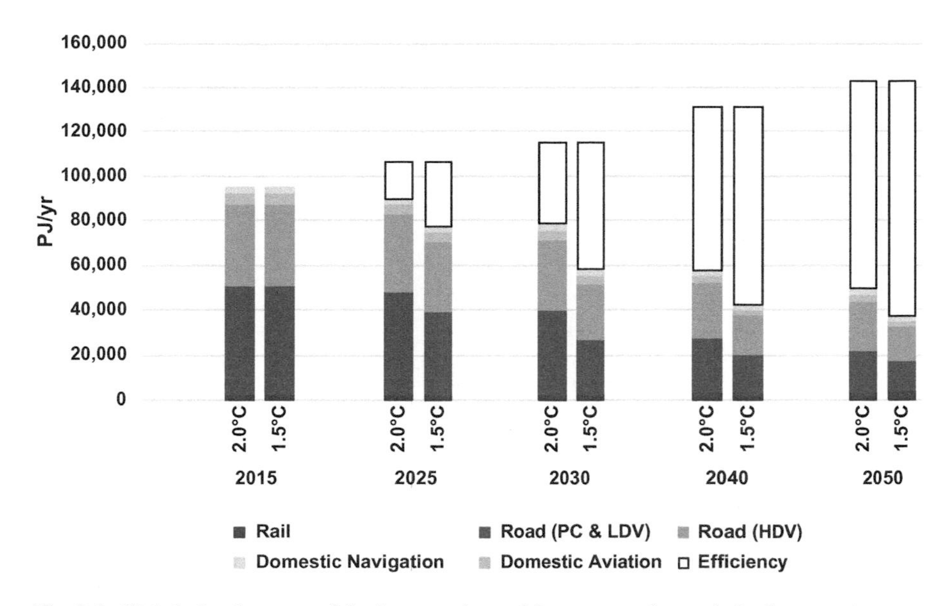

Fig. 8.4 Global: development of final energy demand for transport by mode in the scenarios

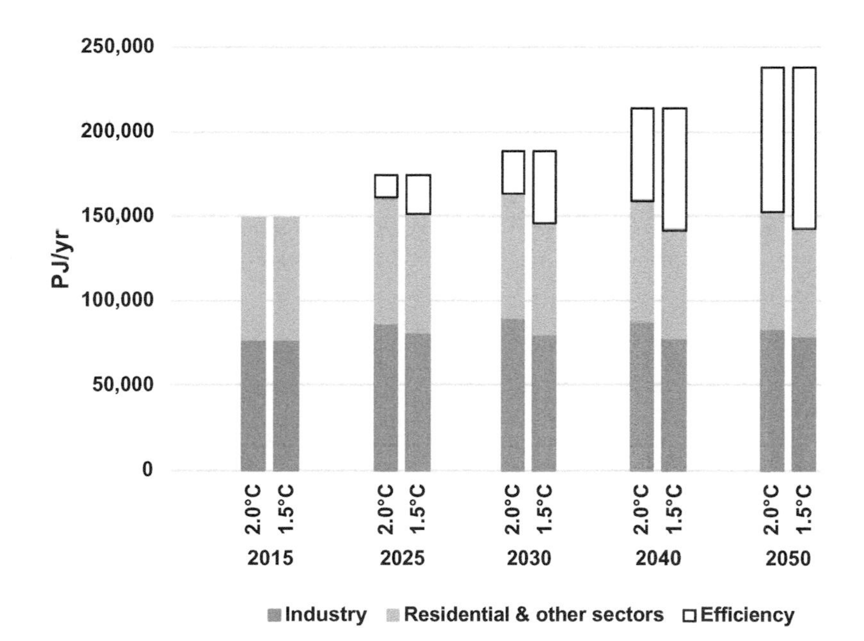

Fig. 8.5 Global: development of heat demand by sector in the scenarios

GW by 2030 and 25,600 GW by 2050. The share of renewable electricity generation in 2030 in the 1.5 °C Scenario is assumed to be 73%. The 1.5 °C Scenario indicates a generation capacity from renewable energy of about 25,700 GW in 2050.

Table 8.1 shows the development of different renewable technologies in the world over time. Figure 8.7 provides an overview of the global power-generation

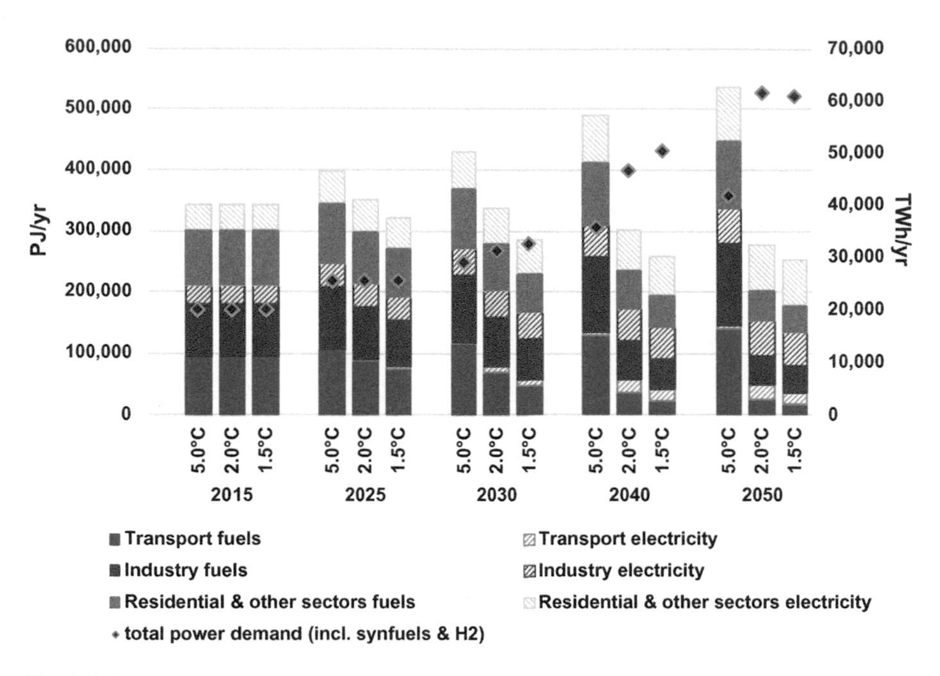

Fig. 8.6 Global: development of the final energy demand by sector in the scenarios

Table 8.1 Global: development of renewable electricity-generation capacity in the scenarios

in GW	(°C)	2015	2025	2030	2040	2050
Hydro	5.0	1202	1420	1558	1757	1951
	2.0	1202	1386	1416	1473	1525
	1.5	1202	1385	1415	1471	1523
Biomass	5.0	112	165	195	235	290
	2.0	112	301	436	617	770
	1.5	112	350	498	656	798
Wind	5.0	413	880	1069	1395	1790
	2.0	413	1582	2901	5809	7851
	1.5	413	1912	3673	6645	7753
Geothermal	5.0	14	20	26	41	62
	2.0	14	49	125	348	557
	1.5	14	53	147	356	525
PV	5.0	225	785	1031	1422	2017
	2.0	225	2194	4158	8343	12,306
	1.5	225	2829	5133	10,017	12,684
CSP	5.0	4	13	20	39	64
	2.0	4	69	361	1346	2062
	1.5	4	92	474	1540	1990
Ocean	5.0	0	1	3	9	22
	2.0	0	22	82	307	512
	1.5	0	22	80	295	450
Total	5.0	1971	3285	3902	4899	6195
	2.0	1971	5604	9478	18,243	25,584
	1.5	1971	6644	11,420	20,980	25,723

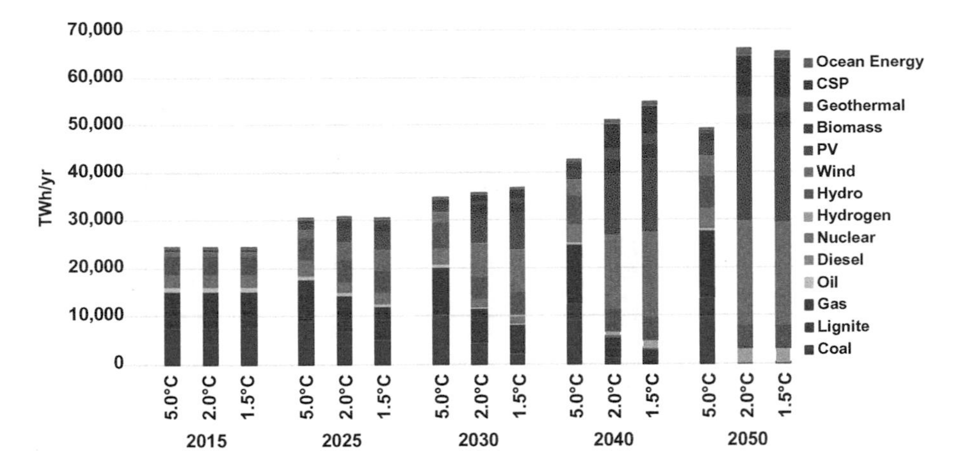

Fig. 8.7 Global: development of electricity-generation structure in the scenarios

structure. From 2020 onwards, the continuing growth of wind and photovoltaic (PV), up to 7850 GW and 12,300 GW, respectively, will be complemented by up to 2060 GW of solar thermal generation, and limited biomass, geothermal, and ocean energy in the 2.0 °C Scenario. Both the 2.0 °C Scenario and 1.5 °C Scenario will lead to a high proportion of variable power generation (PV, wind, and ocean) of 38% and 46%, respectively, by 2030 and 64% and 65%, respectively, by 2050.

8.1.4 Global: Future Costs of Electricity Generation

Figure 8.8 shows the development of the electricity-generation and supply costs over time, including the CO_2 emission costs, in all scenarios. The calculated average electricity generation costs in 2015 (referring to full costs) were around 6 ct/kWh. In the 5.0 °C case, the generation costs will increase until 2050, when they reach 10.6 ct/kWh. The generation costs will also increase in the 2.0 °C and 1.5 °C Scenarios until 2030, when they will reach 9 ct/kWh, and then drop to 7 ct/kWh by 2050. In both alternative scenarios, the generation costs will be around 3.5 ct/kWh lower than in the 5.0 °C Scenario by 2050. Note that these estimates of generation costs do not take into account integration costs such as power grid expansion, storage, or other load-balancing measures.

In the 5.0 °C case, the growth in demand and increasing fossil fuel prices will cause the total electricity supply costs to increase from today's $1560 billion/year to around $5500 billion/year in 2050. In both alternative scenarios, the total supply costs will be $5050 billion/year in 2050. Therefore, the long-term costs for electricity supply in both alternative scenarios are about 8% lower than in the 5.0 °C Scenario as a result of the estimated generation costs and the electrification of heating and mobility.

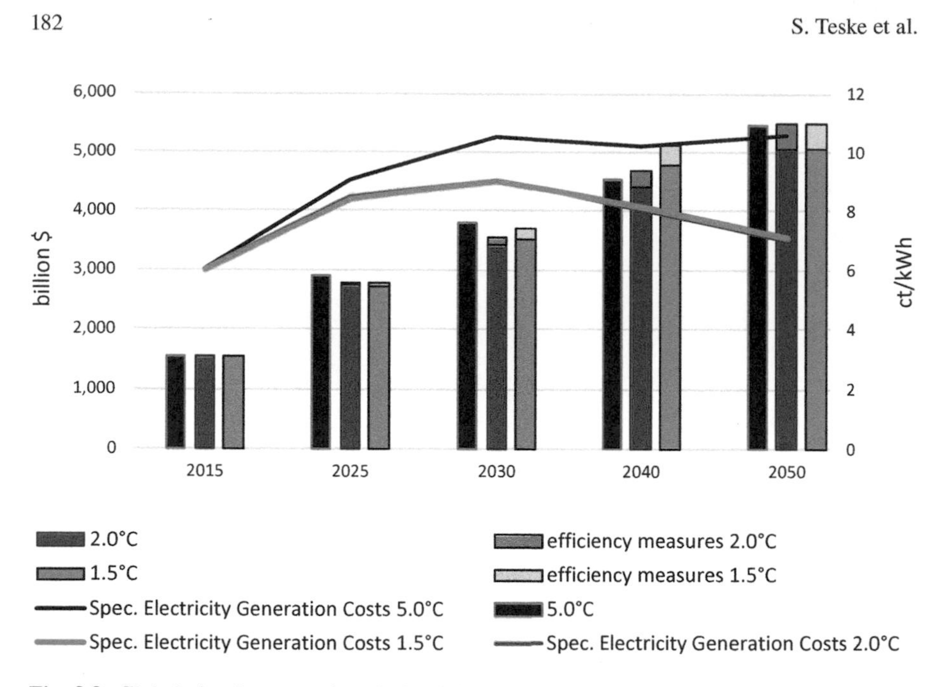

Fig. 8.8 Global: development of total electricity supply costs and specific electricity generation costs in the scenarios

Compared with these results, the generation costs (without including CO_2 emission costs) will increase in the 5.0 °C case to only 7.9 ct/kWh. The generation costs will increase in the 2.0 °C Scenario until 2030 to 7.7 ct/kWh and to a maximum of 8.1 ct/kWh in the 1.5 °C Scenario. Between 2030 and 2050, the costs will decrease to 7 ct/kWh. In the 2.0 °C Scenario, the generation costs will be, at maximum, 0.1 ct/kWh higher than in the 5.0 °C Scenario and this will occur in 2040. In the 1.5 °C Scenario, the generation costs will be, at maximum, 0.5 ct/kWh higher than in the 5.0 °C Scenario, again by around 2040. In 2050, the generation costs in the alternative scenarios will be 0.8–0.9 ct/kWh lower than in the 5.0 °C case. If the CO_2 costs are not considered, the total electricity supply costs in the 5.0 °C case will rise to about $4150 billion/year in 2050.

8.1.5 Global: Future Investments in the Power Sector

In the 2.0 °C Scenario, around $49,000 billion in investment will be required for power generation between 2015 and 2050—including for additional power plants to produce hydrogen and synthetic fuels and for the plant replacement costs at the end of their economic lifetimes. This value will be equivalent to approximately $1360 billion per year on average, and is $28,600 billion more than in the 5.0 °C case ($20,400 billion). An investment of around $51,000 billion for power generation will be required between 2015 and 2050 in the 1.5 °C Scenario. On average, this will be an investment of $1420 billion per year. In the 5.0 °C Scenario, the

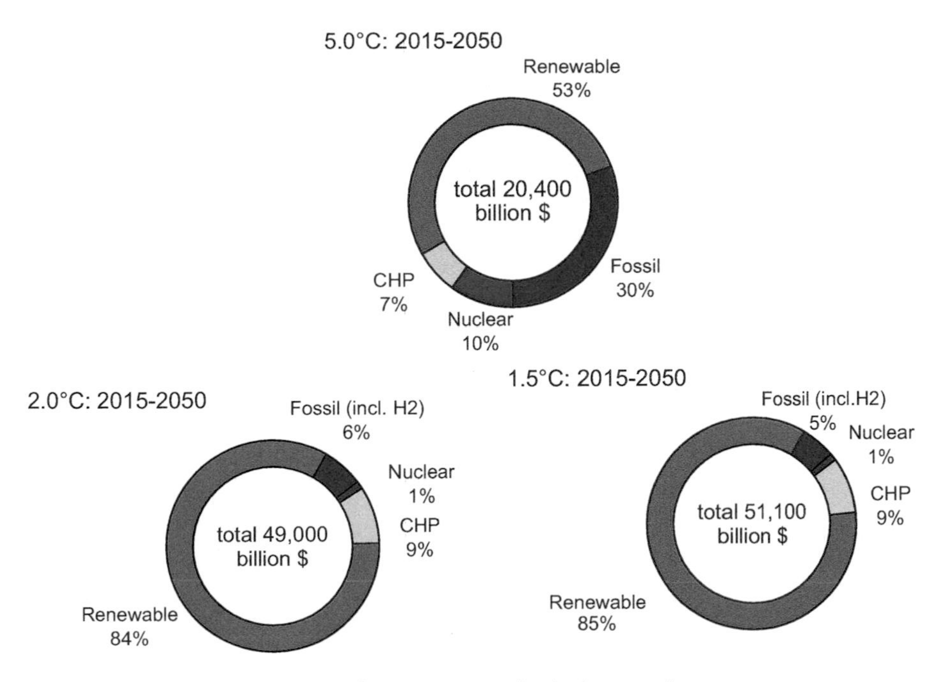

Fig. 8.9 Global: investment shares for power generation in the scenarios

investment in conventional power plants will comprises around 40% of total cumulative investments, whereas approximately 60% will be invested in renewable power generation and co-generation (Fig. 8.9).

However, in the 2.0 °C (1.5 °C) Scenario, the world will shift almost 94% (95%) of its total energy investment to renewables and co-generation. By 2030, the fossil fuel share of the power sector investment will predominantly focus on gas power plants that can also be operated with hydrogen.

Because renewable energy has no fuel costs, other than biomass, the cumulative fuel cost savings in the 2.0 °C Scenario will reach a total of $26,300 billion in 2050, equivalent to $730 billion per year. Therefore, the total fuel cost savings in the 2.0 °C Scenario will be equivalent to 90% of the additional energy investments compared to the 5.0 °C Scenario. The fuel cost savings in the 1.5 °C Scenario will add up to $28,800 billion, or $800 billion per year.

8.1.6 Global: Energy Supply for Heating

The final energy demand for heating will increase in the 5.0 °C Scenario by 59%, from 151 EJ/year in 2015 to around 240 EJ/year in 2050. In the 2.0 °C Scenario, energy efficiency measures will help to reduce the energy demand for heating by 36% in 2050, relative to that in the 5.0 °C Scenario, and by 40% in the 1.5 °C Scenario. Today, renewables supply around 20% of the global final energy demand for heating. The main contribution is from biomass. Renewable energy will provide 42% of the world's total heat demand in 2030 in the 2.0 °C Scenario and 56% in the 1.5 °C Scenario. In both scenarios, renewables will provide 100% of the total heat demand in 2050.

Figure 8.10 shows the development of different technologies for heating worldwide over time, and Table 8.2 provides the resulting renewable heat supply for all scenarios. Until 2030, biomass will remain the main contributor. In the long-term, the growing use of solar, geothermal, and environmental heat will lead to a biomass share in total heating of 33% in the 2.0 °C Scenario and 30% in the 1.5 °C Scenario.

Heat from renewable hydrogen will further reduce the dependence on fossil fuels in both scenarios. Hydrogen consumption in 2050 will be around 15,900 PJ/year in

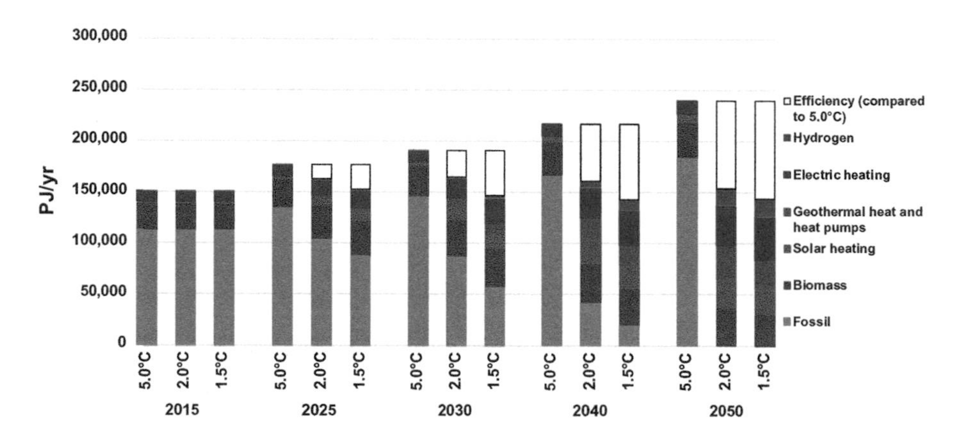

Fig. 8.10 Global: development of heat supply by energy carrier in the scenarios

Table 8.2 Global: development of renewable heat supply in the scenarios (excluding the direct use of electricity)

in PJ/year	(°C)	2015	2025	2030	2040	2050
Biomass	5.0	25,470	27,643	28,878	31,568	34,564
	2.0	25,470	32,078	35,134	38,187	37,536
	1.5	25,470	33,493	36,927	36,385	30,151
Solar heating	5.0	1246	2091	2754	4353	6220
	2.0	1246	6485	12,720	23,329	27,312
	1.5	1246	7656	14,153	21,665	24,725
Geothermal heat and heat pumps	5.0	563	804	925	1293	1823
	2.0	563	4212	8956	21,115	33,123
	1.5	563	4615	10,288	20,031	29,123
Hydrogen	5.0	0	0	0	0	0
	2.0	0	193	508	5670	15,877
	1.5	0	180	1769	10,461	17,173
Total	5.0	27,278	30,538	32,557	37,214	42,608
	2.0	27,278	42,967	57,318	88,301	113,848
	1.5	27,278	45,944	63,137	88,542	101,172

the 2.0 °C Scenario and 17,200 PJ/year in the 1.5 °C Scenario. The direct use of electricity for heating will also increase by a factor of 4.2–4.5 between 2015 and 2050 and will have a final share of 26% in 2050 in the 2.0 °C Scenario and 30% in the 1.5 °C Scenario (Table 8.2).

8.1.7 Global: Future Investments in the Heating Sector

The roughly estimated investments in renewable heating technologies up to 2050 will amount to around $13,230 billion in the 2.0 °C Scenario (including investments for plant replacement after their economic lifetimes)—approximately $368 billion per year. The largest share of this investment is assumed to be for heat pumps (around $5700 billion), followed by solar collectors and geothermal heat use. The 1.5 °C Scenario assumes an even faster expansion of renewable technologies. However, the lower heat demand (compared with the 2.0 °C Scenario) will result in a lower average annual investment of around $344 billion per year (Table 8.3, Fig. 8.11).

8.1.8 Global: Transport

The energy demand in the transport sector will increase in the 5.0 °C Scenario by 50% by 2050, from around 97,200 PJ/year in 2015 to 145,700 PJ/year in 2050. In the 2.0 °C Scenario, assumed technical, structural, and behavioural changes will reduce the energy demand by 66% (96,000 PJ/year) by 2050 compared with the 5.0 °C Scenario. Additional modal shifts, technology switches, and a reduction in

Table 8.3 Global: installed capacities for renewable heat generation in the scenarios

in GW	(°C)	2015	2025	2030	2040	2050
Biomass	5.0	10,215	10,180	9938	9423	8997
	2.0	10,215	10,202	9456	7875	5949
	1.5	10,215	10,418	9568	7073	4141
Geothermal	5.0	5	7	7	8	4
	2.0	5	85	181	492	656
	1.5	5	101	200	433	551
Solar heating	5.0	378	615	781	1175	1652
	2.0	378	1685	3198	5722	6575
	1.5	378	1993	3555	5286	5964
Heat pumps	5.0	89	126	144	199	270
	2.0	89	497	906	1821	2857
	1.5	89	514	967	1726	2430
Total[a]	**5.0**	**10,688**	**10,928**	**10,871**	**10,805**	**10,923**
	2.0	**10,688**	**12,469**	**13,741**	**15,910**	**16,036**
	1.5	**10,688**	**13,026**	**14,290**	**14,517**	**13,086**

[a]Excluding direct electric heating

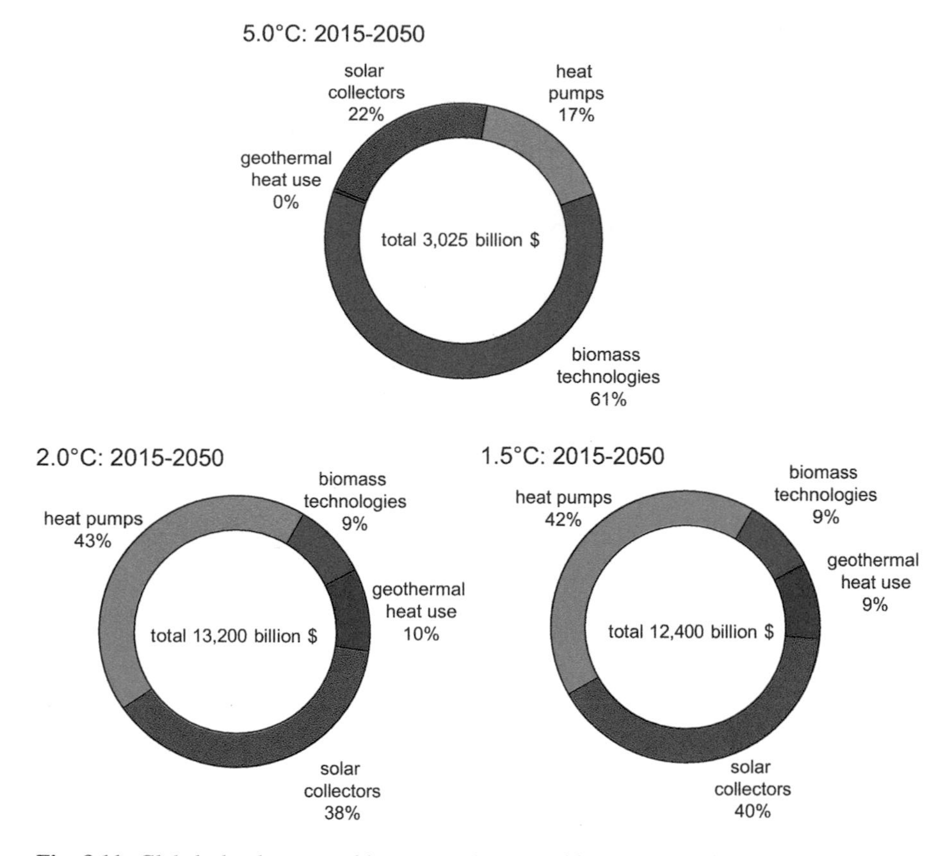

Fig. 8.11 Global: development of investment in renewable heat-generation technologies in the scenarios

the transport demand will lead to even higher energy savings in the 1.5 °C Scenario of 74% (or 108,000 PJ/year) in 2050 compared with the 5.0 °C case (Table 8.4, Fig. 8.12).

By 2030, electricity will provide 12% (2700 TWh/year) of the transport sector's total energy demand in the 2.0 °C Scenario, whereas in 2050, the share will be 47% (6500 TWh/year). In 2050, around 8430 PJ/year of hydrogen will be used in the transport sector, as a complementary renewable option. In the 1.5 °C Scenario, the annual electricity demand will be about 5200 TWh in 2050. The 1.5 °C Scenario also assumes a hydrogen demand of 6850 PJ/year by 2050.

Biofuel use is limited in the 2.0 °C Scenario to a maximum of around 12,000 PJ/year Therefore, by around 2030, synthetic fuels based on power-to-liquid will be introduced, with a maximum amount of 5820 PJ/year in 2050. Because of the lower overall energy demand by transport, biofuel use will be reduced in the 1.5 °C Scenario to a maximum of 10,000 PJ/year The maximum synthetic fuel demand will amount to 6300 PJ/year.

Table 8.4 Global: projection of transport energy demand by mode in the scenarios

in PJ/year	(°C)	2015	2025	2030	2040	2050
Rail	5.0	2705	2708	2814	3024	3199
	2.0	2705	2875	3149	3520	3960
	1.5	2705	2932	3119	3559	4087
Road	5.0	85,169	94,755	102,556	116,449	127,758
	2.0	85,169	79,975	68,660	48,650	40,089
	1.5	85,169	67,579	48,949	34,055	28,859
Domestic aviation	5.0	4719	6544	7745	9080	9176
	2.0	4719	4732	4239	3291	2640
	1.5	4719	4461	3612	2361	1845
Domestic navigation	5.0	2130	2304	2392	2537	2663
	2.0	2130	2303	2388	2512	2601
	1.5	2130	2301	2383	2506	2601
Total	5.0	94,723	106,310	115,506	131,091	142,796
	2.0	94,723	89,886	78,436	57,973	49,290
	1.5	94,723	77,274	58,063	42,482	37,392

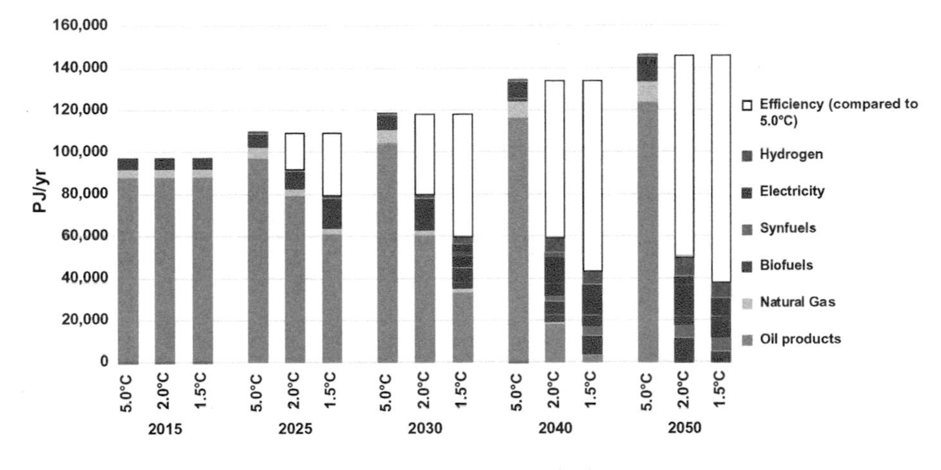

Fig. 8.12 Global: final energy consumption by transport in the scenarios

8.1.9 Global: Development of CO_2 Emissions

In the 5.0 °C Scenario, the annual global energy-related CO_2 emissions will increase by 40%, from 31,180 Mt. in 2015 to more than 43,500 Mt. in 2050. The stringent mitigation measures in both alternative scenarios will cause annual emissions to fall to 7070 Mt. in 2040 in the 2.0 °C Scenario and to 2650 Mt. in the 1.5 °C Scenario, with further reductions to almost zero by 2050. In the 5.0 °C Scenario, the cumulative CO_2 emissions from 2015 until 2050 will add up to 1388 Gt. In contrast, in the 2.0 °C and 1.5 °C Scenarios, the cumulative emissions for the period 2015–2050 will be 587 Gt and 450 Gt, respectively.

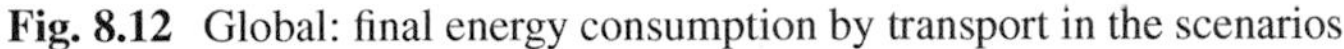

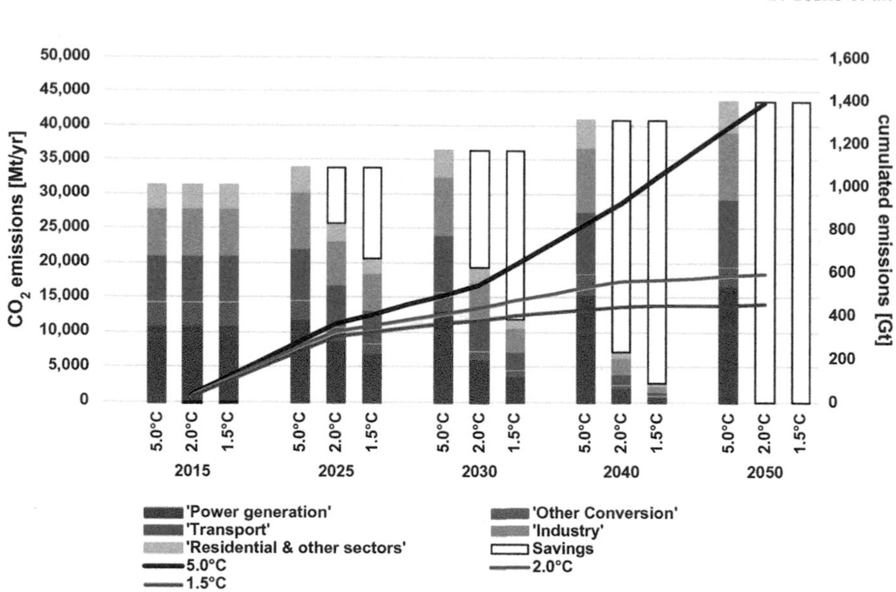

Fig. 8.13 Global: development of CO_2 emissions by sector and cumulative CO_2 emissions (since 2015) in the scenarios ('Savings' = lower than in the 5.0 °C Scenario)

Thus, the cumulative CO_2 emissions will decrease by 58% in the 2.0 °C Scenario and by 68% in the 1.5 °C Scenario compared with the 5.0 °C case. A rapid reduction in annual emissions will occur in both alternative scenarios. In the 2.0 °C Scenario, the reduction will be greatest in 'Power generation', followed by the 'Residential and other' and 'Transport' sectors (Fig. 8.13).

8.1.10 Global: Primary Energy Consumption

The levels of primary energy consumption based on the assumptions discussed above in the three scenarios are shown in Fig. 8.14. In the 2.0 °C Scenario, the primary energy demand will decrease by 21%, from around 556 EJ/year in 2015 to 439 EJ/year in 2050. Compared with the 5.0 °C Scenario, the overall primary energy demand will decrease by 48% by 2050 in the 2.0 °C Scenario (5.0 °C: 837 EJ in 2050). In the 1.5 °C Scenario, the primary energy demand will be even lower (412 EJ in 2050) due to the lower final energy demand and lower conversion losses.

Both the 2.0 °C and 1.5 °C Scenarios aim to rapidly phase-out coal and oil. This will cause renewable energy to have a primary energy share of 35% in 2030 and 92% in 2050 in the 2.0 °C Scenario. In the 1.5 °C Scenario, renewables will have a primary energy share of more than 92% in 2050 (this will includes non-energy consumption, which will still include fossil fuels). Nuclear energy will be phased-out in both the 2.0 °C and 1.5 °C Scenarios. The cumulative primary energy consumption of natural gas in the 5.0 °C Scenario will be 5580 EJ, the cumulative coal consump-

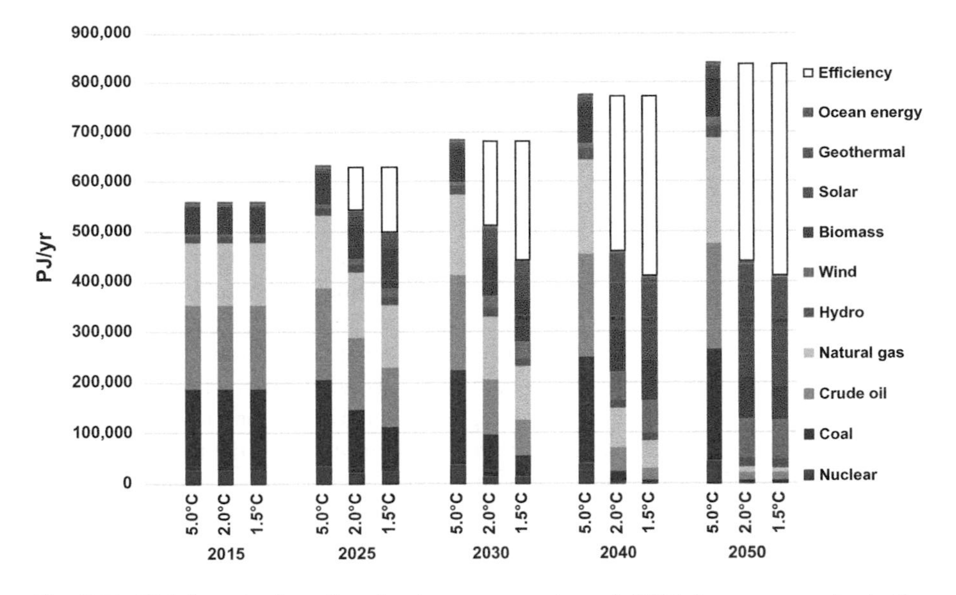

Fig. 8.14 Global: projection of total primary energy demand (PED) by energy carrier in the scenarios

tion will be about 6360 EJ, and the crude oil consumption will be 6380 EJ. In the 2.0 °C Scenario, the cumulative gas demand will amount to 3140 EJ, the cumulative coal demand to 2340 EJ, and the cumulative oil demand to 2960 EJ. Even lower fossil fuel use will be achieved in the 1.5 °C Scenario: 2710 EJ for natural gas, 1570 EJ for coal, and 2230 EJ for oil.

8.2 Global: Bunker Fuels

Bunker fuels for international aviation and navigation are separate categories in the energy statistics. Their use and related emissions are not usually directly allocated to the regional energy balances. However, they contribute significantly to global greenhouse gas (GHG) emissions and pose great challenges regarding their substitution with low-carbon alternatives. In 2015, the annual bunker fuels consumption was in the order of 16,000 PJ, of which 7400 PJ was for aviation and 8600 PJ for navigation. Between 2009 and 2015, bunker fuel consumption increased by 13%. The annual CO_2 emissions from bunker fuels accounted for 1.3 Gt in 2015, approximately 4% of global energy-related CO_2 emissions. In the 5.0 °C Scenario, the development of the final energy demand for bunker fuels is assumed to be that of the IEA World Energy Outlook 2017 Current Policies scenario. This will lead to a further increase of 120% in the demand for bunker fuels until 2050 compared with that in the base year, 2015. Because no substitution with 'green' fuels is assumed, CO_2 emissions will rise by the same order of magnitude.

Although the use of hydrogen and electricity in aviation is technically feasible (at least for regional transport) and synthetic gas use in navigation is an additional option under discussion, this analysis uses a conservative approach and assumes that bunker fuels are only replaced by biofuels or synthetic liquid fuels. Figure 8.15 shows the 5.0 °C and two alternative bunker scenarios, which are defined in consistency to the scenarios for each world region. For the 2.0 °C and 1.5 °C Scenarios, we assume the limited use of sustainable biomass potentials and the complementary central production of power-to-liquid synfuels. In the 2.0 °C Scenario, this production is assumed to take place in three world regions: Africa, the Middle East, and OECD Pacific (especially Australia), where synfuel generation for export is expected to be the most economic. The 1.5 °C Scenario requires even faster decarbonisation, and therefore follows a more ambitious low-energy pathway. This will lead to a faster build-up of the power-to-liquid infrastructure in all regions, which in the long term, will also be used for limited 'regional' bunker fuel production to maintain the utilization of the existing infrastructure. Therefore, the production of bunker fuels is assumed to occur in more regions, with lower exports from the supply regions mentioned above, in the 2.0 °C Scenario. Another assumption is that, consistent with the regional 1.5 °C Scenarios, the biomass consumption for energy supply will decrease in the long term, whereas power-to-liquid will continue to increase as the main option for international aviation and navigation. Finally, the expansion of the power-to-liquid infrastructure for the generation of bunker fuel will be closely associated with the assumed development of regional synthetic fuel demand and generation for transportation in each world region. Figure 8.15 also shows the resulting cumulative CO_2 emissions from bunker fuel consumption between 2015 and 2050, which amount to around 70 Gt in the 5.0 °C case, 30 Gt in the 2.0 °C Scenario, and 21 Gt in the 1.5 °C Scenario. Table 8.5 provides more-detailed data for the three bunker fuel scenarios.

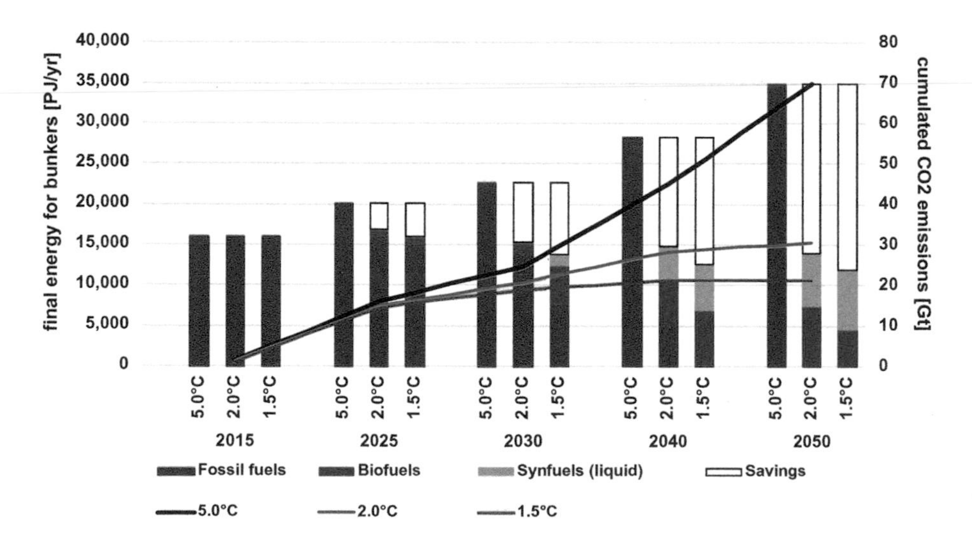

Fig. 8.15 Global: scenario of bunker fuel demand for aviation and navigation and the resulting cumulative CO_2 emissions

Table 8.5 Global: projection of bunker fuel demands for aviation and navigation by fuel in the scenarios

World bunkers 5.0 °C scenario	Unit	2015	2020	2025	2030	2035	2040	2045	2050
Total final energy consumption	**PJ/year**	**15,985**	**17,976**	**20,090**	**22,593**	**25,443**	**28,293**	**31,462**	**34,987**
thereof aviation	PJ/year	7408	8385	9431	10,674	12,097	13,537	15,148	16,950
thereof navigation	PJ/year	8576	9591	10,658	11,919	13,346	14,756	16,314	18,037
fossil fuels	PJ/year	15,985	17,976	20,090	22,593	25,443	28,293	31,462	34,987
biofuels	PJ/year	0	0	0	0	0	0	0	0
synthetic liquid fuels	PJ/year	0	0	0	0	0	0	0	0
Primary energy demand									
crude oil	PJ/year	17,663	19,754	21,956	24,558	27,506	30,423	33,650	37,220
CO_2 emissions	**Mt/year**	**1296**	**1450**	**1611**	**1802**	**2018**	**2232**	**2468**	**2730**
World bunkers 2.0 °C Scenario	**unit**	**2015**	**2020**	**2025**	**2030**	**2035**	**2040**	**2045**	**2050**
Total final energy consumption	**PJ/year**	**15,985**	**17,538**	**16,836**	**15,274**	**15,053**	**14,826**	**14,483**	**14,014**
thereof aviation	PJ/year	7408	8594	8418	7713	7602	7487	7314	7077
thereof navigation	PJ/year	8576	8944	8418	7561	7451	7339	7169	6937
fossil fuels	PJ/year	15,985	17,270	16,180	13,748	10,537	5189	3621	0
biofuels	PJ/year	0	268	657	1526	3146	5417	6381	7430
synthetic liquid fuels	PJ/year	0	0	0	0	1370	4220	4481	6584
Assumed regional structure of synthetic bunker production									
Africa	PJ/year	0	0	0	0	846	2607	2768	4067
Middle East	PJ/year	0	0	0	0	183	564	598	879
OECD Pacific	PJ/year	0	0	0	0	341	1050	1115	1638
Primary energy demand									
crude oil	PJ/year	17,663	18,978	17,683	14,943	11,391	5580	3872	0
biomass	PJ/year	0	400	952	2150	4369	7420	8623	9907
RES electricity demand for PtL	TWh/year	0	0	0	0	961	2880	3058	4375

(continued)

Table 8.5 (continued)

World bunkers 5.0 °C scenario	Unit	2015	2020	2025	2030	2035	2040	2045	2050
CO_2 emissions	**Mt/year**	**1296**	**1391**	**1296**	**1095**	**835**	**409**	**284**	**0**
World bunkers 1.5 °C Scenario	**unit**	**2015**	**2020**	**2025**	**2030**	**2035**	**2040**	**2045**	**2050**
Total final energy consumption	**PJ/year**	**15,985**	**17,538**	**15,995**	**13,747**	**12,795**	**12,602**	**12,311**	**11,912**
thereof aviation	PJ/year	7408	8594	7997	6942	6462	6364	6217	6016
thereof navigation	PJ/year	8576	8944	7997	6805	6334	6238	6094	5896
fossil fuels	PJ/year	15,985	17,538	15,179	7836	2559	0	0	0
biofuels	PJ/year	0	0	816	4536	6398	6931	5540	4527
synthetic liquid fuels	PJ/year	0	0	0	1375	3839	5671	6771	7385
Assumed regional structure of synthetic bunker production									
Africa	PJ/year	0	0	0	717	2002	2863	3093	2882
Middle East	PJ/year	0	0	0	155	433	619	669	873
OECD Pacific	PJ/year	0	0	0	289	836	1265	1622	1697
OECD North America	PJ/year	0	0	0	213	568	798	924	977
OECD Europe	PJ/year	0	0	0	0	0	126	262	557
Eurasia	PJ/year	0	0	0	0	0	0	200	400
Primary energy demand									
crude oil	PJ/year	17,663	19,273	16,589	8517	2766	0	0	0
biomass	PJ/year	0	0	1182	6389	8885	9495	7486	6035
RES electricity demand for PtL	TWh/year	0	0	0	964	2693	3870	4621	4896
CO_2 emissions	**Mt/year**	**1296**	**1413**	**1216**	**624**	**203**	**0**	**0**	**0**

The production of synthetic fuels will cause significant additional electricity demand and a corresponding expansion of the renewable power generation capacities. In the case of liquid bunker fuels, these additional renewable power generation capacities will amount to 1100 GW in the 2.0 °C Scenario and more than 1200 GW in the 1.5 °C Scenario if a flexible utilization rate of 4000 full-load hours per year can be achieved. However, such a situation will require high amounts of electrolyser capacity and hydrogen storage to allow not only flexibility in the power system, but also high utilization rates of the downstream synthesis processes (e.g., via Fischer-Tropsch plants). Other options for renewable synthetic fuel production are solar thermal chemical processes, which directly use high-temperature solar heat.

8.3 Global: Utilization of Solar and Wind Potential

The economic potential, under space constraints, of utility solar PV, concentrated solar power (CSP), and onshore wind was analysed with the methodology described in Sect. 3.3 of Chap. 3.

The 2.0 °C Scenario utilizes only a fraction of the available economic potential of the assumed suitable land for utility-scale solar PV and concentrated solar power plants. This estimate does not include solar PV roof-top systems, which have significant additional potential. India (2.0 °C) will have the highest solar utilization rate of 8.5%, followed by Europe (2.0 °C) and the Middle East (2.0 °C), with 5.9% and 4.6%, respectively.

Onshore wind potential has been utilized to a larger extent than solar potential. In the 2.0 °C Scenario, space-constrained India will utilize more than half of onshore wind, followed by Europe with 20%. This wind potential excludes offshore wind, which has significant potential but the mapping for the offshore wind potential was beyond the scope of this analysis (Table 8.6).

The 1.5 °C Scenario is based on the accelerated deployment of all renewables and the more ambitious implementation of efficiency measures. Therefore, the total installed capacity of solar and wind generators by 2050 is not necessarily larger than it is in the 2.0 °C Scenario, and the utilization rate is in the same order of magnitude. The increased deployment of renewable capacity in OECD Pacific (Australia), the Middle East, and OECD North America (USA) will be due to the production of synthetic bunker fuels from hydrogen to supply global transport energy for international shipping and aviation.

Table 8.6 Economic potential within a space-constrained scenario and utilization rates for the 2.0 °C and 1.5 °C scenarios

		Installed capacity by 2050				Utilization rate			Installed capacity by 2050		Utilization rate	
	SOLAR	2.0 °C		1.5 °C		2.0 °C	1.5 °C	WIND	2.0 °C	1.5 °C	2.0 °C	1.5 °C
Economic Potential within available space	Tech. space potential	PV	CSP	PV	CSP			Onshore wind				
	[GW]	[GW]				[%]		[GW]	[GW]		[%]	
OECD North America	445,954	1688	208	1816	236	0.4%	0.5%	86,846	847	833	1.0%	1.0%
Latin America	148,664	317	66	425	79	0.3%	0.3%	29,736	220	237	0.7%	0.8%
OECD Europe	14,364	793	54	918	57	5.9%	6.8%	2873	577	636	20.1%	22.1%
Middle East	24,451	881	252	742	216	4.6%	3.9%	470	455	434	96.8%	92.4%
Africa	914,180	767	247	930	257	0.1%	0.1%	190,711	485	509	0.3%	0.3%
Eurasia	Not available	658	22	657	34			Not available	564	544		
Non-OECD-Asia	44,064	1065	274	1005	224	3.0%	2.8%	4740	515	506	10.9%	10.7%
India	1323	1257	209	1129	209	8.5%	7.7%	1974	1139	983	57.7%	49.8%
China	176,916	1756	762	1772	614	1.4%	1.3%	17,848	1180	1345	6.6%	7.5%
OECD Pacific	124,178	665	57	745	67	0.6%	0.7%	24,447	244	303	1.0%	1.2%

8.4 Global: Power Sector Analysis

The long-term global and regional energy results were used to conduct a detailed power sector analysis with the methodology described in Sect. 3.5 of Chap. 3. Both the 2.0 °C and 1.5 °C Scenarios rely on high shares of variable solar and wind generation. The aim of the power sector analysis was to gain insight into the power system stability for each region (subdivided into up to eight sub-regions) and to gauge the extent to which power grid interconnections, dispatch generation services, and storage technologies are required. The results presented in this chapter are projections calculated from publicly available data. Detailed load curves for some of the sub-regions and countries discussed in this chapter were not available and, in some cases, the relevant information is classified. Therefore, the outcomes of the [R]E 24/7 model are estimates and require further research with more detailed localized data, especially regarding the available power grid infrastructure. Furthermore, power sector projections for developing countries, especially in Africa and Asia, assume unilateral access to energy services for the residential sector by 2050, and they require transmission and distribution grids in regions where there are none at the time of writing. Further research—in cooperation with local utilities and government representatives—is required to develop a more detailed understanding of power infrastructure needs.

8.4.1 Global: Development of Power Plant Capacities

The size of the global market for renewable power plants will increase significantly under the 2.0 °C Scenario. The annual market for solar PV power must increase from close to 100 GW in 2017 (REN21-GSR 2018) by a factor of 4.5 to an average of 454 GW by 2030. The onshore wind market must expand to 172 GW by 2025, about three times higher than in 2017 (REN21-GSR 2018). The offshore wind market will continue to increase in importance within the renewable power sector. By 2050, offshore wind installations will increase to 32 GW annually—11 times higher than in 2017 (GWEC 2018). Concentrated solar power plants will play an important role in dispatchable solar electricity generation for the supply of bulk power, especially for industry, and will provide secured capacity to power systems. By 2030, the annual CSP market must increase to 78 GW, compared with 3 GW in 2020 and only 0.1 GW in 2017 (REN21-GSR2018) (Table 8.7).

In the 1.5 °C Scenario, the phase-out of coal and lignite power plants is accelerated and a total capacity of 618 GW—equivalent to approximately 515 power stations[1]—must end operation by 2025. The replacement power must come from a variety of renewable power generators, both variable and dispatchable. The annual market for solar PV must be around 30% higher in 2050 than it was in 2025, as in the 2.0 °C Scenario. While the onshore wind market also has an accelerated trajectory

[1]Assumption: average size of one coal power plant side contains multiple generation blocks, with a total of 1200 MW on average for each location.

Table 8.7 World: average annual change in the installed power plant capacity

Global power generation: average annual change of installed capacity [GW/a]	2015–2025		2026–2035		2036–2050	
	2.0 °C	1.5 °C	2.0 °C	1.5 °C	2.0 °C	1.5 °C
Hard coal	2	−107	−96	−119	−68	−12
Lignite	−25	−34	−16	−9	−3	−1
Gas	41	70	44	72	−199	−28
Hydrogen-gas	1	17	12	57	240	246
Oil/diesel	−18	−32	−29	−28	−6	−2
Nuclear	−15	−27	−23	−24	−7	−10
Biomass	24	40	26	29	25	21
Hydro	19	10	7	7	7	8
Wind (onshore)	121	307	273	333	242	158
Wind (offshore)	16	64	75	91	64	45
PV (roof top)	170	413	368	437	399	324
PV (utility scale)	57	138	123	146	133	108
Geothermal	5	16	22	24	28	23
Solar thermal power plants	9	57	93	109	102	85
Ocean energy	4	10	20	20	28	23
Renewable fuel based co-generation	13	31	27	31	25	20

under the 1.5 °C Scenario as well, the offshore wind market is assumed to be almost identical to that in the 2.0 °C pathway because of the longer lead times for these projects. The same is assumed for CSP plants, which are utility-scale projects and significantly higher deployment seems unlikely in the time remaining until 2025.

8.4.2 Global: Utilization of Power-Generation Capacities

On a global scale, in the 2.0 °C and 1.5 °C Scenarios, the shares of variable renewable power generation will increase from 4% in 2015 to 39% and 47%, respectively, by 2030, and to 64% and 60%, respectively, by 2050. The reason for the variations in the two cases is their different assumptions regarding efficiency measures, which may lead to lower overall demand in the 1.5 °C Scenario than in the 2.0 °C Scenario. During the same period, dispatchable renewables—CSP plants, biofuel generation, geothermal energy, and hydropower—will remain around 32% until 2030 on a global average and decrease slightly to 29% in the 2.0 °C Scenario (and to 27% in the 1.5 °C Scenario) by 2050. The shares of dispatchable conventional generation—mainly coal, oil, gas, and nuclear—will decline from a global average of 60% in 2015 to only 14% in 2040. By 2050, the remaining dispatchable conventional gas power plants will have been converted to operate with hydrogen and synthetic fuels, to avoid stranded investments and to achieve higher quantities of dispatch power capacity. Table 8.8 shows the increasing shares of variable renewable power

Table 8.8 Global: power system shares by technology group

Power generation structure in 10 world regions		2.0 °C			1.5 °C		
World		Variable renewables	Dispatch renewables	Dispatch fossil	Variable renewables	Dispatch renewables	Dispatch fossil
OECD North America	2015	7%	35%	58%	7%	41%	52%
	2030	48%	30%	23%	59%	27%	15%
	2050	68%	19%	13%	68%	21%	11%
Latin America	2015	3%	63%	34%	3%	62%	35%
	2030	24%	51%	25%	36%	61%	3%
	2050	39%	45%	16%	40%	46%	13%
Europe	2015	15%	47%	38%	15%	47%	38%
	2030	44%	44%	12%	51%	39%	10%
	2050	67%	28%	4%	69%	27%	4%
Middle East	2015	0%	12%	88%	0%	13%	87%
	2030	51%	19%	31%	56%	18%	27%
	2050	81%	19%	0%	70%	16%	13%
Africa	2015	2%	26%	73%	2%	17%	81%
	2030	47%	21%	32%	52%	13%	35%
	2050	73%	27%	0%	64%	15%	21%
Eurasia	2015	1%	35%	63%	1%	35%	63%
	2030	36%	43%	21%	40%	46%	14%
	2050	69%	23%	7%	65%	25%	10%
Non-OECD Asia	2015	1%	35%	64%	1%	35%	64%
	2030	26%	35%	39%	36%	34%	30%
	2050	52%	28%	19%	55%	28%	17%
India	2015	4%	32%	64%	4%	32%	64%
	2030	45%	26%	29%	60%	21%	19%
	2050	72%	27%	1%	58%	26%	16%
China	2015	6%	35%	59%	6%	21%	73%
	2030	30%	24%	46%	39%	30%	31%
	2050	49%	47%	5%	49%	42%	9%
OECD Pacific	2015	4%	34%	61%	4%	34%	61%
	2030	40%	31%	30%	45%	29%	27%
	2050	71%	26%	2%	64%	22%	14%
Global average	2015	4%	35%	60%	4%	34%	62%
	2030	39%	32%	29%	47%	32%	21%
	2050	64%	29%	7%	60%	27%	13%

Note: Variable renewable generation shares in long term energy pathways and power sector analysis differ due to different calculation methods. The power sector analysis results are based on the sum of up to eight sub-regional simulations, while the long term energy pathway is based on the regional average generation excluding variations in solar and wind resources within that region

generation—solar PV and wind power—under the 2.0 °C and 1.5 °C Scenarios over the entire modelling period. The main difference between the two scenarios is the time horizon until variable renewable power generation is implemented, with more rapid implementation in the 1.5 °C Scenario. Again, increased variable shares—mainly in the USA, the Middle East region, and Australia—will produce synthetic fuels for the export market, as fuel for both renewable power plants and the transport sector.

Table 8.9 provides an overview of the capacity factor developments by technology group for the 2.0 °C and 1.5 °C Scenarios. The operational hours shown are the result of [R]E 24/7 modelling under the *'Dispatch case'*, which assumes that the highest priority is given to the dispatch of power from variable sources, followed by dispatchable renewables. Conventional power generation will only provide power for electricity demand that cannot be met by renewables and storage technologies. Only imports via interconnections will be assigned a lower priority than conventional power. The reason that interconnections are placed last in the supply cascade is the high level of uncertainty about whether these interconnections can actually be implemented in time. Experience with power grid projects—especially transmission lines—indicates that planning and construction can take many years or fail entirely, leaving large-scale utility-based renewable power projects unbuilt.

Table 8.9 Global: capacity factors for *variable* and *dispatchable* power generation

Utilization of variable and dispatchable power generation:		2015	2020	2020	2030	2030	2040	2040	2050	2050
World			2.0 °C	1.5 °C	2.0 °C	1.5 °C	2.0 °C	1.5 °C	2.0 °C	1.5 °C
Capacity factor – average	[%/yr]	49.5%	37%	37%	33%	31%	34%	30%	33%	31%
Limited dispatchable: fossil and nuclear	[%/yr]	58.7%	34%	34%	24%	16%	25%	10%	17%	9%
Limited dispatchable: renewable	[%/yr]	36.9%	45%	45%	42%	36%	58%	31%	39%	34%
Dispatchable: fossil	[%/yr]	42.9%	28%	28%	19%	15%	33%	15%	19%	19%
Dispatchable: renewable	[%/yr]	43.1%	56%	56%	54%	47%	42%	39%	51%	43%
Variable: renewable	[%/yr]	14.6%	14%	14%	28%	26%	28%	26%	29%	27%

On the global level, the average capacity factor across all power-generation technologies is around 45%. For this analysis, we created five different power plant categories based on their current usual operation times and areas of use:

- **Limited dispatchable fossil and nuclear power plants**: coal, lignite, and nuclear power plants with limited ability to respond to changes in demand. These power plants are historically categorized as 'baseload power plants'. Power systems dominated by renewable energy usually contain high proportions of variable generation, and therefore quick reaction times (to ramp up and down) are required. Limited dispatchable power plants cannot deliver these services and are therefore being phased-out.
- **Limited dispatchable renewable systems** are CSP plants with integrated storage and co-generation systems with renewable fuels (including geothermal heat). They cannot respond quickly enough to adapt to minute-by-minute changes in demand, but can still be used as dispatch power plants for 'day ahead' planning.
- **Dispatchable fossil fuel power plants** are gas power plants that have very quick reaction times and therefore provide valid power system services.
- **Dispatchable renewable power plants** are hydropower plants (although they are dependent on the climatic conditions in the region where the plant is used), biogas power plants, and former gas power plants converted to hydrogen and/or synthetic fuel. This technology group is responsible for most of the required load-balancing services and is vital for the stability of the power system, as storage systems, interconnections, and, if possible, demand-side management.
- **Variable renewables** are solar PV plants, onshore and offshore wind farms, and ocean energy generators. A sub-category of ocean energy plants—tidal energy plants—is very predictable.

Table 8.9 shows the development of the utilization of limited and fully dispatchable power generators—both fossil and renewable fuels—and variable power generation. In the 2.0 °C Scenario, conventional power generation in the baseload mode—currently with an annual operation time of around 6000 h per year or more—will decline sharply after 2030 and the annual operation time will be halved, whereas medium-load and dispatch power plants will predominate. The system share of dispatchable renewables will remain around 45%–50% throughout the entire modelling period.

8.4.3 Global: Development of Load, Generation, and Residual Load

Table 8.10 shows the development of the maximum and average loads for the 10 world regions, the average and maximum power generation in each region in megawatts, and the residual loads under both alternative scenarios. The residual load in

Table 8.10 Global: load, generation, and residual load development

Power generation structure in 10 world regions		2.0 °C				1.5 °C			
World		Max demand [GW]	Max generation [GW]	Max residual load [GW]	Max load development (Base 2020) [GW]	Max demand [GW]	Max generation [GW]	Max residual load [GW]	Max load development (Base 2020) [GW]
OECD North America	2020	753	723	57	100%	755	989	58	100%
	2030	864	1159	145	115%	919	1532	194	122%
	2050	1356	2779	469	180%	1362	2900	496	180%
Latin America	2020	218	214	30	100%	218	274	18	100%
	2030	343	377	74	157%	312	418	25	143%
	2050	533	601	154	244%	550	696	122	252%
OECD Europe	2020	574	584	121	100%	574	583	125	100%
	2030	620	718	95	108%	639	936	104	111%
	2050	862	1530	417	150%	900	1727	448	157%
Middle East	2020	174	181	−29	100%	174	180	−26	100%
	2030	229	297	−20	132%	237	346	−13	136%
	2050	551	1164	−67	317%	522	1018	−161	300%
Africa	2020	164	125	47	100%	164	135	37	100%
	2030	280	261	101	171%	296	305	105	181%
	2050	875	1363	647	534%	915	1562	412	559%
Eurasia	2020	257	163	107	100%	257	171	106	100%
	2030	316	332	147	123%	330	416	139	129%
	2050	630	1338	271	245%	632	1296	275	246%

Non-OECD Asia	2020	248	135	122	100%	248	133	124	100%
	2030	415	389	256	167%	423	465	296	171%
	2050	935	1459	728	377%	841	1394	656	339%
India	2020	288	266	44	100%	288	249	61	100%
	2030	493	624	112	171%	491	861	148	170%
	2050	1225	1880	854	425%	1207	1865	558	419%
China	2020	957	935	74	100%	953	946	57	100%
	2030	1233	1249	173	129%	1219	1613	179	128%
	2050	1967	2724	1415	206%	1990	3203	−609	209%
OECD Pacific	2020	354	322	47	100%	354	318	47	100%
	2030	308	468	21	87%	318	544	36	90%
	2050	410	997	196	116%	471	1140	173	133%

this analysis is the load remaining after variable renewable power generation. Negative values indicate that the power generation from solar and wind exceeds the actual load and must be exported to other regions, stored, or curtailed. In each region, the average generation should be on the same level as the average load. The maximum loads and maximum generations shown do not usually occur at the same time, so surplus production of electricity can appear and this should be exported or stored as much as possible. In rare individual cases, solar or wind generation plants can also temporarily reduce their output to a lower load, or some plants can be shut down. Any reduced generation from solar and wind in response to low demand is defined as *curtailment*.

Figure 8.16 illustrates the development of the maximum loads across all 10 world regions under the 2.0 °C and 1.5 °C Scenarios. The most significant increase appears in Africa, where the maximum load surges over the entire modelling period by 534% in response to favourable economic development and increased access to

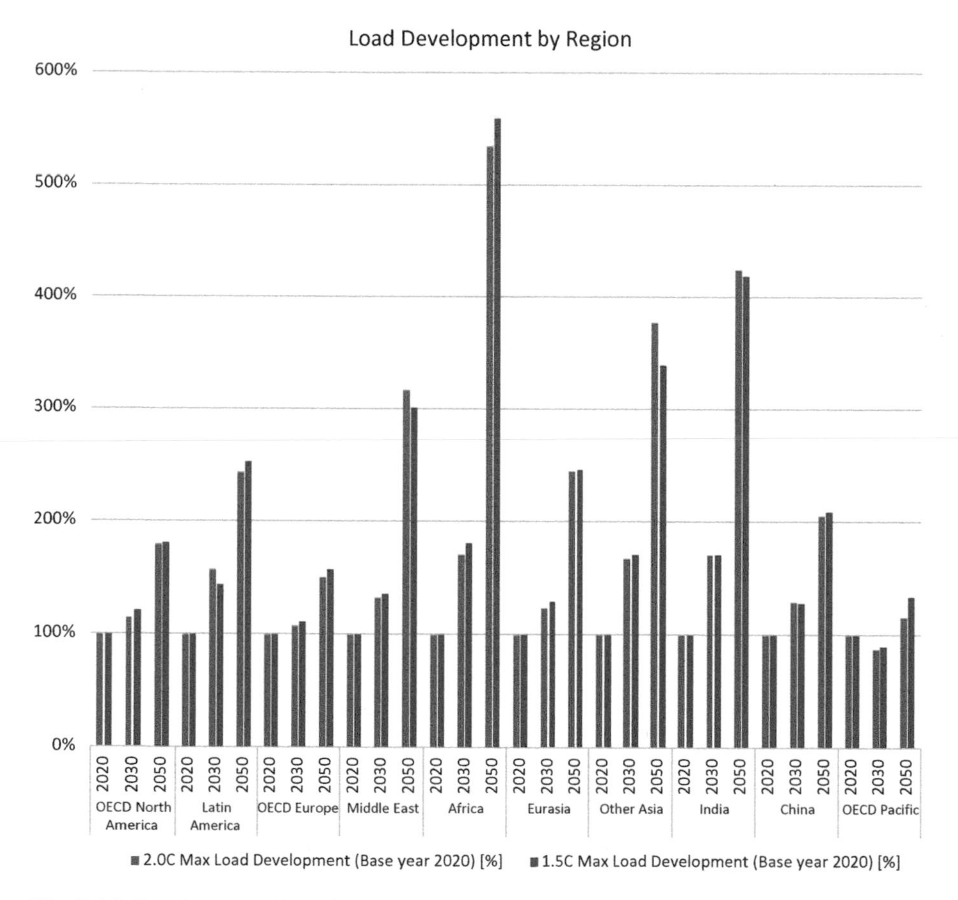

Fig. 8.16 Development of maximum load in 10 world regions in 2020, 2030, and 2050 in the 2.0 °C and 1.5 °C scenarios

energy services by households. In OECD Pacific, efficiency measures will lead to a reduction in the maximum load to 87% of the base year value by 2030 and will increase to 116% by 2050 with the expansion of electric mobility and the increased electrification of the process heat supply in the industry sector. The 1.5 °C Scenario has slightly higher loads in response to the accelerated electrification of the industry, heating, and business sectors, except in three regions (the Middle East, India, and Non OECD Asia), where early action on efficiency measures will lead to an overall lower demand at the end of the modelling period, with the same GDP and population growth rates.

8.4.4 Global System-Relevant Technologies—Storage and Dispatch

The global results of introducing system-relevant technologies are shown in Table 8.8. The first part of this section documents the required power plant markets, the changes and configurations of power-generation systems, and the development of loads in response to high electrification rates in the industry, heating, and transport sectors. The next step in the analysis documents the storage and dispatch demands and possible technology utilization. It is important to note that the results presented here are not cost-optimized. The mixture of battery storage and pumped hydropower plants with hydrogen- and synthetic-fuel-based dispatch power plants presented here represents only one option of many.

Significant simplification is required for the computer simulations of large regions, to reduce the data volumes (and calculation times) or simply because there is not yet any data, because several regions still have no electricity infrastructure in place. Detailed modelling requires access to detailed data. Although the modelling tools used for this analysis could be used to develop significantly more-detailed regional analyses, this is beyond the scope of this research.

The basic concept for the management of power system generation is based on four principles:

1. Diversity;
2. Flexibility;
3. Inter-sectorial connectivity;
4. Resource efficiency.

Diversity in the locally deployed renewable power-generation structure. For example, a combination of onshore and offshore wind with solar PV and CSP plants will reduce storage and dispatch demands.

Flexibility involves a significant number of fast-reacting dispatch power plants operated with fuels produced from renewable electricity (hydrogen and synthetic fuels). The existing gas infrastructure can be further utilized to avoid stranded investments, and the actual fuel production can also be used—with some technical limitations—for load management, which again will reduce the need for storage technologies.

Inter-sectorial connectivity involves the connection of the heating (including process heat) and transport sectors. Neither the transport sector nor the heating sector will undergo complete electrification. To supply industrial process heat, the capacity of co-generation plants—operated with bio-, geothermal, or hydrogen fuels—will be increased by a factor of 2.5 in the 1.5 °C Scenario. Co-generation heating systems with heat storage capacities and heat pumps operated with renewable electricity will lead to more flexibility in the management of both load and demand. However, an analysis of the full potential for these heating systems was not within the scope of this project, so they have not been included in the modelling. Further research with localized data is required.

Resource efficiency In addition to energy and GHG modelling, a resource assessment of selected metals has been undertaken (see Chap. 11). A variety of technologies—especially storage technologies—can be used to reduce the pressure on resource requirements, namely for cobalt and lithium for batteries and electric mobility and the silver required for solar technologies. Therefore, the choice of storage technologies has taken the specific requirements for metals into account.

Table 8.11 shows the storage volumes (in GWh per year) required to avoid the curtailment of variable renewable power generation and the utilization of storage capacities for batteries and pumped hydro for charging with variable renewable electricity in the calculated scenarios. The total storage throughput, including the hydrogen production and the amount of hydrogen-based dispatch power plants, is also shown.

Pumped hydropower will remain the backbone of the storage concept until 2030, when batteries start to overtake pumped hydropower by volume. The model does not distinguish between different battery or pumped hydro technologies. Hydrogen-based dispatch will remain the largest contributor to systems services after 2030 until the end of the modelling period.

8.4.5 Global: Required Storage Capacities for the Stationary Power Sector

The world market for storage and dispatch technologies and services will increase significantly in the 2.0 °C Scenario. The annual market for new hydro pump storage plants will grow on average by 6 GW per year to a total capacity of 244 GW in 2030. During the same period, the total installed capacity for batteries will grow to 12 GW, requiring an annual market of 1 GW. Between 2030 and 2050, the energy service sector for storage and storage technologies must accelerate further. The

Table 8.11 Global: storage and dispatch

Storage and dispatch		2.0 °C					1.5 °C				
		Required to avoid curtailment	Utilization battery -charge-	Utilization PSH -charge-	Total (incl. H2)	Dispatch H2	Required to avoid curtailment	Utilization battery -charge-	Utilization PSH -charge-	Total (incl. H2)	Dispatch H2
World		[GWh/year]	[GWh/year]	[GWh/ year]	[GWh/ year]	[GWh/ year]	[GWh/year]	[GWh/year]	[GWh/ year]	[GWh/ year]	[GWh/ year]
OECD North America	2020	0	0	0	0	0	0	0	0	0	0
	2030	62,369	341	192	1065	11,181	243,235	243,235	475	2405	11,181
	2050	853,401	21,805	868	45,331	238,730	999,704	999,704	924	46,766	238,730
Latin America	2020	0	0	0	0	0	0	0	0	0	0
	2030	0	0	0	0	34	1207	1207	99	318	34
	2050	1314	640	34	1347	127,226	30,526	30,526	621	12,875	127,226
OECD Europe	2020	0	0	0	0	0	0	0	0	0	0
	2030	6238	315	5265	11,161	60,223	38,504	38,504	20,566	42,827	60,223
	2050	212,060	30,546	58,368	177,632	814,585	301,234	301,234	72,812	215,641	814,585
Middle East	2020	0	0	0	0	0	0	0	0	0	0
	2030	18,088	2	943	1890	0	44,945	44,945	1469	2943	0
	2050	752,882	109	4636	9180	0	554,222	554,222	4371	8618	0
Africa	2020	0	0	0	0	0	0	0	0	0	0
	2030	4877	118	2244	4726	0	11,264	11,264	2672	5591	0
	2050	367,201	6514	8977	30,974	212,902	585,423	585,423	9282	31,210	212,902
Eurasia	2020	0	0	0	0	0	0	0	0	0	0
	2030	736	1	169	341	14,106	6031	6031	644	1295	14,106
	2050	296,490	948	8396	18,661	401,044	249,984	249,984	7258	16,303	401,044

(continued)

Table 8.11 (continued)

Storage and dispatch		2.0 °C					1.5 °C				
		Required to avoid curtailment	Utilization battery -charge-	Utilization PSH -charge-	Total (incl. H2)	Dispatch H2	Required to avoid curtailment	Utilization battery -charge-	Utilization PSH -charge-	Total (incl. H2)	Dispatch H2
World		[GWh/year]	[GWh/year]	[GWh/ year]	[GWh/ year]	[GWh/ year]	[GWh/year]	[GWh/year]	[GWh/ year]	[GWh/ year]	[GWh/ year]
Non-OECD Asia	2020	0	0	0	0	0	0	0	0	0	0
	2030	137	2	15	34	0	6848	6848	311	646	0
	2050	171,973	2478	2261	9465	386,454	228,160	228,160	2943	8789	386,454
India	2020	0	0	0	0	0	0	0	0	0	0
	2030	59,399	52	2983	6069	1759	182,561	182,561	8577	17,487	1759
	2050	372,809	2125	6715	17,678	28,113	437,884	437,884	6595	17,199	28,113
China	2020	0	0	0	0	0	0	0	0	0	0
	2030	1102	19	394	827	2582	45,217	45,217	7266	14,957	2582
	2050	102,042	57,483	2966	120,899	623,254	264,729	264,729	20,885	60,022	623,254
OECD Pacific	2020	16	0	0	0	0	16	16	0	0	0
	2030	84,079	623	4601	10,403	831	146,440	146,440	6688	14,855	831
	2050	654,287	70,404	14,815	170,431	81,215	760,962	760,962	14,865	169,093	81,215
Total global	2020	16	0	0	0	0	16	0	0	0	0
	2030	237,026	1474	16,808	36,517	90,716	726,252	2945	48,767	103,323	90,716
	2050	3,784,459	193,051	108,037	601,598	2,913,522	4,412,827	153,528	140,555	586,516	2,913,522

battery market must grow by an annual installation rate of 22 GW, and as a result, it will overtake the global capacity of pumped hydro between 2040 and 2050. The conversion of the gas infrastructure from natural gas to hydrogen and synthetic fuels will start slowly between 2020 and 2030, with the conversion of power plants with an annual capacity of around 2 GW. However, after 2030, the transformation of the global gas industry to hydrogen will accelerate significantly, with a total of 197 GW of gas power plants and gas co-generation capacity converted each year. In parallel, the average capacity factor for gas and hydrogen plants will decrease from 29% (2578 h/year) in 2030 to 11% (975 h/year) by 2050, turning the gas sector from a supply-driven to a service-driven industry.

At around 2030, the 1.5 °C Scenario requires more storage throughput than does the 2.0 °C Scenario, but storage demands for the two scenarios will be equal at the end of the modelling period. It is assumed that this higher throughput can be managed with equally high installed capacities, leading to annual capacity factors for battery and hydro pump storage of around 5–6% by 2050 (Table 8.12).

Table 8.13 shows the average global investment costs for the battery and hydro pump storage capacities in the 2.0 °C and 1.5 °C Scenarios. Both pathways have equal storage capacities and cost projections, especially for batteries, but are highly uncertain in the years beyond 2025. Therefore, the costs are only estimates and require research.

8.5 OECD North America

8.5.1 OECD North America: Long-Term Energy Pathways

8.5.1.1 OECD North America: Final Energy Demand by Sector

Combining the assumptions for population growth, GDP growth, and energy intensity will result in the development pathways for OECD North America's final energy demand shown in Fig. 8.17 under the 5.0 °C, 2.0 °C, and 1.5 °C Scenarios. Under the 5.0 °C Scenario, the total final energy demand will increase by 10% from the current 70,500 PJ/year to 77,800 PJ/year in 2050. In the 2.0 °C Scenario, the final energy demand will decrease by 47% compared with current consumption and will reach 37,300 PJ/year by 2050. The final energy demand in the 1.5 °C Scenario will reach 33,700 PJ, 52% below the 2015 demand level. In the 1.5 °C Scenario, the final energy demand in 2050 will be 10% lower than in the 2.0 °C Scenario. The electricity demand for 'classical' electrical devices (without power-to-heat or e-mobility) will decrease from 4230 TWh/year in 2015 to 3340 TWh/year (2.0 °C) or 2950 TWh/year (1.5 °C) by 2050. Compared with the 5.0 °C case (6050 TWh/year in 2050), the efficiency measures in the 2.0 °C and 1.5 °C Scenarios will save a maximum of 2710 TWh/year and 3100 TWh/year, respectively.

Electrification will lead to a significant increase in the electricity demand by 2050. The 2.0 °C Scenario will require approximately 1400 TWh/year of electricity

Table 8.12 Required increases in storage capacities until 2050

		Global storage and H2-dispatch market volume under 2 scenarios							
		Batteries			Pumped hydro			Hydrogen -production + dispatch	
		[Through-put]	Cumulative capacity	Storage technology share	[Through-put]	Cumulative capacity	Storage technology share	[Through-put]	Cumulative capacity
		[GWh/year]	[GW]	[%]	[GWh/year]	[GW]	[%]	[GWh/year]	[GW]
2015		No data	2	1	No data	153	99		No data
2030	2.0 °C	1474	12	8	16,808	244	92	90,716	35
2030	1.5 °C	2945	13	6	48,767	255	94	351,496	137
2050	2.0 °C	193,051	347	64	108,037	267	36	2,913,522	2990
2050	1.5 °C	153,528	340	52	140,555	278	48	2,075,533	3423

Table 8.13 Estimated average global investment costs for batty and hydro pump storage

Estimated storage investment costs (In $ billion)	2015–2020	Average annual	2021–2030	Average annual	2031–2040	Average annual	2041–2050	Average annual	2015–2050	Average annual
Storage										
Battery	4.8	0.967	44.5	4.4	148.1	14.8	655.8	65.6	853.3	24.4
Hydro pump storage	0	0	38.7	3.9	42.7	4.3	47.2	4.7	128.6	3.7
Total	4.8	0.967	83.2	8.3	190.8	19.1	703.0	70.3	981.9	28.1

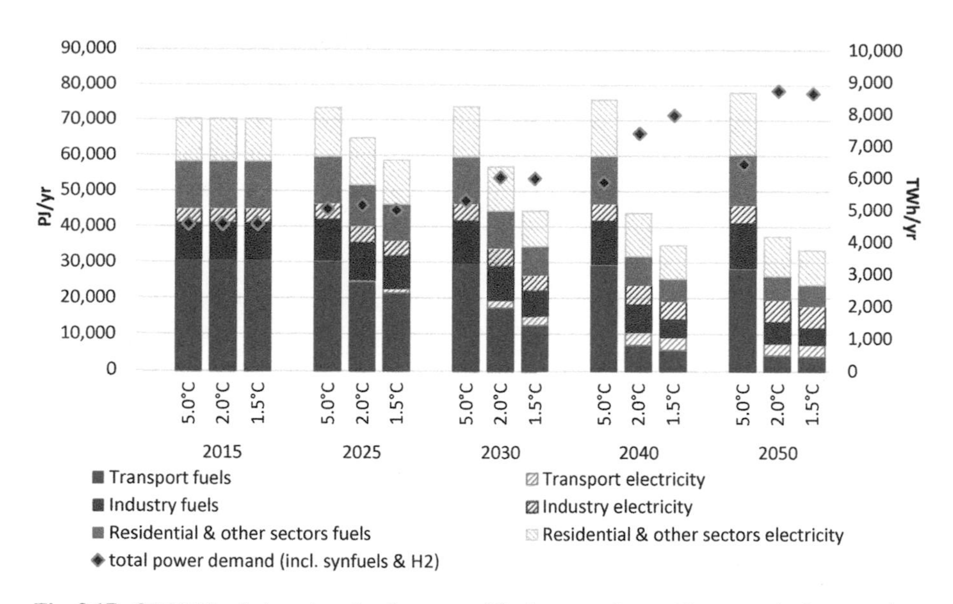

Fig. 8.17 OECD North America: development of final energy demand by sector in the scenarios

for electric heaters and heat pumps, and in the transport sector, it will require approximately 3300 TWh/year for electric mobility. The generation of hydrogen (for transport and high-temperature process heat) and the manufacture of synthetic fuels (mainly for transport) will add an additional power demand of 3000 TWh/year. Therefore, the gross power demand will rise from 5300 TWh/year in 2015 to 9500 TWh/year in 2050 in the 2.0 °C Scenario, 30% higher than in the 5.0 °C case. In the 1.5 °C Scenario, the gross electricity demand will increase to a maximum of 9400 TWh/year in 2050 for similar reasons.

The efficiency gains in the heating sector will be similar in magnitude to those in the electricity sector. Under the 2.0 °C and 1.5 °C Scenarios, a final energy consumption equivalent to about 7000 PJ/year and 9400 PJ/year, respectively, will be avoided by 2050 through efficiency gains compared with the 5.0 °C Scenario.

8.5.1.2 OECD North America: Electricity Generation

The development of the power system is characterized by a dynamically growing renewable energy market and an increasing proportion of total power from renewable sources. In the 2.0 °C Scenario, 100% of the electricity produced in OECD North America will come from renewable energy sources by 2050. 'New' renewables—mainly wind, solar, and geothermal energy—will contribute 85% of the total electricity generated. Renewable electricity's share of the total production will be 68% by 2030 and 89% by 2040. The installed capacity of renewables will reach about 1880 GW by 2030 and 3810 GW by 2050. In the 1.5 °C Scenario, the share of renewable electricity generation in 2030 is assumed to be 84%. The 1.5 °C Scenario projects a generation capacity from renewable energy of about 3920 GW in 2050.

Table 8.14 shows the development of the installed capacities of different renewable technologies in OECD North America over time. Figure 8.18 provides an overview of the overall power-generation structure in OECD North America. From 2020 onwards, the continuing growth of wind and PV—to 1090 GW and 2130 GW, respectively—is complemented by up to 210 GW of solar thermal generation, as well as limited biomass, geothermal, and ocean energy, in the 2.0 °C Scenario. Both the 2.0 °C and 1.5 °C Scenarios will lead to a high proportion of variable power generation (PV, wind, and ocean) of 49% and 59%, respectively, by 2030, and 73% and 74%, respectively, by 2050.

Table 8.14 OECD North America: development of renewable electricity generation capacity in the scenarios

in GW	Case	2015	2025	2030	2040	2050
Hydro	5.0 °C	194	202	207	216	217
	2.0 °C	194	199	202	206	206
	1.5 °C	194	199	202	203	203
Biomass	5.0 °C	22	25	27	30	35
	2.0 °C	22	27	32	42	52
	1.5 °C	22	35	39	43	45
Wind	5.0 °C	87	157	172	197	253
	2.0 °C	87	323	540	812	1092
	1.5 °C	87	358	656	924	1059
Geothermal	5.0 °C	5	5	6	9	12
	2.0 °C	5	6	9	23	37
	1.5 °C	5	5	8	25	37
PV	5.0 °C	29	133	162	220	358
	2.0 °C	29	534	991	1419	2129
	1.5 °C	29	659	1097	1783	2269
CSP	5.0 °C	2	2	3	4	12
	2.0 °C	2	22	87	168	209
	1.5 °C	2	39	148	257	236
Ocean	5.0 °C	0	0	1	2	4
	2.0 °C	0	3	15	59	85
	1.5 °C	0	2	13	52	66
Total	5.0 °C	338	523	577	678	891
	2.0 °C	338	1115	1878	2729	3810
	1.5 °C	338	1298	2163	3288	3916

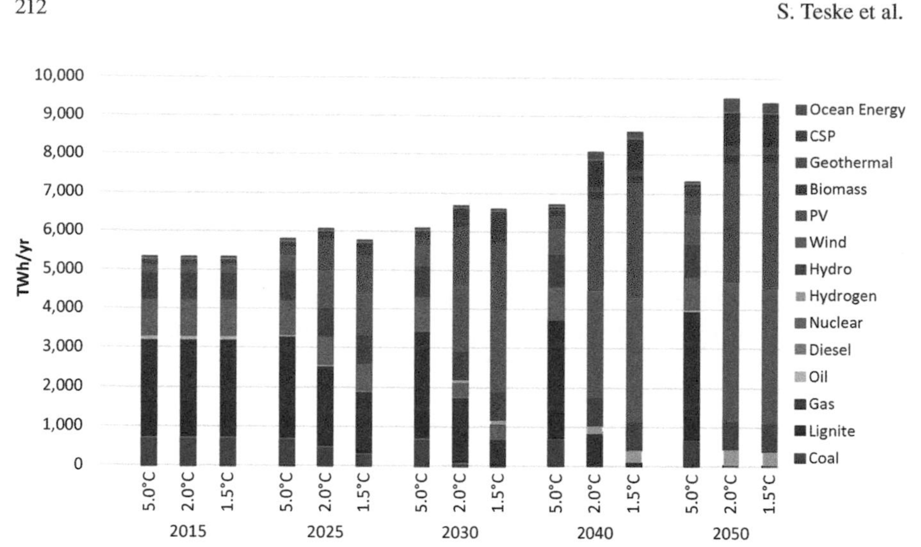

Fig. 8.18 OECD North America: development of electricity-generation structure in the scenarios

8.5.1.3 OECD North America: Future Costs of Electricity Generation

Figure 8.19 shows the development of the electricity-generation and supply costs over time, including CO_2 emission costs, in all scenarios. The calculated electricity-generation costs in 2015 (referring to full costs) were around 4.9 ct/kWh. In the 5.0 °C case, the generation costs will increase until 2050, when they reach 10.1 ct/kWh. The generation costs in the 2.0 °C Scenario will increase in a similar way until 2030, when they reach 8.3 ct/kWh, and then drop to 6.8 ct/kWh by 2050. In the 1.5 °C Scenario, they will increase to 8.8 ct/kWh and then drop to 7.1 ct/kWh by 2050. In the 2.0 °C Scenario, the generation costs in 2050 are 3.3 ct/kWh lower than in the 5.0 °C case. In the 1.5 °C Scenario, the generation costs in 2050 are 3.1 ct/kWh lower than in the 5.0 °C case. Note that these estimates of generation costs do not take into account integration costs such as power grid expansion, storage, or other load-balancing measures.

Under the 5.0 °C case, the growth in demand and increasing fossil fuel prices will result in an increase in total electricity supply costs from today's $270 billion/year to more than $760 billion/year in 2050. In both alternative scenarios, the total supply costs in 2050 will be around $690 billion/year The long-term costs for electricity supply in 2050 will be 8%–9% lower than in the 5.0 °C Scenario as a result of the estimated generation costs and the electrification of heating and mobility.

Compared with these results, the generation costs when the CO_2 emission costs are not considered will increase in the 5.0 °C case to 7.5 ct/kWh. In the 2.0 °C Scenario, they will increase until 2030, when they reach 7.3 ct/kWh, and then drop to 6.8 ct/kWh by 2050. In the 1.5 °C Scenario, they will increase to 8.4 ct/kWh in 2030, and then drop to 7.1 ct/kWh by 2050. In the 2.0 °C Scenario, the generation costs will be, at maximum, 1 ct/kWh higher than in the 5.0 °C case, and this will

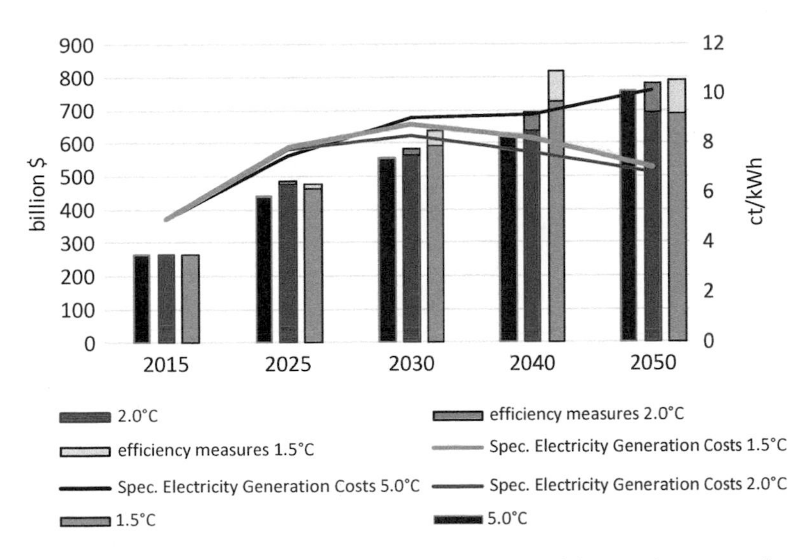

Fig. 8.19 OECD North America: development of total electricity supply costs and specific electricity-generation costs in the scenarios

occur in 2030. In the 1.5 °C Scenario, compared with the 5.0 °C Scenario, the maximum difference in generation costs will be 2 ct/kWh in 2030. If the CO_2 costs are not considered, the total electricity supply costs in the 5.0 °C case will increase to $570 billion/year in 2050.

8.5.1.4 OECD North America: Future Investments in the Power Sector

An investment of around $7600 billion will be required for power generation between 2015 and 2050 in the 2.0 °C Scenario—including additional power plants for the production of hydrogen and synthetic fuels and investments in plant replacement after the end of their economic lifetimes. This value is equivalent to approximately $211 billion per year on average, which is $4400 billion more than in the 5.0 °C case ($3200 billion). In the 1.5 °C Scenario, an investment of around $8180 billion for power generation will be required between 2015 and 2050. On average, this is an investment of $227 billion per year. In the 5.0 °C Scenario, the investment in conventional power plants will be around 48% of the total cumulative investments, whereas approximately 52% will be invested in renewable power generation and co-generation (Fig. 8.20). However, under the 2.0 °C (1.5 °C) Scenario, OECD North America will shift almost 93% (93%) of its entire investment to renewables and co-generation. By 2030, the fossil fuel share of the power sector investment will mainly focus on gas power plants that can also be operated with hydrogen.

Because renewable energy has no fuel costs, other than biomass, the cumulative fuel cost savings in the 2.0 °C Scenario will reach a total of $3240 billion in 2050, equivalent to $90 billion per year. Therefore, the total fuel cost savings will be

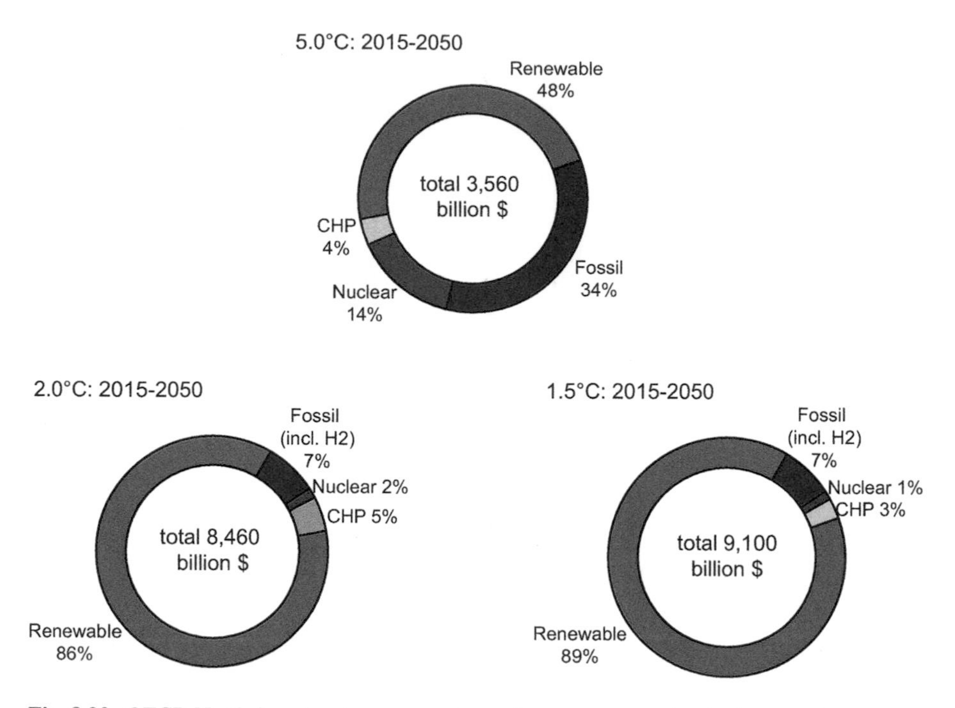

Fig. 8.20 OECD North America: investment shares for power generation in the scenarios

equivalent to 70% of the total additional investments compared to the 5.0 °C Scenario. The fuel cost savings in the 1.5 °C Scenario will add up to $3910 billion, or $109 billion per year.

8.5.1.5 OECD North America: Energy Supply for Heating

The final energy demand for heating will increases in the 5.0 °C Scenario by 32%, from 19,700 PJ/year in 2015 to 26,000 PJ/year in 2050. Energy efficiency measures will help to reduce the energy demand for heating by 27% by 2050 in the 2.0 °C Scenario relative to the 5.0 °C case, and by 36% in the 1.5 °C Scenario. Today, renewables supply around 11% of OECD North America's final energy demand for heating, with the main contribution from biomass. Renewable energy will provide 38% of OECD North America's total heat demand in 2030 in the 2.0 °C Scenario and 61% in the 1.5 °C Scenario. In both scenarios, renewables will provide 100% of the total heat demand in 2050.

Figure 8.21 shows the development of different technologies for heating in OECD North America over time, and Table 8.15 provides the resulting renewable heat supply for all scenarios. Until 2030, biomass will remain the main contributor. The growing use of solar, geothermal, and environmental heat will lead, in the long term, to a biomass share of 25% under the 2.0 °C Scenario and 19% under the 1.5 °C Scenario. Heat from renewable hydrogen will further reduce the dependence on fossil fuels under both scenarios. Hydrogen consumption in 2050 will be around

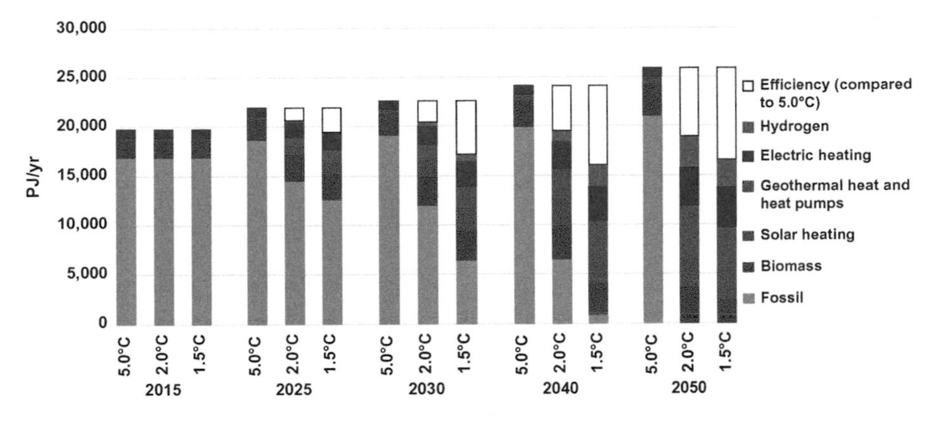

Fig. 8.21 OECD North America: development of heat supply by energy carrier in the scenarios

Table 8.15 OECD North America: development of renewable heat supply in the scenarios (excluding the direct use of electricity)

in PJ/year	Case	2015	2025	2030	2040	2050
Biomass	5.0 °C	1868	2142	2334	2787	3279
	2.0 °C	1868	2758	3019	3493	3686
	1.5 °C	1868	2707	3149	3191	2378
Solar heating	5.0 °C	107	210	277	451	695
	2.0 °C	107	887	1772	2639	2962
	1.5 °C	107	1290	2169	2839	3128
Geothermal heat and heat pumps	5.0 °C	17	17	18	18	19
	2.0 °C	17	875	1378	3031	5257
	1.5 °C	17	1076	2185	3463	4152
Hydrogen	5.0 °C	0	0	0	0	0
	2.0 °C	0	144	276	1014	3045
	1.5 °C	0	22	677	2100	2666
Total	5.0 °C	1991	2369	2629	3256	3994
	2.0 °C	1991	4664	6445	10,176	14,949
	1.5 °C	1991	5095	8180	11,592	12,324

3000 PJ/year in the 2.0 °C Scenario and 2700 PJ/year in the 1.5 °C Scenario. The direct use of electricity for heating will also increase by a factor of 4.6–4.9 between 2015 and 2050 and will have a final energy share of 21% in 2050 in the 2.0 °C Scenario and 26% in the 1.5 °C Scenario.

8.5.1.6 OECD North America: Future Investments in the Heating Sector

The roughly estimated investments in renewable heating technologies up to 2050 will amount to around $2580 billion in the 2.0 °C Scenario (including investments for plant replacement after their economic lifetimes) or approximately $72 billion

Table 8.16 OECD North America: installed capacities for renewable heat generation in the scenarios

in GW	Case	2015	2025	2030	2040	2050
Biomass	5.0 °C	292	315	330	366	411
	2.0 °C	292	381	387	355	272
	1.5 °C	292	360	384	334	179
Geothermal	5.0 °C	0	0	0	0	0
	2.0 °C	0	17	30	44	52
	1.5 °C	0	34	57	82	109
Solar heating	5.0 °C	29	58	76	124	191
	2.0 °C	29	232	466	697	780
	1.5 °C	29	331	557	728	793
Heat pumps	5.0 °C	3	3	3	3	3
	2.0 °C	3	123	188	393	677
	1.5 °C	3	143	292	479	568
Total[a]	5.0 °C	324	375	410	494	605
	2.0 °C	324	752	1071	1489	1781
	1.5 °C	324	868	1290	1622	1649

[a]Excluding direct electric heating

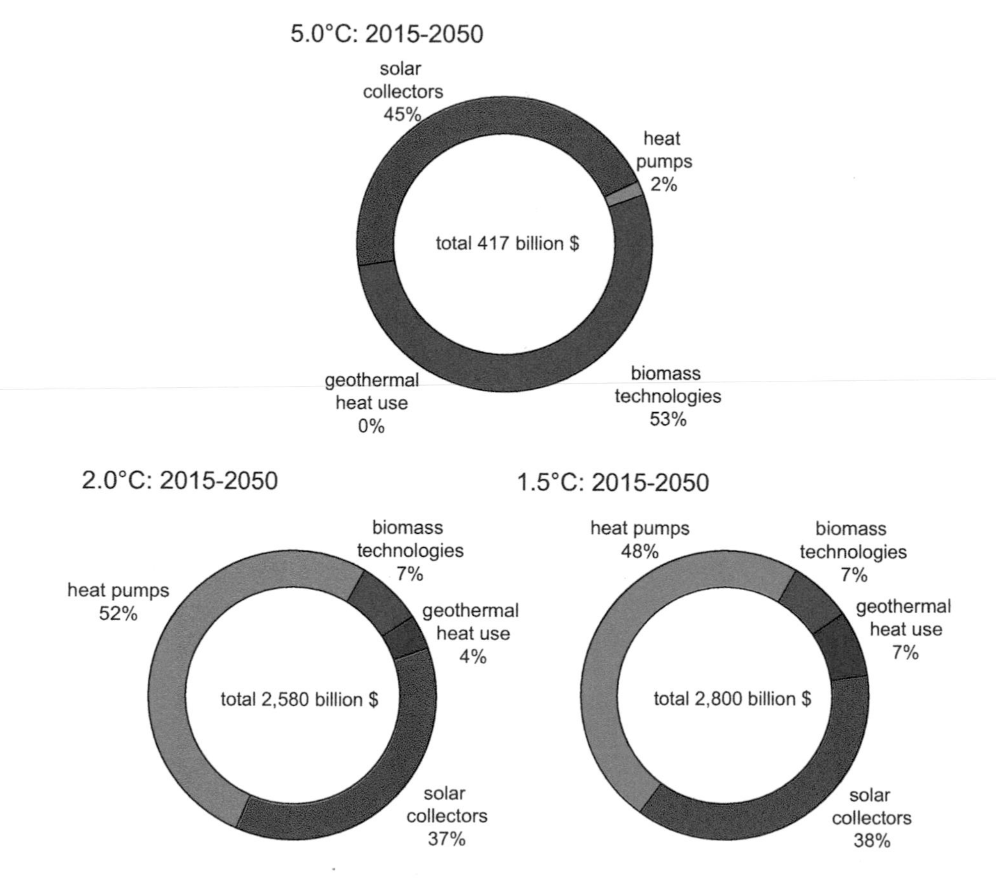

Fig. 8.22 OECD North America: development of investments in renewable heat generation technologies in the scenarios

per year. The largest share of investment in OECD North America is assumed to be for heat pumps (around $1300 billion), followed by solar collectors and biomass technologies. The 1.5 °C Scenario assumes an even faster expansion of renewable technologies, resulting in a lower average annual investment of around $78 billion per year (Table 8.16, Fig. 8.22).

8.5.1.7 OECD North America: Transport

Energy demand in the transport sector in OECD North America is expected to decrease by 8% in the 5.0 °C Scenario, from around 31,000 PJ/year in 2015 to 28,600 PJ/year in 2050. In the 2.0 °C Scenario, assumed technical, structural, and behavioural changes will save 73% (20,970 PJ/year) in 2050 compared with the 5.0 °C case. Additional modal shifts, technology switches, and a reduction in transport demand will lead to even higher energy savings in the 1.5 °C Scenario, of 74% (or 21,100 PJ/year) in 2050 compared with the 5.0 °C case (Table 8.17, Fig. 8.23).

By 2030, electricity will provide 11% (620 TWh/year) of the transport sector's total energy demand in the 2.0 °C Scenario, and in 2050, the share will be 44% (930 TWh/year). In 2050, up to 2090 PJ/year of hydrogen will be used in the transport sector as a complementary renewable option. In the 1.5 °C Scenario, the annual electricity demand will be 1030 TWh in 2050. The 1.5 °C Scenario also assumes a hydrogen demand of 2020 PJ/year by 2050.

Biofuel use is limited in the 2.0 °C Scenario to a maximum of 2540 PJ/year Therefore, around 2030, synthetic fuels based on power-to-liquid will be introduced, with a maximum amount of 270 PJ/year in 2050. Because the reduction in

Table 8.17 OECD North America: projection of the transport energy demand by mode in the scenarios

in PJ/year	Case	2015	2025	2030	2040	2050
Rail	5.0 °C	674	628	609	570	529
	2.0 °C	674	660	655	523	516
	1.5 °C	674	743	730	773	806
Road	5.0 °C	26,686	25,691	24,838	24,222	23,414
	2.0 °C	26,686	21,257	15,933	7731	5124
	1.5 °C	26,686	18,612	11,973	6717	5251
Domestic aviation	5.0 °C	2421	2978	3274	3398	3186
	2.0 °C	2421	2309	2026	1530	1242
	1.5 °C	2421	2167	1689	1063	840
Domestic navigation	5.0 °C	461	482	493	514	535
	2.0 °C	461	481	489	489	473
	1.5 °C	461	479	484	483	473
Total	5.0 °C	30,241	29,779	29,214	28,704	27,664
	2.0 °C	30,241	24,707	19,104	10,273	7354
	1.5 °C	30,241	22,000	14,875	9036	7370

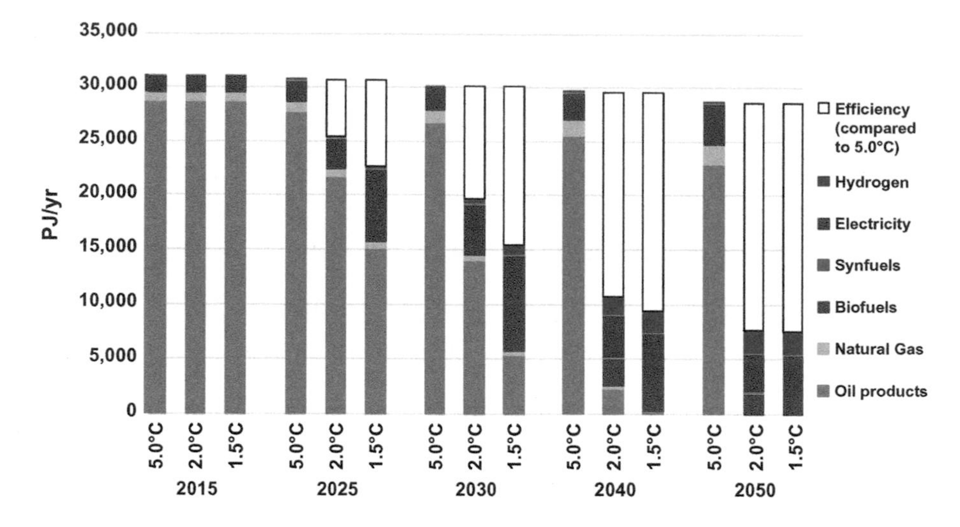

Fig. 8.23 OECD North America: final energy consumption by transport in the scenarios

fossil fuel for transport will be faster, biofuel use will increase in the 1.5 °C Scenario to a maximum of 5900 PJ/year. The demand for synthetic fuels will decrease to zero by 2050 in the 1.5 °C Scenario because of the lower overall energy demand by transport.

8.5.1.8 OECD North America: Development of CO_2 Emissions

In the 5.0 °C Scenario, OECD North America's annual CO_2 emissions will decrease by 9% from 6170 Mt. in 2015 to 5612 Mt. in 2050. Stringent mitigation measures in both the alternative scenarios will lead to reductions in annual emissions to 930 Mt. in 2040 in the 2.0 °C Scenario and to 120 Mt. in the 1.5 °C Scenario, with further reductions to almost zero by 2050. In the 5.0 °C case, the cumulative CO_2 emissions from 2015 until 2050 will add up to 216 Gt. In contrast, in the 2.0 °C and 1.5 °C Scenarios, the cumulative emissions for the period from 2015 until 2050 will be 99 Gt and 72 Gt, respectively.

Therefore, the cumulative CO_2 emissions will decrease by 54% in the 2.0 °C Scenario and by 67% in the 1.5 °C Scenario compared with the 5.0 °C case. A rapid decrease in the annual emissions will occur under both alternative scenarios. In the 2.0 °C Scenario, the reduction will be greatest in 'Power generation', followed by the 'Transport' and 'Residential and other' sectors (Fig. 8.24).

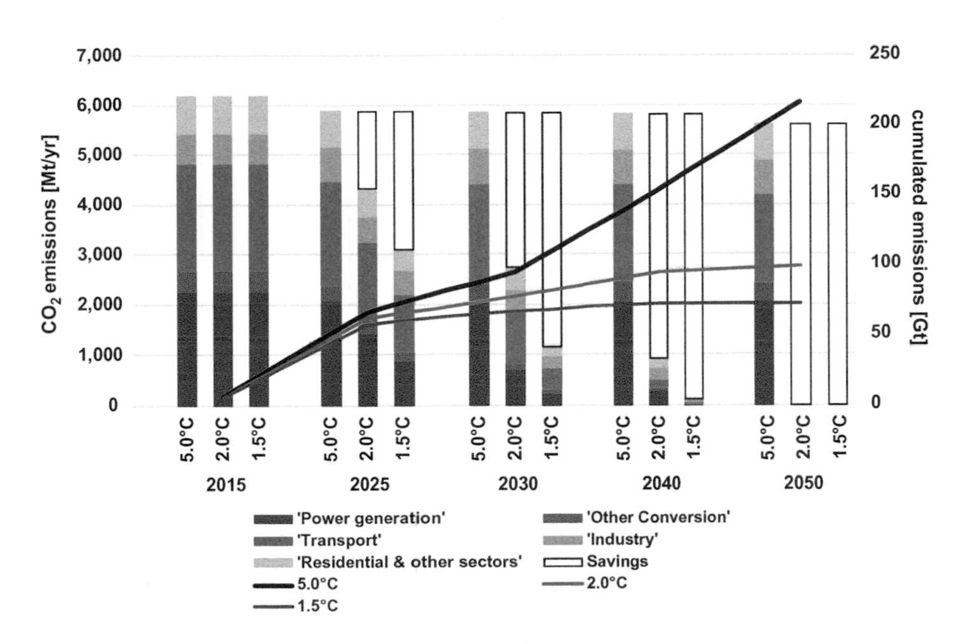

Fig. 8.24 OECD North America: development of CO_2 emissions by sector and cumulative CO_2 emissions (after 2015) in the scenarios ('Savings' = reduction compared with the 5.0 °C Scenario)

8.5.1.9 OECD North America: Primary Energy Consumption

Taking into account the assumptions discussed above, the levels of primary energy consumption under the three scenarios are shown in Fig. 8.25. In the 2.0 °C Scenario, the primary energy demand will decrease by 46%, from around 111,600 PJ/year in 2015 to 60,600 PJ/year in 2050. Compared with the 5.0 °C Scenario, the overall primary energy demand will decrease by 50% by 2050 in the 2.0 °C Scenario (5.0 °C: 121,000 PJ in 2050). In the 1.5 °C Scenario, the primary energy demand will be even lower (56,600 PJ in 2050) because the final energy demand and conversion losses will be lower.

Both the 2.0 °C and 1.5 °C Scenarios aim to rapidly phase-out coal and oil. As a result, renewable energy will have a primary energy share of 34% in 2030 and 91% in 2050 in the 2.0 °C Scenario. In the 1.5 °C Scenario, renewables will have a pri-

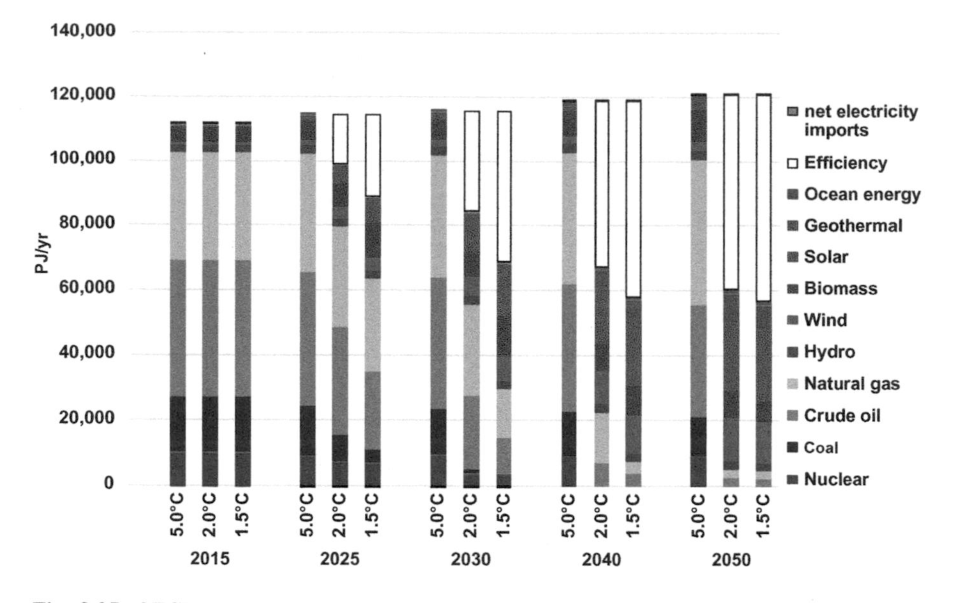

Fig. 8.25 OECD North America: projection of total primary energy demand (PED) by energy carrier in the scenarios (including electricity import balance)

mary share of more than 91% in 2050 (including non-energy consumption, which will still include fossil fuels). Nuclear energy will be phased-out by 2040 under both the 2.0 °C and the 1.5 °C Scenarios. The cumulative primary energy consumption of natural gas in the 5.0 °C case will add up to 1290 EJ, the cumulative coal consumption to about 470 EJ, and the crude oil consumption to 1300 EJ. In contrast, in the 2.0 °C Scenario, the cumulative gas demand will amount to 730 EJ, the cumulative coal demand to 120 EJ, and the cumulative oil demand to 640 EJ. Even lower cumulative fossil fuel use will be achieved in the 1.5 °C Scenario: 480 EJ for natural gas, 80 EJ for coal, and 440 EJ for oil.

8.5.2 Regional Results: Power Sector Analysis

The key results for all 10 world regions and their sub-regions are presented in this section, with standardized tables to make them comparable for the reader. Regional differences and particularities are summarized. It is important to note that the electricity loads for the sub-regions—which are in several cases also countries—are calculated (see Chap. 3) and are not real measured values. When information was available, the model results are compared with published peak loads and adapted as far as possible. However, deviations of 10% or more for the base year are in the range of the probability. The interconnection capacities between sub-regions are simplified assumptions based on current practices in liberalized power markets, and include cross-border trade (e.g., between Canada and the USA) (C2ES 2017) or within the European Union (EU). The EU set a target of 10% interconnection capacity between their member states in 2002 (EU-EG 2017). The interconnection capacities for sub-regions that are not geographically connected are set to zero for the entire modelling period, even when there is current discussion about the implementation of new interconnections, such as for the ASEAN Power Grid (ASEAN-CE 2018).

8.5.3 OECD North America: Power Sector Analysis

The OECD North America region includes Canada, the USA, and Mexico, and therefore contains more than one large electricity market. Although the power sector is liberalized in all three countries, the state of implementation and the market rules in place vary significantly. Even within the USA, each state has different market rules and grid regulations. Therefore, the calculated scenarios assume the priority dispatch of all renewables and priority grid connections for new renewable power plants, and a streamlined process for required construction permits. The power sector analysis for all regions is based on technical, not political, considerations.

8.5.3.1 OECD North America: Development of Power Plant Capacities

The size of the renewable power market in OECD North America will increase significantly in the 2.0 °C Scenario. The annual market for solar PV must increase from 22.76 GW in 2020 by a factor of 5 to an average of 95 GW by 2030. The onshore wind market must expand to 35 GW by 2025, an increase from around 13 GW 5 years earlier. By 2050, offshore wind generation will increase to 9.7 GW annually, by a factor of 7 compared with the base year (2015). Concentrated solar power plants will play an important role in dispatchable solar electricity generation to supply bulk power, especially for industry and industrial process heat. The annual market in 2030 will increase to 16 GW, compared with 1.7 GW in 2020. The 1.5 °C Scenario accelerates both the phase-out of fossil-fuel-based power generation and the deployment of renewables—mainly solar PV and wind in the first decade—about 5–7 years faster than the 2.0 °C Scenario (Table 8.18).

8.5.3.2 OECD North America: Utilization of Power-Generation Capacities

Table 8.19 shows the increasing shares of variable renewable power generation across all North American regions. Whereas Alaska and Canada are dominated by variable wind power generation, Mexico and the sunny mid-west of the USA have significant contributions from CSP. Solar PV is distributed evenly across the entire

Table 8.18 OECD North America: average annual change in installed power plant capacity

Power generation: average annual change of installed capacity [GW/a]	2015–2025		2026–2035		2036–2050	
	2.0 °C	1.5 °C	2.0 °C	1.5 °C	2.0 °C	1.5 °C
Hard coal	−7	−16	−6	−8	−4	0
Lignite	−14	−18	−7	0	0	0
Gas	6	9	12	1	−55	−4
Hydrogen-gas	1	10	4	24	55	39
Oil/diesel	−5	−7	−3	−4	−1	0
Nuclear	−4	−9	−10	−10	0	−1
Biomass	1	2	1	1	1	0
Hydro	−5	−3	0	0	0	2
Wind (onshore)	24	48	36	36	24	19
Wind (offshore)	2	19	11	19	10	3
PV (roof top)	39	94	64	68	61	55
PV (utility scale)	13	31	21	23	20	18
Geothermal	0	0	1	1	2	2
Solar thermal power plants	3	18	15	18	6	4
Ocean energy	1	2	4	4	4	3
Renewable fuel based co-generation	1	2	2	2	2	0

region. Onshore and offshore wind penetration is highest in rural areas, whereas solar roof-top power generation is highest in suburban regions where roof space and electricity demand from residential buildings correlate best. The south-west of the USA will have the highest share of variable renewables—mainly solar PV for residual homes and office buildings, connected to battery systems. There are no structural differences between the 2.0 °C and 1.5 °C Scenarios, except faster implementation in the latter. It is assumed that all regions will have an interconnection capacity of 20% of the regional average load, with which to exchange renewable and dispatch electricity to neighbouring regions.

Capacity factors for the five generation types and the resulting average utilization are shown in Table 8.20. Compared with the global average, North America will start with a capacity factor for limited dispatchable generation of about 10% over the global average. By 2050, the average capacity factor across all power-generation types will be 29% for both scenarios. A low average capacity factor requires flexible power plants and a power market framework that incentivizes them.

8.5.3.3 OECD North America: Development of Load, Generation, and Residual Load

Table 8.21 shows the development of the maximum load, generation, and resulting residual load (the load remaining after variable renewable generation). With increased shares of variable solar PV and wind power, the minimum residual load

Table 8.19 OECD North America and sub-regions: power system shares by technology group

Power generation structure and interconnection		2.0 °C				1.5 °C			
OECD North America		Variable RE	Dispatch RE	Dispatch fossil	Inter-connection	Variable RE	Dispatch RE	Dispatch fossil	Inter-connection
USA Alaska	2015	4%	35%	61%	10%				
	2030	29%	31%	40%	15%	36%	30%	34%	15%
	2050	42%	23%	35%	20%	42%	26%	32%	20%
Canada West	2015	6%	35%	59%	10%				
	2030	43%	30%	27%	15%	53%	28%	19%	15%
	2050	63%	21%	16%	20%	63%	23%	14%	20%
Canada East	2015	7%	35%	59%	10%				
	2030	45%	30%	25%	15%	56%	27%	16%	15%
	2050	66%	21%	13%	20%	66%	23%	11%	20%
USA North East	2015	7%	35%	58%	10%				
	2030	47%	31%	22%	15%	58%	28%	14%	15%
	2050	69%	20%	11%	20%	69%	22%	9%	20%
USA North West	2015	4%	35%	61%	10%				
	2030	36%	32%	32%	15%	47%	30%	23%	15%
	2050	59%	23%	18%	20%	59%	25%	16%	20%
USA South West	2015	7%	35%	58%	10%				
	2030	53%	28%	19%	15%	64%	25%	11%	15%
	2050	73%	17%	10%	20%	73%	18%	8%	20%
USA South East	2015	8%	35%	58%	10%				
	2030	53%	28%	19%	15%	63%	25%	12%	15%
	2050	71%	18%	11%	20%	71%	20%	9%	20%

(continued)

Table 8.19 (continued)

Power generation structure and interconnection		2.0 °C				1.5 °C			
OECD North America		Variable RE	Dispatch RE	Dispatch fossil	Inter-connection	Variable RE	Dispatch RE	Dispatch fossil	Inter-connection
Mexico	2015	5%	35%	61%	10%				
	2030	37%	30%	32%	15%	46%	28%	26%	15%
	2050	56%	23%	22%	20%	55%	25%	19%	20%
OECD North America	2015	7%	35%	58%					
	2030	48%	30%	23%		59%	27%	15%	
	2050	68%	19%	13%		68%	21%	11%	

Table 8.20 OECD North America: capacity factors by generation type

Utilization of variable and dispatchable power generation:		2015	2020	2020	2030	2030	2040	2040	2050	2050
OECD North America			2.0 °C	1.5 °C	2.0 °C	1.5 °C	2.0 °C	1.5 °C	2.0 °C	1.5 °C
Capacity Factor – average	[%/yr]	53.1%	35%	33%	29%	28%	34%	28%	29%	29%
Limited dispatchable: fossil and nuclear	[%/yr]	68.6%	40%	10%	28%	2%	20%	6%	10%	10%
Limited dispatchable: renewable	[%/yr]	45.9%	46%	57%	37%	39%	59%	36%	36%	35%
Dispatchable: fossil	[%/yr]	39.7%	23%	21%	11%	5%	30%	8%	12%	11%
Dispatchable: renewable	[%/yr]	44.0%	52%	68%	49%	52%	47%	44%	49%	45%
Variable: renewable	[%/yr]	18.9%	12%	12%	25%	26%	34%	27%	28%	28%

can become negative. If this happens, the surplus generation can either be exported to other regions, stored, or curtailed. The export of load to other regions requires transmission lines. If the theoretical utilization rate of transmission cables (= interconnection) exceeds 100%, the transport capacity must be increased. We assume that the entire load need not be exported, and that surplus generation capacities can be curtailed because interconnections are costly and require a certain level of utilization to make them economically viable. An analysis of the economic viability of new interconnections and their optimal transmission capacities is beyond the scope of this research project.

In Alaska in the 2.0 °C Scenario, for example, generation and demand are balanced in 2020 and 2030, but peak generation is substantially higher than demand in 2050. In the 1.5 °C Scenario, a significant level of overproduction is achieved by 2030. In the two scenarios, the surplus peak generation is equally high. These results have been calculated under the assumption that surplus generation will be stored in a cascade of batteries and pumped-storage hydroelectricity (PSH) or used to produce hydrogen and/or synthetic fuels. Therefore, the maximal interconnection requirements shown in this chapter represent the maximum surplus generation capacity. To avoid curtailment, these overcapacities have mainly been used for hydrogen production. Therefore, Alaska could remain an energy exporter but switch from oil to wind-generated synthetic gas and/or hydrogen.

Table 8.22 provides an overview of the calculated storage and dispatch power requirements by sub-region. To store or export the entire electricity output during

Table 8.21 OECD North America: load, generation, and residual load development

Power generation structure		2.0 °C				1.5 °C			
OEC D North America		Max demand [GW]	Max generation [GW]	Max Residual Load [GW]	Max interconnection requirements [GW]	Max demand [GW]	Max generation [GW]	Max residual load [GW]	Max interconnection requirements [GW]
NA – USA Alaska	2020	1.4	1.4	0.0		1.4	18.8	0.1	
	2030	1.5	1.5	0.1	0	1.6	13.5	0.3	12
	2050	2.4	11.8	0.5	9	2.4	11.5	0.5	9
NA – Canada West	2020	21.1	21.1	0.0		21.2	34.0	0.3	
	2030	23.0	31.2	5.7	3	24.5	39.8	4.6	11
	2050	37.2	73.1	15.2	21	37.3	76.4	15.3	24
NA – Canada East	2020	53.0	53.0	0.0		53.1	117.3	0.8	
	2030	58.0	88.0	14.6	15	61.6	117.5	15.3	40
	2050	94.3	213.7	41.2	78	94.6	223.0	41.0	87
NA – USA North East	2020	258.6	243.6	29.9		259.5	273.2	21.8	
	2030	288.5	355.7	47.7	20	304.2	468.8	63.5	101
	2050	433.0	853.7	175.3	246	434.6	891.6	176.7	280
NA – USA North West	2020	25.6	25.6	0.0		25.7	81.1	2.2	
	2030	28.5	30.6	5.9	0	30.1	40.8	6.0	5
	2050	42.5	74.3	16.0	16	42.7	77.7	16.1	19

NA – USA South West	2020	109.4	109.1	4.6		109.8	167.5	9.3	
	2030	121.8	163.0	11.8	29	128.5	208.8	20.0	60
	2050	181.8	384.2	38.3	164	182.4	402.3	42.0	178
NA – USA South East	2020	217.7	217.7	0.4		217.4	232.1	15.3	
	2030	255.8	372.6	38.0	79	270.9	490.7	64.7	155
	2050	393.3	890.9	102.6	395	394.5	927.6	122.3	411
Mexico	2020	66.6	51.3	22.3					
	2030	87.2	116.1	21.3	8	97.6	151.9	19.8	35
	2050	171.9	277.1	80.5	25	173.3	289.7	81.9	34

Table 8.22 OECD North America: storage and dispatch service requirements

Storage and dispatch		2.0 °C					1.5 °C				
OECD North America		Required to avoid curtailment [GWh/year]	Utilization battery -through-put- [GWh/year]	Utilization PSH -through-put- [GWh/year]	Total storage demand (incl. H2) [GWh/year]	Dispatch hydrogen-based [GWh/year]	Required to avoid curtailment [GWh/year]	Utilization battery -through-put- [GWh/year]	Utilization PSH -through-put- [GWh/year]	Total Storage demand (incl. H2) [GWh/year]	Dispatch Hydrogen-based [GWh/year]
USA Alaska	2020	0	0	0	0	0	0	0	0	0	0
	2030	11	0	0	0	136	68	1	1	2	136
	2050	328	38	1	39	542	407	41	1	42	542
Canada West	2020	0	0	0	0	0	0	0	0	0	0
	2030	1011	14	7	21	1957	4078	31	18	49	1957
	2050	14,665	1044	34	1078	7776	17,557	1100	38	1137	7776
Canada East	2020	0	0	0	0	0	0	0	0	0	0
	2030	3014	38	20	58	4482	13,352	82	53	135	4482
	2050	42,780	2545	91	2636	18,129	50,077	2623	97	2720	18,129
USA North East	2020	0	0	0	0	0	0	0	0	0	0
	2030	9092	148	73	221	17,290	50,047	404	239	643	17,290
	2050	212,448	13,990	509	14,499	60,398	252,243	14,457	546	15,004	60,398
USA North West	2020	0	0	0	0	0	0	0	0	0	0
	2030	90	4	1	5	2394	1854	26	13	39	2394
	2050	11,806	1013	33	1046	8707	14,933	1085	37	1122	8707
USA South West	2020	0	0	0	0	0	0	0	0	0	0
	2030	10,722	121	68	189	6370	47,636	238	172	410	6370
	2050	172,771	6661	301	6962	22,741	201,316	6894	316	7210	22,741

USA South East	2020	0	0	0	0	0	0	0	0	0	0
	2030	35,827	320	195	516	15,281	115,409	579	402	981	15,281
	2050	372,747	15,600	690	16,290	53,958	429,227	15,734	725	16,459	53,958
Mexico	2020	0	0	0	0	0	0	0	0	0	0
	2030	2604	37	18	55	7792	10,790	95	52	147	7792
	2050	25,855	2706	75	2781	32,716	33,945	2985	86	3071	32,716
OECD North America	2020	0	0	0	0	0	0	0	0	0	0
	2030	62,369	682	384	1065	55,702	243,235	1456	949	2405	55,702
	2050	853,401	43,597	1735	45,331	204,967	999,704	44,919	1846	46,766	204,967

each production peak would require significant additional investment. Therefore, it is assumed that not all surplus solar and wind generation must be stored, and that up to 5% (in 2030) and 10% (in 2050) of the annual production can be curtailed without significant economic disadvantage. We assume that regions with favourable wind and solar potentials, and advantages regarding available space, will use their overcapacities to export electricity via transmission lines and/or to produce synthetic and/or hydrogen fuels.

The southern part of the USA will achieve a significant solar PV share by 2050 and storage demand will be highest in this region. Storage and dispatch demand will increase in all sub-regions between 2025 and 2035. Before 2025, storage demand will be zero in all regions.

8.6 Latin America

8.6.1 Latin America: Long-Term Energy Pathways

8.6.1.1 Latin America: Final Energy Demand by Sector

Combining the assumptions on population growth, GDP growth, and energy intensity will produce the future development pathways for Latin America's final energy demand shown in Fig. 8.26 for the 5.0 °C, 2.0 °C, and 1.5 °C Scenarios. Under the 5.0 °C Scenario, the total final energy demand will increase by 70% from the current 19,200 PJ/year to 32,600 PJ/year in 2050. In the 2.0 °C Scenario, the final energy demand will decrease by 11% compared with current consumption and will reach 17,000 PJ/year by 2050. The final energy demand in the 1.5 °C Scenario will fall to 15,800 PJ in 2050, 18% below the 2015 demand. In the 1.5 °C Scenario, the final energy demand in 2050 will be 7% lower than in the 2.0 °C Scenario. The electricity demand for 'classical' electrical devices (without power-to-heat or e-mobility) will increase from 740 TWh/year in 2015 to around 1560 TWh/year in 2050 in both alternative scenarios, around 300 TWh/year lower than in the 5.0 °C Scenario (1860 TWh/year in 2050).

Electrification will lead to a significant increase in the electricity demand by 2050. In the 2.0 °C Scenario, the electricity demand for heating will be about 600 TWh/year due to electric heaters and heat pumps, and in the transport sector an increase of approximately 1700 TWh/year will be caused by electric mobility. The generation of hydrogen (for transport and high-temperature process heat) and the manufacture of synthetic fuels (mainly for transport) will add an additional power demand of 600 TWh/year. The gross power demand will thus increase from 1300 TWh/year in 2015 to 3500 TWh/year in 2050 in the 2.0 °C Scenario, 25% higher than in the 5.0 °C case. In the 1.5 °C Scenario, the gross electricity demand will increase to a maximum of 3800 TWh/year in 2050.

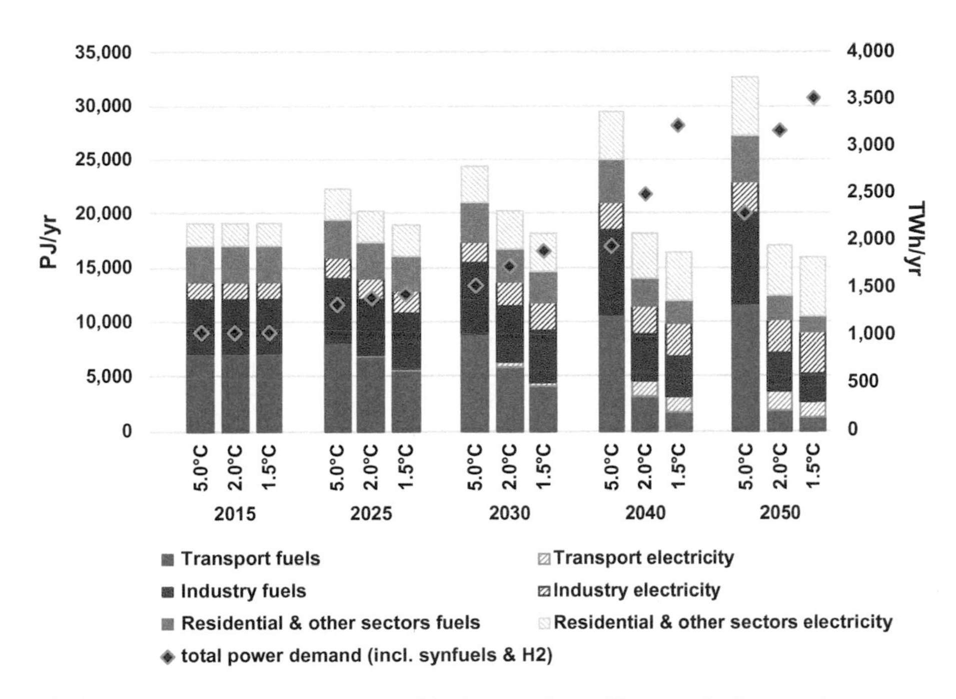

Fig. 8.26 Latin America: development of final energy demand by sector in the scenarios

Efficiency gains in the heating sector could be even larger than in the electricity sector. Under the 2.0 °C and 1.5 °C Scenarios, a final energy consumption equivalent to about 4300 PJ/year will be avoided through efficiency gains in both scenarios by 2050 compared with the 5.0 °C Scenario.

8.6.1.2 Latin America: Electricity Generation

The development of the power system is characterized by a dynamically growing renewable energy market and an increasing proportion of total power coming from renewable sources. By 2050, 100% of the electricity produced in Latin America will come from renewable energy sources in the 2.0 °C Scenario. 'New' renewables—mainly wind, solar, and geothermal energy—will contribute 63% of the total electricity generation. Renewable electricity's share of the total production will be 87% by 2030 and 96% by 2040. The installed capacity of renewables will reach about 530 GW by 2030 and 1030 GW by 2050. The share of renewable electricity generation in 2030 in the 1.5 °C Scenario will be 91%. In the 1.5 °C Scenario, the generation capacity from renewable energy will be approximately 1210 GW in 2050.

Table 8.23 shows the development of different renewable technologies in Latin America over time. Figure 8.27 provides an overview of the overall power-generation structure in Latin America. From 2020 onwards, the continuing growth of wind and PV, up to 230 GW and 410 GW, respectively, will be complemented by up to 60 GW solar thermal generation, as well as limited biomass, geothermal, and

Table 8.23 Latin America: development of renewable electricity-generation capacity in the scenarios

in GW	Case	2015	2025	2030	2040	2050
Hydro	5.0 °C	161	200	222	269	302
	2.0 °C	161	180	180	183	184
	1.5 °C	161	180	180	183	184
Biomass	5.0 °C	18	23	25	29	34
	2.0 °C	18	43	57	75	89
	1.5 °C	18	43	61	75	81
Wind	5.0 °C	11	31	38	50	66
	2.0 °C	11	56	95	199	234
	1.5 °C	11	67	134	272	285
Geothermal	5.0 °C	1	1	2	3	4
	2.0 °C	1	3	5	12	18
	1.5 °C	1	3	5	12	15
PV	5.0 °C	2	14	19	29	42
	2.0 °C	2	108	175	295	409
	1.5 °C	2	133	237	529	537
CSP	5.0 °C	0	1	1	2	3
	2.0 °C	0	4	20	51	63
	1.5 °C	0	4	20	76	78
Ocean	5.0 °C	0	0	0	0	4
	2.0 °C	0	1	2	20	37
	1.5 °C	0	1	2	20	30
Total	5.0 °C	193	270	306	382	456
	2.0 °C	193	395	534	834	1034
	1.5 °C	193	432	640	1167	1209

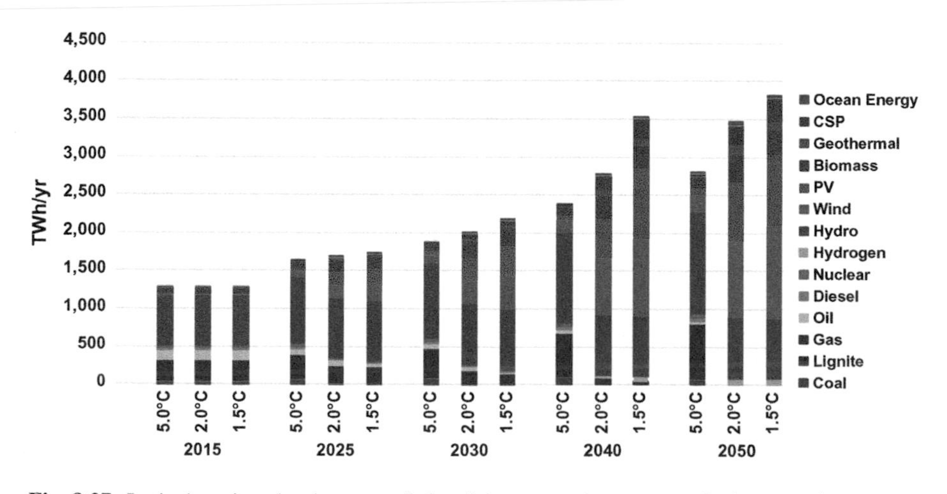

Fig. 8.27 Latin America: development of electricity-generation structure in the scenarios

ocean energy, in the 2.0 °C Scenario. Both the 2.0 °C and 1.5 °C Scenarios will lead to a high proportion of variable power generation (PV, wind, and ocean) of 31% and 39%, respectively, by 2030, and 52% and 57%, respectively, by 2050.

8.6.1.3 Latin America: Future Costs of Electricity Generation

Figure 8.28 shows the development of the electricity-generation and supply costs over time, including CO_2 emission costs, under all scenarios. The calculated electricity-generation costs in 2015 (referring to full costs) were around 4.5 ct/kWh. In the 5.0 °C case, the generation costs will increase until 2050, when they reach 8.3 ct/kWh. The generation costs in the 2.0 °C Scenario will increase until 2030, when they reach 7 ct/kWh, and will then drop to 5.9 ct/kWh by 2050. In the 1.5 °C Scenario, they will increase to 6.7 ct/kWh, and then drop to 5.6 ct/kWh by 2050. In the 2.0 °C Scenario, the generation costs in 2050 will be 2.4 ct/kWh lower than in the 5.0 °C case. In the 1.5 °C Scenario, the maximum difference in generation costs will be 2.6 ct/kWh in 2050. Note that these estimates of generation costs do not take into account integration costs such as power grid expansion, storage, or other load-balancing measures.

In the 5.0 °C case, the growth in demand and increasing fossil fuel prices will result in an increase in total electricity supply costs from today's $70 billion/year to more than $240 billion/year in 2050. In the 2.0 °C Scenario, the total supply costs will be $230 billion/year and in the 1.5 °C Scenario, they will be $240 billion/year in 2050. The long-term costs for electricity supply will be more than 5% lower in

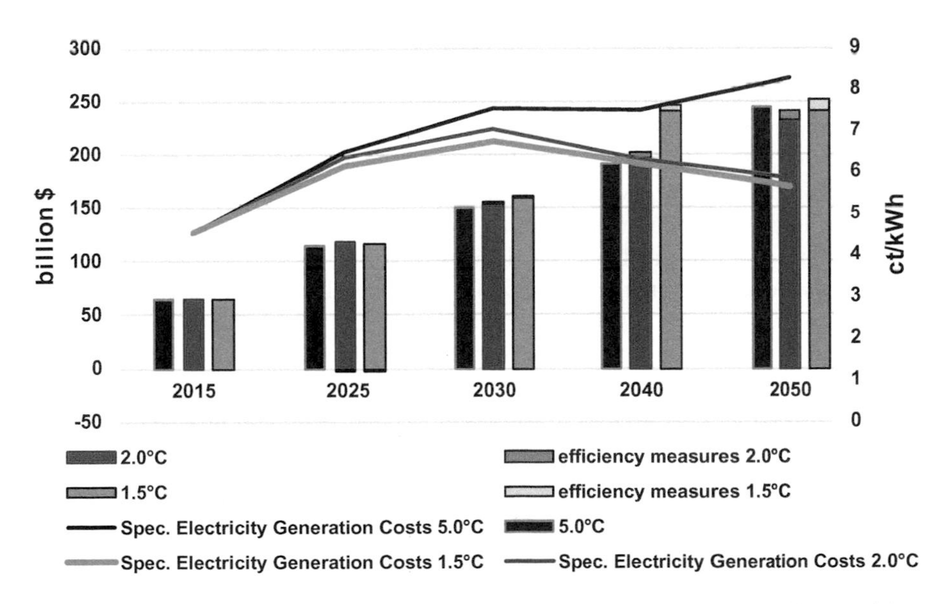

Fig. 8.28 Latin America: development of total electricity supply costs and specific electricity-generation costs in the scenarios

the 2.0 °C Scenario than in the 5.0 °C Scenario as a result of the estimated generation costs and the electrification of heating and mobility. Further electrification and synthetic fuel generation in the 1.5 °C Scenario will result in total power generation costs that are similar to the 5.0 °C case.

Compared with these results, the generation costs when the CO_2 emission costs are not considered will increase in the 5.0 °C case to 7.1 ct/kWh. In the 2.0 °C Scenario, they will increase until 2030, when they will reach 6.6 ct/kWh, and then drop to 5.9 ct/kWh by 2050. In the 1.5 °C Scenario, they will increase to 6.5 ct/kWh and then drop to 5.6 ct/kWh by 2050. In the 2.0 °C Scenario, the generation costs will be maximum, at 0.25 ct/kWh higher than in the 5.0 °C case, in 2030 (0.1 ct/kWh in the 1.5 °C Scenario). The generation costs in 2050 will again be lower in the alternative scenarios than in the 5.0 °C case: 1.2 ct/kWh in the 2.0 °C Scenario and 1.5 ct/kWh in the 1.5 °C Scenario. If CO_2 costs are not considered, the total electricity supply costs in the 5.0 °C case will increase to about $210 billion/year in 2050.

8.6.1.4 Latin America: Future Investments in the Power Sector

An investment of about $1920 billion will be required for power generation between 2015 and 2050 in the 2.0 °C Scenario, including additional power plants for the production of hydrogen and synthetic fuels and investments in plant replacement after the ends of their economic lives. This value is equivalent to approximately $53 billion per year, on average, which is $880 billion more than in the 5.0 °C case ($1040 billion). An investment of around $2190 billion for power generation will be required between 2015 and 2050 in the 1.5 °C Scenario. On average, this will be an investment of $61 billion per year. Under the 5.0 °C Scenario, the investment in conventional power plants will be around 25% of the total cumulative investments, whereas approximately 75% will be invested in renewable power generation and co-generation (Fig. 8.29).

However, under the 2.0 °C (1.5 °C) Scenario, Latin America will shift almost 94% (95%) of its entire investment to renewables and co-generation. By 2030, the fossil fuel share of the power sector investment will predominantly focus on gas power plants that can also be operated with hydrogen.

Because renewable energy has no fuel costs, other than biomass, the cumulative fuel cost savings in the 2.0 °C Scenario will reach a total of $820 billion in 2050, equivalent to $23 billion per year. Therefore, the total fuel cost savings will be equivalent to 90% of the total additional investments compared to the 5.0 °C Scenario. The fuel cost savings in the 1.5 °C Scenario will add up to $900 billion, or $25 billion per year.

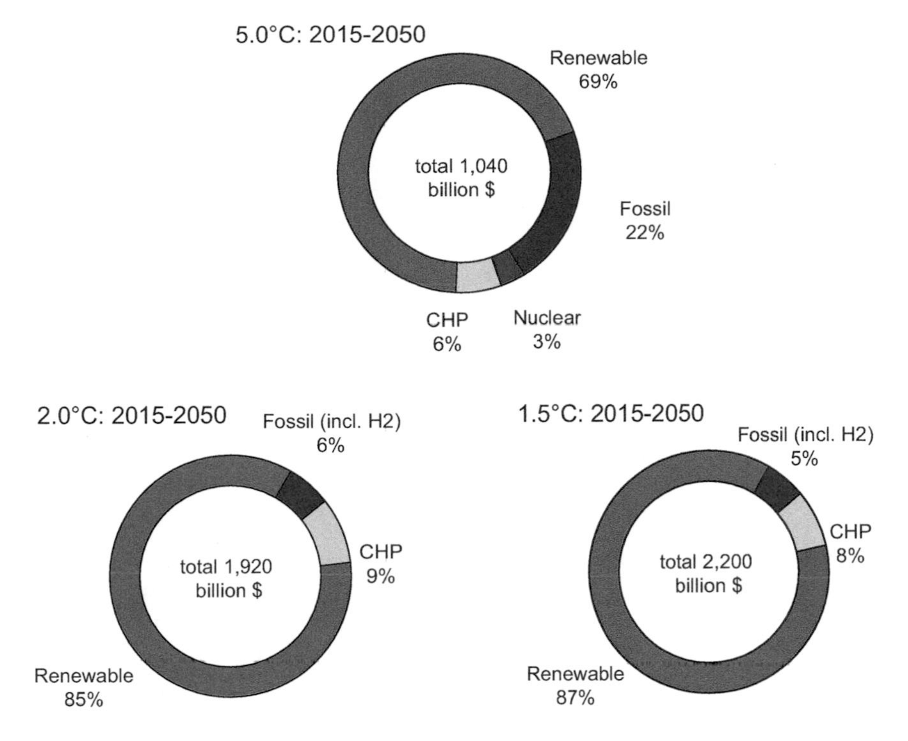

Fig. 8.29 Latin America: investment shares for power generation in the scenarios

8.6.1.5 Latin America: Energy Supply for Heating

The final energy demand for heating will increase in the 5.0 °C Scenario by 72%, from 7800 PJ/year in 2015 to 13,300 PJ/year in 2050. In the 2.0 °C and 1.5 °C Scenarios, energy efficiency measures will help to reduce the energy demand for heating by 32% in 2050, relative to that in the 5.0 °C Scenario. Today, renewables supply around 42% of Latin America's final energy demand for heating, with the main contribution from biomass. Renewable energy will provide 68% of Latin America's total heat demand in 2030 in the 2.0 °C Scenario and 75% in the 1.5 °C Scenario. In both scenarios, renewables will provide 100% of the total heat demand in 2050.

Figure 8.30 shows the development of different technologies for heating in Latin America over time, and Table 8.24 provides the resulting renewable heat supply for all scenarios. Biomass will remain the main contributor. The growing use of solar, geothermal, and environmental heat will supplement mainly fossil fuels. This will lead in the long term to a biomass share of 65% under the 2.0 °C Scenario and 50% under the 1.5 °C Scenario.

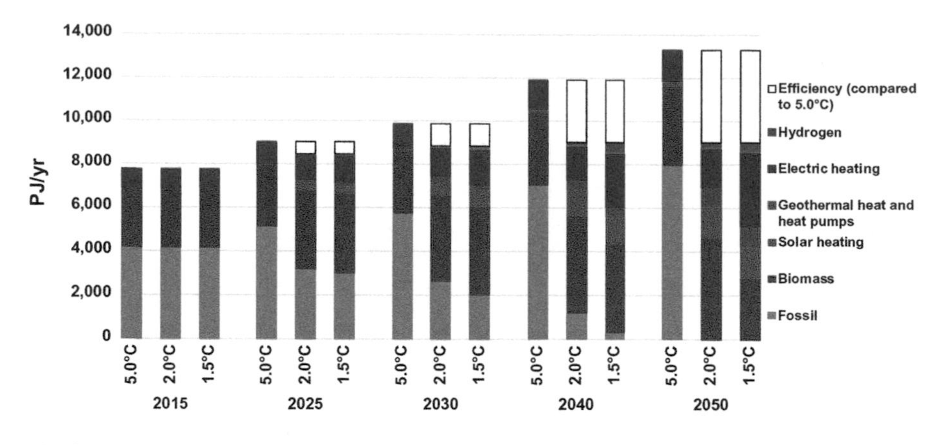

Fig. 8.30 Latin America: development of heat supply by energy carrier in the scenarios

Table 8.24 Latin America: development of renewable heat supply in the scenarios (excluding the direct use of electricity)

in PJ/year	Case	2015	2025	2030	2040	2050
Biomass	5.0 °C	2684	2760	2888	3335	3622
	2.0 °C	2684	3550	3895	4412	4654
	1.5 °C	2684	3632	4007	4023	2767
Solar heating	5.0 °C	32	64	88	146	227
	2.0 °C	32	394	712	1217	1418
	1.5 °C	32	394	783	1265	1445
Geothermal heat and heat pumps	5.0 °C	0	0	0	0	0
	2.0 °C	0	133	206	458	910
	1.5 °C	0	133	204	452	930
Hydrogen	5.0 °C	0	0	0	0	0
	2.0 °C	0	0	4	169	220
	1.5 °C	0	0	88	473	404
Total	5.0 °C	2715	2824	2976	3480	3849
	2.0 °C	2715	4077	4817	6255	7202
	1.5 °C	2715	4159	5082	6213	5546

Heat from renewable hydrogen will further reduce the dependence on fossil fuels in both scenarios. Hydrogen consumption in 2050 will be around 200 PJ/year in the 2.0 °C Scenario and 400 PJ/year in the 1.5 °C Scenario. The direct use of electricity for heating will also increase by a factor of 2–4 between 2015 and 2050 and will attain a final energy share of 20% in 2050 in the 2.0 °C Scenario and 39% in the 1.5 °C Scenario.

8.6.1.6 Latin America: Future Investments in the Heating Sector

The roughly estimated investments in renewable heating technologies up to 2050 will amount to around \$580 billion in the 2.0 °C Scenario (including investments in plant replacement after their economic lifetimes), or approximately \$16 billion per year. The largest share of investment in Latin America is assumed to be for solar collectors (more than \$200 billion), followed by biomass technologies and heat pumps. The 1.5 °C Scenario assumes an even faster expansion of renewable technologies, but due to the lower heat demand, the average annual investment will again be around \$16 billion per year (Fig. 8.31, Table 8.25).

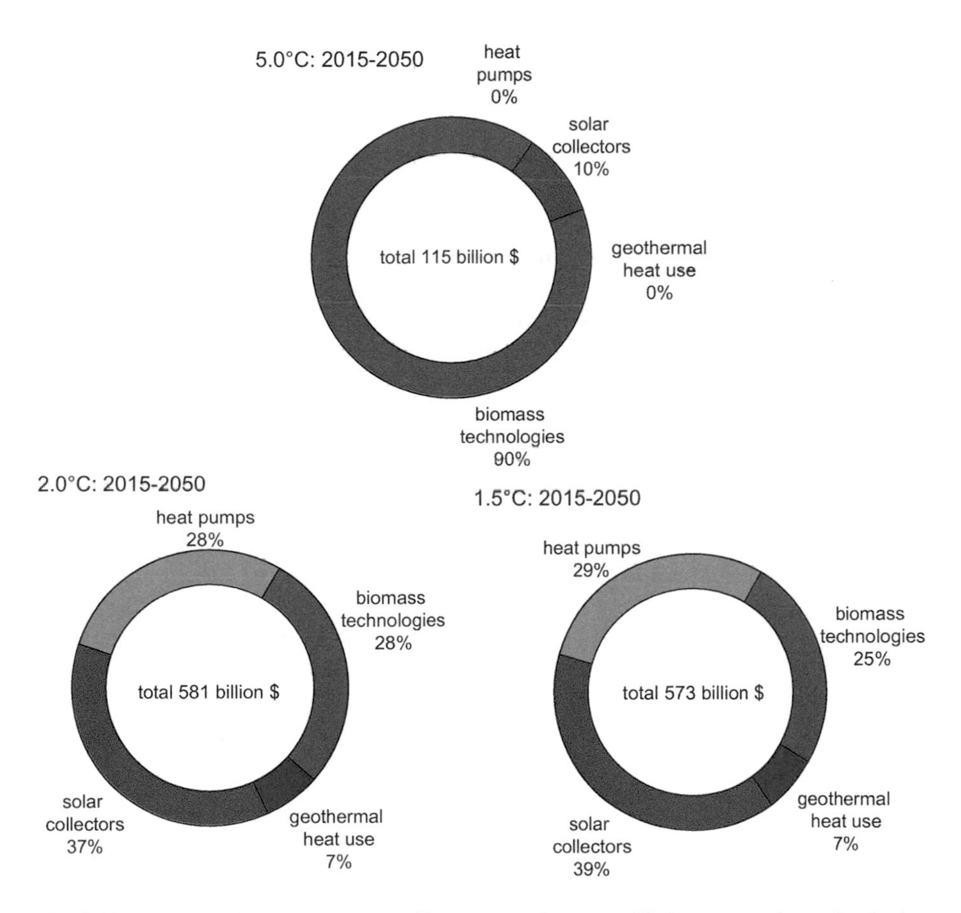

Fig. 8.31 Latin America: development of investments for renewable heat generation technologies in the scenarios

Table 8.25 Latin America: installed capacities for renewable heat generation in the scenarios

in GW	Case	2015	2025	2030	2040	2050
Biomass	5.0 °C	549	531	536	552	542
	2.0 °C	549	730	742	657	603
	1.5 °C	549	770	752	513	179
Geothermal	5.0 °C	0	0	0	0	0
	2.0 °C	0	2	4	12	16
	1.5 °C	0	2	4	12	17
Solar heating	5.0 °C	7	15	20	34	52
	2.0 °C	7	91	164	281	327
	1.5 °C	7	91	181	292	333
Heat pumps	5.0 °C	0	0	0	0	0
	2.0 °C	0	13	18	36	88
	1.5 °C	0	13	18	36	89
Total[a]	5.0 °C	556	546	556	585	594
	2.0 °C	556	835	929	986	1034
	1.5 °C	556	876	955	853	619

[a]Excluding direct electric heating

8.6.1.7 Latin America: Transport

Energy demand in the transport sector in Latin America is expected to increase by 63% under the 5.0 °C Scenario, from around 7100 PJ/year in 2015 to 11,700 PJ/year in 2050. In the 2.0 °C Scenario, assumed technical, structural, and behavioural changes will save 69% (8090 PJ/year) by 2050 relative to the 5.0 °C Scenario. Additional modal shifts, technology switches, and a reduction in transport demand will lead to even greater energy savings in the 1.5 °C Scenario of 77% (or 9040 PJ/year) in 2050 compared with the 5.0 °C case (Table 8.26, Fig. 8.32).

By 2030, electricity will provide 6% (110 TWh/year) of the transport sector's total energy demand under the 2.0 °C Scenario, whereas in 2050, the share will be 47% (470 TWh/year). In 2050, up to 480 PJ/year of hydrogen will be used in the transport sector as a complementary renewable option. In the 1.5 °C Scenario, the annual electricity demand will be 390 TWh in 2050. The 1.5 °C Scenario also assumes a hydrogen demand of 430 PJ/year by 2050.

Biofuel use is limited in the 2.0 °C Scenario to a maximum of 1340 PJ/year Around 2030, synthetic fuels based on power-to-liquid will be introduced, with a maximum of 190 PJ/year by 2050. Due to the lower overall energy demand in transport, biofuel use will be reduced in the 1.5 °C Scenario to a maximum of 1030 PJ/year The maximum synthetic fuel demand will amount to 350 PJ/year.

8.6.1.8 Latin America: Development of CO_2 Emissions

In the 5.0 °C Scenario, Latin America's annual CO_2 emissions will increase by 48%, from 1220 Mt. in 2015 to 1806 Mt. in 2050. The stringent mitigation measures in both alternative scenarios will cause the annual emissions to fall to 240 Mt. in

Table 8.26 Latin America: projection of transport energy demand by mode in the scenarios

in PJ/year	Case	2015	2025	2030	2040	2050
Rail	5.0 °C	90	110	122	145	163
	2.0 °C	90	113	133	157	192
	1.5 °C	90	130	145	163	224
Road	5.0 °C	6662	7486	8102	9754	10,610
	2.0 °C	6662	6424	5799	4107	3112
	1.5 °C	6662	5196	3971	2744	2161
Domestic aviation	5.0 °C	211	348	453	593	638
	2.0 °C	211	228	213	175	139
	1.5 °C	211	218	196	137	104
Domestic navigation	5.0 °C	101	104	107	113	117
	2.0 °C	101	104	107	113	117
	1.5 °C	101	104	107	113	117
Total	5.0 °C	7064	8047	8783	10,605	11,529
	2.0 °C	7064	6868	6251	4551	3559
	1.5 °C	7064	5648	4419	3157	2605

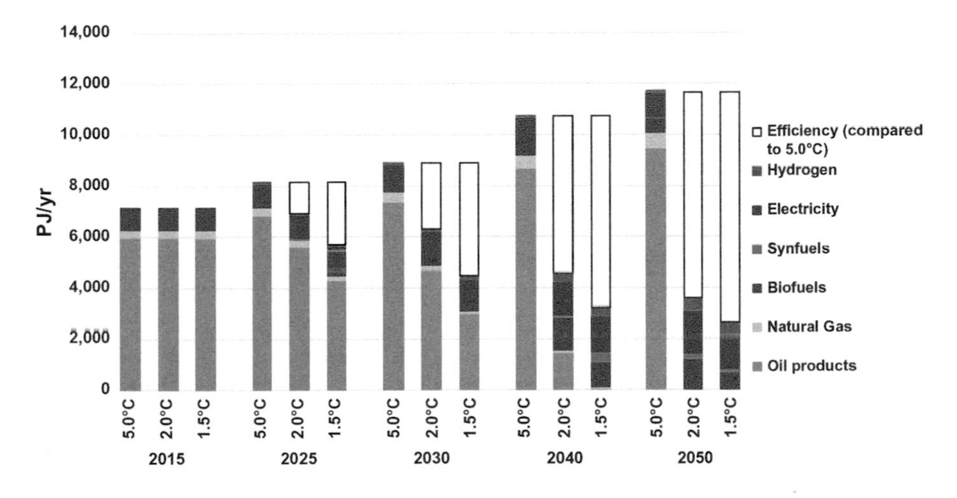

Fig. 8.32 Latin America: final energy consumption by transport in the scenarios

2040 in the 2.0 °C Scenario and to 50 Mt. in the 1.5 °C Scenario, with further reductions to almost zero by 2050. In the 5.0 °C case, the cumulative CO_2 emissions from 2015 until 2050 will add up to 56 Gt. In contrast, in the 2.0 °C and 1.5 °C Scenarios, the cumulative emissions for the period from 2015 until 2050 will be 21 Gt and 17 Gt, respectively.

Therefore, the cumulative CO_2 emissions will decrease by 63% in the 2.0 °C Scenario and by 70% in the 1.5 °C Scenario compared with the 5.0 °C case. A rapid reduction in annual emissions will occur in both alternative scenarios. In the 2.0 °C Scenario, the reduction will be greatest in 'Power generation', followed by the 'Residential and other' and 'Industry' sectors (Fig. 8.33).

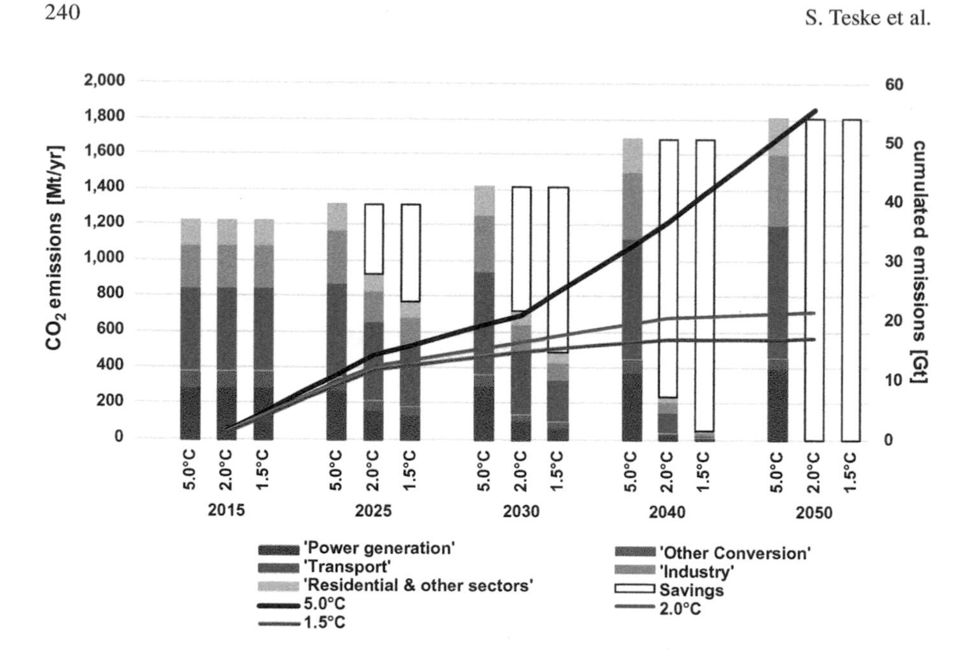

Fig. 8.33 Latin America: development of CO_2 emissions by sector and cumulative CO_2 emissions (after 2015) in the scenarios ('Savings' = reduction compared with the 5.0 °C Scenario)

8.6.1.9 Latin America: Primary Energy Consumption

The levels of primary energy consumption under the three scenarios when the assumptions discussed above are taken into account are shown in Fig. 8.34. In the 2.0 °C Scenario, the primary energy demand will decrease by 2%, from around 28,400 PJ/year in 2015 to 27,900 PJ/year in 2050. Compared with the 5.0 °C Scenario, the overall primary energy demand will decrease by 38% in 2050 in the 2.0 °C Scenario (5.0 °C: 45000 PJ in 2050). In the 1.5 °C Scenario, the primary energy demand will be even lower (25,700 PJ in 2050) because the final energy demand and conversion losses will be lower.

Both the 2.0 °C and 1.5 °C Scenarios aim to rapidly phase-out coal and oil. This will cause renewable energy to have a primary energy share of 55% in 2030 and 94% in 2050 in the 2.0 °C Scenario. In the 1.5 °C Scenario, renewables will also have a primary energy share of more than 94% in 2050 (including non-energy consumption, which will still include fossil fuels). Nuclear energy will be phased-out by 2035 under both the 2.0 °C and the 1.5 °C Scenarios. The cumulative primary energy consumption of natural gas in the 5.0 °C case will add up to 290 EJ, the cumulative coal consumption to about 60 EJ, and the crude oil consumption to 460 EJ. In contrast, in the 2.0 °C Scenario, the cumulative gas demand will amount to 130 EJ, the cumulative coal demand to 20 EJ, and the cumulative oil demand to 200

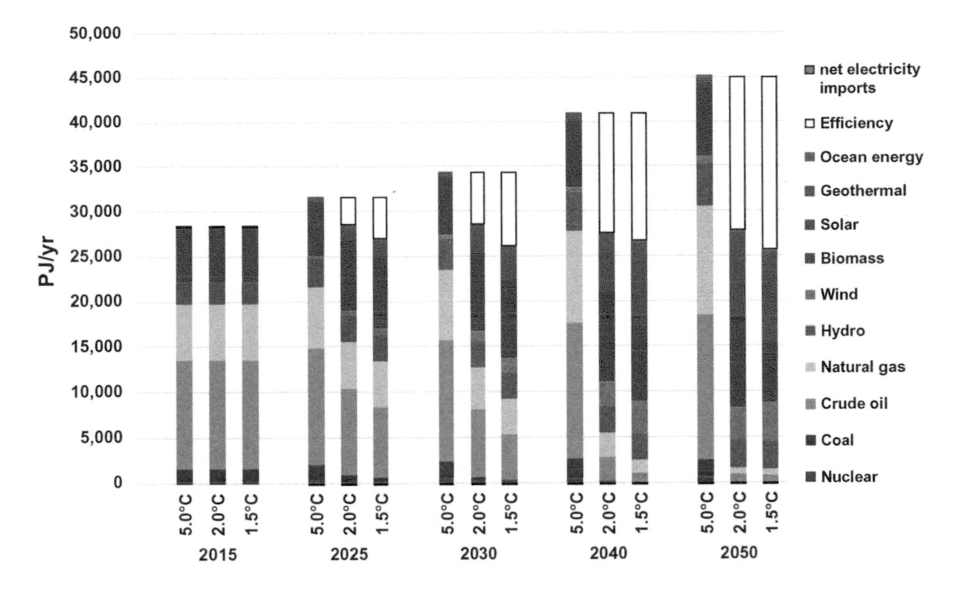

Fig. 8.34 Latin America: projection of total primary energy demand (PED) by energy carrier in the scenarios (including electricity import balance)

EJ. Even lower fossil fuel use will be achieved in the 1.5 °C Scenario: 110 EJ for natural gas, 10 EJ for coal, and 150 EJ for oil.

8.6.2 Latin America: Power Sector Analysis

The Latin American region is extremely diverse. It borders Mexico in the north and its southern tip is in the South Pacific. It also includes all the Caribbean islands and Central America. The power-generation situation is equally diverse, and the sub-regional breakdown tries to reflect this diversity to some extent. In the Caribbean, which contains 28 island nations and more than 7000 islands, the calculated storage demand will almost certainly be higher than the region's average, because a regional power exchange grid between the islands seems impractical. To calculate the detailed storage demand, island-specific analyses would be required, as has recently been done for Barbados (Hohmeyer 2015). The mainland of South America has been subdivided into the large economic centres of Chile, Argentina, and Brazil, and Central America and the northern part of South America have been clustered into two parts.

8.6.2.1 Latin America: Development of Power Plant Capacities

The most important future renewable technologies for Latin America are solar PV and onshore wind, followed by CSP (which will be especially suited to the Atacama Desert in Chile) and offshore wind, mainly in the coastal areas of Brazil and Argentina. The annual market for solar PV must increase from 6.5 GW in 2020 by a factor of three to an average of 15.5 GW by 2030 under the 2.0 °C Scenario and to around 23 GW under the 1.5 °C Scenario. The onshore wind market in the 1.5 °C Scenario must increase to 15 GW by 2025, compared with the average annual onshore wind market of around 3 GW between 2014 and 2017 (GWEC 2018). By 2050, offshore wind will have increased to a moderate annual new installation capacity of around 2–3 GW from 2025 to 2050 in both scenarios. Concentrated solar power plants will be limited to the desert regions of South America, especially Chile. The market for biofuels for electricity generation will play an important role in all agricultural areas, including the Caribbean and Central America, where most geothermal resources are located (Table 8.27).

8.6.2.2 Latin America: Utilization of Power-Generation Capacities

Table 8.28 shows that our modelling assumes that for the entire modelling period, there will be no interconnection capacity between the Caribbean, Central America, and South America, whereas the interconnection capacity in the rest of South America will increase to 15% by 2030 and to 20% by 2050. The shares of variable

Table 8.27 Latin America: average annual change in installed power plant capacity

Latin Power Generation: average annual change of installed capacity [GW/a]	2015–2025		2026–2035		2036–2050	
	2.0 °C	1.5 °C	2.0 °C	1.5 °C	2.0 °C	1.5 °C
Hard coal	0	−1	0	−1	−1	0
Lignite	0	0	0	0	0	0
Gas	4	2	1	6	−9	5
Hydrogen-gas	0	1	1	4	11	14
Oil/diesel	−1	−4	−4	−3	0	0
Nuclear	0	0	0	0	0	0
Biomass	3	5	3	4	4	3
Hydro	2	0	0	0	0	0
Wind (onshore)	5	11	11	17	6	3
Wind (offshore)	0	1	2	2	3	2
PV (roof top)	9	18	14	25	9	8
PV (utility scale)	3	6	5	8	3	3
Geothermal	0	1	1	1	1	1
Solar thermal power plants	0	2	4	5	2	3
Ocean energy	0	0	1	1	2	2
Renewable fuel based co-generation	1	2	2	2	2	1

Table 8.28 Latin America: power system shares by technology group

Power generation structure and interconnection		2.0 °C				1.5 °C			
Latin America		Variable RE	Dispatch RE	Dispatch fossil	Inter-connection	Variable RE	Dispatch RE	Dispatch fossil	Inter-connection
Caribbean	2015	3%	63%	34%	0%				
	2030	25%	62%	12%	0%	25%	62%	12%	0%
	2050	44%	53%	3%	0%	44%	53%	3%	0%
Central America	2015	2%	64%	35%	0%				
	2030	21%	64%	14%	0%	21%	64%	14%	0%
	2050	40%	58%	2%	0%	40%	58%	2%	0%
North L. America	2015	2%	64%	34%	10%				
	2030	20%	41%	39%	15%	20%	41%	39%	15%
	2050	30%	40%	30%	20%	30%	40%	30%	20%
Central L. America	2015	1%	64%	36%	10%				
	2030	16%	52%	32%	15%	16%	52%	32%	15%
	2050	29%	49%	22%	20%	29%	49%	22%	20%
Brazil	2015	4%	63%	33%	10%				
	2030	30%	54%	16%	15%	30%	54%	16%	15%
	2050	47%	44%	8%	20%	47%	44%	8%	20%
Uruguay	2015	2%	61%	37%	10%				
	2030	21%	57%	22%	15%	21%	57%	22%	15%
	2050	37%	52%	11%	20%	37%	52%	11%	20%
Argentina	2015	2%	62%	36%	10%				
	2030	19%	42%	38%	15%	19%	42%	38%	15%
	2050	31%	40%	29%	20%	31%	40%	29%	20%

(continued)

Table 8.28 (continued)

Power generation structure and interconnection		2.0 °C				1.5 °C			
Latin America		Variable RE	Dispatch RE	Dispatch fossil	Inter-connection	Variable RE	Dispatch RE	Dispatch fossil	Inter-connection
Chile	2015	2%	64%	35%	10%				
	2030	18%	45%	37%	15%	18%	45%	37%	15%
	2050	33%	47%	19%	20%	33%	47%	19%	20%
Latin America	2015	3%	63%	34%					
	2030	24%	51%	25%		24%	51%	25%	
	2050	39%	45%	16%		39%	45%	16%	

Table 8.29 Latin America: capacity factors by generation type

Utilization of variable and dispatchable power generation:		2015	2020	2020	2030	2030	2040	2040	2050	2050
Latin America			2.0 °C	1.5 °C	2.0 °C	1.5 °C	2.0 °C	1.5 °C	2.0 °C	1.5 °C
Capacity factor – average	[%/yr]	48.9%	31%	25%	36%	21%	41%	18%	34%	24%
Limited dispatchable: fossil and nuclear	[%/yr]	73.4%	14%	3%	17%	0%	45%	0%	13%	4%
Limited dispatchable: renewable	[%/yr]	26.0%	53%	48%	46%	19%	56%	23%	47%	33%
Dispatchable: fossil	[%/yr]	53.2%	24%	11%	31%	2%	37%	6%	31%	11%
Dispatchable: renewable	[%/yr]	45.6%	37%	28%	46%	26%	43%	25%	46%	35%
Variable: renewable	[%/yr]	12.2%	12%	12%	21%	14%	31%	15%	22%	15%

renewables are almost identical in the 2.0 °C and 1.5 °C Scenarios. The lowest rates of variable renewables are in central South America and Central America because the onshore wind potential is limited by average wind speeds that are lower than elsewhere. Compared with all the other world regions, Latin America has the highest share of dispatchable renewables, mainly attributable to existing hydropower plants.

Compared with other regions of the world, Latin America currently has a small fleet of coal and nuclear power plants, but they are operated with a high capacity factor (Table 8.29). The dispatch order for all world regions in all cases is assumed to be the same, to make the results comparable. Therefore, the capacity factors of these dispatch power plants (mainly gas) will increase at the expense of those for coal and nuclear power plants, which explains the rapid reduction in the capacity factor in 2020. Therefore, this effect is the result of the assumed dispatch order, rather than of an increase in variable power generation.

8.6.2.3 Latin America: Development of Load, Generation and Residual Load

The sub-regions of Latin America are highly diverse in their geographic features and population densities, so the maximum loads in the different sub-regions vary widely. Table 8.30 shows that the sub-region with the smallest calculated maximum load is Uruguay, with only 2.3 GW, which seems realistic because the maximum

Table 8.30 Latin America: load, generation, and residual load development

Power generation structure		2.0 °C			1.5 °C				
		Max demand	Max generation	Max residual load	Max interconnection requirements	Max demand	Max generation	Max residual load	Max interconnection requirements
Latin America		[GW]	[GW]	[GW]	[GW]	[GW]	[GW]	[GW]	[GW]
Caribbean	2020	14.9	4.9	10.4		14.8	7.4	8.1	
	2030	23.4	23.1	8.9	0	21.0	27.5	1.5	5
	2050	36.6	38.6	14.8	0	36.9	48.4	6.9	5
Central America	2020	13.1	13.0	3.3		13.1	20.0	0.5	
	2030	20.7	23.6	7.8	0	18.8	31.8	1.4	12
	2050	33.3	42.6	14.3	0	33.2	47.3	7.2	7
North Latin America	2020	37.5	41.5	1.4		37.4	67.9	1.4	
	2030	59.1	76.7	7.1	11	53.1	80.9	5.0	23
	2050	92.9	117.2	15.8	9	94.7	108.3	24.9	0
Central South America	2020	16.8	9.9	6.9		16.8	14.7	2.1	
	2030	26.2	29.4	5.7	0	24.1	39.9	2.3	14
	2050	42.0	46.3	11.3	0	42.9	59.5	11.5	5
Brazil	2020	99.0	96.4	5.7		98.9	102.3	4.7	
	2030	153.8	150.1	38.4	0	140.7	145.2	9.9	0
	2050	241.0	247.5	74.1	0	250.7	306.1	45.5	10

Uruguay	2020	2.3	2.9	0.4		2.3	4.4	0.1	
	2030	3.4	4.0	1.1	0	3.1	5.3	0.2	2
	2050	4.9	6.6	1.7	0	5.1	7.8	1.0	2
Argentina	2020	25.5	26.2	1.0		25.5	35.7	1.0	
	2030	40.1	176.4	3.1	133	36.6	176.4	3.6	136
	2050	56.4	71.8	14.0	2	59.4	82.7	18.2	5
Chile	2020	9.3	19.2	0.4					
	2030	16.5	21.0	1.7	3	15.0	23.5	1.4	7
	2050	26.1	30.7	7.7	0	27.7	35.5	7.2	1

load was 1.7 GW in 2012 according to IDB (2013). Brazil, Uruguay's direct neighbour, has the largest load of close to 100 GW, which will increase by a factor of 2.5 to around 250 GW by 2050 under both scenarios. Brazil's maximum generation will increase accordingly, without significant overproduction peaks. The calculated maximum increase in interconnection required is only 10 GW. In Argentina, peak generation matches peak demand because Argentina has one of the best wind resources in the world in Patagonia. Surplus wind power can either be exported after a significant increase in transmission capacity or, as assumed in our scenario, it can be used to produce synthetic and hydrogen fuels.

Table 8.31 provides an overview of the calculated storage and dispatch power requirements by sub-region. As indicated in the introduction to the Latin America results, the storage requirements for the Caribbean might be high because the region cannot exchange solar or wind electricity with other sub-regions. However, all other sub-regions contain either several countries or larger provinces, so they are more suited to the integration of variable electricity. Compared with other world regions, Latin America has one of the lowest storage capacities and one of the lowest needs for additional dispatch. This is because the region's installed capacity of hydropower is high. However, this research does not include a water resource assessment for hydropower plants. Droughts may increase the demand for storage and/or hydrogen dispatch.

8.7 OECD Europe

8.7.1 OECD Europe: Long-Term Energy Pathways

8.7.1.1 OECD Europe: Final Energy Demand by Sector

Combining the assumptions on population growth, GDP growth, and energy intensity produces the future development pathways for OECD Europe's final energy demand shown in Fig. 8.35 for the 5.0 °C, 2.0 °C, and 1.5 °C Scenarios. In the 5.0 °C Scenario, the total final energy demand will increase by 9%, from the current 46,000 PJ/year to 50,000 PJ/year in 2050. In the 2.0 °C Scenario, the final energy demand will decrease by 39% compared with current consumption and will reach 28,000 PJ/year by 2050. The final energy demand in the 1.5 °C Scenario will reach 25,200 PJ, 45% below the 2015 demand. In the 1.5 °C Scenario, the final energy demand in 2050 will be 10% lower than in the 2.0 °C Scenario. The electricity demand for 'classical' electrical devices (without power-to-heat or e-mobility) will decrease from 2300 TWh/year in 2015 to 2040 TWh/year by 2050 in both alternative scenarios. Compared with the 5.0 °C case (3200 TWh/year in 2050), the efficiency measures implemented in the 2.0 °C and 1.5 °C Scenarios will save 1160 TWh/year in 2050.

Table 8.31 Latin America: storage and dispatch service requirements in the 2.0 °C and 1.5 °C Scenarios

Storage and dispatch		2.0 °C					1.5 °C				
		Required to avoid curtailment	Utilization battery -through-put-	Utilization PSH -through-put-	Total storage demand (incl. H2)	Dispatch hydrogen-based	Required to avoid curtailment	Utilization battery -through-put-	Utilization PSH -through-put-	Total storage demand (incl. H2)	Dispatch hydrogen-based
Latin America		[GWh/year]	[GWh/year]	[GWh/year]	[GWh/year]	[GWh/year]	[GWh/year]	[GWh/year]	[GWh/year]	[GWh/year]	[GWh/year]
Caribbean	2020	0	0	0	0	0	0	0	0	0	0
	2030	0	0	0	0	6	81	6	11	17	0
	2050	100	46	3	49	15,282	1816	534	59	594	1808
Central America	2020	0	0	0	0	0	0	0	0	0	0
	2030	0	0	0	0	5	57	5	8	13	0
	2050	34	47	2	49	15,010	1462	560	59	619	5843
North Latin America	2020	0	0	0	0	0	0	0	0	0	0
	2030	0	0	0	0	0	3	1	1	1	0
	2050	0	0	0	0	7086	1047	633	57	690	0
Central L. America	2020	0	0	0	0	0	0	0	0	0	0
	2030	0	0	0	0	3	82	9	14	23	0
	2050	36	41	1	42	16,031	2768	1032	104	1136	40
Brazil	2020	0	0	0	0	0	0	0	0	0	0
	2030	0	0	0	0	19	774	83	138	221	0
	2050	475	666	27	693	63,131	18,024	6977	769	7746	1103
Uruguay	2020	77	0	0	0	0	511	0	0	0	0
	2030	0	0	0	0	0	20	1	2	3	0
	2050	42	20	2	22	1591	279	78	9	86	65

(continued)

Table 8.31 (continued)

Storage and dispatch		2.0 °C					1.5 °C				
		Required to avoid curtailment	Utilization battery -through-put-	Utilization PSH -through-put-	Total storage demand (incl. H2)	Dispatch hydrogen-based	Required to avoid curtailment	Utilization battery -through-put-	Utilization PSH -through-put-	Total storage demand (incl. H2)	Dispatch hydrogen-based
Latin America		[GWh/year]	[GWh/year]	[GWh/year]	[GWh/year]	[GWh/year]	[GWh/year]	[GWh/year]	[GWh/year]	[GWh/year]	[GWh/year]
Argentina	2020	0	0	0	0	0	0	0	0	0	0
	2030	0	0	0	0	0	177	14	23	37	0
	2050	617	446	32	478	315	4969	1727	180	1908	0
Chile	2020	2	0	0	0	0	2669	1	0	1	0
	2030	0	0	0	0	1	13	1	2	3	0
	2050	10	14	1	15	8781	162	91	7	97	58
Latin America	2020	79	0	0	0	0	3180	2	0	2	0
	2030	0	0	0	0	34	1207	121	197	318	1
	2050	1314	1279	68	1347	127,226	30,526	11,633	1243	12,875	8917

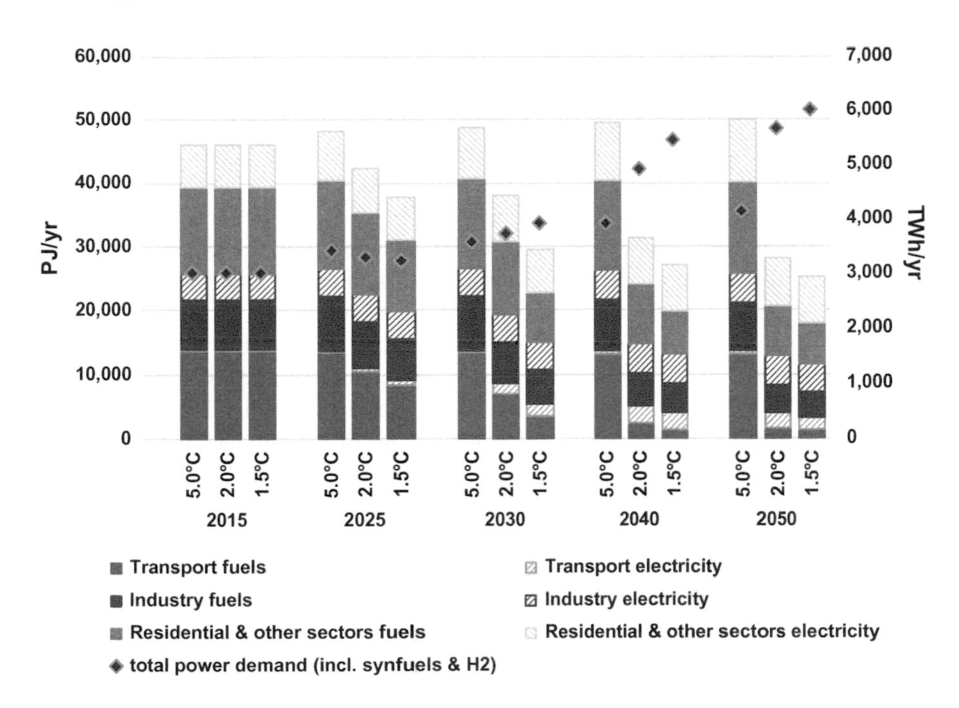

Fig. 8.35 OECD Europe: development in three scenarios

Electrification will cause a significant increase in the electricity demand by 2050. In the 2.0 °C Scenario, the electricity demand for heating will increase to approximately 1300 TWh/year due to electric heaters and heat pumps, and in the transport sector, the demand will increase to approximately 2600 TWh/year in response to increased electric mobility. The generation of hydrogen (for transport and high-temperature process heat) and the manufacture of synthetic fuels (mainly for transport) will add an additional power demand of 1600 TWh/year The gross power demand will thus rise from 3600 TWh/year in 2015 to 6000 TWh/year by 2050 in the 2.0 °C Scenario, 28% higher than in the 5.0 °C case. In the 1.5 °C Scenario, the gross electricity demand will increase to a maximum of 6400 TWh/year by 2050.

Efficiency gains could be even larger in the heating sector than in the electricity sector. Under the 2.0 °C and 1.5 °C Scenarios, a final energy consumption equivalent to about 6200 PJ/year and 8200 PJ/year, respectively, are avoided by efficiency gains by 2050 compared with the 5.0 °C Scenario.

8.7.1.2 OECD Europe: Electricity Generation

The development of the power system is characterized by a dynamically growing renewable energy market and an increasing proportion of total power from renewable sources. By 2050, 100% of the electricity produced in OECD Europe will come from renewable energy sources in the 2.0 °C Scenario. 'New' renewables—mainly wind, solar, and geothermal energy—will contribute 75% of the total electricity generation. Renewable electricity's share of the total production will be 68% by 2030 and 89% by 2040. The installed capacity of renewables will reach about 1200 GW by 2030 and 2270 GW by 2050. The share of renewable electricity generation in 2030 in the 1.5 °C Scenario is assumed to be 74%. The 1.5 °C Scenario will have a generation capacity from renewable energy of approximately 2480 GW in 2050.

Table 8.32 shows the development of different renewable technologies in OECD Europe over time. Figure 8.36 provides an overview of the overall power-generation structure in OECD Europe. From 2020 onwards, the continuing growth of wind and PV, up to 790 GW and 1000 GW, respectively, will be complemented by generation from biomass (ca. 110 GW) CSP and ocean energy (more than 50 GW each), in the 2.0 °C Scenario. Both the 2.0 °C and 1.5 °C Scenarios will lead to high proportions of variable power generation (PV, wind, and ocean) of 38% and 45%, respectively, by 2030 and 67% and 68%, respectively, by 2050.

Table 8.32 OECD Europe: development of renewable electricity-generation capacity in the scenarios

in GW	Case	2015	2025	2030	2040	2050
Hydro	5.0 °C	207	224	231	238	248
	2.0 °C	207	218	219	221	225
	1.5 °C	207	218	219	221	225
Biomass	5.0 °C	40	51	56	60	65
	2.0 °C	40	78	105	115	113
	1.5 °C	40	84	111	113	113
Wind	5.0 °C	138	216	254	296	347
	2.0 °C	138	279	409	655	787
	1.5 °C	138	299	468	778	847
Geothermal	5.0 °C	2	3	3	3	4
	2.0 °C	2	6	11	27	39
	1.5 °C	2	6	11	27	39
PV	5.0 °C	95	137	157	172	191
	2.0 °C	95	264	422	745	996
	1.5 °C	95	364	598	1028	1151
CSP	5.0 °C	2	3	4	7	11
	2.0 °C	2	7	17	38	54
	1.5 °C	2	7	22	48	57
Ocean	5.0 °C	0	1	1	4	8
	2.0 °C	0	7	16	42	53
	1.5 °C	0	7	16	42	53
Total	5.0 °C	484	635	706	780	873
	2.0 °C	484	859	1198	1842	2267
	1.5 °C	484	985	1444	2256	2485

8.7.1.3 OECD Europe: Future Costs of Electricity Generation

Figure 8.37 shows the development of the electricity-generation and supply costs over time, including the CO_2 emission costs, in all scenarios. The calculated electricity generation costs in 2015 (referring to full costs) were around 7 ct/kWh. In the 5.0 °C case, generation costs will increase until 2050, when they will reach 10.4 ct/

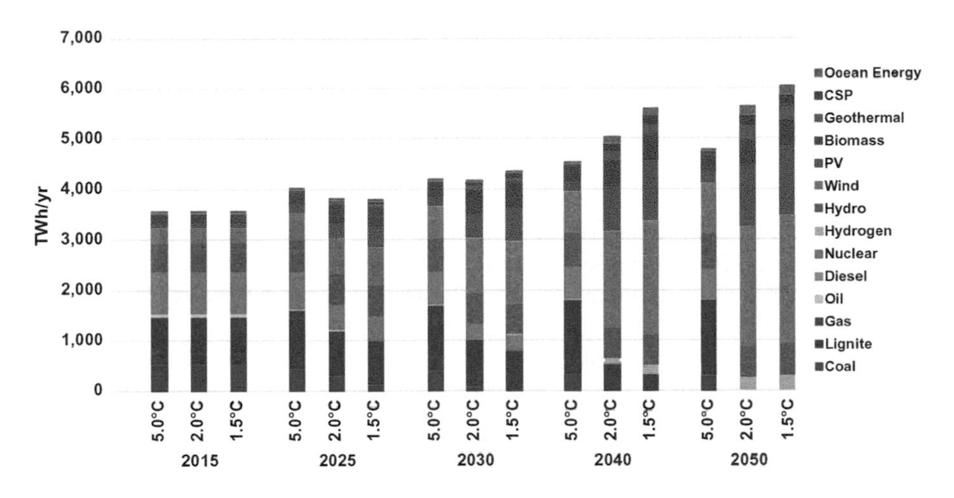

Fig. 8.36 OECD Europe: development of electricity-generation structure in the scenarios

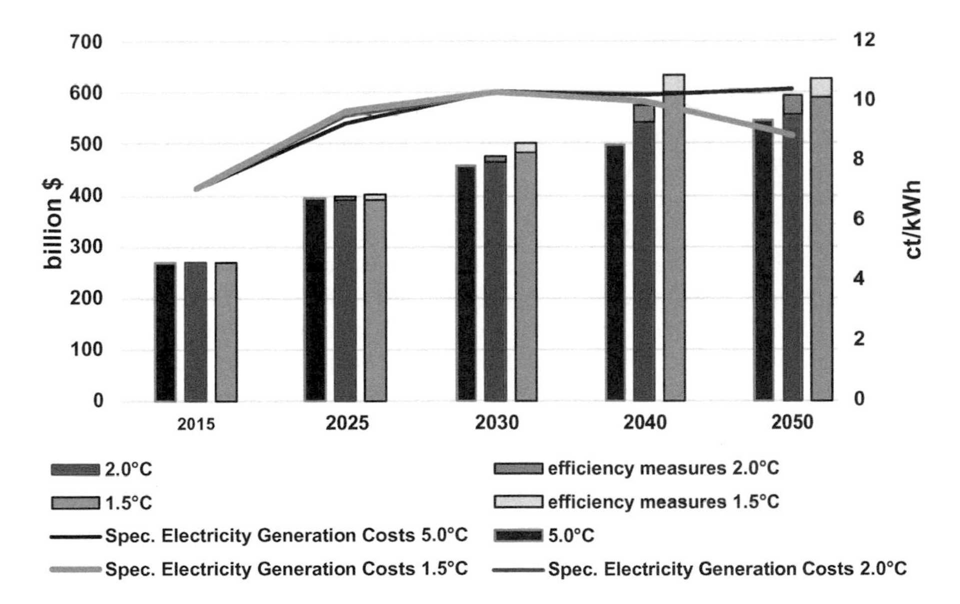

Fig. 8.37 OECD Europe: development of total electricity supply costs and specific electricity-generation costs in the scenarios

kWh. The generation costs in both alternative scenarios will increase until 2030, when they will reach 10.3 ct/kWh, and they will drop by 2050 to 8.9 ct/kWh and 8.8 ct/kWh, respectively, 1.5–1.6 ct/kWh lower than in the 5.0 °C case. Note that these estimates of generation costs do not take into account integration costs such as power grid expansion, storage, or other load-balancing measures.

In the 5.0 °C case, the growth in demand and increasing fossil fuel prices will result in an increase in total electricity supply costs from today's $270 billion/year to more than $550 billion/year in 2050. In the 2.0 °C Scenario, the total supply costs will be $560 billion/year and in the 1.5 °C Scenario, they will be $590 billion/year The long-term costs for electricity supply will be more than 2% higher in the 2.0 °C Scenario than in the 5.0 °C Scenario as a result of the estimated generation costs and the electrification of heating and mobility. Further electrification and synthetic fuel generation in the 1.5 °C Scenario will result in total power generation costs that are 8% higher than in the 5.0 °C case.

Compared with these results, the generation costs when the CO_2 emission costs are not considered will increase in the 5.0 °C Scenario to 8.8 ct/kWh by 2050. In the 2.0 °C Scenario, they will increase until 2030 when they reach 9.5 ct/kWh, and then drop to 8.9 ct/kWh by 2050. In the 1.5 °C Scenario, they will increase to 9.7 ct/kWh, and then drop to 8.8 ct/kWh by 2050. In the 2.0 °C Scenario, the generation costs will reach a maximum of 1 ct/kWh higher than in the 5.0 °C case in 2030. In the 1.5 °C Scenario, the maximum difference in generation costs compared with the 5.0 °C Scenario will be 1.2 ct/kWh, which will occur in 2040. If the CO_2 costs are not considered, the total electricity supply costs in the 5.0 °C case will rise to about $470 billion/year in 2050.

8.7.1.4 OECD Europe: Future Investments in the Power Sector

An investment of around $4900 billion will be required for power generation between 2015 and 2050 in the 2.0 °C Scenario—including additional power plants for the production of hydrogen and synthetic fuels and investments to replace plants at the ends of their economic lives. This value is equivalent to approximately $136 billion per year on average, which is $2150 billion more than in the 5.0 °C case ($2750 billion). An investment of around $5340 billion for power generation will be required between 2015 and 2050 under the 1.5 °C Scenario. On average, this will be an investment of $148 billion per year. In the 5.0 °C Scenario, investment in conventional power plants will be around 26% of the total cumulative investments, whereas approximately 74% will be invested in renewable power generation and co-generation (Fig. 8.38).

However, in the 2.0 °C (1.5 °C) Scenario, OECD Europe will shift almost 96% (97%) of its entire investments to renewables and co-generation. By 2030, the fossil fuel share of the power sector investments will predominantly focus on gas power plants that can also be operated with hydrogen.

Because renewable energy has no fuel costs, other than biomass, the cumulative fuel cost savings in the 2.0 °C Scenario will reach a total of $2340 billion in 2050,

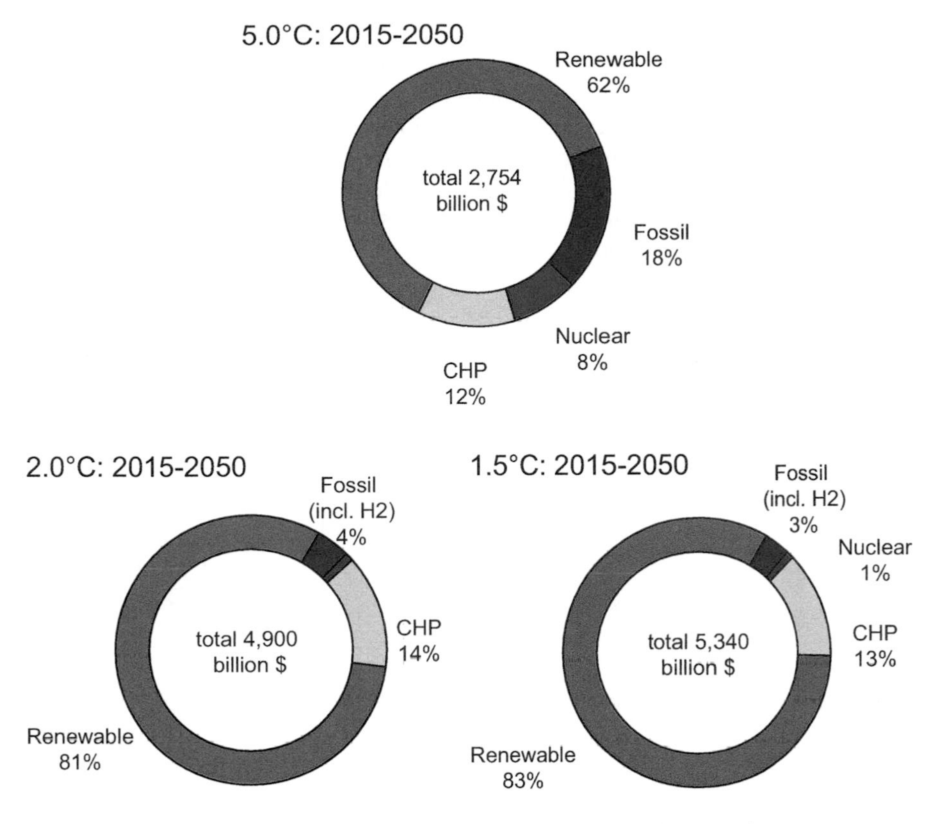

Fig. 8.38 OECD Europe: investment shares for power generation in the scenarios

equivalent to $65 billion per year. Therefore, the total fuel cost savings will be equivalent to 110% of the total additional investments compared to the 5.0 °C Scenario. The fuel cost savings in the 1.5 °C Scenario will add up to $2600 billion, or $72 billion per year.

8.7.1.5 OECD Europe: Energy Supply for Heating

The final energy demand for heating will increase in the 5.0 °C Scenario by 16%, from 20,600 PJ/year in 2015 to 24,000 PJ/year in 2050. Energy efficiency measures will help to reduce the energy demand for heating by 26% in 2050 in the 2.0 °C Scenario relative to that in the 5.0 °C case, and by 34% in the 1.5 °C Scenario. Today, renewables supply around 19% of OECD Europe's final energy demand for heating, with the main contribution from biomass. Renewable energy will provide 44% of OECD Europe's total heat demand in 2030 under the 2.0 °C Scenario and 53% under the 1.5 °C Scenario. In both scenarios, renewables will provide 100% of the total heat demand in 2050.

Figure 8.39 shows the development of different technologies for heating in OECD Europe over time, and Table 8.33 provides the resulting renewable heat supply for all scenarios. Up to 2030, biomass will remain the main contributor. The growing use of solar, geothermal, and environmental heat will lead in the long term to a biomass share of 27% in the 2.0 °C Scenario and 28% in the 1.5 °C Scenario.

Heat from renewable hydrogen will further reduce the dependence on fossil fuels in both scenarios. Hydrogen consumption in 2050 will be around 1900 PJ/year in the 2.0 °C Scenario and 2200 PJ/year in the 1.5 °C Scenario. The direct use of electricity for heating will also increase by a factor of 1.5–1.6 between 2015 and 2050, and will have a final energy share of 22% in 2050 in the 2.0 °C Scenario and 23% in the 1.5 °C Scenario.

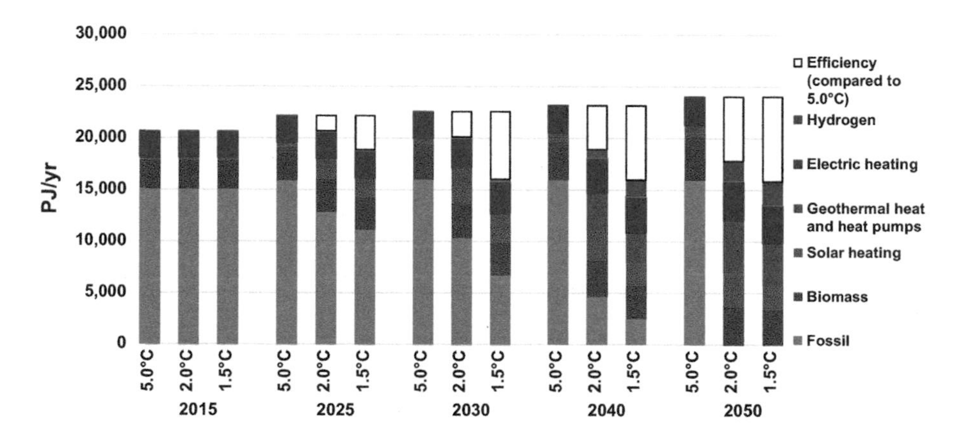

Fig. 8.39 OECD Europe: development of heat supply by energy carrier in the scenarios

Table 8.33 OECD Europe: development of renewable heat supply in the scenarios (excluding the direct use of electricity)

in PJ/year	Case	2015	2025	2030	2040	2050
Biomass	5.0 °C	2681	3115	3343	3713	4153
	2.0 °C	2681	3109	3295	3483	3772
	1.5 °C	2681	3046	3096	3220	3433
Solar heating	5.0 °C	119	216	251	345	454
	2.0 °C	119	1043	1788	2904	3243
	1.5 °C	119	1013	1464	2182	2327
Geothermal heat and heat pumps	5.0 °C	203	291	336	479	717
	2.0 °C	203	968	1731	3572	5080
	1.5 °C	203	878	1430	2933	4147
Hydrogen	5.0 °C	0	0	0	0	0
	2.0 °C	0	0	1	788	1895
	1.5 °C	0	0	162	1595	2227
Total	5.0 °C	3003	3623	3931	4537	5325
	2.0 °C	3003	5121	6815	10,748	13,989
	1.5 °C	3003	4937	6152	9930	12,134

8.7.1.6 OECD Europe: Future Investments in the Heating Sector

The roughly estimated investments in renewable heating technologies up to 2050 will amount to around $2410 billion in the 2.0 °C Scenario (including investments for plant replacement at the ends of their economic lifetimes), or approximately $67 billion per year. The largest share of investments in OECD Europe is assumed to be for heat pumps (around $1200 billion), followed by solar collectors ($1080 billion). The 1.5 °C Scenario assumes an even faster expansion of renewable technologies. However, the lower heat demand (compared with the 2.0 °C Scenario) will results in a lower average annual investment of around $51 billion per year (Fig. 8.40, Table 8.34).

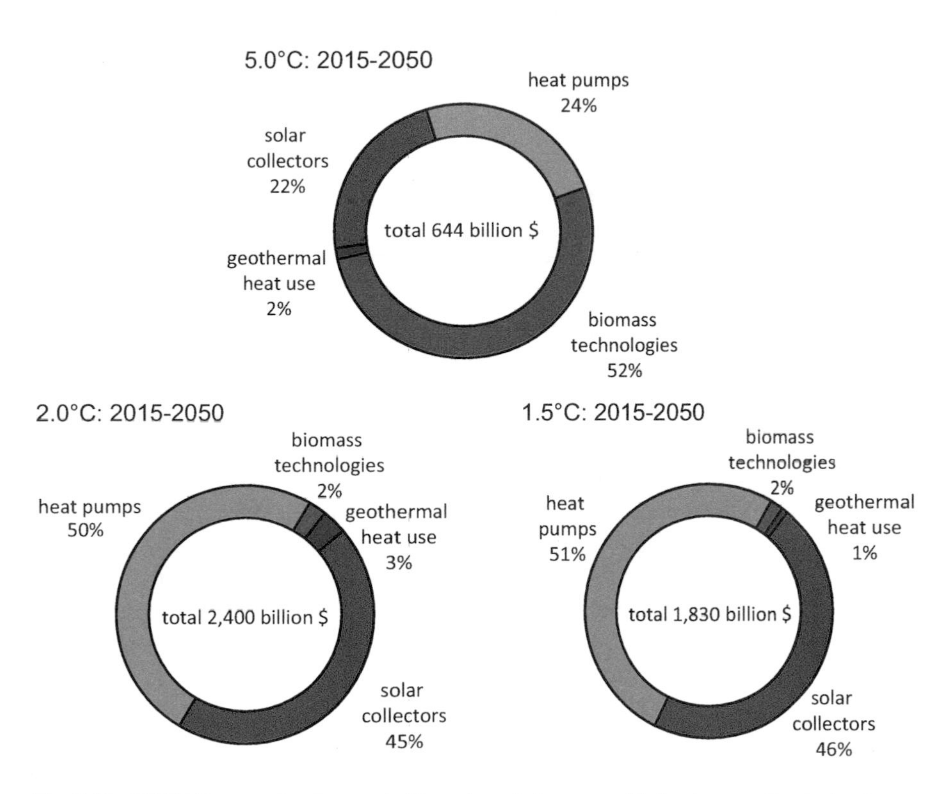

Fig. 8.40 OECD Europe: development of investments for renewable heat-generation technologies in the scenarios

Table 8.34 OECD Europe: installed capacities for renewable heat generation in the scenarios

in GW	Case	2015	2025	2030	2040	2050
Biomass	5.0 °C	434	467	486	507	519
	2.0 °C	434	407	339	293	289
	1.5 °C	434	381	276	256	242
Geothermal	5.0 °C	5	7	7	7	3
	2.0 °C	5	15	24	49	48
	1.5 °C	5	14	16	21	11
Solar heating	5.0 °C	36	65	76	104	137
	2.0 °C	36	298	510	790	885
	1.5 °C	36	291	423	624	685
Heat pumps	5.0 °C	29	40	46	62	84
	2.0 °C	29	134	228	417	566
	1.5 °C	29	121	183	336	444
Total[a]	5.0 °C	504	579	615	681	744
	2.0 °C	504	855	1101	1548	1789
	1.5 °C	504	807	897	1237	1383

[a]Excluding direct electric heating

8.7.1.7 OECD Europe: Transport

Energy demand in the transport sector in OECD Europe is expected to decrease by 3% in the 5.0 °C Scenario, from around 14,000 PJ/year in 2015 to 13,600 PJ/year in 2050. In the 2.0 °C Scenario, assumed technical, structural, and behavioural changes will save 69% (9460 PJ/year) by 2050 compared with the 5.0 °C Scenario. Additional modal shifts, technology switches, and a reduction in the transport demand will lead to even higher energy savings in the 1.5 °C Scenario of 76% (or 10,300 PJ/year) in 2050 compared with the 5.0 °C case (Table 8.35, Fig. 8.41).

By 2030, electricity will provide 18% (430 TWh/year) of the transport sector's total energy demand in the 2.0 °C Scenario, whereas in 2050, the share will be 64% (740 TWh/year). In 2050, up to 840 PJ/year of hydrogen will be used in the transport sector as a complementary renewable option. In the 1.5 °C Scenario, the annual electricity demand will be 580 TWh in 2050. The 1.5 °C Scenario also assumes a hydrogen demand of 730 PJ/year by 2050.

Biofuel use is limited in the 2.0 °C Scenario to a maximum of 600 PJ/year Therefore, around 2030, synthetic fuels based on power-to-liquid will be introduced, with a maximum amount of 130 PJ/year in 2050. Biofuel use will be reduced in the 1.5 °C Scenario to a maximum of 590 PJ/year. The maximum synthetic fuel demand will reach 170 PJ/year.

8.7.1.8 OECD Europe: Development of CO_2 Emissions

In the 5.0 °C Scenario, OECD Europe's annual CO_2 emissions will decrease by 15% from 3400 Mt. in 2015 to 2876 Mt. in 2050. The stringent mitigation measures in both alternative scenarios will cause the annual emissions to fall to 570 Mt. in

Table 8.35 OECD Europe: projection of the transport energy demand by mode in the scenarios

in PJ/year	Case	2015	2025	2030	2040	2050
Rail	5.0 °C	323	334	335	337	344
	2.0 °C	323	362	409	509	643
	1.5 °C	323	383	458	453	400
Road	5.0 °C	13,087	12,699	12,633	12,529	12,464
	2.0 °C	13,087	10,163	7540	4196	3097
	1.5 °C	13,087	8197	4404	3215	2556
Domestic aviation	5.0 °C	300	397	448	485	474
	2.0 °C	300	294	254	182	142
	1.5 °C	300	273	198	105	82
Domestic navigation	5.0 °C	227	236	240	248	259
	2.0 °C	227	236	240	247	258
	1.5 °C	227	236	240	247	258
Total	5.0 °C	13,938	13,665	13,656	13,598	13,541
	2.0 °C	13,938	11,055	8443	5134	4140
	1.5 °C	13,938	9090	5300	4020	3296

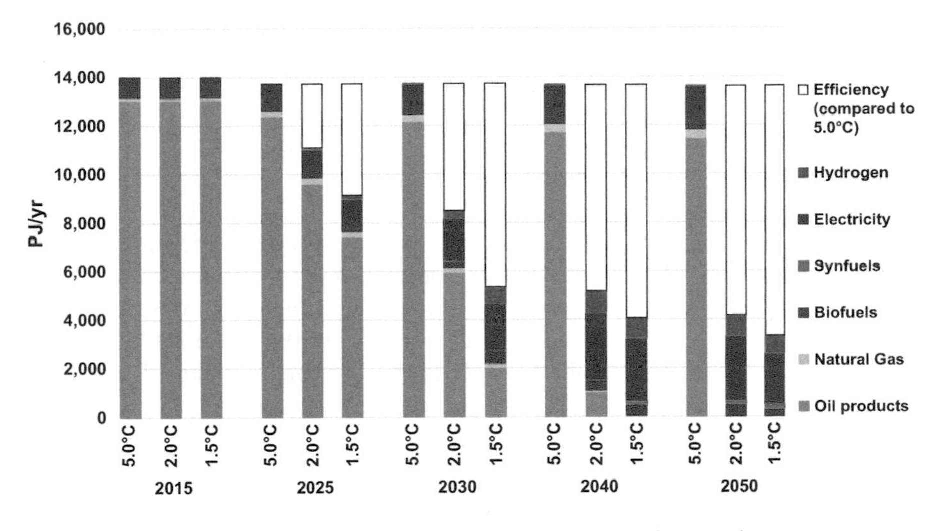

Fig. 8.41 OECD Europe: final energy consumption by transport in the scenarios

2040 in the 2.0 °C Scenario and to 270 Mt. in the 1.5 °C Scenario, with further reductions to almost zero by 2050. In the 5.0 °C case, the cumulative CO_2 emissions from 2015 until 2050 will add up to 116 Gt. In contrast, in the 2.0 °C and 1.5 °C Scenarios, the cumulative emissions for the period from 2015 until 2050 will be 55 Gt and 44 Gt, respectively.

Therefore, the cumulative CO_2 emissions will decrease by 53% in the 2.0 °C Scenario and by 62% in the 1.5 °C Scenario compared with the 5.0 °C case. A rapid reduction in the annual emissions will occur in both alternative scenarios. In the 2.0 °C Scenario, this reduction will be greatest in 'Power generation', followed by the 'Transport' and the 'Residential and other' sectors (Fig. 8.42).

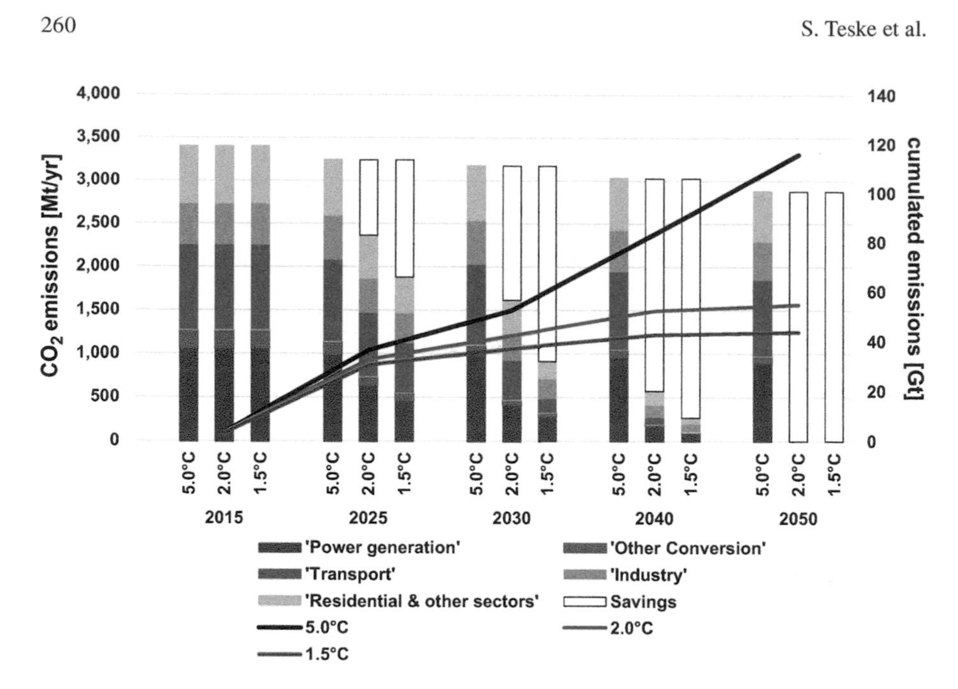

Fig. 8.42 OECD Europe: development of CO_2 emissions by sector and cumulative CO_2 emissions (after 2015) in the scenarios ('Savings' = reduction compared with the 5.0 °C Scenario)

8.7.1.9 OECD Europe: Primary Energy Consumption

The levels of primary energy consumption in the three scenarios when the assumptions discussed above are taken into account are shown in Fig. 8.43. In the 2.0 °C Scenario, the primary energy demand will decrease by 44%, from around 71,200 PJ/year in 2015 to 40,100 PJ/year in 2050. Compared with the 5.0 °C Scenario, the overall primary energy demand will decrease by 43% by 2050 in the 2.0 °C Scenario (5.0 °C: 70,700 PJ in 2050). In the 1.5 °C Scenario, the primary energy demand will be even lower (39,000 PJ in 2050) because the final energy demand and conversion losses will be lower.

Both the 2.0 °C and 1.5 °C Scenarios aim to rapidly phase-out coal and oil. This will cause renewable energy to have primary energy shares of 39% in 2030 and 92% in 2050 in the 2.0 °C Scenario. In the 1.5 °C Scenario, renewables will have a primary energy share of more than 92% in 2050 (including non-energy consumption, which will still include fossil fuels). Nuclear energy will be phased-out by 2040 under both the 2.0 °C and the 1.5 °C Scenarios. The cumulative primary energy consumption of natural gas in the 5.0 °C case will add up to 670 EJ, the cumulative coal consumption to about 300 EJm, and the crude oil consumption to 660 EJ. In contrast, in the 2.0 °C case, the cumulative gas demand will amount to 420 EJ, the cumulative coal demand to 100 EJ, and the cumulative oil demand to 320 EJ. Even lower fossil fuel use will be achieved in the 1.5 °C Scenario: 340 EJ for natural gas, 70 EJ for coal, and 240 EJ for oil.

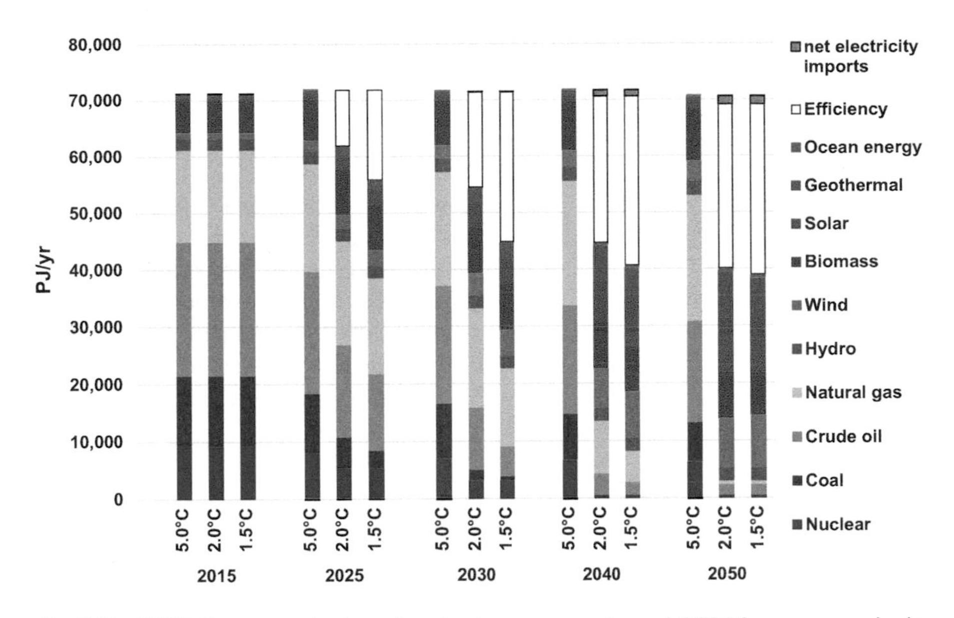

Fig. 8.43 OECD Europe: projection of total primary energy demand (PED) by energy carrier in the scenarios (including electricity import balance)

8.7.2 OECD Europe: Power Sector Analysis

The European power sector is liberalized across the EU and cross-border trade in electricity has a long tradition and is very well documented. The European Network of Transmission System Operators for Electricity (ENTSO-E) publishes detailed data about the annual cross-border trade (ENTSO-E 2018) and produces the *Ten-Year-Network Development Plan (TYNDP)*, which aims to integrate 60% renewable electricity by 2040 (TYNDP 2016). While the extent to which the power sector is liberalised and open for competition for generation and supply varies significantly across the EU, at the time of the writing of this book all 28-member states had renewable electricity and energy efficiency targets and policies to implement them. However, the OECD Europe region covers not only the EU but also neighbouring countries such as Norway, Switzerland and Turkey, which are not members of the EU, but are connected to the EU grid and are also involved in the cross-border electricity trade. The region also includes Iceland, Malta, and a significant number of islands in the coastal waters of the European continent and the Mediterranean Sea. The storage demand for all the islands and island nations cannot be calculated with a regional approach, and doing so was beyond the scope of this research. Israel is also part of OECD Europe in the IEA world regions used for this analysis. However, because of its geographic position, and to reflect current and possible future interconnections with its neighbours, Israel has been taken out of the energy balance of OECD Europe and integrated into the Middle East region.

8.7.2.1 OECD Europe: Development of Power Plant Capacities

The annual market for solar PV must increase from 11 GW in 2020 by a factor of 2 to an average of 40 GW by 2030. The onshore wind market must expand to 18 GW by 2025 under the 2.0 °C Scenario. This is only a minor increase on the average European wind market of 10–14 GW between 2009 and 2016 and 16.8 GW in 2017. However, the 1.5 °C Scenario requires that the size of the onshore wind market double between 2020 and 2025. The offshore wind market for both scenarios is similar and must increase from 3 GW (GWEC 2018) in 2017 to around 10 GW per year throughout the entire modelling period until 2050. All European lignite power plants will have stopped operations by 2035, and the last hard coal power plant will have gone offline by 2040 under the 2.0 °C Scenario. The 1.5 °C pathway requires the phase-out 5 years earlier (Table 8.36).

8.7.2.2 OECD Europe: Utilization of Power-Generation Capacities

The UK, Ireland, and the Iberian Peninsula are the least interconnected sub-regions of OECD Europe, and they already have relatively high shares of variable renewables, as shown in Table 8.37.

Table 8.37 shows that the Nordic countries, especially Norway and Sweden, have very high shares of hydropower, including pumped hydropower. Therefore, an increased interconnection capacity with other sub-regions by 2030 will contribute to the integration of larger shares of wind and solar in other European regions.

Table 8.36 OECD Europe: average annual change in installed power plant capacity

OECD Europe power generation: average annual change of installed capacity [GW/a]	2015–2025		2026–2035		2036–2050	
	2.0 °C	1.5 °C	2.0 °C	1.5 °C	2.0 °C	1.5 °C
Hard coal	−5	−9	−8	−4	0	0
Lignite	−5	−6	−3	−2	0	0
Gas	2	1	0	−5	−22	−19
Hydrogen-gas	0	1	2	6	14	14
Oil/diesel	−7	−5	−1	−2	0	0
Nuclear	−6	−9	−6	−6	−2	−2
Biomass	5	7	4	3	1	1
Hydro	1	0	0	0	0	0
Wind (onshore)	13	28	22	32	13	10
Wind (offshore)	4	9	10	11	8	8
PV (roof top)	16	43	30	42	25	21
PV (utility scale)	5	14	10	14	8	7
Geothermal	0	1	2	2	2	2
Solar thermal power plants	1	2	2	4	2	2
Ocean energy	1	2	3	3	2	2
Renewable fuel based co-generation	3	6	4	4	1	1

Table 8.37 OECD Europe: power system shares by technology group

Power generation structure and interconnection		2.0 °C				1.5 °C			
OECD Europe		Variable RE	Dispatch RE	Dispatch fossil	Inter-connection	Variable RE	Dispatch RE	Dispatch fossil	Inter-connection
Central	2015	12%	47%	41%	20%				
	2030	38%	47%	15%	20%	45%	43%	12%	20%
	2050	62%	33%	5%	20%	64%	31%	5%	20%
UK & Islands	2015	25%	47%	28%	10%				
	2030	63%	31%	6%	20%	71%	25%	5%	20%
	2050	84%	15%	2%	20%	85%	13%	2%	20%
Iberian Peninsula	2015	26%	47%	26%	10%				
	2030	67%	30%	3%	20%	76%	22%	3%	20%
	2050	86%	13%	1%	20%	88%	12%	1%	20%
Balkans + Greece	2015	17%	47%	35%	10%				
	2030	53%	42%	6%	20%	60%	35%	5%	20%
	2050	73%	24%	3%	20%	74%	23%	3%	20%
Baltic	2015	15%	47%	38%	10%				
	2030	44%	45%	12%	20%	50%	40%	10%	20%
	2050	67%	29%	4%	20%	68%	28%	4%	20%
Nordic	2015	13%	47%	39%	10%				
	2030	39%	46%	14%	20%	46%	43%	11%	20%
	2050	65%	31%	4%	20%	67%	29%	4%	20%

(continued)

Table 8.37 (continued)

Power generation structure and interconnection		2.0 °C				1.5 °C			
OECD Europe		Variable RE	Dispatch RE	Dispatch fossil	Inter-connection	Variable RE	Dispatch RE	Dispatch fossil	Inter-connection
Turkey	2015	10%	47%	42%	5%				
	2030	35%	48%	17%	5%	40%	44%	16%	5%
	2050	59%	35%	6%	5%	60%	34%	6%	5%
OECD Europe Central	2015	15%	47%	38%					
	2030	44%	44%	12%		51%	39%	10%	
	2050	67%	28%	4%		69%	27%	4%	

Table 8.38 OECD Europe: capacity factors by generation type

Utilization of variable and dispatchable power generation:		2015	2020	2020	2030	2030	2040	2040	2050	2050
World			2.0 °C	1.5 °C	2.0 °C	1.5 °C	2.0 °C	1.5 °C	2.0 °C	1.5 °C
Capacity factor – average	[%/yr]	45.2%	37%	37%	48%	44%	35%	36%	39%	38%
Limited dispatchable: fossil and nuclear	[%/yr]	57.5%	14%	14%	3%	2%	19%	1%	20%	9%
Limited dispatchable: renewable	[%/yr]	54.0%	60%	60%	52%	48%	60%	39%	41%	40%
Dispatchable: fossil	[%/yr]	32.0%	20%	20%	7%	7%	30%	10%	15%	16%
Dispatchable: renewable	[%/yr]	43.7%	67%	67%	67%	61%	39%	49%	52%	50%
Variable: renewable	[%/yr]	22.5%	22%	22%	40%	38%	29%	35%	36%	35%

Across the EU, it is assumed that the average interconnection capacities will increase to 20% of the regional peak load.

Both alternative scenarios assume that limited dispatchable power generation—namely coal, lignite, and nuclear—will not have priority dispatch and will be last in the dispatch queue. Therefore, the average calculated capacity factor will decrease from 57.5% in 2015 to only 14% in 2020, as shown in Table 8.38.

Table 8.38 shows that by 2020, most of the installed coal and nuclear capacity will not be required to secure power supply. Instead, dispatchable renewable power plants will fill the gap and their capacity factors will increase.

8.7.2.3 OECD Europe: Development of Load, Generation, and Residual Load

The loads of the European sub-regions will not increase until 2030 in the two alternative scenarios, as shown in Table 8.39. The only exception is Turkey, which will have a constantly increasing load. This is attributed to Turkey's assumed economic development and increasing per capita electricity demand, which is currently lower than in most EU countries (WB-DB 2018). The calculated load will increase in all sub-regions between 2030 and 2050 due to the increased deployment of electric mobility. Central Europe has a very high requirement for increased transmission interconnection—or storage, see Table 8.40—because of increases in variable generation, including offshore wind in the North Sea and Baltic Sea. Central Europe,

Table 8.39 OECD Europe: load, generation, and residual load development

Power generation structure		2.0 °C				1.5 °C			
		Max demand	Max generation	Max residual load	Max interconnection requirements	Max demand	Max generation	Max residual load	Max interconnection requirements
OECD Europe		[GW]	[GW]	[GW]	[GW]	[GW]	[GW]	[GW]	[GW]
Central	2020	328.9	322.3	73.6		328.9	322.3	77.8	
	2030	350.7	397.4	44.5	2	360.3	520.4	47.4	113
	2050	491.9	842.8	243.1	108	511.2	954.3	259.0	184
UK & Islands	2020	66.1	73.5	33.2		66.1	73.4	33.1	
	2030	71.6	87.6	21.0	0	73.7	112.9	23.4	16
	2050	98.0	187.9	51.5	38	102.2	210.7	55.1	53
Iberian Peninsula	2020	47.0	56.1	10.3		47.0	56.1	10.3	
	2030	50.8	62.3	7.3	4	52.6	80.8	7.9	20
	2050	70.8	133.2	31.7	31	74.3	149.4	34.6	41
Balkans + Greece	2020	37.9	38.2	1.4		37.9	37.9	1.4	
	2030	39.5	49.3	6.3	4	41.6	63.1	6.8	15
	2050	55.6	105.4	24.1	26	59.8	117.8	27.5	30
Baltic	2020	4.6	4.5	0.1		4.6	4.5	0.1	
	2030	4.9	6.1	0.7	1	5.1	7.9	0.7	2
	2050	6.8	13.1	3.2	3	7.2	14.7	3.5	4
Nordic	2020	52.0	50.8	1.3		52.0	50.8	1.3	
	2030	54.4	65.9	8.7	3	55.2	86.0	10.4	20
	2050	71.0	140.3	30.0	39	72.6	158.5	31.0	55
Turkey	2020	37.5	38.5	0.8		37.5	38.2	0.8	
	2030	48.4	49.1	6.9	0	50.8	64.4	7.5	6
	2050	68.2	107.4	33.1	6	73.0	121.5	37.4	11

Table 8.40 OECD Europe: storage and dispatch service requirements

Storage and dispatch		2.0 °C					1.5 °C				
		Required to avoid curtailment	Utilization battery -through-put-	Utilization PSH -through-put-	Total storage demand (incl. H2)	Dispatch hydrogen-based	Required to avoid curtailment	Utilization battery -through-put-	Utilization PSH -through-put-	Total storage demand (incl. H2)	Dispatch hydrogen-based
OECD Europe		[GWh/year]	[GWh/year]	[GWh/year]	[GWh/year]	[GWh/year]	[GWh/year]	[GWh/year]	[GWh/year]	[GWh/year]	[GWh/year]
Central	2020	0	0	0	0	0	0	0	0	0	0
	2030	425	67	728	796	38,043	6947	515	7996	8511	139,501
	2050	59,495	28,998	32,425	61,423	546,511	99,134	35,542	48,679	84,222	549,376
UK & Islands	2020	0	0	0	0	0	0	0	0	0	0
	2030	3419	293	5808	6101	4148	12,239	440	13,977	14,417	13,195
	2050	57,089	9507	34,158	43,665	41,134	72,011	9738	38,301	48,039	40,932
Iberian Peninsula	2020	0	0	0	0	0	0	0	0	0	0
	2030	1688	186	2763	2949	2712	12,555	407	11,672	12,079	8127
	2050	52,580	7952	27,526	35,478	22,000	69,483	8273	30,928	39,201	22,448
Balkans + Greece	2020	0	0	0	0	0	0	0	0	0	0
	2030	523	62	895	957	3274	3699	172	3996	4168	11,349
	2050	19,794	5717	10,649	16,366	39,208	25,680	6267	12,033	18,300	42,798
Baltic	2020	0	0	0	0	0	0	0	0	0	0
	2030	27	2	41	42	482	190	7	174	181	1775
	2050	1071	360	542	902	6365	1504	413	677	1090	6636
Nordic	2020	0	0	0	0	0	0	0	0	0	0
	2030	149	16	274	291	6276	2111	95	2237	2332	23,031
	2050	14,144	4425	6905	11,330	80,577	22,171	5219	9360	14,580	78,294

(continued)

Table 8.40 (continued)

Storage and dispatch		2.0 °C					1.5 °C				
		Required to avoid curtailment	Utilization battery -through-put-	Utilization PSH -through-put-	Total storage demand (incl. H2)	Dispatch hydrogen-based	Required to avoid curtailment	Utilization battery -through-put-	Utilization PSH -through-put-	Total storage demand (incl. H2)	Dispatch hydrogen-based
OECD Europe		[GWh/year]	[GWh/year]	[GWh/year]	[GWh/year]	[GWh/year]	[GWh/year]	[GWh/year]	[GWh/year]	[GWh/year]	[GWh/year]
Turkey	2020	0	0	0	0	0	0	0	0	0	0
	2030	8	4	21	25	5287	762	72	1067	1139	20,038
	2050	7887	4120	4348	8467	78,788	11,251	4744	5467	10,211	82,142
OECD Europe	2020	0	0	0	0	0	0	0	0	0	0
	2030	6238	630	10,531	11,161	60,223	38,504	1710	41,118	42,827	217,016
	2050	212,060	61,078	116,554	177,632	814,585	301,234	70,196	145,445	215,641	822,626

the Iberian Peninsula, and the UK have the highest storage demands, as shown in Table 8.40. This corresponds to the calculated results for increased interconnections. To avoid curtailment, renewably produced hydrogen will be used to store surplus generation for dispatch when required. Finding the optimal mix of battery capacity, pumped hydro capacity, hydrogen production, and expansion of transmission capacity was beyond the scope of this analysis, and further research is required on this issue.

8.8 Africa

8.8.1 Africa: Long-Term Energy Pathways

8.8.1.1 Africa: Final Energy Demand by Sector

The development pathways for Africa's final energy demand when the assumptions on population growth, GDP growth, and energy intensity are combined are shown in Fig. 8.44 for the 5.0 °C, 2.0 °C, and 1.5 °C Scenarios. In the 5.0 °C Scenario, the total final energy demand will increase by 103% from the current 23,200 PJ/year to 47,100 PJ/year in 2050. In the 2.0 °C Scenario, the final energy demand will increase at a much slower rate, by 39% compared with current consumption, and will reach

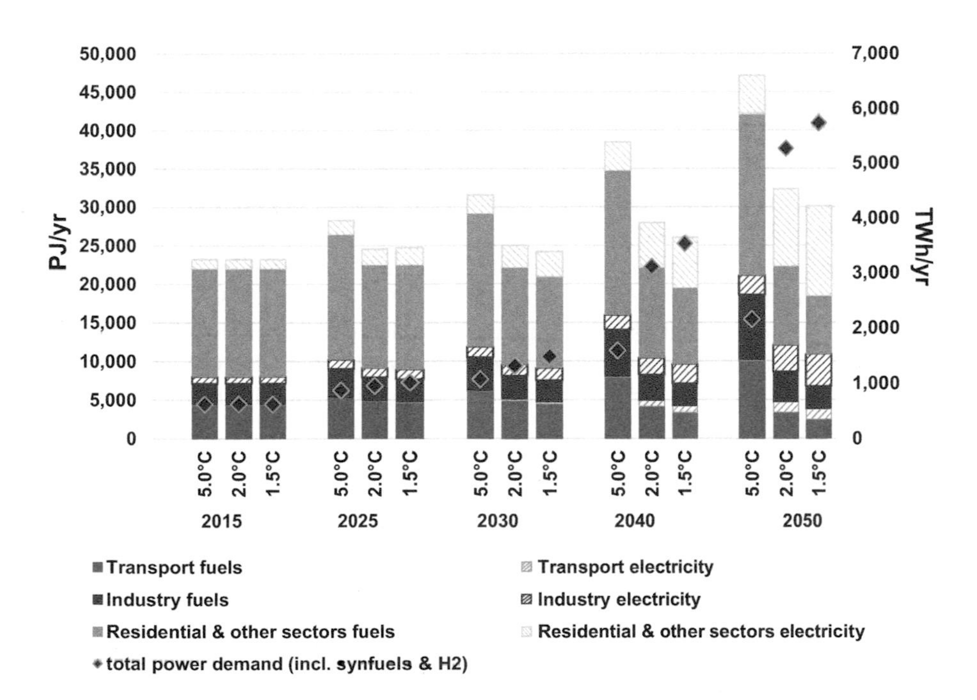

Fig. 8.44 Africa: development of final energy demand by sector in the scenarios

32,300 PJ/year by 2050. The final energy demand under the 1.5 °C Scenario will reach 30,100 PJ, 30% above the 2015 demand level. In the 1.5 °C Scenario, the final energy demand in 2050 will be 7% lower than in the 2.0 °C Scenario. The electricity demand for 'classical' electrical devices (without power-to-heat or e-mobility) will increase from 540 TWh/year in 2015 to around 2590 TWh/year in 2050 in both alternative scenarios, which will be 590 TWh/year higher than in the 5.0 °C case. Although efficiency measures will reduce the specific energy consumption by appliances, the scenarios consider higher consumption to achieve higher living standards.

Electrification will lead to a significant increase in the electricity demand by 2050. In the 2.0 °C Scenario, the electricity demand for heating will increase to approximately 1200 TWh/year due to electric heaters and heat pumps, and in the transport sector, the demand will increase to approximately 1300 TWh/year in response to increased electric mobility. The generation of hydrogen (for transport and high-temperature process heat) and the manufacture of synthetic fuels (mainly for transport) will add an additional power demand of 1100 TWh/year The gross power demand will thus increase from 800 TWh/year in 2015 to 5700 TWh/year in 2050 in the 2.0 °C Scenario, 119% higher than in the 5.0 °C case. In the 1.5 °C Scenario, the gross electricity demand will increase to a maximum of 6300 TWh/year in 2050.

The efficiency gains in the heating sector could be even larger than in the electricity sector. In the 2.0 °C and 1.5 °C Scenarios, a final energy consumption equivalent to about 3600 PJ/year is avoided through efficiency gains by 2050 compared with the 5.0 °C Scenario.

8.8.1.2 Africa: Electricity Generation

The development of the power system is characterized by a dynamically growing renewable energy market and an increasing proportion of total power from renewable sources. By 2050, 100% of the electricity produced in Africa will come from renewable energy sources in the 2.0 °C Scenario. 'New' renewables—mainly wind, solar, and geothermal energy—will contribute 92% of the total electricity generation. Renewable electricity's share of total production will be 61% by 2030 and 96% by 2040. The installed capacity of renewables will reach about 360 GW by 2030 and 2040 GW by 2050. In the 1.5 °C Scenario, the share of renewable electricity generation in 2030 is assumed to be 73%. The 1.5 °C Scenario will have a generation capacity from renewable energy of approximately 2280 GW in 2050.

Table 8.41 shows the development of different renewable technologies in Africa over time. Figure 8.45 provides an overview of the overall power-generation structure in Africa. From 2020 onwards, the continuing growth of wind and PV, up to 610 GW and 980 GW, respectively, will be complemented by up to 230 GW of solar thermal generation, as well as limited biomass, geothermal, and ocean energy, in the 2.0 °C Scenario. Both the 2.0 °C and 1.5 °C Scenarios will lead to high proportions of variable power generation (PV, wind, and ocean) of 40% and 49%, respectively, by 2030, and 71% by 2050.

Table 8.41 Africa: development of renewable electricity-generation capacity in the scenarios

in GW	Case	2015	2025	2030	2040	2050
Hydro	5.0 °C	28	47	58	84	117
	2.0 °C	28	46	49	51	54
	1.5 °C	28	46	48	51	54
Biomass	5.0 °C	1	2	4	8	13
	2.0 °C	1	8	17	33	48
	1.5 °C	1	8	25	42	72
Wind	5.0 °C	3	11	14	20	29
	2.0 °C	3	42	132	415	609
	1.5 °C	3	87	197	453	633
Geothermal	5.0 °C	1	2	3	7	14
	2.0 °C	1	7	16	33	64
	1.5 °C	1	7	16	33	64
PV	5.0 °C	2	17	27	52	89
	2.0 °C	2	38	134	611	983
	1.5 °C	2	70	166	757	1162
CSP	5.0 °C	0	2	3	10	17
	2.0 °C	0	0	1	80	235
	1.5 °C	0	2	19	108	257
Ocean	5.0 °C	0	0	0	0	0
	2.0 °C	0	2	10	20	43
	1.5 °C	0	2	10	20	43
Total	5.0 °C	35	81	110	180	279
	2.0 °C	35	144	359	1243	2036
	1.5 °C	35	223	481	1464	2284

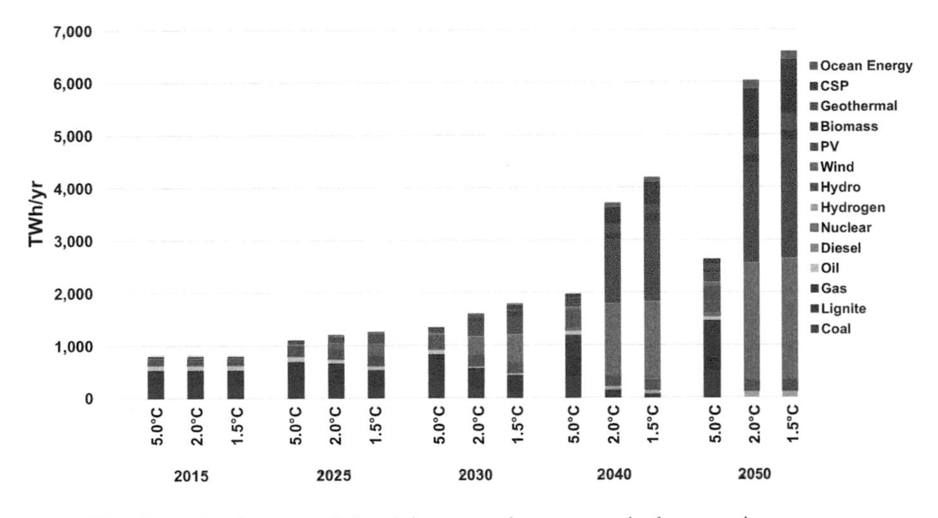

Fig. 8.45 Africa: development of electricity-generation structure in the scenarios

8.8.1.3 Africa: Future Costs of Electricity Generation

Figure 8.46 shows the development of the electricity-generation and supply costs over time, including the CO_2 emission costs, in all scenarios. The calculated electricity-generation costs in 2015 (referring to full costs) were around 5.4 ct/kWh. In the 5.0 °C case, generation costs will increase until 2030, when they reach 11 ct/kWh, and will then stabilize at 10.8 ct/kWh by 2050. In the 2.0 °C and 1.5 °C Scenarios, the generation costs will increase until 2030, when they reach 8.4 ct/kWh and 8.2 ct/kWh, respectively. They will then drop to 5.6 ct/kWh by 2050 in both scenarios, 5.2 ct/kWh lower than in the 5.0 °C case. Note that these estimates of generation costs do not take into account integration costs such as power grid expansion, storage, or other load-balancing measures.

In the 5.0 °C case, the growth in demand and increasing fossil fuel prices will cause the total electricity supply costs to increase from today's $40 billion/year to more than $290 billion/year in 2050. In the 2.0 °C Scenario, the total supply costs will be $350 billion/year, and in the 1.5 °C Scenario, they will be $380 billion/year The long-term costs of electricity supply will be more than 23% higher under the 2.0 °C Scenario than under the 5.0 °C Scenario as a result of the estimated generation costs and the electrification of heating and mobility. Further electrification and synthetic fuel generation in the 1.5 °C Scenario will result in total power generation costs that are 34% higher than in the 5.0 °C case.

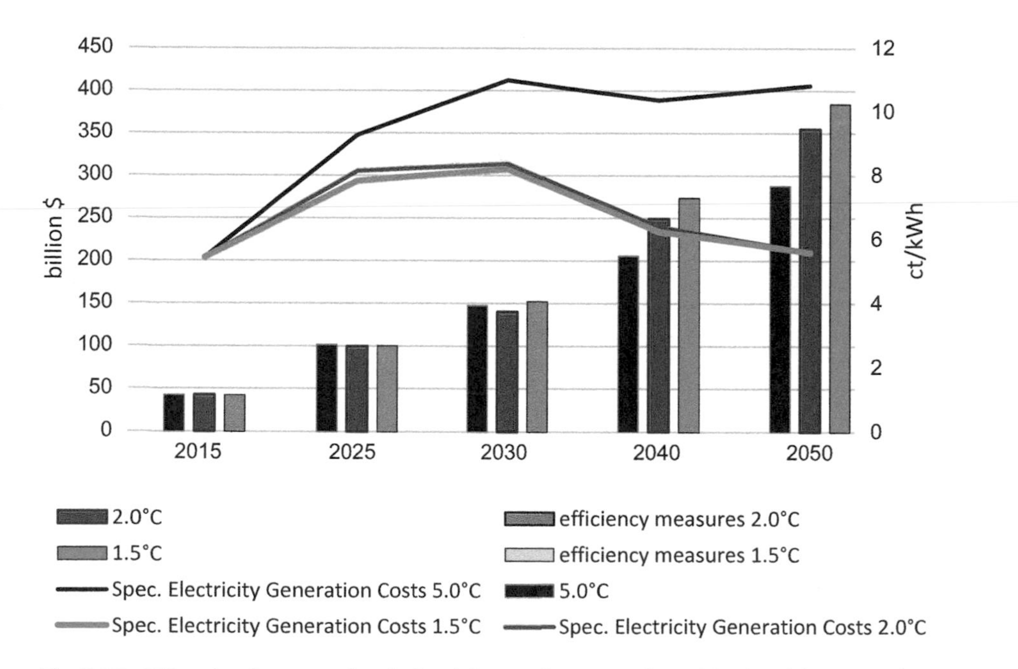

Fig. 8.46 Africa: development of total electricity supply costs and specific electricity-generation costs in the scenarios

Compared with these results, the generation costs when the CO_2 emission costs are not considered will increase in the 5.0 °C case to 8.1 ct/kWh. In the 2.0 °C Scenario, they will increase until 2030, when they reach 6.8 ct/kWh, and then drop to 5.6 ct/kWh by 2050. In the 1.5 °C Scenario, they will increase to 7.2 ct/kWh and then drop to 5.6 ct/kWh by 2050. Therefore, the generation costs in both alternative scenarios are, at maximum, 2.5 ct/kWh lower than in the 5.0 °C case. If the CO_2 costs are not considered, the total electricity supply costs in the 5.0 °C case will increase to about $220 billion/year in 2050.

8.8.1.4 Africa: Future Investments in the Power Sector

An investment of around $3500 billion will be required for power generation between 2015 and 2050 in the 2.0 °C Scenario—including additional power plants for the production of hydrogen and synthetic fuels and investments in plant replacement at the ends of their economic lives. This value is equivalent to approximately $97 billion per year, on average, and is $2590 billion more than in the 5.0 ° C case ($910 billion). An investment of around $3910 billion for power generation will be required between 2015 and 2050 in the 1.5 °C Scenario. On average, this is an investment of $109 billion per year. In the 5.0 °C Scenario, the investment in conventional power plants will be around 45% of the total cumulative investments, and approximately 55% will be invested in renewable power generation and co-generation (Fig. 8.47).

However, in the 2.0 °C (1.5 °C) Scenario, Africa will shift almost 93% (94%) of its entire investments to renewables and co-generation. By 2030, the fossil fuel share of power sector investments will focus predominantly on gas power plants that can also be operated with hydrogen.

Because renewable energy has no fuel costs, other than biomass, the cumulative fuel cost savings in the 2.0 °C Scenario will reach a total of $1510 billion in 2050, equivalent to $42 billion per year. Therefore, the total fuel cost savings will be equivalent to 60% of the total additional investments compared to the 5.0 °C Scenario. The fuel cost savings in the 1.5 °C Scenario will add up to $1610 billion, or $45 billion per year.

8.8.1.5 Africa: Energy Supply for Heating

The final energy demand for heating will increase in the 5.0 °C Scenario by 166%, from 7600 PJ/year in 2015 to 20,200 PJ/year in 2050. Energy efficiency measures will help to reduce the energy demand for heating by 18% in 2050 in both alternative scenarios, relative to the 5.0 °C case. Today, renewables supply around 61% of Africa's final energy demand for heating, with the main contribution from biomass. Renewable energy will provide 71% of Africa's total heat demand in 2030 under the 2.0 °C Scenario and 79% under the 1.5 °C Scenario. In both scenarios, renewables will provide 100% of the total heat demand from renewable energy in 2050.

Figure 8.48 shows the development of different technologies for heating in Africa over time, and Table 8.42 provides the resulting renewable heat supply for all scenarios. Biomass will remain the main contributor. The growing use of solar, geothermal, and environmental heat will lead, in the long term, to a reduced biomass share of 51% in the 2.0 °C Scenario and 40% in the 1.5 °C Scenario.

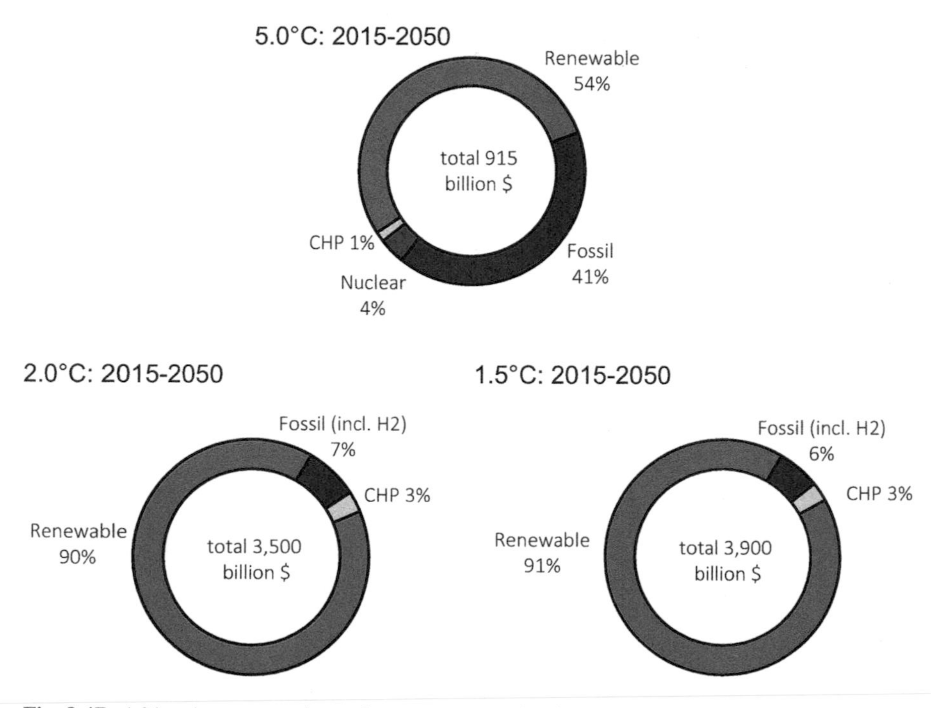

Fig. 8.47 Africa: investment shares for power generation in the scenarios

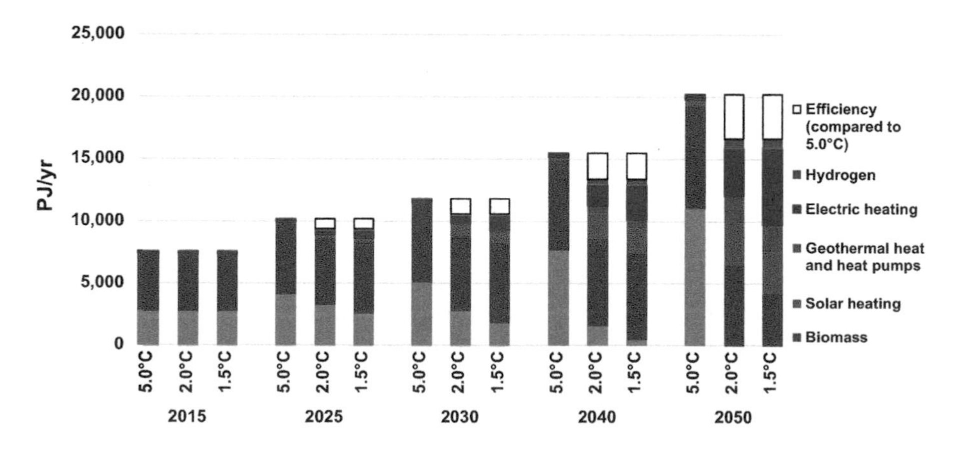

Fig. 8.48 Africa: development of heat supply by energy carrier in the scenarios

Table 8.42 Africa: development of renewable heat supply in the scenarios (excluding the direct use of electricity)

in PJ/year	Case	2015	2025	2030	2040	2050
Biomass	5.0 °C	4586	5761	6317	7211	8203
	2.0 °C	4586	5308	6047	7039	6551
	1.5 °C	4586	5748	6448	6938	4222
Solar heating	5.0 °C	7	37	86	228	481
	2.0 °C	7	204	786	2066	3416
	1.5 °C	7	203	783	2109	3416
Geothermal heat and heat pumps	5.0 °C	0	0	0	0	0
	2.0 °C	0	86	215	559	2106
	1.5 °C	0	86	213	591	2106
Hydrogen	5.0 °C	0	0	0	0	0
	2.0 °C	0	0	0	397	720
	1.5 °C	0	0	0	429	720
Total	5.0 °C	4593	5797	6404	7440	8684
	2.0 °C	4593	5598	7047	10,061	12,793
	1.5 °C	4593	6037	7444	10,067	10,464

Heat from renewable hydrogen will further reduce the dependence on fossil fuels in both scenarios. Hydrogen consumption in 2050 will be around 720 PJ/year in both the 2.0 °C Scenario and 1.5 °C Scenario. The direct use of electricity for heating will also increase by a factor of 21–34 between 2015 and 2050, and will attain a final energy share of 23% in 2050 in the 2.0 °C Scenario and 37% in the 1.5 °C Scenario.

8.8.1.6 Africa: Future Investments in the Heating Sector

The roughly estimated investments in renewable heating technologies up to 2050 will amount to around $790 billion in the 2.0 °C Scenario (including investments in plant replacement after their economic lifetimes), or approximately $22 billion per year. The largest share of investment in Africa is assumed to be for heat pumps (around $370 billion), followed by solar collectors and biomass technologies. The 1.5 °C Scenario assumes an even faster expansion of renewable technologies. However, the lower heat demand (compared with the 2.0 °C Scenario) will result in a lower average annual investment of around $21 billion per year (Table 8.43, Fig. 8.49).

Table 8.43 Africa: installed capacities for renewable heat generation in the scenarios

in GW	Case	2015	2025	2030	2040	2050
Biomass	5.0 °C	3655	4036	4100	3973	3870
	2.0 °C	3655	3276	3063	2792	2251
	1.5 °C	3655	3562	3069	2440	1307
Geothermal	5.0 °C	0	0	0	0	0
	2.0 °C	0	5	9	15	37
	1.5 °C	0	5	8	15	37
Solar heating	5.0 °C	1	7	16	44	92
	2.0 °C	1	39	150	396	654
	1.5 °C	1	39	150	404	654
Heat pumps	5.0 °C	0	0	0	0	0
	2.0 °C	0	3	16	51	227
	1.5 °C	0	3	16	54	227
Total[a]	5.0 °C	3656	4043	4116	4017	3962
	2.0 °C	3656	3324	3239	3253	3169
	1.5 °C	3656	3610	3244	2912	2225

[a]Excluding direct electric heating

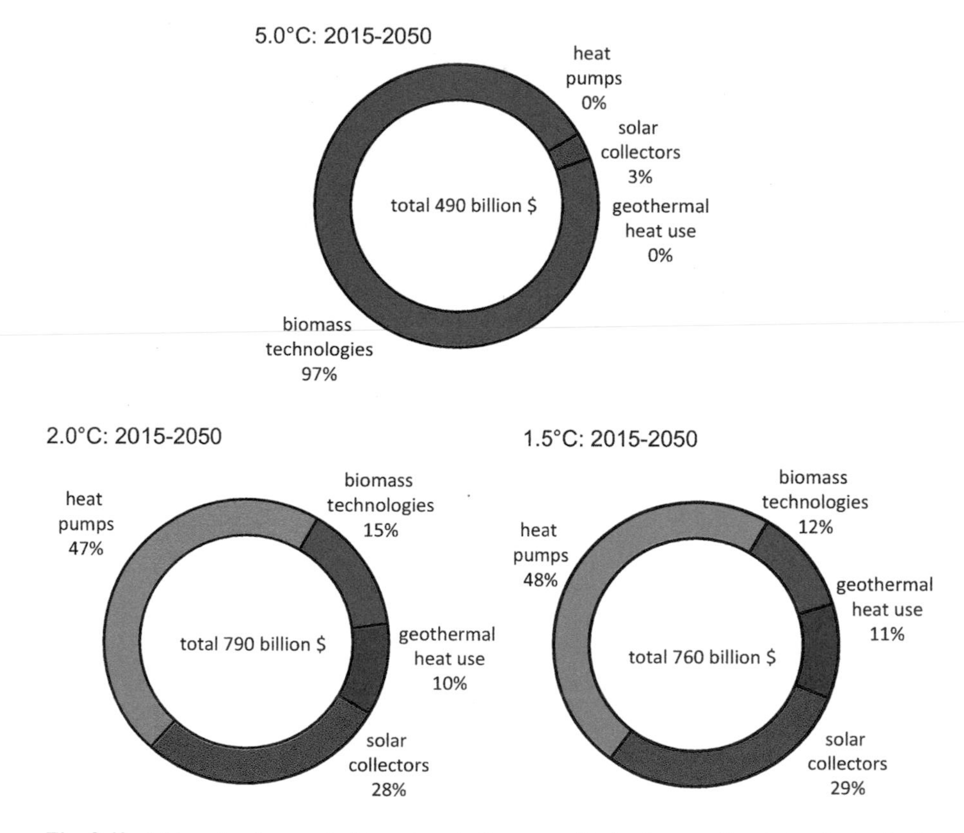

Fig. 8.49 Africa: development of investments for renewable heat-generation technologies in the scenarios

8.8.1.7 Africa: Transport

The energy demand in the transport sector in Africa is expected to increase by 131% in the 5.0 °C Scenario, from around 4400 PJ/year in 2015 to 10,100 PJ/year in 2050. In the 2.0 °C Scenario, assumed technical, structural, and behavioural changes will save 53% (5410 PJ/year) by 2050 compared with the 5.0 °C Scenario. Additional modal shifts, technology switches, and a reduction in the transport demand will lead to even higher energy savings in the 1.5 °C Scenario of 63% (or 6360 PJ/year) in 2050 compared with the 5.0 °C case (Table 8.44, Fig. 8.50).

By 2030, electricity will provide 4% (50 TWh/year) of the transport sector's total energy demand in the 2.0 °C Scenario, whereas by 2050, the share will be 28% (370 TWh/year). In 2050, up to 410 PJ/year of hydrogen will be used in the transport sector as a complementary renewable option. In the 1.5 °C Scenario, the annual electricity demand will be 360 TWh in 2050. The 1.5 °C Scenario also assumes a hydrogen demand of 340 PJ/year by 2050.

Biofuel use is limited in the 2.0 °C Scenario to a maximum of 2300 PJ/year. Therefore, around 2030, synthetic fuels based on power-to-liquid will be introduced, with a maximum amount of 700 PJ/year in 2050. With the lower overall energy demand by transport, biofuel use will be reduced in the 1.5 °C Scenario to a maximum of 1700 PJ/year The maximum synthetic fuel demand will amount to 470 PJ/year.

8.8.1.8 Africa: Development of CO_2 Emissions

In the 5.0 °C Scenario, Africa's annual CO_2 emissions will increase by 126%, from 1140 Mt. in 2015 to 2585 Mt. in 2050. The stringent mitigation measures in both alternative scenarios will cause annual emissions to fall to 400 Mt. in 2040 in the

Table 8.44 Africa: projection of transport energy demand by mode in the scenarios

in PJ/year	Case	2015	2025	2030	2040	2050
Rail	5.0 °C	46	52	58	67	74
	2.0 °C	46	58	71	96	110
	1.5 °C	46	69	88	125	186
Road	5.0 °C	4182	5000	5812	7522	9635
	2.0 °C	4182	4688	4828	4651	4488
	1.5 °C	4182	4493	4422	3925	3482
Domestic aviation	5.0 °C	105	159	198	256	272
	2.0 °C	105	114	110	90	71
	1.5 °C	105	110	102	74	54
Domestic navigation	5.0 °C	32	35	37	40	44
	2.0 °C	32	35	37	40	44
	1.5 °C	32	35	37	40	44
Total	5.0 °C	4366	5246	6105	7885	10,027
	2.0 °C	4366	4895	5045	4877	4714
	1.5 °C	4366	4707	4648	4164	3765

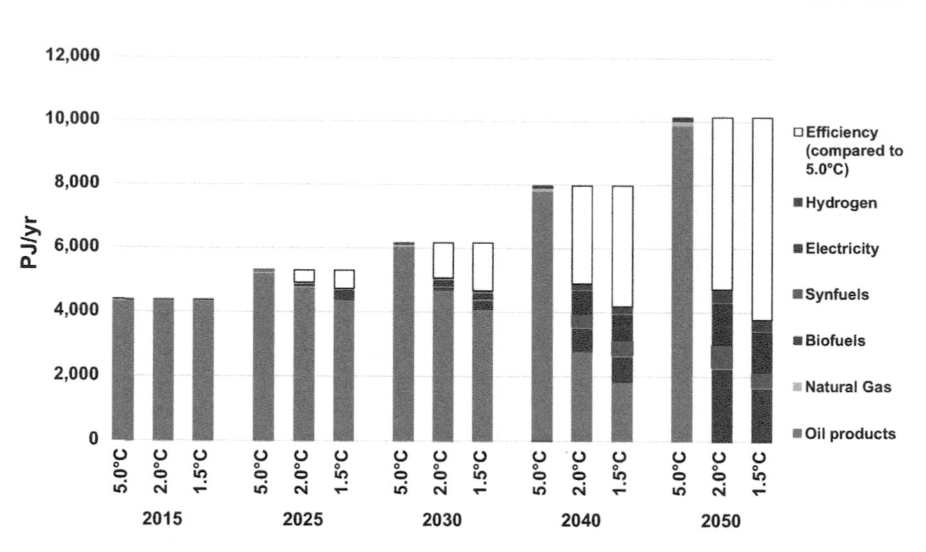

Fig. 8.50 Africa: final energy consumption by transport in the scenarios

2.0 °C Scenario and to 200 Mt. in the 1.5 °C Scenario, with further reductions to almost zero by 2050. In the 5.0 °C case, the cumulative CO_2 emissions from 2015 until 2050 will add up to 66 Gt. In contrast, in the 2.0 °C and 1.5 °C Scenarios, the cumulative emissions for the period from 2015 until 2050 will be 27 Gt and 22 Gt, respectively.

Therefore, the cumulative CO_2 emissions will decrease by 59% in the 2.0 °C Scenario and by 67% in the 1.5 °C Scenario compared with the 5.0 °C case. A rapid reduction in annual emissions will occur in both alternative scenarios. In the 2.0 °C Scenario, this reduction will be greatest in 'Power generation', followed by the 'Industry' and 'Residential and other' sectors (Fig. 8.51).

8.8.1.9 Africa: Primary Energy Consumption

The levels of primary energy consumption in the three scenarios when the assumptions discussed above are taken into account are shown in Fig. 8.52. In the 2.0 °C Scenario, the primary energy demand will increase by 50% from around 33,200 PJ/year in 2015 to around 50,000 PJ/year in 2050. Compared with the 5.0 °C Scenario, the overall primary energy demand will decrease by 26% by 2050 in the 2.0 °C Scenario (5.0 °C: 67700 PJ in 2050). In the 1.5 °C Scenario, the primary energy demand will be even lower (48,000 PJ in 2050) because the final energy demand and conversion losses will be lower.

Both the 2.0 °C and 1.5 °C Scenarios aim to rapidly phase-out coal and oil. This will cause renewable energy to have a primary energy share of 56% in 2030 and 98% in 2050 in the 2.0 °C Scenario. In the 1.5 °C Scenario, renewables will have a primary energy share of more than 98% in 2050 (including non-energy consumption, which will still include fossil fuels). Nuclear energy will be phased-out by

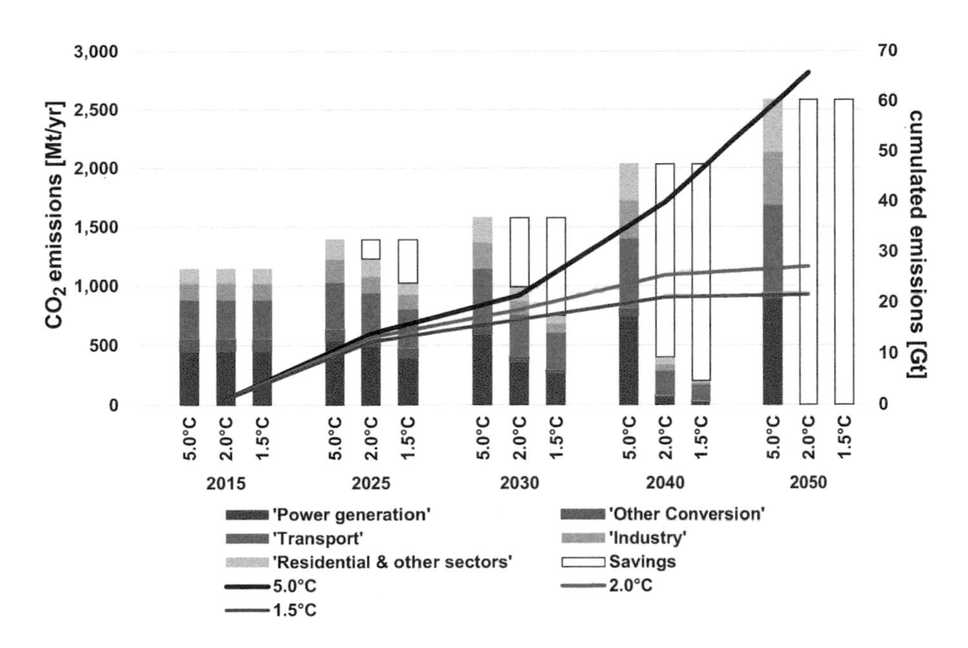

Fig. 8.51 Africa: development of CO_2 emissions by sector and cumulative CO_2 emissions (after 2015) in the scenarios ('Savings' = reduction compared with the 5.0 °C Scenario)

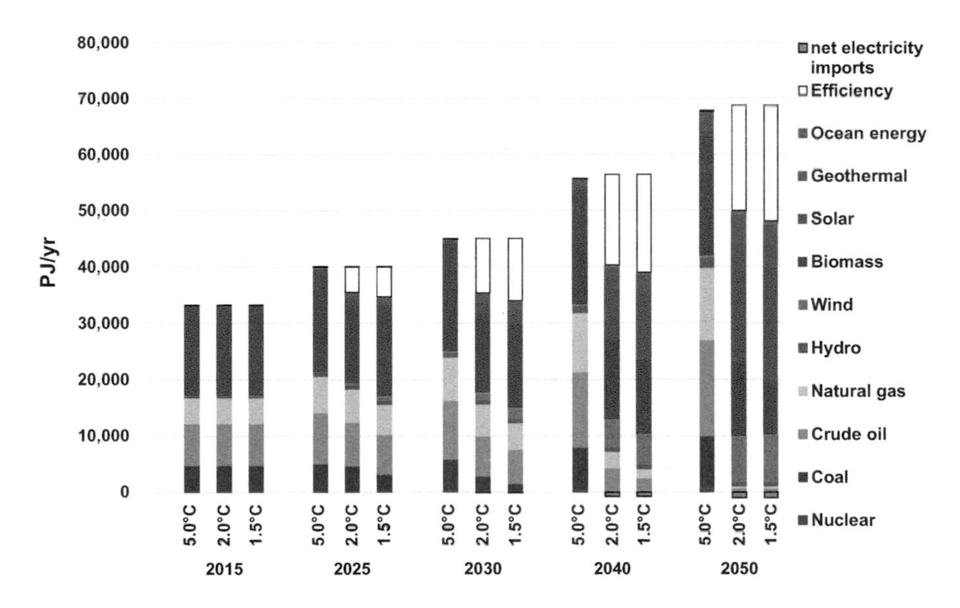

Fig. 8.52 Africa: projection of total primary energy demand (PED) by energy carrier in the scenarios (including electricity import balance)

2035 under both the 2.0 °C Scenario and 1.5 °C Scenario. The cumulative primary energy consumption of natural gas in the 5.0 °C case will add up to 290 EJ, the cumulative coal consumption to about 210 EJ, and the crude oil consumption to 390 EJ. In contrast, in the 2.0 °C Scenario, the cumulative gas demand will amount to 130 EJ, the cumulative coal demand to 70 EJ, and the cumulative oil demand to 180 EJ. Even lower fossil fuel use will achieved in the 1.5 °C Scenario: 110 EJ for natural gas, 50 EJ for coal, and 150 EJ for oil.

8.8.2 *Africa: Power Sector Analysis*

The African continent has 54 countries and its geographic, economic, and climatic diversity are significant. Its regional breakdown into sub-regions tries to reflect this diversity, but still requires a level of simplification. There is no pan-African power grid yet, although it is currently under discussion. The African Clean Energy Corridor (ACEC) is the most prominent regional initiative and aims to connect the Eastern Africa Power Pool (EAPP) with the Southern Africa Power Pool (SAPP). It was politically endorsed in January 2014 at the Assembly of the International Renewable Energy Agency (IRENA 2014).

8.8.2.1 Africa: Development of Power Plant Capacities

In 2050, Africa's most important renewable power-generation technology in both scenarios will be solar PV. In the 1.5 °C Scenario, solar PV will provide just over 40% of the total generation capacity, followed by onshore wind (with 24%), hydrogen power (15%), and CSP plants (located in the desert regions), with 10% of the total capacity. All other renewable power plant technologies will have only 2%–3% shares. The 2.0 °C Scenario will arrive at similar capacities by 2050, although the transition times in the two scenarios differ. Africa must build up solar PV and onshore wind markets equal to the market sizes in China in 2017: 50 GW of solar PV installation (REN21-GSR2018) and 23 GW of onshore wind (GWEC 2018). The market for CSP plants must reach about 1 GW per year by 2025, increasing rapidly to 3 GW per year in 2029 and 15 GW per year in 2035 (Table 8.45).

8.8.2.2 Africa: Utilization of Power-Generation Capacities

Africa's sub-regions are assumed to have an interconnection capacity of 5% at the beginning of the calculation period (2015). This capacity is not required for any exchange of variable electricity production, because currently, shares are only at or below 2% of the total generation capacity (Table 8.46). However, the variable generation capacity will increase rapidly towards 2030. We assume that the interconnection capacity between sub-regions will increase and that initiatives such as the African Clean Energy Corridor (ACEC) will be implemented successfully.

Table 8.45 Africa: average annual change in installed power plant capacity

Africa power generation: average annual change of installed capacity [GW/a]	2015–2025		2026–2035		2036–2050	
	2.0 °C	1.5 °C	2.0 °C	1.5 °C	2.0 °C	1.5 °C
Hard coal	2	0	−2	−7	−4	0
Lignite	0	0	0	0	0	0
Gas	6	3	10	16	13	14
Hydrogen-gas	0	0	1	3	15	32
Oil/diesel	−1	−2	−2	−2	−1	−1
Nuclear	0	0	0	0	0	0
Biomass	1	3	2	3	2	3
Hydro	2	1	1	1	0	0
Wind (onshore)	5	20	21	21	23	21
Wind (offshore)	0	2	5	10	7	4
PV (roof top)	3	12	29	31	41	48
PV (utility scale)	1	4	10	10	14	16
Geothermal	1	2	2	2	3	3
Solar thermal power plants	0	2	4	9	18	16
Ocean energy	0	1	1	1	3	3
Renewable fuel based co-generation	1	2	2	2	1	1

The development of average capacity factors for each generation type will follow the same trend as in most world regions. Table 8.47 shows the significant drop in the capacity factors of limited dispatchable power plants under the 1.5 °C Scenario.

8.8.2.3 Africa: Development of Load, Generation, and Residual Load

Table 8.48 shows that under the 2.0 °C Scenario, the transmission capacities need not exceed the assumed 25% interconnection capacity. If the exchange capacity between Africa's sub-regions is 20%—as calculated under the 1.5 °C Scenario—additional capacity will be required. Therefore, a 25% interconnection capacity seems a good target for high renewable penetration scenarios in Africa. The load in all sub-regions—from North Africa to South Africa—will increase significantly. The greatest increase is calculated for Southern Africa, with the load increasing by a factor of 7, followed by Central Africa (a factor of 6.5), East Africa (6), West Africa (5.5), and North Africa (4). The load increase in the Republic of South Africa will follow the patterns of other industrialized countries, more than doubling, due mainly to increases in electric mobility. The load increases in other parts of Africa will be first and foremost due to universal access to energy services for all households and favourable economic development.

Table 8.49 provides an overview of the calculated storage and dispatch power requirements by African sub-region. East and West Africa will require the highest battery capacity, due to the very high share of solar PV battery systems in rural and residential areas with low power grid availability. Like the Middle East, Africa is

Table 8.46 Africa: power system shares by technology group

Power generation structure and interconnection		2.0 °C				1.5 °C			
		Variable RE	Dispatch RE	Dispatch fossil	Inter-connection	Variable RE	Dispatch RE	Dispatch fossil	Inter-connection
North Africa	2015	2%	25%	73%	5%				
	2030	56%	23%	21%	20%	60%	8%	32%	5%
	2050	75%	25%	0%	25%	61%	10%	29%	20%
West Africa	2015	1%	26%	73%	5%				
	2030	38%	24%	38%	20%	41%	18%	41%	5%
	2050	67%	33%	0%	25%	63%	23%	14%	20%
Central Africa	2015	0%	26%	74%	5%				
	2030	20%	29%	50%	20%	19%	30%	52%	5%
	2050	42%	58%	0%	25%	39%	44%	17%	20%
East Africa	2015	2%	26%	72%	5%				
	2030	50%	22%	28%	20%	59%	10%	31%	5%
	2050	75%	25%	0%	25%	68%	13%	18%	20%
Southern Africa	2015	1%	25%	73%	5%				
	2030	46%	20%	34%	20%	52%	17%	31%	5%
	2050	81%	19%	0%	25%	70%	12%	17%	20%
South Africa	2015	2%	25%	73%	5%				
	2030	63%	0%	36%	20%	54%	8%	38%	5%
	2050	67%	33%	0%	25%	49%	9%	42%	20%
Africa	2015	2%	26%	73%					
	2030	47%	21%	32%		52%	13%	35%	
	2050	73%	27%	0%		64%	15%	21%	

Table 8.47 Africa: capacity factors by generation type

Utilization of Variable and Dispatchable power generation:		2015	2020	2020	2030	2030	2040	2040	2050	2050
Africa			2.0 °C	1.5 °C	2.0 °C	1.5 °C	2.0 °C	1.5 °C	2.0 °C	1.5 °C
Capacity factor – average	[%/yr]	**54.7%**	**33%**	**33%**	**29%**	**25%**	**40%**	**23%**	**36%**	**23%**
Limited dispatchable: fossil and nuclear	[%/yr]	69.4%	31%	5%	19%	8%	20%	4%	10%	5%
Limited dispatchable: renewable	[%/yr]	29.7%	52%	32%	35%	24%	51%	17%	36%	17%
Dispatchable: fossil	[%/yr]	49.2%	32%	37%	16%	23%	36%	15%	16%	17%
Dispatchable: renewable	[%/yr]	43.7%	39%	28%	27%	20%	41%	12%	49%	14%
Variable: renewable	[%/yr]	12.2%	12%	12%	38%	28%	34%	27%	35%	27%

Table 8.48 Africa: load, generation, and residual load development

Power generation structure		2.0 °C				1.5 °C			
		Max demand	Max generation	Max residual load	Max interconnection requirements	Max demand	Max generation	Max residual load	Max interconnection requirements
Africa		[GW]	[GW]	[GW]	[GW]	[GW]	[GW]	[GW]	[GW]
North Africa	2020	23.8	19.3	8.4		23.8	20.2	7.8	
	2030	31.0	33.9	4.2	0	34.0	43.6	6.0	4
	2050	99.3	161.9	63.6	0	109.8	186.1	28.5	48
West Africa	2020	38.7	19.5	19.7		38.6	22.9	16.5	
	2030	64.7	56.7	25.3	0	66.5	58.5	24.5	0
	2050	214.4	310.3	164.6	0	216.1	355.7	118.0	22
Central Africa	2020	4.2	3.4	0.8		4.2	3.9	0.3	
	2030	8.2	7.3	2.6	0	8.5	7.7	2.6	0
	2050	27.0	38.6	26.4	0	27.3	46.8	26.6	0
East Africa	2020	44.0	34.8	11.9		44.0	39.5	7.0	
	2030	86.5	75.0	30.0	0	88.5	82.9	28.5	0
	2050	265.1	369.8	197.4	0	267.1	425.1	101.7	56
Southern Africa	2020	27.8	24.2	4.0		27.7	25.4	2.3	
	2030	67.2	57.9	35.9	0	68.3	74.5	36.6	0
	2050	199.3	359.3	169.9	0	199.6	407.3	111.5	96
South Africa	2020	25.3	23.5	1.7		25.3	23.5	2.7	
	2030	22.4	30.0	3.3	4	30.4	37.5	7.0	0
	2050	70.1	122.9	24.7	28	94.9	141.5	25.4	21

Table 8.49 Africa: storage and dispatch service requirements

Storage and dispatch		2.0 °C					1.5 °C				
		Required to avoid curtailment	Utilization battery -through-put-	Utilization PSH -through-put-	Total storage demand (incl. H2)	Dispatch hydrogen-based	Required to avoid curtailment	Utilization battery -through-put-	Utilization PSH -through-put-	Total storage demand (incl. H2)	Dispatch hydrogen-based
Africa		[GWh/year]	[GWh/year]	[GWh/year]	[GWh/ year]	[GWh/ year]	[GWh/year]	[GWh/year]	[GWh/year]	[GWh/ year]	[GWh/year]
North Africa	2020	0	0	0	0	0	0	0	0	0	0
	2030	1456	44	857	901	0	4611	65	1500	1565	0
	2050	59,499	1959	2904	4864	37,284	77,546	1976	2994	4969	2904
West Africa	2020	0	0	0	0	0	0	0	0	0	0
	2030	0	1	11	12	0	18	10	126	136	0
	2050	62,015	2525	3154	5679	41,842	125,281	2552	3797	6349	10,940
Central Africa	2020	0	0	0	0	0	0	0	0	0	0
	2030	0	0	0	0	0	0	0	0	0	0
	2050	4938	293	298	590	6107	10,557	323	391	714	3879
East Africa	2020	0	0	0	0	0	0	0	0	0	0
	2030	54	45	827	872	0	1960	78	1787	1865	0
	2050	104,983	3467	4976	8444	65,953	182,399	3573	5673	9246	6375
Southern Africa	2020	0	0	0	0	0	0	0	0	0	0
	2030	609	33	640	673	0	3268	49	1053	1102	0
	2050	110,532	2122	3371	5493	42,521	177,898	2189	3818	6008	19,886
South Africa	2020	0	0	0	0	0	0	0	0	0	0
	2030	2757	113	2155	2268	0	1407	46	877	923	0
	2050	25,233	2659	3245	5904	19,194	11,741	2038	1886	3924	0

(continued)

Table 8.49 (continued)

Storage and dispatch		2.0 °C					1.5 °C				
		Required to avoid curtailment	Utilization battery -through-put-	Utilization PSH -through-put-	Total storage demand (incl. H2)	Dispatch hydrogen-based	Required to avoid curtailment	Utilization battery -through-put-	Utilization PSH -through-put-	Total storage demand (incl. H2)	Dispatch hydrogen-based
Africa		[GWh/year]	[GWh/year]	[GWh/year]	[GWh/ year]	[GWh/ year]	[GWh/year]	[GWh/year]	[GWh/year]	[GWh/ year]	[GWh/year]
Africa	2020	0	0	0	0	0	0	0	0	0	0
	2030	4877	237	4489	4726	0	11,264	248	5343	5591	0
	2050	367,201	13,026	17,948	30,974	212,902	585,423	12,651	18,558	31,210	43,984

one of the global renewable fuel production regions and it is assumed that all subregions of Africa have equal amounts of energy export potential. However, a more detailed examination of export energy is required, which is beyond the scope of this project.

8.9 The Middle East

8.9.1 The Middle East: Long-Term Energy Pathways

8.9.1.1 The Middle East: Final Energy Demand by Sector

The future development pathways for the Middle East's final energy demand when the assumptions on population growth, GDP growth, and energy intensity are combined are shown in Fig. 8.53 for the 5.0 °C, 2.0 °C, and 1.5 °C Scenarios. In the 5.0 °C Scenario, the total final energy demand will increase by 133% from the current 17,100 PJ/year to around 40,000 PJ/year in 2050. In the 2.0 °C Scenario, the final energy demand will decrease by 8% compared with current consumption and will reach 15,800 PJ/year by 2050. The final energy demand in the 1.5 °C Scenario will reach 13,600 PJ, 20% below the 2015 demand level. In the 1.5 °C Scenario, the final energy demand in 2050 will be 14% lower than in the 2.0 °C Scenario. The

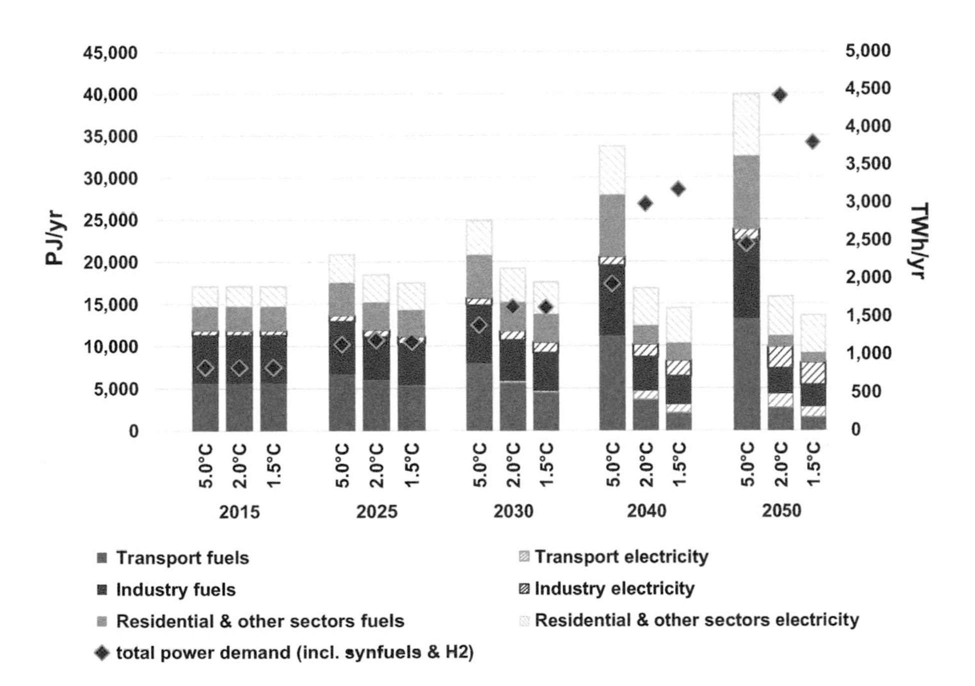

Fig. 8.53 Middle East: development of the final energy demand by sector in the scenarios

electricity demand for 'classical' electrical devices (without power-to-heat or e-mobility) will increase from 650 TWh/year in 2015 to 1230 TWh/year (2.0 °C) and 1160 TWh/year (1.5 °C) by 2050. Compared with the 5.0 °C case (2330 TWh/year in 2050), the efficiency measures in the 2.0 °C and 1.5 °C Scenarios will save a maximum of 1100 TWh/year and 1170 TWh/year, respectively.

Electrification will lead to a significant increase in the electricity demand. In the 2.0 °C Scenario, the electricity demand for heating will rise to approximately 800 TWh/year due to electric heaters and heat pumps, and in the transport sector, the demand will rise to approximately 1700 TWh/year due to the increase in electric mobility. The generation of hydrogen (for transport and high-temperature process heat) and the manufacture of synthetic fuels (mainly for transport) will add an additional power demand of 1900 TWh/year. The gross power demand will thus rise from 1100 TWh/year in 2015 to 4700 TWh/year in 2050 in the 2.0 °C Scenario, 57% higher than in the 5.0 °C case. In the 1.5 °C Scenario, the gross electricity demand will increase to a maximum of 4100 TWh/year by 2045.

The efficiency gains could be even larger in the heating sector than in the electricity sector. In the 2.0 °C and 1.5 °C Scenarios, a final energy consumption equivalent to about 10,100 PJ/year and 10,500 PJ/year, respectively, will be avoided through efficiency gains by 2050 compared with the 5.0 °C Scenario.

8.9.1.2 The Middle East: Electricity Generation

The development of the power system is characterized by a dynamically growing renewable energy market and an increasing proportion of total power from renewable sources. By 2050, 100% of the electricity produced in the Middle East will come from renewable energy sources under the 2.0 °C Scenario. 'New' renewables—mainly wind, solar, and geothermal energy—will contribute 96% of the total electricity generation. Renewable electricity's share of the total production will be 49% by 2030 and 91% by 2040. The installed capacity of renewables will reach about 430 GW by 2030 and 1910 GW by 2050. The share of renewable electricity generation in 2030 in the 1.5 °C Scenario is assumed to be 58%. In the 1.5 °C Scenario, the generation capacity from renewable energy will be approximately 1700 GW in 2050.

Table 8.50 shows the development of different renewable technologies in the Middle East over time. Figure 8.54 provides an overview of the overall power-generation structure in the Middle East. From 2020 onwards, the continuing growth of wind and PV, up to 480 GW and 1070 GW, respectively, will be complemented by up to 250 GW of solar thermal generation, as well as limited biomass, geothermal, and ocean energy, in the 2.0 °C Scenario. Both the 2.0 °C Scenario and 1.5 °C Scenario will lead to high proportions of variable power generation (PV, wind, and ocean) of 39% and 46%, respectively, by 2030, and 64% and 66%, respectively, by 2050.

Table 8.50 Middle East: development of renewable electricity-generation capacity in the scenarios

in GW	Case	2015	2025	2030	2040	2050
Hydro	5.0 °C	16	20	22	25	29
	2.0 °C	16	22	22	25	29
	1.5 °C	16	22	22	25	29
Biomass	5.0 °C	0	0	1	3	7
	2.0 °C	0	2	3	4	4
	1.5 °C	0	3	3	4	4
Wind	5.0 °C	0	4	9	23	49
	2.0 °C	0	54	156	371	481
	1.5 °C	0	60	175	432	456
Geothermal	5.0 °C	0	0	0	0	0
	2.0 °C	0	5	7	20	25
	1.5 °C	0	5	7	20	21
PV	5.0 °C	0	7	10	21	40
	2.0 °C	0	76	187	560	1069
	1.5 °C	0	92	236	587	928
CSP	5.0 °C	0	2	3	6	7
	2.0 °C	0	10	43	270	252
	1.5 °C	0	10	47	342	216
Ocean	5.0 °C	0	0	0	0	0
	2.0 °C	0	5	10	40	50
	1.5 °C	0	5	10	40	45
Total	5.0 °C	16	32	45	79	132
	2.0 °C	16	174	427	1290	1911
	1.5 °C	16	197	500	1449	1699

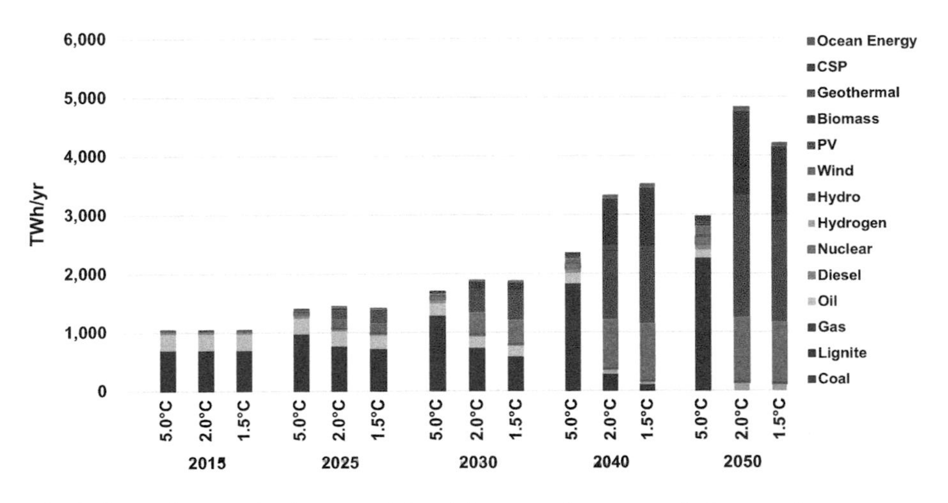

Fig. 8.54 Middle East: development of electricity-generation structure in the scenarios

8.9.1.3 The Middle East: Future Costs of Electricity Generation

Figure 8.55 shows the development of the electricity-generation and supply costs over time, including the CO_2 emission costs, in all scenarios. The calculated electricity-generation costs in 2015 (referring to full costs) were around 7.1 ct/kWh. In the 5.0 °C case, the generation costs will increase until 2030, when they reach 14.8 ct/kWh, and then drop to 13.7 ct/kWh by 2050. The generation costs in the 2.0 °C Scenario will increase until 2030, when they reach 11.1 ct/kWh, and then drop to 6.1 ct/kWh by 2050. In the 1.5 °C Scenario, they will increase to 10.7 ct/kWh, and then drop to 7.3 ct/kWh by 2050. In the 2.0 °C Scenario, the generation costs in 2050 will be 7.6 ct/kWh lower than in the 5.0 °C case. In the 1.5 °C Scenario, the generation costs in 2050 will be 6.4 ct/kWh lower than in the 5.0 °C case. Note that these estimates of generation costs do not take into account integration costs such as power grid expansion, storage, or other load-balancing measures.

In the 5.0 °C case, growth in demand and increasing fossil fuel prices will cause the total electricity supply costs to rise from today's \$70 billion/year to more than \$410 billion/year in 2050. In the 2.0 °C Scenario, the total supply costs will be \$300 billion/year and in the 1.5 °C Scenario, they will be \$310 billion/year. The long-term cost of electricity supply will be more than 27% lower in the 2.0 °C Scenario than in the 5.0 °C Scenario as a result of the estimated generation costs and the electrification of heating and mobility. Further demand reductions in the 1.5 °C Scenario will result in total power-generation costs that are 24% lower than in the 5.0 °C case.

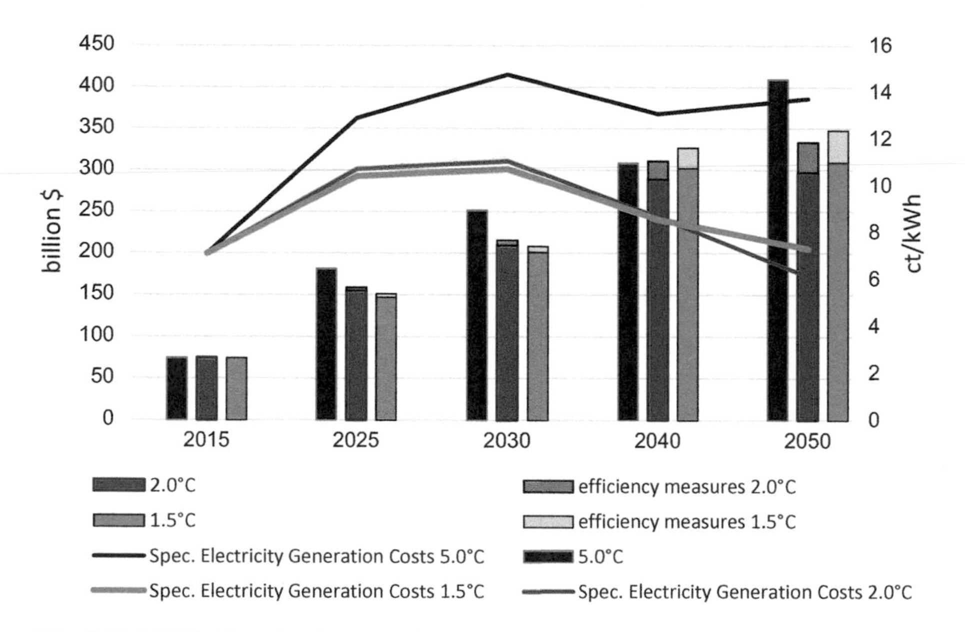

Fig. 8.55 Middle East: development of total electricity supply costs and specific electricity-generation costs in the scenarios

The generation costs without the CO_2 emission costs will increase in the 5.0 °C case to 11.1 ct/kWh by 2030, and then stabilize at 10.8 ct/kWh by 2050. In the 2.0 °C Scenario and the 1.5 °C Scenario, they will increase to a maximum of 9 ct/kWh in 2030, before they drop to 6.1 ct/kWh and 7.3 ct/kWh by 2050, respectively. In the 2.0 °C Scenario, the generation costs will be 4.7 ct/kWh lower than in the 5.0 °C case and this maximum difference will occur in 2050. In the 1.5 °C Scenario, the maximum difference in generation costs compared with the 5.0 °C case will be 3.5 ct/kWh in 2050. If the CO_2 costs are not considered, the total electricity supply costs in the 5.0 °C case will rise to about $320 billion/year by 2050.

8.9.1.4 The Middle East: Future Investments in the Power Sector

An investment of around $3450 billion will be required for power generation between 2015 and 2050 in the 2.0 °C Scenario—including additional power plants for the production of hydrogen and synthetic fuels and investments in plant replacement at the ends of their economic lives. This value will be equivalent to approximately $96 billion per year on average, and this is $2720 billion more than in the 5.0 °C case ($730 billion). An investment of around $3470 billion for power generation will be required between 2015 and 2050 in the 1.5 °C Scenario, or on average, $96 billion per year. In the 5.0 °C Scenario, the investment in conventional power plants will be around 68% of the total cumulative investments, whereas approximately 32% will be invested in renewable power generation and co-generation (Fig. 8.56). However, in both alternative scenarios, the Middle East will shift almost 94% of its entire investments to renewables and co-generation. By 2030, the fossil fuel share of power sector investment will predominantly focus on gas power plants that can also be operated with hydrogen.

Because renewable energy has no fuel costs, other than biomass, the cumulative fuel cost savings in the 2.0 °C Scenario will reach a total of $2900 billion in 2050, equivalent to $81 billion per year. Therefore, the total fuel cost savings will be equivalent to 110% of the total additional investments compared to the 5.0 °C Scenario. The fuel cost savings in the 1.5 °C Scenario will add up to $3100 billion, or $86 billion per year.

8.9.1.5 The Middle East: Energy Supply for Heating

The final energy demand for heating will increase by 139% in the 5.0 °C Scenario, from 7100 PJ/year in 2015 to 17,100 PJ/year in 2050. Energy efficiency measures will help to reduce the energy demand for heating by 59% in 2050 in the 2.0 °C Scenario, relative to the 5.0 °C case, and by 62% in the 1.5 °C Scenario. Today, renewables supply almost none of the Middle East's final energy demand for heating. Renewable energy will provide 23% of the Middle East's total heat demand in 2030 in the 2.0 °C Scenario and 25% in the 1.5 °C Scenario. In both scenarios, renewables will provide 100% of the total heat demand in 2050.

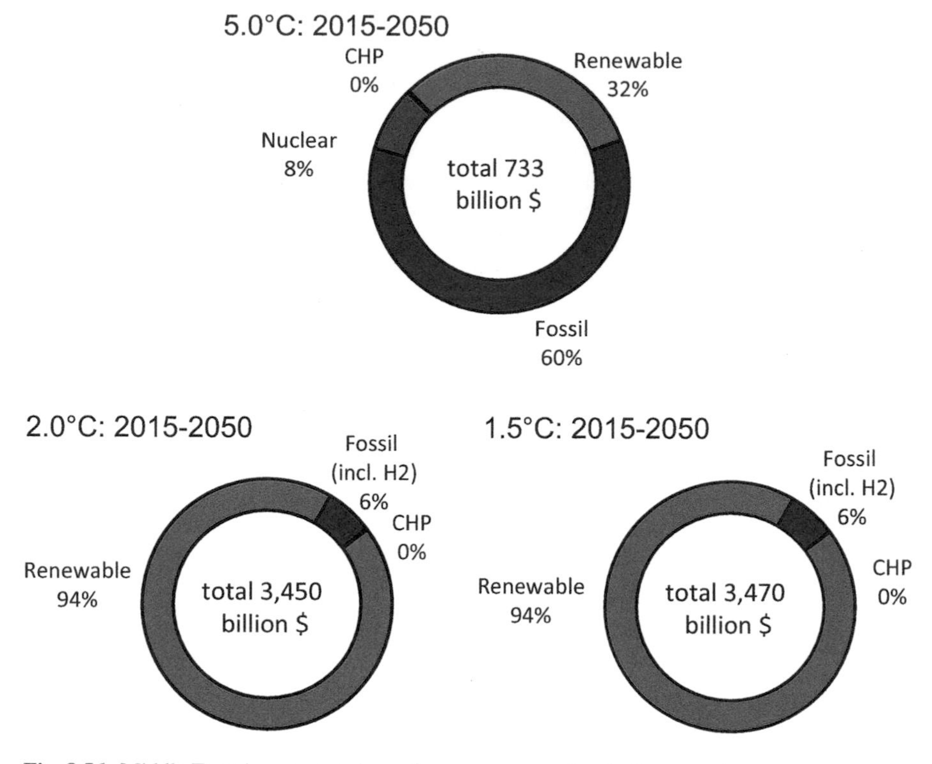

Fig. 8.56 Middle East: investment shares for power generation in the scenarios

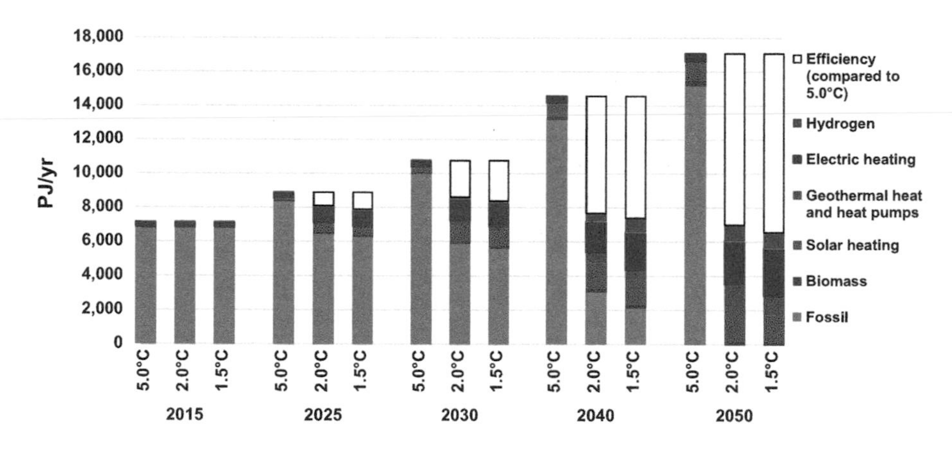

Fig. 8.57 Middle East: development of heat supply by energy carrier in the scenarios

Figure 8.57 shows the development of different technologies for heating in the Middle East over time, and Table 8.51 provides the resulting renewable heat supply for all scenarios. The growing use of solar, geothermal, and environmental heat will

Table 8.51 Middle East: development of renewable heat supply in the scenarios (excluding the direct use of electricity)

in PJ/year	Case	2015	2025	2030	2040	2050
Biomass	5.0 °C	20	56	86	169	291
	2.0 °C	20	101	132	200	196
	1.5 °C	20	92	124	183	155
Solar heating	5.0 °C	8	92	284	778	1113
	2.0 °C	8	404	932	1535	1961
	1.5 °C	8	393	909	1475	1619
Geothermal heat and heat pumps	5.0 °C	0	0	0	0	0
	2.0 °C	0	118	232	565	1387
	1.5 °C	0	115	226	540	1057
Hydrogen	5.0 °C	0	0	0	0	0
	2.0 °C	0	0	51	488	946
	1.5 °C	0	0	48	828	915
Total	5.0 °C	28	149	370	947	1404
	2.0 °C	28	624	1346	2788	4489
	1.5 °C	28	601	1307	3025	3746

supplement electrification, with solar heat becoming the main direct renewable heat source in the 2.0 °C Scenario and 1.5 °C Scenario.

Heat from renewable hydrogen will further reduce the dependence on fossil fuels in both scenarios. Hydrogen consumption in 2050 will be around 950 PJ/year in the 2.0 °C Scenario and 920 PJ/year in the 1.5 °C Scenario. The direct use of electricity for heating will also increase by a factor of 9–10 between 2015 and 2050, and its final energy share will be 36% in 2050 in the 2.0 °C Scenario and 43% in the 1.5 °C Scenario (Fig. 8.57).

8.9.1.6 The Middle East: Future Investments in the Heating Sector

The roughly estimated investments in renewable heating technologies to 2050 will amount to less than $440 billion in the 2.0 °C Scenario (including investments for plant replacement after their economic lifetimes), or approximately $12 billion per year. The largest share of investments in the Middle East is assumed to be for heat pumps (more than $200 billion), followed by solar collectors and geothermal heat use. The 1.5 °C Scenario assumes an even faster expansion of renewable technologies. However, the lower heat demand (compared with the 2.0 °C Scenario) will result in a lower average annual investment of around $10 billion per year (Table 8.52, Fig. 8.58).

Table 8.52 Middle East: installed capacities for renewable heat generation in the scenarios

in GW	Case	2015	2025	2030	2040	2050
Biomass	5.0 °C	4	10	14	25	38
	2.0 °C	4	13	15	18	14
	1.5 °C	4	12	15	17	13
Geothermal	5.0 °C	0	0	0	0	0
	2.0 °C	0	2	8	19	30
	1.5 °C	0	2	8	18	35
Solar heating	5.0 °C	1	17	51	139	198
	2.0 °C	1	72	142	217	252
	1.5 °C	1	71	139	209	206
Heat pumps	5.0 °C	0	0	0	0	0
	2.0 °C	0	12	17	43	122
	1.5 °C	0	12	17	42	76
Total[a]	5.0 °C	6	26	65	164	237
	2.0 °C	6	99	183	297	418
	1.5 °C	6	96	178	286	330

[a]Excluding direct electric heating

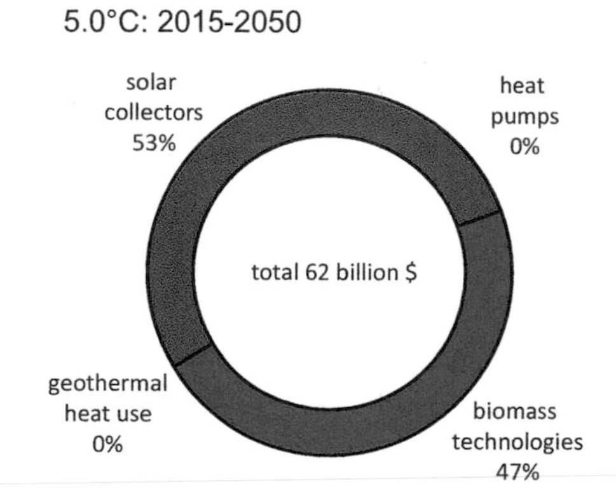

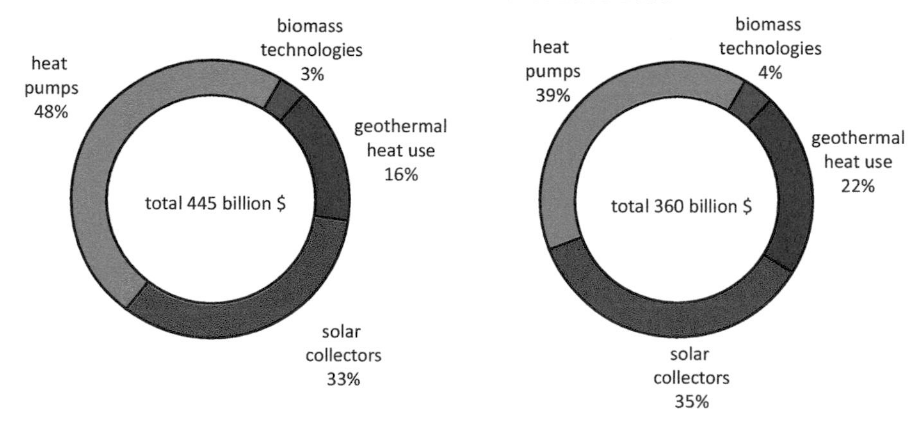

Fig. 8.58 Middle East: development of investments for renewable heat-generation technologies in the scenarios

8.9.1.7 The Middle East: transport

Energy demand in the transport sector in the Middle East is expected to increase in the 5.0 °C Scenario by 133%, from around 5700 PJ/year in 2015 to 13,300 PJ/year in 2050. In the 2.0 °C Scenario, assumed technical, structural, and behavioural changes will save 67% (8860 PJ/year) by 2050 compared with the 5.0 °C Scenario. Additional modal shifts, technology switches, and a reduction in the transport demand will lead to even higher energy savings in the 1.5 °C Scenario of 79% (or 10,400 PJ/year) in 2050 compared with the 5.0 °C case (Table 8.53, Fig. 8.59).

By 2030, electricity will provide 4% (70 TWh/year) of the transport sector's total energy demand in the 2.0 °C Scenario, whereas in 2050, the share will be 39% (480 TWh/year). In 2050, up to 620 PJ/year of hydrogen will be used in the transport sector as a complementary renewable option. In the 1.5 °C Scenario, the annual electricity demand will be 350 TWh in 2050. The 1.5 °C Scenario also assumes a hydrogen demand of 450 PJ/year by 2050.

Biofuel use is limited in the 2.0 °C Scenario to a maximum of 370 PJ/year. Therefore, around 2030, synthetic fuels based on power-to-liquid will be introduced, with a maximum consumption of 1670 PJ/year in 2050. Biofuel use in the 1.5 °C Scenario with have a maximum of 430 PJ/year. The maximum synthetic fuel demand will amount to 920 PJ/year.

Table 8.53 Middle East: projection of transport energy demand by mode in the scenarios

in PJ/year	Case	2015	2025	2030	2040	2050
Rail	5.0 °C	184	38	48	65	75
	2.0 °C	184	64	103	169	157
	1.5 °C	184	89	117	161	194
Road	5.0 °C	5425	6613	7802	10,999	12,992
	2.0 °C	5425	5928	5732	4510	4194
	1.5 °C	5425	5246	4528	2899	2618
Domestic aviation	5.0 °C	57	83	103	136	146
	2.0 °C	57	60	57	47	37
	1.5 °C	57	57	52	36	28
Domestic navigation	5.0 °C	0	0	0	0	0
	2.0 °C	0	0	0	0	0
	1.5 °C	0	0	0	0	0
Total	5.0 °C	5666	6734	7954	11,200	13,213
	2.0 °C	5666	6051	5893	4726	4388
	1.5 °C	5666	5392	4697	3096	2840

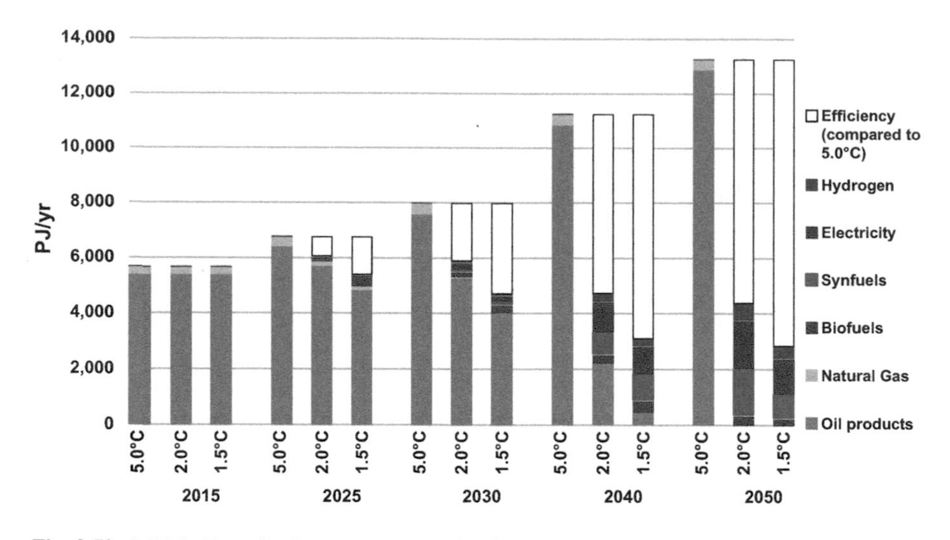

Fig. 8.59 Middle East: final energy consumption by transport in the scenarios

8.9.1.8 The Middle East: Development of CO_2 Emissions

In the 5.0 °C Scenario, the Middle East's annual CO_2 emissions will increase by 76% from 1760 Mt. in 2015 to 3094 Mt. in 2050. The stringent mitigation measures in both alternative scenarios will cause the annual emissions to fall to 510 Mt. in 2040 in the 2.0 °C Scenario and to 220 Mt. in the 1.5 °C Scenario, with further reductions to almost zero by 2050. In the 5.0 °C case, the cumulative CO_2 emissions from 2015 until 2050 will add up to 90 Gt. In contrast, in the 2.0 °C and 1.5 °C Scenarios, the cumulative emissions for the period from 2015 until 2050 will be 38 Gt and 31 Gt, respectively.

Therefore, the cumulative CO_2 emissions will decrease by 58% in the 2.0 °C Scenario and by 66% in the 1.5 °C Scenario compared with the 5.0 °C case. A rapid reduction in annual emissions will occur in both alternative scenarios. In the 2.0 °C Scenario, this reduction will be greatest in 'Industry' followed by the 'Power generation' and 'Transport' sectors (Fig. 8.60).

8.9.1.9 The Middle East: Primary Energy Consumption

The levels of primary energy consumption in the three scenarios when the assumptions discussed above are taken into account are shown in Fig. 8.61. In the 2.0 °C Scenario, the primary energy demand will decrease by 16%, from around 30,300 PJ/year in 2015 to 25,400 PJ/year in 2050. Compared with the 5.0 °C Scenario, the overall primary energy demand will decrease by 59% by 2050 in the 2.0 °C Scenario (5.0 °C: 61,700 PJ in 2050). In the 1.5 °C Scenario, the primary energy demand will be even lower (22,300 PJ in 2050) because the final energy demand and conversion losses will be lower.

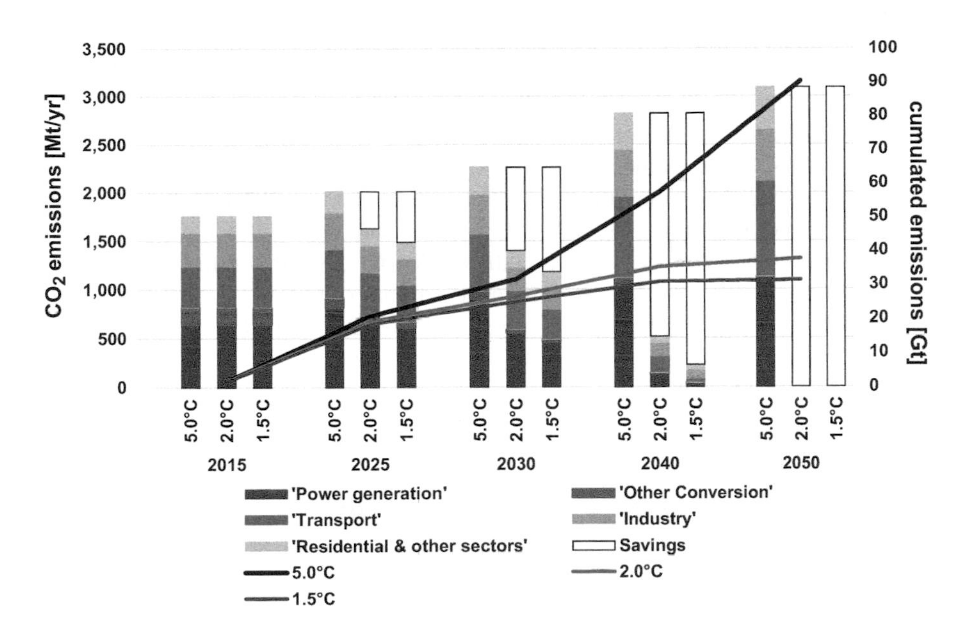

Fig. 8.60 Middle East: development of CO_2 emissions by sector and cumulative CO_2 emissions (after 2015) in the scenarios ('Savings' = reduction compared with the 5.0 °C Scenario)

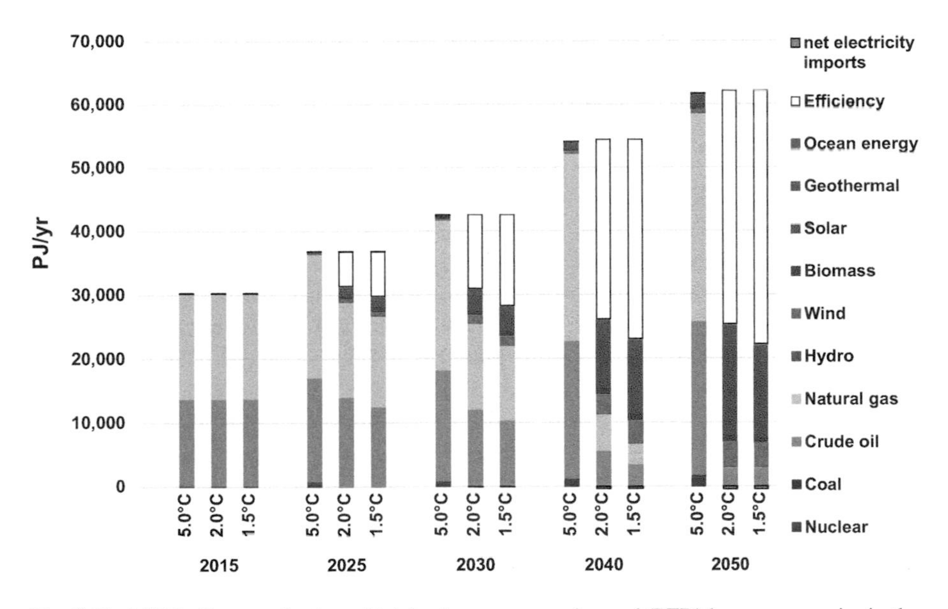

Fig. 8.61 Middle East: projection of total primary energy demand (PED) by energy carrier in the scenarios (including electricity import balance)

Both the 2.0 °C and 1.5 °C Scenarios aim to rapidly phase-out coal and oil. This will cause renewable energy to have a primary energy share of 18% in 2030 and 88% in 2050 in the 2.0 °C Scenario. In the 1.5 °C Scenario, renewables will have a primary energy share of more than 86% in 2050 (including non-energy consumption, which will still include fossil fuels). Nuclear energy will be phased-out in 2035 in both the 2.0 °C and the 1.5 °C Scenarios. The cumulative primary energy consumption of natural gas in the 5.0 °C case will add up to 830 EJ, the cumulative coal consumption to about 10 EJ, and the crude oil consumption to 630 EJ. In the 2.0 °C Scenario, the cumulative gas demand will amount to 330 EJ, the cumulative coal demand to 1 EJ, and the cumulative oil demand to 310 EJ. Even lower fossil fuel use will be achieved in the 1.5 °C Scenario: 280 EJ for natural gas, 0.9 EJ for coal, and 270 EJ for oil.

8.9.2 The Middle East: Power Sector Analysis

The Middle East has significant renewable energy potential. The region's solar radiation is among the highest in the world and it has good wind conditions in coastal areas and in its mountain ranges. The electricity market is fragmented, and policies differ significantly. However, most countries are connected to their neighbours by transmission lines. Saudi Arabia, the geographic centre of the region, has connections to most neighbouring countries. Both the 2.0 °C Scenario and the 1.5 °C Scenario assume that the Middle East will remain a significant player in the energy market, moving from oil and gas to solar, and that it will play an important role in producing synthetic fuels and hydrogen for export.

8.9.2.1 The Middle East: Development of Power Plant Capacities

The overwhelming majority of fossil-fuel-based power generation in the Middle East is from gas-fired power plants. Both scenarios assume that this gas capacity (in GW) will remain on the same level until 2050, but will be converted to hydrogen. The annual market for solar PV must increase to 2.5 GW in 2020 and to 28.5 GW by 2030 in the 2.0 °C Scenario, and to 35 GW in the 1.5 °C Scenario. The onshore wind market must expand to 10 GW by 2025 in both scenarios. This represents a very ambitious target because the market for wind power plants in the Middle East has never been higher than 117 MW (GWEC 2018) (in 2015). Parts of the offshore oil and gas industry can be transitioned into an offshore wind industry. The total capacity assumed for the Middle East by 2050 is 20–25 GW under both scenarios. For comparison, the UK had an installed capacity for offshore wind of 6.8 GW and Germany of 5.4 GW in 2017 (GWEC 2018). The vast solar resources in the Middle

Table 8.54 Middle East: average annual change in installed power plant capacity

Middle East – power generation: average annual change of installed capacity [GW/a]	2015–2025		2026–2035		2036–2050	
	2.0 °C	1.5 °C	2.0 °C	1.5 °C	2.0 °C	1.5 °C
Hard coal	0.0	0.0	0.0	0.0	0.0	0.0
Lignite	0.0	0.0	0.0	0.0	0.0	0.0
Gas	1.5	7.0	1.9	6.2	−19.1	3.0
Hydrogen-gas	0.0	0.3	1.5	1.7	20.3	24.2
Oil/Diesel	−0.1	−4.0	−8.9	−8.1	−0.8	−0.5
Nuclear	−0.1	0.0	−0.1	−0.1	0.0	0.0
Biomass	0.2	0.3	0.2	0.1	0.2	0.0
Hydro	1.0	0.5	0.2	0.2	0.5	0.5
Wind (onshore)	6.5	19.3	28.3	35.5	14.7	7.6
Wind (offshore)	0.2	0.5	0.8	0.8	1.4	1.2
PV (roof top)	7.3	19.0	26.2	29.9	46.4	32.3
PV (utility scale)	2.4	6.3	8.7	10.0	15.5	10.8
Geothermal	0.6	0.8	1.1	1.1	1.0	0.6
Solar thermal power plants	1.3	5.4	13.1	20.3	11.4	3.7
Ocean energy	0.3	1.3	1.3	2.5	1.0	1.7
Renewable fuel based co-generation	0.0	0.0	0.1	0.1	0.0	0.0

East make it suitable for CSP plants—the total capacity by 2050 is calculated to be 252 GW (2.0 °C Scenario), equal to the gas power plant capacity in the Middle East in 2017 (Table 8.54).

8.9.2.2 Middle East: Utilization of Power-Generation Capacities

In 2015, the base year of the scenario calculations, the Middle East had less than 0.5% variable power generation. Table 8.55 shows the rapidly increasing shares of variable renewable power generation across the Middle East. Israel is included in the Middle East region (as opposed to the IEA region used for the long-term scenario) to reflect its current and possible future interconnection with the regional power system. The current interconnection capacity between all sub-regions is assumed to be 5%, increasing to 20% in 2030 and 25% in 2050. Dispatchable renewables will have a stable market share of around 15%–20% over the entire modelling period in the 2.0 °C Scenario and 15%–20% in the 1.5 °C Scenario.

Average capacity factors correspond to the results for the other world regions. Table 8.56 shows that the limited dispatchable fossil and nuclear generation will drop quickly, whereas the significant gas power plant capacity within the region can increase capacity factors to take over their load and reduce carbon emissions at an early stage. The calculation results are attributed to the assumed dispatch order, which prioritizes gas over coal and nuclear.

Table 8.55 Middle East: power system shares by technology group

Power generation structure and interconnection		2.0 °C				1.5 °C			
Middle East		Variable RE	Dispatch RE	Dispatch fossil	Inter-connection	Variable RE	Dispatch RE	Dispatch fossil	Inter-connection
Israel	2015	0%	12%	88%	5%				
	2030	41%	18%	41%	20%	46%	18%	36%	15%
	2050	81%	18%	0%	25%	64%	15%	21%	20%
North-ME	2015	0%	12%	88%	5%				
	2030	50%	20%	30%	20%	55%	19%	26%	15%
	2050	83%	17%	0%	25%	74%	15%	10%	20%
Saudi Arabia-ME	2015	0%	12%	88%	5%				
	2030	51%	17%	32%	20%	55%	17%	28%	15%
	2050	83%	17%	0%	25%	72%	16%	12%	20%
UAE-ME	2015	0%	12%	88%	5%				
	2030	36%	20%	45%	20%	40%	20%	40%	15%
	2050	76%	24%	0%	25%	53%	18%	28%	20%
East-ME	2015	0%	12%	88%	5%				
	2030	42%	20%	38%	20%	47%	21%	32%	15%
	2050	80%	20%	0%	25%	63%	17%	20%	20%
Iraq-ME	2015	0%	12%	88%	5%				
	2030	60%	18%	21%	20%	65%	17%	18%	15%
	2050	82%	18%	0%	25%	76%	16%	7%	20%

Iran-ME	2015	0%	12%	88%	5%				
	2030	57%	19%	24%	20%	62%	18%	21%	15%
	2050	81%	19%	0%	25%	73%	17%	9%	20%
Middle East	2015	0%	12%	88%					
	2030	51%	19%	31%		56%	18%	27%	
	2050	81%	19%	0%		70%	16%	13%	

Table 8.56 Middle East: capacity factors by generation type

Utilization of variable and dispatchable power generation:		2015	2020	2020	2030	2030	2040	2040	2050	2050
Middle East			2.0 °C	1.5 °C	2.0 °C	1.5 °C	2.0 °C	1.5 °C	2.0 °C	1.5 °C
Capacity factor – average	[%/yr]	52.6%	45%	43%	27%	24%	34%	21%	29%	25%
Limited dispatchable: fossil and nuclear	[%/yr]	31.1%	13%	13%	5%	2%	19%	3%	10%	5%
Limited dispatchable: renewable	[%/yr]	26.3%	34%	34%	47%	42%	50%	21%	28%	30%
Dispatchable: fossil	[%/yr]	52.9%	41%	40%	15%	10%	45%	8%	17%	16%
Dispatchable: renewable	[%/yr]	38.9%	83%	83%	66%	57%	43%	20%	36%	38%
Variable: renewable	[%/yr]	6.6%	12%	12%	24%	23%	27%	23%	29%	25%

8.9.2.3 The Middle East: Development of Load, Generation, and Residual Load

The Middle East is assumed to be one of the exporters of solar electricity into the EU, so the calculated solar installation capacities throughout the region will be significantly higher than required for self-supply.

Table 8.57 shows a negative residual load in almost all sub-regions for every year and in both scenarios. This is attributable to substantial oversupply, so the production of renewables is almost constantly higher than the demand. This electricity has been calculated as exports from the Middle East and imports to Europe.

The Middle East will be one of three dedicated renewable energy export regions. These exports are in the form of renewable fuels and electricity. The [R]E 24/7 model does not calculate electricity exchange in 1 h steps between the world regions, and therefore the amount of electricity exported accumulates from year to year. The load curves for the Middle East and European regions are not calculated separately.

Table 8.58 provides an overview of the calculated storage and dispatch power requirements by sub-region in the Middle East. Iran and Saudi Arabia West Africa will require the highest storage capacity, due to the very high share of solar PV systems in residential areas. Like the Africa, the Middle East is one of the global renewable fuel production regions and it is assumed that all sub-regions of the Middle East have equal amounts of energy export potential. However, a more detailed examination of export energy is required, which is beyond the scope of this project.

Table 8.57 Middle East: load, generation, and residual load development

Power generation structure		2.0 °C				1.5 °C			
		Max demand	Max generation	Max residual load	Max interconnection requirements	Max demand	Max generation	Max residual load	Max interconnection requirements
Middle East		[GW]	[GW]	[GW]	[GW]	[GW]	[GW]	[GW]	[GW]
Israel	2020	11.5	11.0	−2.9		11.5	11.9	−2.9	
	2030	14.0	17.3	−1.8	5	14.3	19.9	−1.4	7
	2050	29.7	62.7	−3.7	37	29.3	55.2	−10.3	36
North-ME	2020	33.7	25.7	−12.1		33.8	29.4	−11.6	
	2030	39.7	44.6	−12.1	17	40.6	52.6	−11.2	23
	2050	83.6	196.5	−11.5	124	77.5	172.8	−19.9	115
Saudi Arabia-ME	2020	59.5	45.9	−3.4		59.6	45.4	−2.4	
	2030	72.4	85.9	2.3	11	75.1	99.8	5.6	19
	2050	173.6	380.9	−21.7	229	168.6	334.0	−59.4	225
UAE-ME	2020	21.2	29.8	−0.4		21.2	29.6	1.3	
	2030	26.0	44.1	2.2	16	27.0	50.7	3.4	20
	2050	62.2	120.2	−7.4	65	61.3	105.4	−23.4	67
East-Middle East	2020	12.0	23.3	−2.5		12.0	22.6	−2.6	
	2030	14.8	31.3	−1.2	18	15.1	35.9	−0.8	22
	2050	32.5	63.4	−3.9	35	31.2	55.6	−7.2	32
Iraq-ME	2020	20.1	13.8	−7.6		20.2	13.8	−7.3	
	2030	26.0	30.4	−7.4	12	26.8	35.8	−8.1	17
	2050	64.3	137.5	−7.8	81	57.7	119.9	−20.0	82
Iran-ME	2020	49.4	56.9	−12.2		49.4	56.4	−12.2	
	2030	76.1	88.1	−14.3	26	78.3	103.9	−11.4	37
	2050	188.5	399.1	−22.7	233	174.3	348.0	−40.5	214

Table 8.58 Middle East: storage and dispatch service requirements

Storage and dispatch		2.0 °C					1.5 °C				
		Required to avoid curtailment	Utilization battery -through-put-	Utilization PSH -through-put-	Total storage demand (incl. H2)	Dispatch hydrogen-based	Required to avoid curtailment	Utilization battery -through-put-	Utilization PSH -through-put-	Total storage demand (incl. H2)	Dispatch hydrogen-based
Middle East		[GWh/year]	[GWh/year]	[GWh/year]	[GWh/year]	[GWh/year]	[GWh/year]	[GWh/year]	[GWh/year]	[GWh/year]	[GWh/year]
Israel	2020	0	0	0	0	0	0	0	0	0	0
	2030	29	0	10	10	0	226	0	36	36	8
	2050	24,725	11	379	390	0	14,244	11	320	331	529
North-ME	2020	0	0	0	0	0	0	0	0	0	0
	2030	1164	0	193	194	0	3596	1	355	356	20
	2050	109,498	32	1409	1441	0	84,974	31	1434	1465	1193
Saudi Arabia	2020	0	0	0	0	0	0	0	0	0	0
	2030	3366	1	513	514	0	11,457	2	900	902	39
	2050	231,140	73	2685	2757	0	159,949	74	2429	2503	2624
UAE	2020	0	0	0	0	0	0	0	0	0	0
	2030	9	0	5	5	0	233	0	45	45	17
	2050	35,463	24	679	703	0	17,093	23	507	531	1075

East-ME	2020	0	0	0	0	0	0	0	0	0	0
	2030	2	0	3	3	0	117	0	29	29	9
	2050	21,916	12	410	421	0	12,920	12	350	362	490
Iraq	2020	0	0	0	0	0	0	0	0	0	0
	2030	3941	0	330	330	0	8185	0	446	447	10
	2050	87,343	18	892	910	0	74,252	17	920	937	684
Iran	2020	0	0	0	0	0	0	0	0	0	0
	2030	9576	1	831	833	0	21,130	1	1127	1128	30
	2050	242,799	50	2508	2558	0	190,790	47	2443	2490	2036
Middle East	2020	0	0	0	0	0	0	0	0	0	0
	2030	18,088	4	1886	1890	0	44,945	4	2939	2943	132
	2050	752,882	218	8962	9180	0	554,222	215	8404	8618	8630

8.10 Eastern Europe/Eurasia

8.10.1 Eastern Europe/Eurasia: Long-Term Energy Pathways

8.10.1.1 Eastern Europe/Eurasia: Final Energy Demand by Sector

The future development pathways for Eastern Europe/Eurasia's final energy demand when the assumptions on population growth, GDP growth, and energy intensity are combined are shown in Fig. 8.62 for the 5.0 °C, 2.0 °C, and 1.5 °C Scenarios. In the 5.0 °C Scenario, the total final energy demand will increase by 45%, from the current 25,500 PJ/year to 37,000 PJ/year in 2050. In the 2.0 °C Scenario, the final energy demand will decrease by 25% compared with current consumption and will reach 19,100 PJ/year by 2050. The final energy demand in the 1.5 °C Scenario will reach 17,800 PJ, 30% below the 2015 level. In the 1.5 °C Scenario, the final energy demand in 2050 will be 7% lower than in the 2.0 °C Scenario. The electricity demand for 'classical' electrical devices (without power-to-heat or e-mobility) will increase from 910 TWh/year in 2015 to 1000 TWh/year (2.0 °C) or 940 TWh/year (1.5 °C) by 2050. Compared with the 5.0 °C case (1600 TWh/year in 2050), the efficiency measures in the 2.0 °C and 1.5 °C Scenarios will save a maximum of 600 TWh/year and 660 TWh/year, respectively.

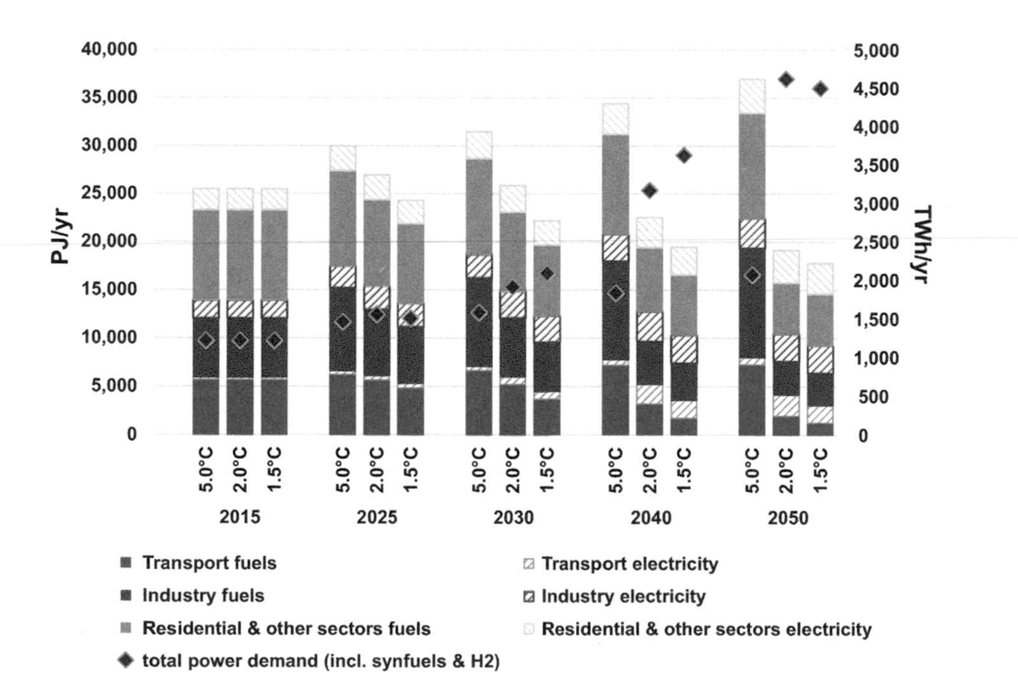

Fig. 8.62 Eastern Europe/Eurasia: development of the final energy demand by sector in the scenarios

Electrification will lead to a significant increase in the electricity demand by 2050. In the 2.0 °C Scenario, the electricity demand for heating will be approximately 700 TWh/year due to electric heaters and heat pumps, and in the transport sector, the electricity demand will be approximately 2300 TWh/year due to increased electric mobility. The generation of hydrogen (for transport and high-temperature process heat) and the manufacture of synthetic fuels (mainly for transport) will add an additional power demand of 2300 TWh/year. Therefore, the gross power demand will rise from 1700 TWh/year in 2015 to 4900 TWh/year in 2050 in the 2.0 °C Scenario, 88% higher than in the 5.0 °C case. In the 1.5 °C Scenario, the gross electricity demand will increase to a maximum of 4800 TWh/year in 2050.

Efficiency gains could be even larger in the heating sector than in the electricity sector. In the 2.0 °C and 1.5 °C Scenarios, a final energy consumption equivalent to more than 10,700 PJ/year is avoided by 2050 compared with the 5.0 °C Scenario through efficiency gains.

8.10.1.2 Eastern Europe/Eurasia: Electricity Generation

The development of the power system is characterized by a dynamically growing renewable energy market and an increasing proportion of total power from renewable sources. By 2050, 100% of the electricity produced in Eastern Europe/Eurasia will come from renewable energy sources in the 2.0 °C Scenario. 'New' renewables—mainly wind, solar, and geothermal energy—will contribute 75% of the total electricity generation. Renewable electricity's share of the total production will be 55% by 2030 and 84% by 2040. The installed capacity of renewables will reach about 560 GW by 2030 and 1900 GW by 2050. The share of renewable electricity generation in 2030 in the 1.5 °C Scenario is assumed to be 66%. In the 1.5 °C Scenario, the generation capacity from renewable energy will be approximately 1870 GW in 2050.

Table 8.59 shows the development of different renewable technologies in Eastern Europe/Eurasia over time. Figure 8.63 provides an overview of the overall power-generation structure in Eastern Europe/Eurasia. From 2020 onwards, the continuing growth of wind and PV, up to 740 GW and 820 GW, respectively, will be complemented by up to 30 GW of solar thermal generation, as well as limited biomass, geothermal, and ocean energy, in the 2.0 °C Scenario. Both the 2.0 °C Scenario and 1.5 °C Scenario will lead to a high proportion of variable power generation (PV, wind, and ocean) of 28% and 32%, respectively, by 2030, and 62% and 61%, respectively, by 2050.

Table 8.59 Eastern Europe/Eurasia: development of renewable electricity-generation capacity in the scenarios

in GW	Case	2015	2025	2030	2040	2050
Hydro	5.0 °C	98	107	112	123	136
	2.0 °C	98	107	112	115	116
	1.5 °C	98	107	112	115	116
Biomass	5.0 °C	1	4	6	9	14
	2.0 °C	1	21	45	64	96
	1.5 °C	1	40	74	86	109
Wind	5.0 °C	6	9	10	17	23
	2.0 °C	6	70	176	469	744
	1.5 °C	6	74	196	531	697
Geothermal	5.0 °C	0	1	1	2	4
	2.0 °C	0	5	12	38	71
	1.5 °C	0	7	21	46	71
PV	5.0 °C	4	5	6	8	10
	2.0 °C	4	108	209	502	817
	1.5 °C	4	132	294	678	821
CSP	5.0 °C	0	0	0	0	0
	2.0 °C	0	0	1	16	33
	1.5 °C	0	0	1	22	34
Ocean	5.0 °C	0	0	0	0	0
	2.0 °C	0	0	1	13	19
	1.5 °C	0	0	1	13	19
Total	5.0 °C	108	126	136	159	186
	2.0 °C	108	310	555	1216	1896
	1.5 °C	108	360	698	1492	1869

8.10.1.3 Eastern Europe/Eurasia: Future Costs of Electricity Generation

Figure 8.64 shows the development of the electricity-generation and supply costs over time, including the CO_2 emission costs, in all scenarios. The calculated electricity-generation costs in 2015 (referring to full costs) were around 4.5 ct/kWh. In the 5.0 °C case, the generation costs will increase until 2050, when they reach 10 ct/kWh. In the 2.0 °C Scenario, the generation costs will increase until 2050, when they will reach 8.6 ct/kWh. In the 1.5 °C Scenario, they will increase to 9.3 ct/kWh, and then drop to 8.8 ct/kWh by 2050. In the 2.0 °C Scenario, the generation costs in 2050 will be 1.4 ct/kWh lower than in the 5.0 °C case. In the 1.5 °C Scenario, the generation costs in 2050 will be 1.1 ct/kWh lower than in the 5.0 °C case. Note that these estimates of generation costs do not take into account integration costs such as power grid expansion, storage, or other load-balancing measures.

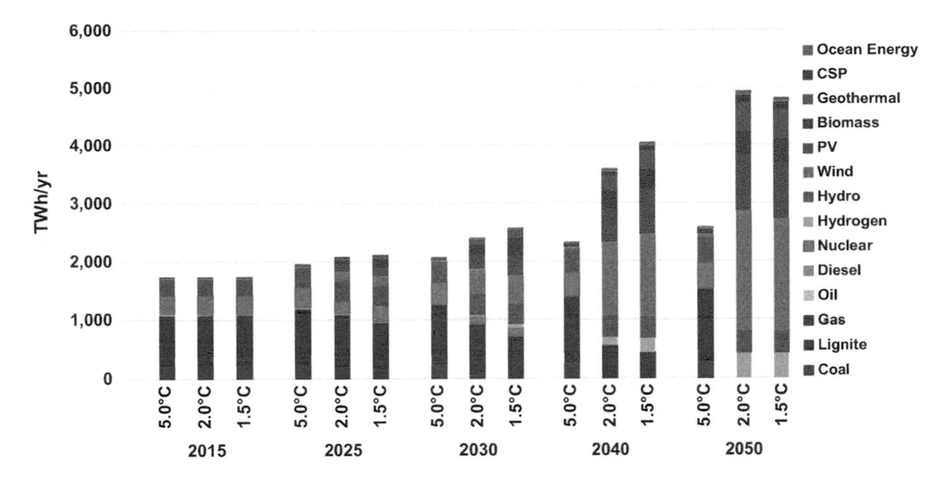

Fig. 8.63 Eastern Europe/Eurasia: development of electricity-generation structure in the scenarios

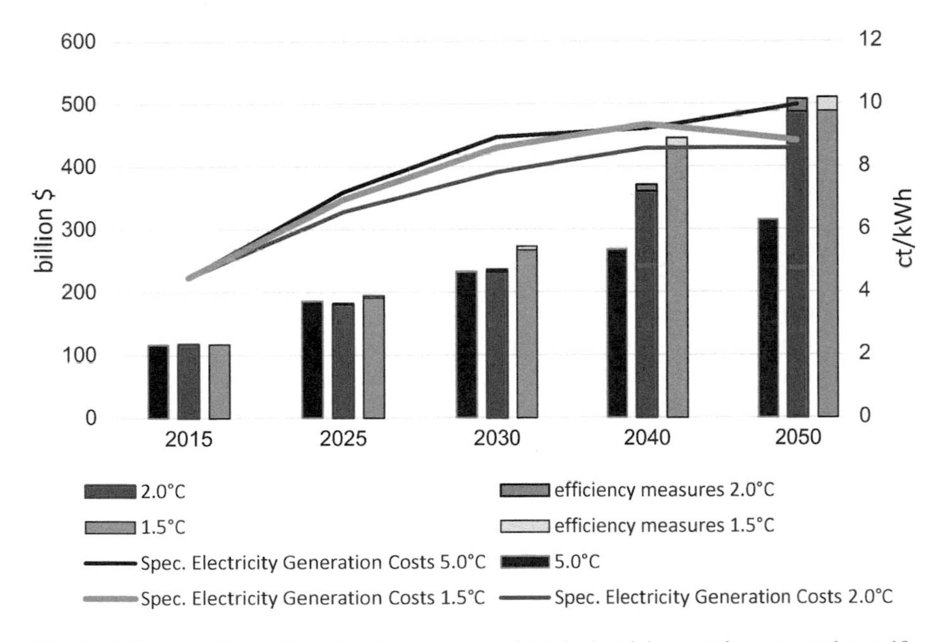

Fig. 8.64 Eastern Europe/Eurasia: development of total electricity supply costs and specific electricity-generation costs in the scenarios

In the 5.0 °C case, the growth of demand and increasing fossil fuel prices will cause the total electricity supply costs to rise from today's $120 billion/year to more than $320 billion/year in 2050. In both alternative scenarios, the total supply costs will be $490 billion/year in 2050. The long-term costs of electricity supply will be more than 54% higher in the 2.0 °C Scenario than in the 5.0 °C Scenario as a result of the estimated generation costs and the electrification of heating and mobility. Further electrification and synthetic fuel generation in the 1.5 °C Scenario will result in total power generation costs that are 55% higher than in the 5.0 °C case.

Compared with these results, the generation costs when the CO_2 emission costs are not considered will increase in the 5.0 °C case to 6.9 ct/kWh. In the 2.0 °C Scenario, the generation costs will increase continuously until 2050, when they reach 8.6 ct/kWh. They will increase to 8.8 ct/kWh in the 1.5 °C Scenario. In the 2.0 °C Scenario, the generation costs will reach a maximum, at 1.7 ct/kWh higher than in the 5.0 °C case, and this will occur in 2050. In the 1.5 °C Scenario, the maximum difference in generation costs compared with the 5.0 °C case will be 2.6 ct/kWh in 2040. The generation costs in 2050 will still be 2 ct/kWh higher than in the 5.0 °C case. If the CO_2 costs are not considered, the total electricity supply costs in the 5.0 °C case will rise to about $240 billion in 2050.

8.10.1.4 Eastern Europe/Eurasia: Future Investments in the Power Sector

An investment of around $3600 billion will be required for power generation between 2015 and 2050 in the 2.0 °C Scenario—including additional power plants for the production of hydrogen and synthetic fuels and investments in plant replacement at the end of their economic lives. This value is equivalent to approximately $100 billion per year on average, and is $2660 billion more than in the 5.0 °C case ($940 billion). An investment of around $3770 billion for power generation will be required between 2015 and 2050 in the 1.5 °C Scenario. On average, this is an investment of $105 billion per year. In the 5.0 °C Scenario, the investment in conventional power plants will be around 40% of the total cumulative investments, whereas approximately 60% will be invested in renewable power generation and co-generation (Fig. 8.65).

However, in the 2.0 °C (1.5 °C) scenario, Eastern Europe/Eurasia will shift almost 97% (98%) of its entire investments to renewables and co-generation. By 2030, the fossil fuel share of the power sector investments will predominantly focus on gas power plants that can also be operated with hydrogen.

Because renewable energy has no fuel costs, other than biomass, the cumulative fuel cost savings in the 2.0 °C Scenario will reach a total of $1730 billion in 2050, equivalent to $48 billion per year. Therefore, the total fuel cost savings will be equivalent to 70% of the total additional investments compared to the 5.0 °C Scenario. The fuel cost savings in the 1.5 °C Scenario will add up to $1900 billion, or $53 billion per year.

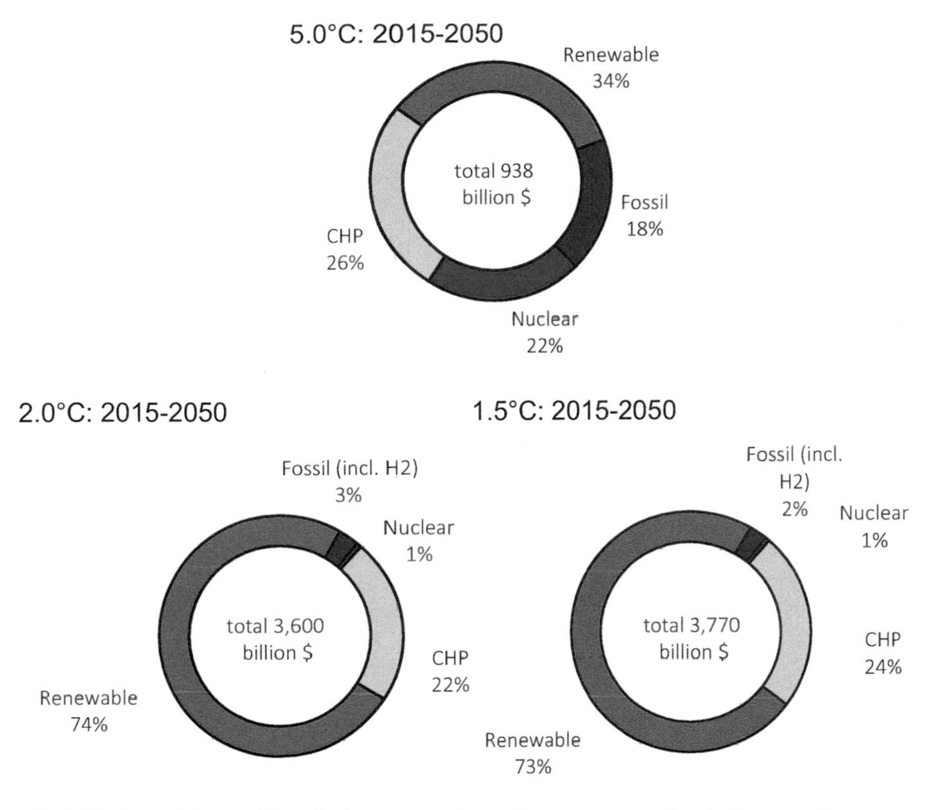

Fig. 8.65 Eastern Europe/Eurasia: investment shares for power generation in the scenarios

8.10.1.5 Eastern Europe/Eurasia: Energy Supply for Heating

The final energy demand for heating will increase in the 5.0 °C Scenario by 46%, from 15,700 PJ/year in 2015 to 22,900 PJ/year in 2050. Energy efficiency measures will help to reduce the energy demand for heating by 47% in 2050 in both alternative scenarios. Today, renewables supply around 4% of Eastern Europe/Eurasia's final energy demand for heating, with the main contribution from biomass. Renewable energy will provide 29% of Eastern Europe/Eurasia's total heat demand in 2030 in the 2.0 °C Scenario and 42% in the 1.5 °C Scenario. In both scenarios, renewables will provide 100% of the total heat demand in 2050.

Figure 8.66 shows the development of different technologies for heating in Eastern Europe/Eurasia over time, and Table 8.60 provides the resulting renewable heat supply for all scenarios. Until 2030, biomass will remain the main contributor. In the long term, the growing use of solar, geothermal, and environmental heat will lead to a biomass share of 28% in both alternative scenarios.

Heat from renewable hydrogen will further reduce the dependence on fossil fuels in both scenarios. Hydrogen consumption in 2050 will be around 1900 PJ/year in the 2.0 °C Scenario and 2000 PJ/year in the 1.5 °C Scenario.

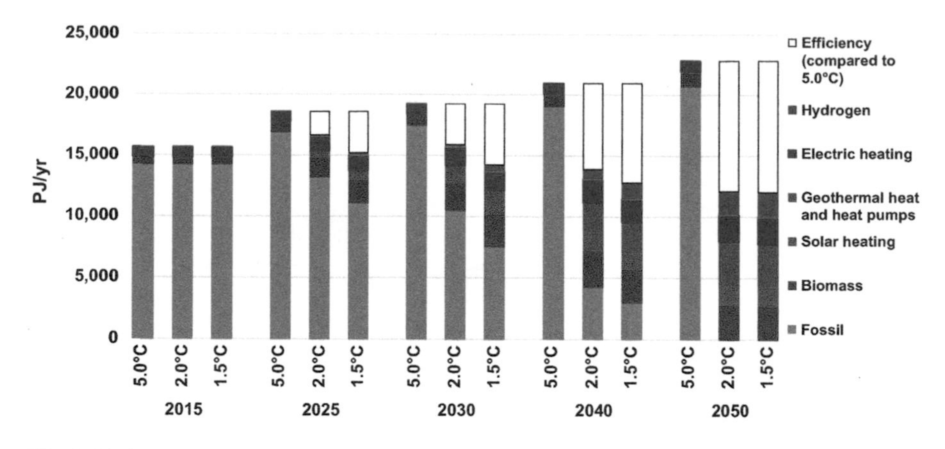

Fig. 8.66 Eastern Europe/Eurasia: development of heat supply by energy carrier in the scenarios

Table 8.60 Eastern Europe/Eurasia: development of renewable heat supply in the scenarios (excluding the direct use of electricity)

in PJ/year	Case	2015	2025	2030	2040	2050
Biomass	5.0 °C	537	810	873	1005	1164
	2.0 °C	537	1604	2199	2971	2819
	1.5 °C	537	1869	2684	2734	2722
Solar heating	5.0 °C	5	10	13	24	41
	2.0 °C	5	277	706	1560	1662
	1.5 °C	5	351	768	1395	1620
Geothermal heat and heat pumps	5.0 °C	6	9	11	15	21
	2.0 °C	6	265	780	2314	3493
	1.5 °C	6	410	1163	2434	3393
Hydrogen	5.0 °C	0	0	0	0	0
	2.0 °C	0	42	152	795	1934
	1.5 °C	0	155	494	1344	2032
Total	5.0 °C	548	829	897	1044	1226
	2.0 °C	548	2187	3837	7640	9908
	1.5 °C	548	2786	5110	7906	9767

The direct use of electricity for heating will also increases by a factor of 2.7 between 2015 and 2050, and its final energy share will be 18% in 2050 in the 2.0 °C Scenario and 19% in the 1.5 °C Scenario.

8.10.1.6 Eastern Europe/Eurasia: Future Investments in the Heating Sector

The roughly estimated investment in renewable heating technologies up to 2050 will amount to around $1070 billion in the 2.0 °C Scenario (including investments in plant replacement after their economic lifetimes), or approximately $30 billion

Table 8.61 Eastern Europe/Eurasia: installed capacities for renewable heat generation in the scenarios

in GW	Case	2015	2025	2030	2040	2050
Biomass	5.0 °C	107	150	157	169	183
	2.0 °C	107	230	249	263	172
	1.5 °C	107	241	252	230	162
Geothermal	5.0 °C	0	0	0	1	1
	2.0 °C	0	14	26	64	61
	1.5 °C	0	12	30	52	54
Solar heating	5.0 °C	1	2	3	5	9
	2.0 °C	1	56	145	330	359
	1.5 °C	1	74	163	300	352
Heat pumps	5.0 °C	1	1	2	2	3
	2.0 °C	1	25	64	184	248
	1.5 °C	1	33	76	175	236
Total[a]	5.0 °C	109	154	162	177	196
	2.0 °C	109	325	483	841	839
	1.5 °C	109	361	522	758	805

[a] Excluding direct electric heating

per year. The largest share of the investments in Eastern Europe/Eurasia is assumed to be for heat pumps (around $490 billion), followed by solar collectors and biomass technologies. The 1.5 °C Scenario assumes an even faster expansion of renewable technologies. However, the lower heat demand (compared with the 2.0 °C Scenario) will result in a lower average annual investment of around $29 billion per year (Table 8.61, Fig. 8.67).

8.10.1.7 Eastern Europe/Eurasia: Transport

Energy demand in the transport sector in Eastern Europe/Eurasia is expected to increase in the 5.0 °C Scenario by 34%, from around 6000 PJ/year in 2015 to 8000 PJ/year in 2050. In the 2.0 °C Scenario, assumed technical, structural, and behavioural changes will save 48% (3840 PJ/year) by 2050 compared with the 5.0 °C Scenario. Additional modal shifts, technology switches, and a reduction in the transport demand will lead to even higher energy savings in the 1.5 °C Scenario of 62% (or 4970 PJ/year) in 2050 compared with the 5.0 °C case (Table 8.62, Fig. 8.68).

By 2030, electricity will provide 14% (240 TWh/year) of the transport sector's total energy demand in the 2.0 °C Scenario, whereas in 2050, the share will be 54% (630 TWh/year). In 2050, up to 410 PJ/year of hydrogen will be used in the transport sector as a complementary renewable option. In the 1.5 °C Scenario, the annual electricity demand will be 510 TWh in 2050. The 1.5 °C Scenario also assumes a hydrogen demand of 330 PJ/year by 2050.

Biofuel use is limited in the 2.0 °C Scenario to a maximum of 720 PJ/year Therefore, around 2030, synthetic fuels based on power-to-liquid will be intro-

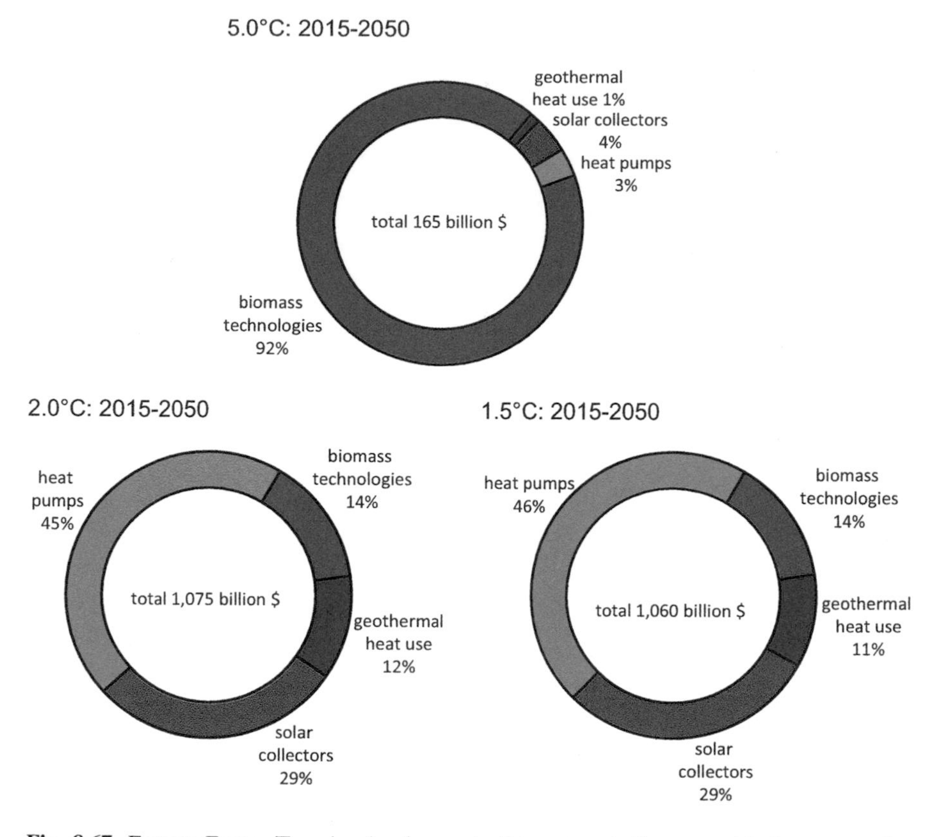

Fig. 8.67 Eastern Europe/Eurasia: development of investments for renewable heat-generation technologies in the scenarios

duced, with a maximum amount of 880 PJ/year in 2050. With the lower overall energy demand in transport, biofuel use will also be reduced in the 1.5 °C Scenario to a maximum of 700 PJ/year The maximum synthetic fuel demand will amount to 540 PJ/year.

8.10.1.8 Eastern Europe/Eurasia: Development of CO_2 Emissions

In the 5.0 °C Scenario, Eastern Europe/Eurasia's annual CO_2 emissions will increase by 14%, from 2420 Mt. in 2015 to 2768 Mt. in 2050. The stringent mitigation measures in both alternative scenarios will cause the annual emissions to fall to 590 Mt. in 2040 in the 2.0 °C Scenario and to 340 Mt. in the 1.5 °C Scenario, with further reductions to almost zero by 2050. In the 5.0 °C case, the cumulative CO_2 emissions from 2015 until 2050 will add up to 95 Gt. In contrast, in the 2.0 °C and 1.5 °C Scenarios, the cumulative emissions for the period from 2015 until 2050 will be 45 Gt and 36 Gt, respectively.

Table 8.62 Eastern Europe/Eurasia: projection of transport energy demand by mode in the scenarios

in PJ/year	Case	2015	2025	2030	2040	2050
Rail	5.0 °C	434	498	528	599	674
	2.0 °C	434	509	544	646	712
	1.5 °C	434	449	470	620	796
Road	5.0 °C	3873	4321	4680	5181	5319
	2.0 °C	3873	4336	4403	3923	3195
	1.5 °C	3873	3593	2963	2346	2016
Domestic aviation	5.0 °C	232	336	403	482	471
	2.0 °C	232	247	228	188	150
	1.5 °C	232	237	207	146	114
Domestic navigation	5.0 °C	34	35	36	38	40
	2.0 °C	34	35	36	38	40
	1.5 °C	34	35	36	38	40
Total	5.0 °C	4573	5191	5647	6301	6504
	2.0 °C	4573	5127	5210	4795	4097
	1.5 °C	4573	4313	3677	3150	2966

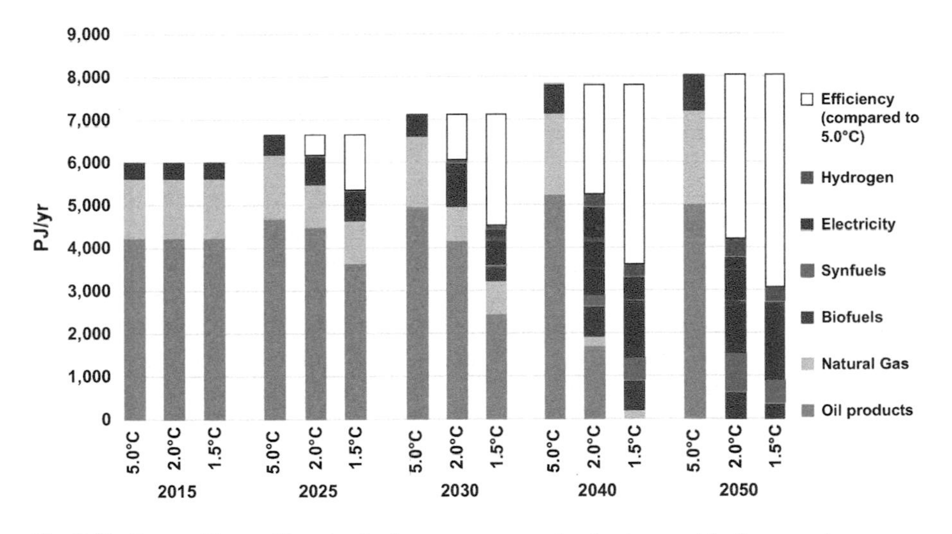

Fig. 8.68 Eastern Europe/Eurasia: final energy consumption by transport in the scenarios

Therefore, the cumulative CO_2 emissions will decrease by 53% in the 2.0 °C Scenario and by 62% in the 1.5 °C Scenario compared with the 5.0 °C case. A rapid reduction in annual emissions will occur in both alternative scenarios. In the 2.0 °C Scenario, this reduction will be greatest in 'Power generation', followed by the 'Residential and other' and 'Industry' sectors (Fig. 8.69).

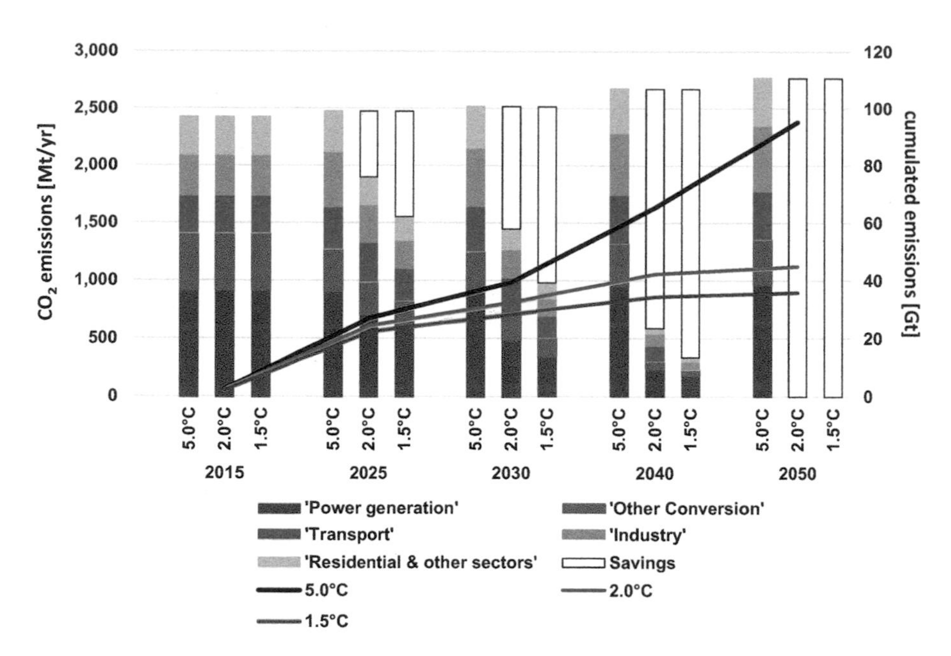

Fig. 8.69 Eastern Europe/Eurasia: development of CO_2 emissions by sector and cumulative CO_2 emissions (after 2015) in the scenarios ('Savings' = reduction compared with the 5.0 °C Scenario)

8.10.1.9 Eastern Europe/Eurasia: Primary Energy Consumption

The levels of primary energy consumption in the three scenarios when the assumptions discussed above are taken into account are shown in Fig. 8.70. In the 2.0 °C Scenario, the primary energy demand will decrease by 25%, from around 46,000 PJ/year in 2015 to 34,600 PJ/year in 2050. Compared with the 5.0 °C Scenario, the overall primary energy demand will decrease by 40% by 2050 in the 2.0 °C Scenario (5.0 °C: 57,700 PJ in 2050). In the 1.5 °C Scenario, the primary energy demand will be even lower (33,600 PJ in 2050) because the final energy demand and conversion losses will be lower.

Both the 2.0 °C and 1.5 °C Scenarios aim to rapidly phase-out coal and oil. This will cause renewable energy to have a primary energy share of 26% in 2030 and 91% in 2050 in the 2.0 °C Scenario. In the 1.5 °C Scenario, renewables will have a primary energy share of more than 90% in 2050 (including non-energy consumption, which will still include fossil fuels). Nuclear energy will be phased-out by 2040 in both the 2.0 °C Scenario and 1.5 °C Scenario. The cumulative primary energy consumption of natural gas in the 5.0 °C case will add up to 840 EJ, the cumulative coal consumption to about 290 EJ, and the crude oil consumption to 340 EJ. In contrast, in the 2.0 °C Scenario, the cumulative gas demand will amount to 510 EJ, the cumulative coal demand to 100 EJ, and the cumulative oil demand to 160 EJ. Even lower fossil fuel use will be achieved in the 1.5 °C Scenario: 450 EJ for natural gas, 70 EJ for coal, and 120 EJ for oil.

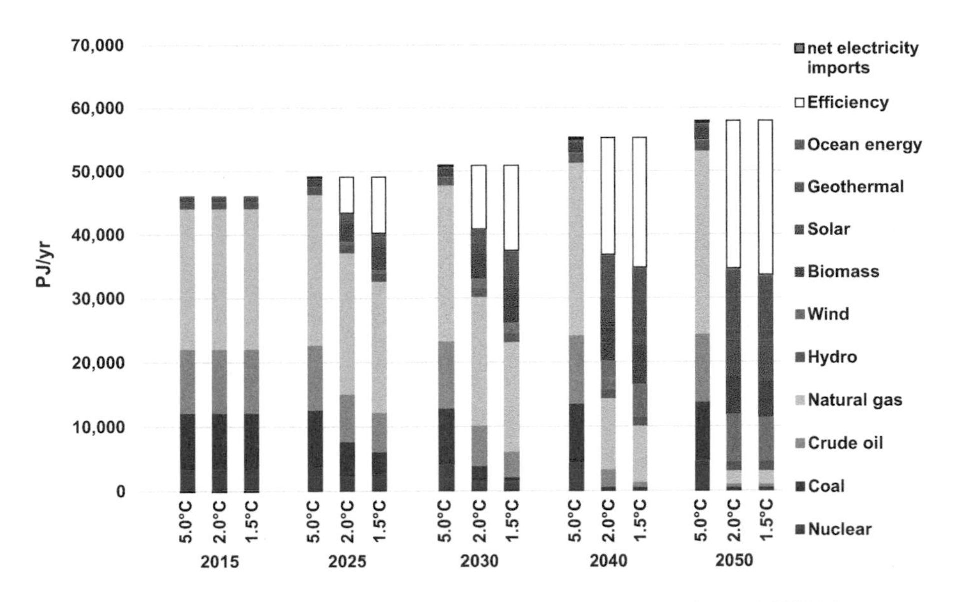

Fig. 8.70 Eastern Europe/Eurasia: projection of total primary energy demand (PED) by energy carrier in the scenarios (including electricity import balance)

8.10.2 Eastern Europe/Eurasia: Power Sector Analysis

This region sits between the strong economic hubs of the EU, China, and India. Russia, by far the largest country within this region, is an important producer of oil and gas, and supplies all surrounding countries. Therefore, Eurasia will be key in future energy developments. Its renewable energy industry is among the smallest in the world, but recent developments indicate growth in both the wind (WPM 3-2018) and solar industries (PVM 3-2018).

8.10.2.1 Eurasia: Development of Power Plant Capacities—2.0 °C Scenario

The northern part of Eurasia and Mongolia have significant wind potential, whereas the southern part, especially in Central Asia, has substantial possibilities for utility-scale solar power plants—both for solar PV and concentrated solar. The annual market for solar PV and onshore wind—as for all other renewable power generation technologies—must develop from a very low MW range in 2017 to a GW market by 2025. Besides solar PV and onshore wind, bioenergy has significant potential in Eurasia, especially in the European part, Russia, and the agricultural regions around the Caspian Sea (Table 8.63).

Table 8.63 Eurasia: average annual change in installed power plant capacity

Eurasia power generation: average annual change of installed capacity [GW/a]	2015–2025		2026–2035		2036–2050	
	2.0 °C	1.5 °C	2.0 °C	1.5 °C	2.0 °C	1.5 °C
Hard coal	−1	−6	−6	−4	0	0
Lignite	−3	−4	−2	−1	0	0
Gas	4	1	0	−2	−17	−5
Hydrogen-gas	0	2	2	4	20	17
Oil/Diesel	−2	−2	−1	−1	0	0
Nuclear	−2	−3	−2	−4	−1	0
Biomass	3	8	3	5	4	2
Hydro	2	1	1	1	0	0
Wind (onshore)	7	20	26	28	24	21
Wind (offshore)	1	3	6	6	11	8
PV (roof top)	9	25	21	32	31	22
PV (utility scale)	3	8	7	11	10	7
Geothermal	1	3	2	4	4	3
Solar thermal power plants	0	0	1	1	1	2
Ocean energy	0	0	1	1	1	1
Renewable fuel based co-generation	2	7	4	7	5	3

8.10.2.2 Eurasia: Utilization of Power-Generation Capacities

Variable power generation starts at almost zero, but increases rapidly to over 30% in most sub-regions of Eurasia, as shown in Table 8.64.

Table 8.64 shows that dispatchable renewables will experience stable market conditions throughout the entire modelling period across the whole region. Both scenarios assume that the interconnections between Eastern Europe and Russia will increase significantly, whereas the power transmission capacities for Kazakhstan, Central Asia, the area around the Caspian Sea, and Mongolia will remain low due to geographic distances.

Compared with other world regions, it will take longer for the capacity factor of the limited dispatchable power plants to drop below economic viability, as shown in Table 8.65.

Table 8.65. The capacity factor of variable renewables will rise by 2030, mainly due to increased deployment of wind and concentrated solar power with storage. The average capacity factor of the power-generation fleet will be around 35% by 2050 and will therefore be on the same level as it was 2015 in both scenarios.

8.10.2.3 Eurasia: Development of Load, Generation, and Residual Load

The modelling of both scenarios predicts small increases in interconnection beyond those assumed to occur by 2030 (see Table 8.64).

Table 8.64 Eurasia: power system shares by technology group

Power generation structure and interconnection		2.0 °C				1.5 °C			
Eurasia		Variable RE	Dispatch RE	Dispatch fossil	Inter-connection	Variable RE	Dispatch RE	Dispatch fossil	Inter-connection
Eastern Europe	2015	1%	35%	63%	5%				
	2030	37%	45%	18%	10%	41%	46%	13%	10%
	2050	70%	22%	7%	20%	66%	24%	10%	20%
Russia	2015	1%	35%	63%	5%				
	2030	35%	43%	22%	5%	39%	47%	14%	5%
	2050	68%	24%	8%	5%	64%	26%	10%	5%
Kazakhstan	2015	2%	35%	63%	5%				
	2030	44%	42%	14%	5%	49%	42%	9%	5%
	2050	80%	16%	4%	5%	77%	18%	5%	5%
Mongolia	2015	2%	35%	63%	5%				
	2030	43%	43%	13%	5%	48%	43%	10%	5%
	2050	74%	20%	6%	10%	71%	22%	8%	10%
West Caspian Sea	2015	1%	35%	63%	5%				
	2030	43%	41%	16%	5%	47%	40%	12%	5%
	2050	77%	17%	6%	10%	72%	19%	9%	10%
East Caspian Sea	2015	1%	35%	63%	5%				
	2030	37%	44%	19%	5%	41%	45%	14%	5%
	2050	71%	22%	7%	10%	67%	24%	10%	10%

(continued)

Table 8.64 (continued)

Power generation structure and interconnection		2.0 °C				1.5 °C			
Eurasia		Variable RE	Dispatch RE	Dispatch fossil	Inter-connection	Variable RE	Dispatch RE	Dispatch fossil	Inter-connection
Central Asia	2015	0%	35%	64%	5%				
	2030	18%	50%	31%	5%	23%	50%	27%	5%
	2050	43%	39%	18%	5%	38%	37%	26%	5%
Eurasia	2015	1%	35%	63%					
	2030	36%	43%	21%		40%	46%	14%	
	2050	69%	23%	7%		65%	25%	10%	

Table 8.65 Eurasia: capacity factors by generation type

Utilization of variable and dispatchable power generation:		2015	2020	2020	2030	2030	2040	2040	2050	2050
Eurasia			2.0 °C	1.5 °C	2.0 °C	1.5 °C	2.0 °C	1.5 °C	2.0 °C	1.5 °C
Capacity factor – average	**[%/yr]**	**36.8%**	**31%**	**40%**	**48%**	**47%**	**34%**	**34%**	**34%**	**34%**
Limited dispatchable: fossil and nuclear	[%/yr]	43.8%	31%	30%	22%	18%	19%	0%	7%	4%
Limited dispatchable: renewable	[%/yr]	39.3%	42%	42%	57%	54%	60%	39%	39%	40%
Dispatchable: fossil	[%/yr]	27.6%	18%	17%	7%	6%	31%	8%	12%	15%
Dispatchable: renewable	[%/yr]	38.7%	48%	73%	73%	68%	41%	49%	50%	51%
Variable: renewable	[%/yr]	10.5%	11%	11%	40%	39%	25%	32%	32%	33%

Table 8.64. However, after 2030, significant increases will be required by 2050, especially in Russia. The export of renewable electricity can also take place via existing gas pipelines with power-to-gas technologies. Between 2030 and 2050, the loads for all regions will double, due to the increased electrification of the heating, industry, and transport sectors (Table 8.66).

In Eurasia, the main storage technology for both scenarios is pumped hydro, whereas hydrogen plays a major role for the grid integration of variable generation (Table 8.67). Hydrogen production can also be used for load management, although not for short peak loads. Due to the technical and economic limitations associated with the increased interconnection via transmission lines and pumped hydro storage systems, curtailment will be higher than the scenario target (a maximum of 10% by 2050). For Eastern Europe, Kazakhstan, Mongolia, and the East Caspian Sea, the calculated curtailment will be between 10% and 14%, whereas the West Caspian Region will have the highest curtailment of 19% in the 2.0 °C Scenario and 17% in the 1.5 °C Scenario. Further research and optimization are required.

Table 8.66 Eurasia: load, generation, and residual load development

Power generation structure		2.0 °C				1.5 °C			
		Max demand	Max generation	Max residual load	Max interconnection requirements	Max demand	Max generation	Max residual load	Max interconnection requirements
Eurasia		[GW]	[GW]	[GW]	[GW]	[GW]	[GW]	[GW]	[GW]
Eastern Europe	2020	32.9	30.8	3.9		32.9	33.0	4.6	
	2030	38.1	45.0	12.4	0	40.8	56.2	14.2	1
	2050	77.2	179.6	30.9	71	77.5	174.3	31.6	65
Russia	2020	172.7	95.8	83.6		172.7	100.6	81.8	
	2030	214.2	218.6	103.5	0	221.4	275.2	94.9	0
	2050	428.3	887.6	191.7	268	429.2	859.0	194.4	235
Kazakhstan	2020	14.7	18.8	0.9		14.7	17.9	0.9	
	2030	17.9	18.6	8.3	0	18.9	23.1	7.7	0
	2050	34.3	74.5	14.1	26	34.4	72.2	14.4	23
Mongolia	2020	1.7	2.0	0.1		1.7	2.0	0.1	
	2030	2.0	2.3	0.9	0	2.1	2.9	0.9	0
	2050	3.7	8.5	1.2	4	3.7	8.4	1.2	3
West Caspian Sea	2020	10.7	6.2	4.6		10.7	6.9	4.2	
	2030	12.5	13.9	6.4	0	13.4	17.3	5.9	0
	2050	24.3	55.8	9.8	22	24.4	54.1	10.0	20
East Caspian Sea	2020	21.6	7.5	14.2		21.6	7.8	13.8	
	2030	25.2	28.2	12.7	0	26.9	35.0	12.5	0
	2050	50.0	113.4	18.6	45	50.2	109.7	19.1	40
Central Asia	2020	2.5	2.3	0.2		2.5	2.3	0.2	
	2030	6.0	5.8	2.8	0	6.7	6.5	3.0	0
	2050	12.0	18.2	4.2	2	12.1	18.2	4.4	2

Table 8.67 Eurasia: storage and dispatch service requirements

Storage and dispatch		2.0 °C					1.5 °C				
		Required to avoid curtailment	Utilization battery -through-put-	Utilization PSH -through-put-	Total storage demand (incl. H2)	Dispatch hydrogen-based	Required to avoid curtailment	Utilization battery -through-put-	Utilization PSH -through-put-	Total storage demand (incl. H2)	Dispatch hydrogen-based
Eurasia		[GWh/year]	[GWh/year]	[GWh/year]	[GWh/year]	[GWh/year]	[GWh/year]	[GWh/year]	[GWh/year]	[GWh/year]	[GWh/year]
Eastern Europe	2020	0	0	0	0	0	0	0	0	0	0
	2030	373	1	137	138	1720	1674	2	317	319	5920
	2050	52,516	274	2626	2900	49,057	43,933	267	2303	2570	49,858
Russia	2020	0	0	0	0	0	0	0	0	0	0
	2030	36	0	41	41	9711	2290	3	517	520	33,707
	2050	147,854	1132	9342	10,474	282,100	123,490	1049	7895	8944	287,188
Kazakhstan	2020	0	0	0	0	0	0	0	0	0	0
	2030	7	0	7	7	690	281	1	84	85	2223
	2050	28,094	133	1444	1577	13,192	23,926	127	1271	1398	13,544
Mongolia	2020	0	0	0	0	0	0	0	0	0	0
	2030	24	0	11	11	78	131	0	25	25	258
	2050	3177	17	152	169	1997	2938	16	139	155	1971
West Caspian Sea	2020	0	0	0	0	0	0	0	0	0	0
	2030	163	0	78	79	472	882	1	173	174	1558
	2050	30,281	96	1207	1303	12,025	26,053	94	1120	1214	12,088
East Caspian Sea	2020	0	0	0	0	0	0	0	0	0	0
	2030	134	0	65	65	1125	773	1	170	170	3759
	2050	32,074	202	1785	1988	30,493	27,253	195	1580	1775	30,852

(continued)

Table 8.67 (continued)

Storage and dispatch		2.0 °C					1.5 °C				
		Required to avoid curtailment	Utilization battery -through-put-	Utilization PSH -through-put-	Total storage demand (incl. H2)	Dispatch hydrogen-based	Required to avoid curtailment	Utilization battery -through-put-	Utilization PSH -through-put-	Total storage demand (incl. H2)	Dispatch hydrogen-based
Eurasia		[GWh/year]	[GWh/year]	[GWh/year]	[GWh/year]	[GWh/year]	[GWh/year]	[GWh/year]	[GWh/year]	[GWh/year]	[GWh/year]
Central Asia	2020	0	0	0	0	0	0	0	0	0	0
	2030	0	0	0	0	309	0	0	1	1	1090
	2050	2495	39	211	250	12,181	2391	39	207	245	12,037
Eurasia Eastern Europe	2020	0	0	0	0	0	0	0	0	0	0
	2030	736	2	339	341	14,106	6031	7	1287	1295	48,516
	2050	296,490	1894	16,767	18,661	401,044	249,984	1788	14,515	16,303	407,537

8.11 Non-OECD Asia

8.11.1 Non-OECD Asia: Long-Term Energy Pathways

8.11.1.1 Non-OECD Asia: Final Energy Demand by Sector

The future development pathways for Non-OECD Asia's final energy demand when the assumptions on population growth, GDP growth, and energy intensity are combined are shown in Fig. 8.71 for the 5.0 °C, 2.0 °C, and 1.5 °C Scenarios. In the 5.0 °C Scenario, the total final energy demand will increase by 111% from the current 24,500 PJ/year to 51,800 PJ/year in 2050. In the 2.0 °C Scenario, the final energy demand will increase at a much lower rate by 16% compared with current consumption, and will reach 28,300 PJ/year by 2050. The final energy demand in the 1.5 °C Scenario will reach 25,700 PJ, 5% above the 2015 demand. In the 1.5 °C Scenario, the final energy demand in 2050 will be 9% lower than in the 2.0 °C Scenario. The electricity demand for 'classical' electrical devices (without power-to-heat or e-mobility) will increase from 830 TWh/year in 2015 to 2480 TWh/year in 2050 in both alternative scenarios. Compared with the reference case (3880 TWh/year in 2050), the efficiency measures in the 2.0 °C and 1.5 °C scenarios will save 1400 TWh/year in 2050.

Electrification will lead to a significant increase in the electricity demand by 2050. In the 2.0 °C Scenario, the electricity demand for heating will be approxi-

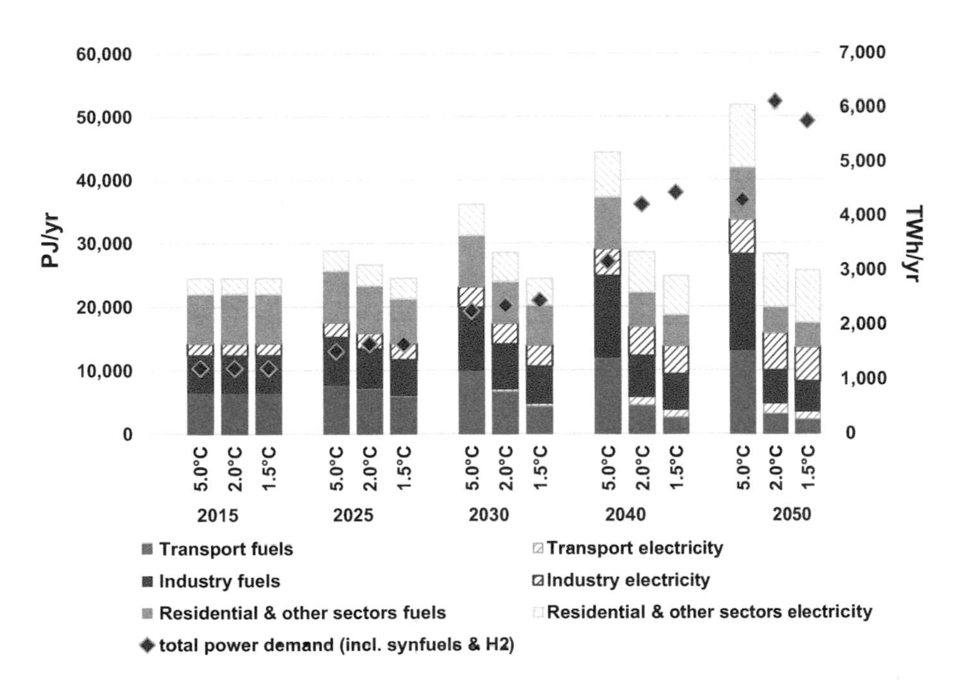

Fig. 8.71 Non-OECD Asia: development of the final energy demand by sector in the scenarios

mately 1500 TWh/year due to electric heaters and heat pumps, and in the transport sector, the electricity demand will be approximately 1700 TWh/year due to electric mobility. The generation of hydrogen (for transport and high-temperature process heat) and the manufacture of synthetic fuels (mainly for transport) will add an additional power demand of 1700 TWh/year. Therefore, the gross power demand will rise from 1400 TWh/year in 2015 to 6400 TWh/year in 2050 in the 2.0 °C Scenario, 33% higher than in the 5.0 °C case. In the 1.5 °C Scenario, the gross electricity demand will increase to a maximum of 6000 TWh/year in 2050.

The efficiency gains in the heating sector could be even larger than those in the electricity sector. In the 2.0 °C and 1.5 °C Scenarios, a final energy consumption equivalent to about 6900 PJ/year and 8100 PJ/year, respectively, will be avoided by 2050 compared with the 5.0 °C Scenario, through efficiency gains.

8.11.1.2 Non-OECD Asia: Electricity Generation

The development of the power system is characterized by a dynamically growing renewable energy market and an increasing proportion of total power from renewable sources. By 2050, 100% of the electricity produced in Non-OECD Asia will come from renewable energy sources in the 2.0 °C Scenario. 'New' renewables—mainly wind, solar, and geothermal energy—will contribute 87% of the total electricity generation. Renewable electricity's share of the total production will be 59% by 2030 and 87% by 2040. The installed capacity of renewables will reach about 610 GW by 2030 and 2430 GW by 2050. The share of renewable electricity generation in 2030 in the 1.5 °C Scenario is assumed to be 74%. In the 1.5 °C Scenario, the generation capacity from renewable energy will be approximately 2320 GW in 2050.

Table 8.68 shows the development of different renewable technologies in Non-OECD Asia over time. Figure 8.72 provides an overview of the overall power-generation structure in Non-OECD Asia. From 2020 onwards, the continuing growth of wind and PV up to 635 GW and 1280 GW, respectively, will be complemented by up to 275 GW solar thermal generation, as well as limited biomass, geothermal, and ocean energy in the 2.0 °C Scenario. Both the 2.0 °C Scenario and 1.5 °C Scenario will lead to a high proportion of variable power generation (PV, wind, and ocean) of 34% and 48%, respectively, by 2030, and 64% and 66%, respectively, by 2050.

8.11.1.3 Non-OECD Asia: Future Costs of Electricity Generation

Figure 8.73 shows the development of the electricity-generation and supply costs over time, including the CO_2 emission costs, in all scenarios. The calculated electricity generation costs in 2015 (referring to full costs) were around 5.2 ct/kWh. In the 5.0 °C case, the generation costs will increase until 2050, when they reach 11.7 ct/kWh. The generation costs will increase in the 2.0 °C Scenario until 2030, when they will reach 8.1 ct/kWh, and will drop to 6.3 ct/kWh by 2050. In the 1.5 °C

Table 8.68 Non-OECD Asia: development of renewable electricity-generation capacity in the scenarios

in GW	Case	2015	2025	2030	2040	2050
Hydro	5.0 °C	63	85	124	151	183
	2.0 °C	63	86	86	90	91
	1.5 °C	63	86	86	90	91
Biomass	5.0 °C	7	10	17	22	31
	2.0 °C	7	19	19	30	36
	1.5 °C	7	19	20	31	39
Wind	5.0 °C	2	5	17	32	54
	2.0 °C	2	53	148	389	635
	1.5 °C	2	98	229	458	631
Geothermal	5.0 °C	3	4	6	8	10
	2.0 °C	3	6	23	50	63
	1.5 °C	3	7	26	47	54
PV	5.0 °C	3	9	26	44	70
	2.0 °C	3	107	287	806	1282
	1.5 °C	3	157	396	907	1256
CSP	5.0 °C	0	0	0	0	0
	2.0 °C	0	5	45	134	275
	1.5 °C	0	5	45	110	224
Ocean	5.0 °C	0	0	0	0	0
	2.0 °C	0	0	2	20	50
	1.5 °C	0	0	2	15	30
Total	5.0 °C	78	113	191	257	348
	2.0 °C	78	276	610	1518	2432
	1.5 °C	78	373	804	1658	2325

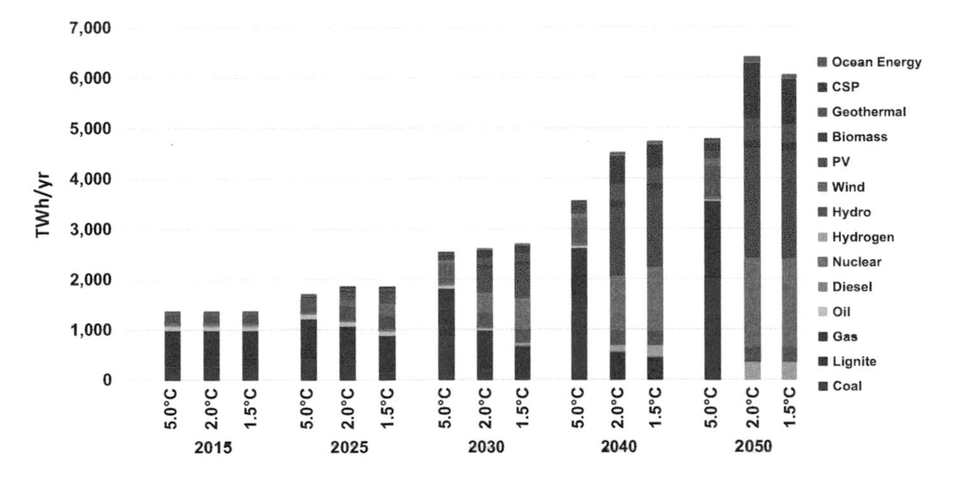

Fig. 8.72 Non-OECD Asia: development of electricity-generation structure in the scenarios

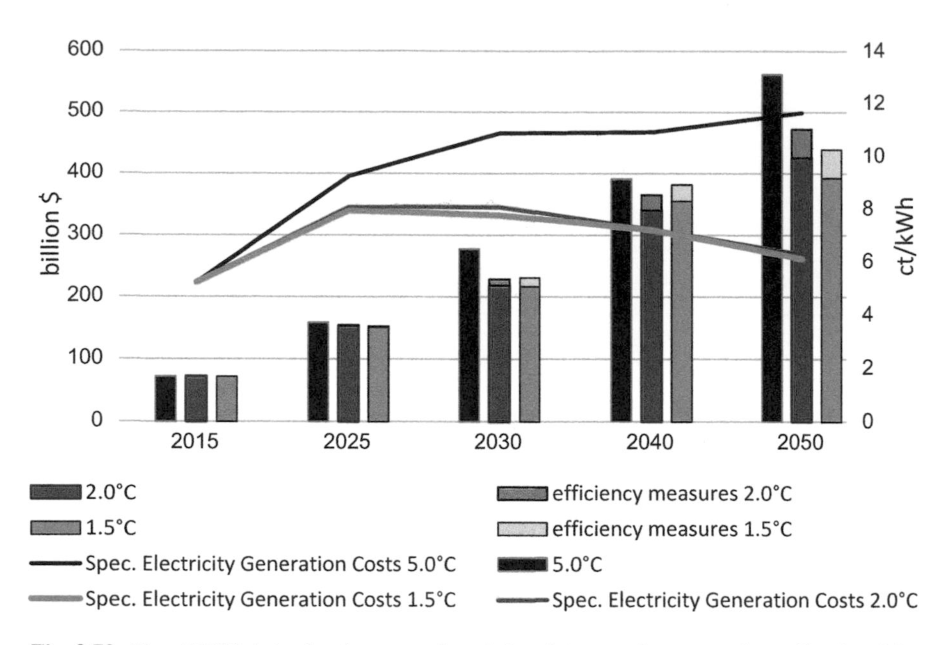

Fig. 8.73 Non-OECD Asia: development of total electricity supply costs and specific electricity generation costs in the scenarios

Scenario, they will increase to 7.9 ct/kWh, and drop to 6.1 ct/kWh by 2050. In both alternative scenarios, the generation costs in 2050 will be around 5.5 ct/kWh lower than in the 5.0 °C case. Note that these estimates of generation costs do not take into account integration costs such as power grid expansion, storage, or other load-balancing measures.

In the 5.0 °C case, the growth in demand and increasing fossil fuel prices will cause the total electricity supply costs to rise from today's $70 billion/year to more than $560 billion/year in 2050. In the 2.0 °C Scenario, the total supply costs will be $430 billion/year and in the 1.5 °C Scenario they will be $390 billion/year. The long-term costs for electricity supply will be more than 24% lower in the 2.0 °C Scenario than in the 5.0 °C Scenario as a result of the estimated generation costs and the electrification of heating and mobility. Further reductions in demand in the 1.5 °C Scenario will result in total power generation costs that are 30% lower than in the 5.0 °C case.

Compared with these results, the generation costs when the CO_2 emission costs are not considered will increase in the 5.0 °C case to 7.4 ct/kWh. In the 2.0 °C Scenario, they still increase until 2030, when they reach 6.5 ct/kWh, and then drop to 6.3 ct/kWh by 2050. In the 1.5 °C Scenario, they will increase to 6.9 ct/kWh and then drop to 6.1 ct/kWh by 2050. In the 2.0 °C case, the generation costs will be maximum in 2050, and 1.1 ct/kWh lower than in the 5.0 °C, whereas they will be 1.3 ct/kWh in the 1.5 °C Scenario. If the CO_2 costs are not considered, the total electricity supply costs in the 5.0 °C case will increase to about $360 billion/year in 2050.

8.11.1.4 Non-OECD Asia: Future Investments in the Power Sector

An investment of \$4030 billion will be required for power generation between 2015 and 2050 in the 2.0 °C Scenario—including investment in additional power plants for the production of hydrogen and synthetic fuels and investments in plant replacement at the end of their economic lifetimes. This value is equivalent to approximately \$112 billion per year on average, and is \$2660 billion more than in the 5.0 °C case (\$1370 billion). An investment of around \$3950 billion for power generation will be required between 2015 and 2050 in the 1.5 °C Scenario. On average, this is an investment of \$110 billion per year. In the 5.0 °C Scenario, the investment in conventional power plants will be around 55% of the total cumulative investments, whereas approximately 45% will be invested in renewable power generation and co-generation (Fig. 8.74).

However, in the 2.0 °C (1.5 °C) Scenario, Non-OECD Asia will shift almost 93% (95%) of its entire investment to renewables and co-generation. By 2030, the fossil fuel share of power sector investment will predominantly focus on gas power plants that can also be operated with hydrogen.

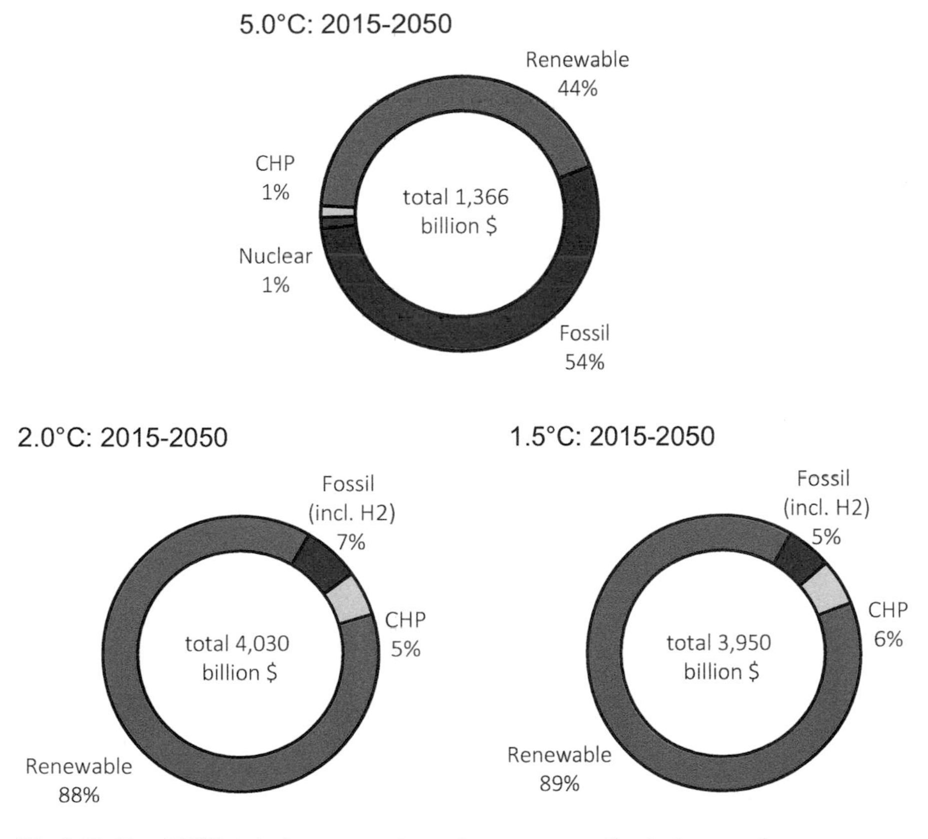

Fig. 8.74 Non-OECD Asia: investment shares for power generation in the scenarios

Because renewable energy has no fuel costs, other than biomass, the cumulative fuel cost savings in the 2.0 °C Scenario will reach a total of $2610 billion in 2050, equivalent to $73 billion per year. Therefore, the total fuel cost savings will be equivalent to 98% of the total additional investments compared to the 5.0 °C Scenario. The fuel cost savings in the 1.5 °C Scenario will add up to $2770 billion, or $77 billion per year.

8.11.1.5 Non-OECD Asia: Energy Supply for Heating

The final energy demand for heating will increase by 103% in the 5.0 °C scenario, from 10,800 PJ/year in 2015 to 21,900 PJ/year in 2050. Energy efficiency measures will help to reduce the energy demand for heating by 32% by 2050 in the 2.0 °C Scenario, relative to the 5.0 °C case, and by 37% in the 1.5 °C Scenario. Today, renewables supply around 43% of Non-OECD Asia's final energy demand for heating, with the main contribution from biomass. Renewable energy will provide 57% of Non-OECD Asia's total heat demand in 2030 in the 2.0 °C Scenario and 70% in the 1.5 °C Scenario. In both scenarios, renewables will provide 100% of the total heat demand in 2050.

Figure 8.75 shows the development of different technologies for heating in Non-OECD Asia over time, and Table 8.69 provides the resulting renewable heat supply for all scenarios. Up to 2030, biomass remains the main contributor. In the long term, the growing use of solar, geothermal, and environmental heat will lead to a biomass share of 40% in the 2.0 °C Scenario and 38% in the 1.5 °C Scenario. The heat from renewable hydrogen will further reduce the dependence on fossil fuels in both scenarios. The hydrogen consumption in 2050 will be around 900 PJ/year in the 2.0 °C Scenario and 1300 PJ/year in the 1.5 °C Scenario. The direct use of electricity for heating will also increase by a factor of 5-5.7 between 2015 and 2050. Energy for heating will have a final energy share of 34% in 2050 in the 2.0 °C Scenario and 32% in the 1.5 °C Scenario.

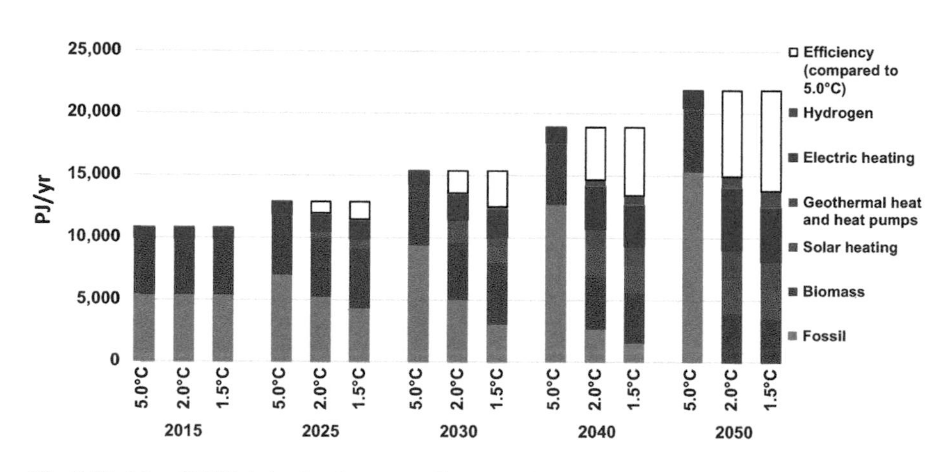

Fig. 8.75 Non-OECD Asia: development of heat supply by energy carrier in the scenarios

Table 8.69 Non-OECD Asia: development of renewable heat supply in the scenarios (excluding the direct use of electricity)

in PJ/year	Case	2015	2025	2030	2040	2050
Biomass	5.0 °C	4459	4800	4787	4878	4919
	2.0 °C	4459	4680	4529	4232	3948
	1.5 °C	4459	4772	4890	4054	3549
Solar heating	5.0 °C	4	12	33	70	128
	2.0 °C	4	401	1129	2252	2723
	1.5 °C	4	509	1221	2141	2389
Geothermal heat and heat pumps	5.0 °C	0	0	0	0	0
	2.0 °C	0	141	740	1563	2410
	1.5 °C	0	262	839	1587	2198
Hydrogen	5.0 °C	0	0	0	0	0
	2.0 °C	0	0	0	454	862
	1.5 °C	0	0	133	735	1274
Total	5.0 °C	4464	4811	4821	4948	5047
	2.0 °C	4464	5222	6398	8501	9942
	1.5 °C	4464	5542	7083	8516	9411

8.11.1.6 Non-OECD Asia: Future Investments in the Heating Sector

The roughly estimated investments in renewable heating technologies up to 2050 will amount to around $1120 billion in the 2.0 °C Scenario (including investments for the replacement of plants after their economic lifetimes), or approximately $31 billion per year. The largest share of investment in Non-OECD Asia is assumed to be for solar collectors (around $480 billion), followed by heat pumps and geothermal heat use. The 1.5 °C Scenario assumes an even faster expansion of renewable technologies. However, the lower heat demand (compared with the 2.0 °C Scenario) will results in a lower average annual investment of around $28 billion per year (Table 8.70, Fig. 8.76).

8.11.1.7 Non-OECD Asia: Transport

The energy demand in the transport sector in Non-OECD Asia is expected to increase in 2015 in the 5.0 °C Scenario from around 6500 PJ/year by 102% to 13,200 PJ/year in 2050. In the 2.0 °C Scenario, assumed technical, structural, and behavioural changes will save 63% (8320 PJ/year) by 2050 compared to the 5.0 °C Scenario. Additional modal shifts, technology switches, and a reduction in the transport demand will lead to even higher energy savings in the 1.5 °C Scenario of 73% (or 9660 PJ/year) by 2050 compared to the 5.0 °C case (Table 8.71, Fig. 8.77).

By 2030, electricity will provide 6% (120 TWh/year) of the transport sector's total energy demand in the 2.0 °C Scenario, whereas in 2050, the share will be 36% (480 TWh/year). In 2050, up to 650 PJ/year of hydrogen will be used in the trans-

Table 8.70 Non-OECD Asia: installed capacities for renewable heat generation in the scenarios

	Case	2015	2025	2030	2040	2050
Biomass	5.0 °C	1886	1925	1767	1610	1459
	2.0 °C	1886	1850	1557	1150	821
	1.5 °C	1886	1829	1693	1084	713
Geothermal	5.0 °C	0	0	0	0	0
	2.0 °C	0	4	18	51	73
	1.5 °C	0	4	15	44	64
Solar heating	5.0 °C	1	3	10	20	37
	2.0 °C	1	114	321	639	772
	1.5 °C	1	145	349	609	678
Heat pumps	5.0 °C	0	0	0	0	0
	2.0 °C	0	13	58	103	159
	1.5 °C	0	27	70	110	144
Total[a]	5.0 °C	1888	1928	1777	1631	1496
	2.0 °C	1888	1981	1954	1944	1825
	1.5 °C	1888	2004	2127	1847	1598

[a] Excluding direct electric heating

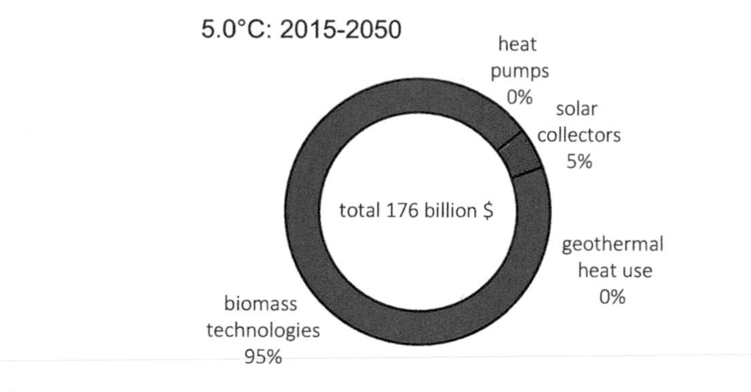

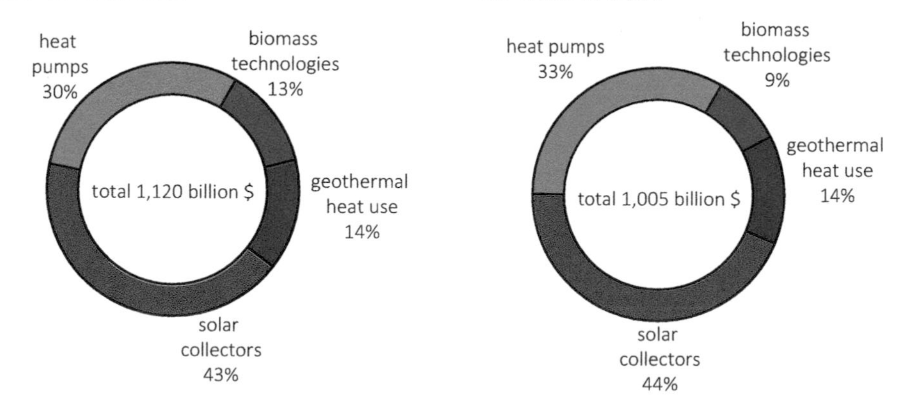

Fig. 8.76 Non-OECD Asia: development of investments for renewable heat-generation technologies in the scenarios

Table 8.71 Non-OECD Asia: projection of transport energy demand by mode in the scenarios

in PJ/year	Case	2015	2025	2030	2040	2050
Rail	5.0 °C	76	81	81	83	83
	2.0 °C	76	96	116	158	183
	1.5 °C	76	115	124	148	212
Road	5.0 °C	6023	7139	9256	11,061	12,181
	2.0 °C	6023	6694	6489	5251	4245
	1.5 °C	6023	5493	4217	3258	2903
Domestic aviation	5.0 °C	225	353	447	581	621
	2.0 °C	225	240	220	180	143
	1.5 °C	225	230	200	139	108
Domestic navigation	5.0 °C	196	216	227	246	267
	2.0 °C	196	216	227	246	267
	1.5 °C	196	216	227	246	267
Total	5.0 °C	6521	7789	10,010	11,970	13,153
	2.0 °C	6521	7246	7051	5834	4838
	1.5 °C	6521	6053	4769	3791	3489

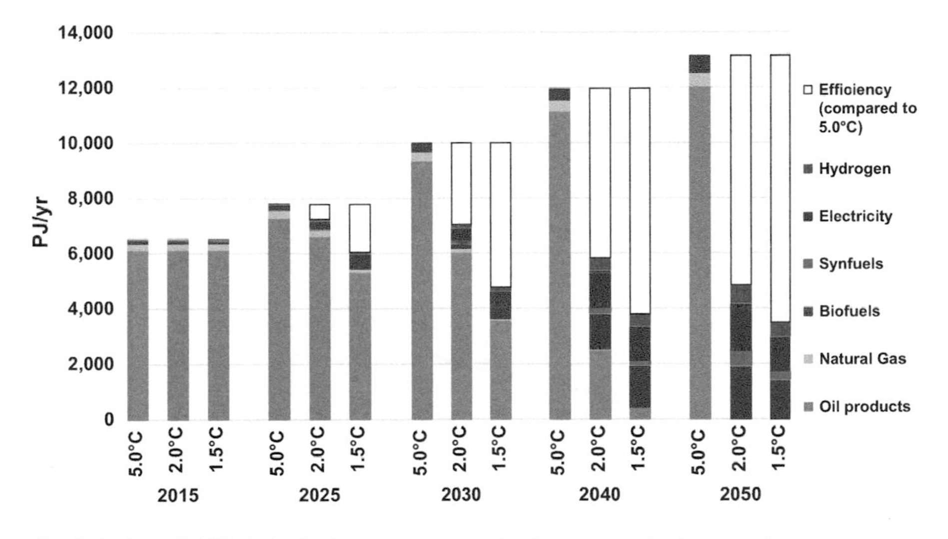

Fig. 8.77 Non-OECD Asia: final energy consumption by transport in the scenarios

port sector as a complementary renewable option. In the 1.5 °C Scenario, the annual electricity demand will be 350 TWh in 2050. The 1.5 °C Scenario also assumes a hydrogen demand of 500 PJ/year by 2050.

Biofuel use is limited in the 2.0 °C Scenario to a maximum of 1940 PJ/year Therefore, around 2030, synthetic fuels based on power-to-liquid will be introduced, with a maximum amount of 530 PJ/year in 2050. Due to the lower overall energy demand in transport, biofuel use will be reduced in the 1.5 °C Scenario to a maximum of 1540 PJ/year. The maximum synthetic fuel demand will amount to 280 PJ/year.

8.11.1.8 Non-OECD Asia: Development of CO_2 Emissions

In the 5.0 °C Scenario, Non-OECD Asia's annual CO_2 emissions will increase by 160%, from 1880 Mt. in 2015 to 4880 Mt. in 2050. The stringent mitigation measures in both alternative scenarios will cause the annual emissions to fall to 630 Mt. in 2040 in the 2.0 °C Scenario and to 330 Mt. in the 1.5 °C Scenario, with further reductions to almost zero by 2050. In the 5.0 °C case, the cumulative CO_2 emissions from 2015 until 2050 will add up to 121 Gt. In contrast, in the 2.0 °C and 1.5 °C Scenarios, the cumulative emissions for the period from 2015 until 2050 will be 42 Gt and 32 Gt, respectively.

Therefore, the cumulative CO_2 emissions will decrease by 65% in the 2.0 °C Scenario and by 74% in the 1.5 °C Scenario compared with the 5.0 °C case. A rapid reduction in annual emissions will occur in both alternative scenarios. In the 2.0 °C Scenario, this reduction will be greatest in 'Power generation', followed by the 'Residential and other' and 'Industry' sectors (Fig. 8.78).

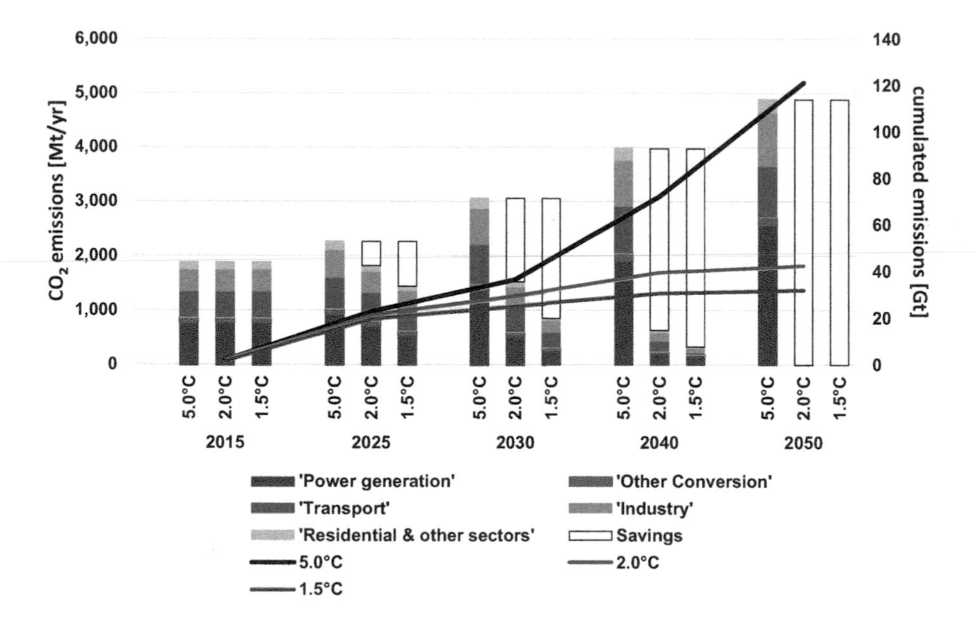

Fig. 8.78 Non-OECD Asia: development of CO_2 emissions by sector and cumulative CO_2 emissions (after 2015) in the scenarios ('Savings' = reduction compared with the 5.0 °C Scenario)

8.11.1.9 Non-OECD Asia: Primary Energy Consumption

The levels of primary energy consumption in the three scenarios when the assumptions discussed above are taken into account are shown in Fig. 8.79. In the 2.0 °C Scenario, the primary energy demand will increase by 13%, from around 38,100 PJ/year in 2015 to 43,200 PJ/year. Compared with the 5.0 °C Scenario, the overall primary energy demand will decrease by 47% by 2050 in the 2.0 °C Scenario (5.0 °C: 81600 PJ in 2050). In the 1.5 °C Scenario, the primary energy demand will be even lower (39,300 PJ in 2050) because the final energy demand and conversion losses will be lower.

Both the 2.0 °C Scenario and 1.5 °C Scenario aim to rapidly phase-out coal and oil. This will cause renewable energy to have a primary energy share of 40% in 2030 and 93% in 2050 in the 2.0 °C Scenario. In the 1.5 °C Scenario, renewables will have a primary energy share of more than 92% in 2050 (including non-energy consumption, which will still include fossil fuels). Nuclear energy will be phased out by 2045 in both the 2.0 °C Scenario and 1.5 °C Scenario. The cumulative primary energy consumption of natural gas in the 5.0 °C case will add up to 430 EJ, the cumulative coal consumption to about 530 EJ, and the crude oil consumption to 580

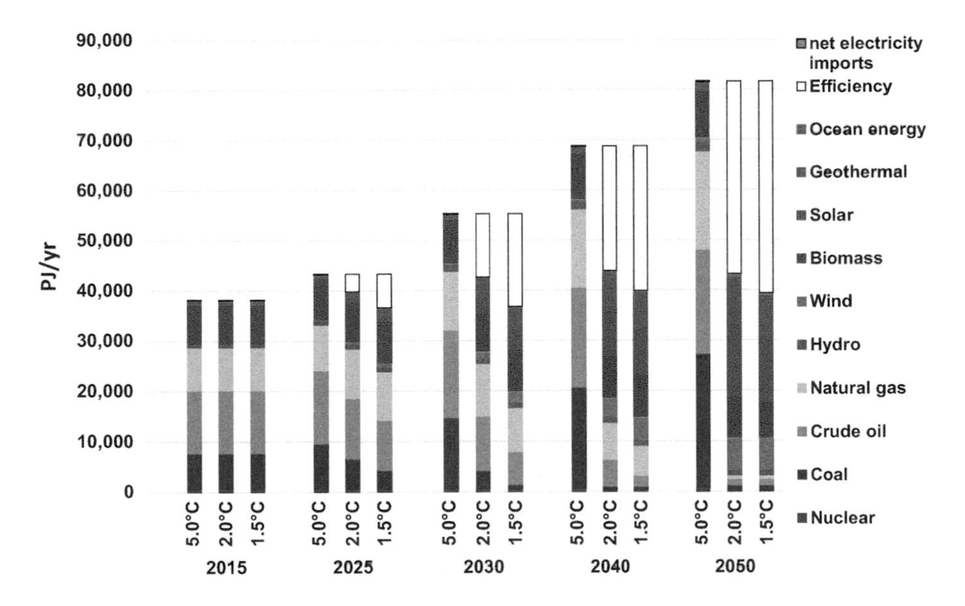

Fig. 8.79 Non-OECD Asia: projection of total primary energy demand (PED) by energy carrier in the scenarios (including electricity import balance)

EJ. In contrast, in the 2.0 °C Scenario, the cumulative gas demand will amount to 260 EJ, the cumulative coal demand to 120 EJ, and the cumulative oil demand to 270 EJ. Even lower fossil fuel use will be achieved in the 1.5 °C Scenario: 230 EJ for natural gas, 70 EJ for coal, and 190 EJ for oil.

8.11.2 Non-OECD Asia: Power Sector Analysis

Non-OECD Asia is the most heterogeneous region of all IEA world energy regions because it includes not only all the ASEAN countries (ASEAN 2018) of South East Asia, but also central and south Asian nations, as well all 16 Pacific Island states. As for the Caribbean Islands, a power system assessment—especially with regard to possible storage demand—that examines all Pacific Island states together rather than individually, is not sufficient to provide the actual required storage demand. However, with this is in mind, the ratio of solar PV generation to storage requirements does provide some indication. A specific assessment for each of the Pacific Island states is required, but is beyond the scope of this study. Indonesia and the Philippines are selected as sub-regions because they are island states with some interconnection between islands.

8.11.2.1 Non-OECD Asia: Development of Power Plant Capacities

Non-OECD Asia's renewable power market can be subdivided into the following categories: technologies for small and medium islands (mainly solar PV–battery systems, mini-hydro and small-scale bioenergy systems); and utility-scale solar and onshore wind for all major economies in mainland Asia or on the large islands of the Philippines and Indonesia. Several countries in this region are on the Pacific Ring of Fire and have significant geothermal energy resources. The annual market for geothermal power plants is one of the world's largest, with a projected 3–4 GW each year for almost two decades between 2025 and 2045 in both scenarios (Table 8.72).

8.11.2.2 Non-OECD Asia: Utilization of Power-Generation Capacities

Due to the geographic diversity and wide distribution of all sub-regions of the Non-OECD Asia region, it is assumed that there are no interconnection capacities available, and that there will not be any at the end of the modelling period (Table 8.73).

Table 8.72 Non-OECD Asia: average annual change in installed power plant capacity

Non-OECD-Asia power generation: average annual change of installed capacity [GW/a]	2015–2025		2026–2035		2036–2050	
	2.0 °C	1.5 °C	2.0 °C	1.5 °C	2.0 °C	1.5 °C
Hard coal	2	−6	−7	−4	−1	0
Lignite	−2	−4	−1	−2	0	0
Gas	4	10	19	14	−26	−22
Hydrogen-gas	0	1	0	6	33	24
Oil/diesel	0	−5	−4	−5	−1	0
Nuclear	0	0	0	0	0	0
Biomass	2	1	1	1	1	1
Hydro	3	2	0	0	0	0
Wind (onshore)	4	21	20	24	26	20
Wind (offshore)	3	7	6	7	5	4
PV (roof top)	10	36	40	47	50	37
PV (utility scale)	3	12	13	16	17	12
Geothermal	0	3	4	4	2	1
Solar thermal power plants	1	6	9	8	17	13
Ocean energy	0	0	1	1	3	2
Renewable fuel based co-generation	**1**	**2**	**1**	**1**	**1**	**1**

In both scenarios, variable power generation will jump from only 1% today to around 25% in all sub-regions, whereas dispatchable renewables will remain stable at around 25%–30% until 2050.

Compared with other world regions, the capacity factors for limited dispatchable fossil and nuclear energy will remain relatively high until 2030, as shown in Table 8.74. The time required for variable power generation to replace fossil and nuclear generation will be greater than it is in other regions. In the 1.5 °C Scenario, all coal capacities across the region will be phased out by 2030, except for 4 GW (equivalent to 4–5 power plants), which will be off-line 5 years later.

8.11.2.3 Non-OECD Asia: Development of Load, Generation, and Residual Load

Because both scenarios were calculated under the assumption that there are no interconnection capacities at the sub-regional level, more dispatch capacity will be deployed. Table 8.75 shows that only Asia North-West and Asia South-West will require some interconnection to avoid curtailment. The development of the maximum load, generation, and the resulting residual load—the load remaining after

Table 8.73 Non-OECD Asia: power system shares by technology group

Power generation structure and interconnection		2.0 °C				1.5 °C			
		Variable RE	Dispatch RE	Dispatch Fossil	Inter-connection	Variable RE	Dispatch RE	Dispatch Fossil	Inter-connection
Asia West: Pakistan, Afghanistan, Nepal, Bhutan	2015	1%	35%	63%	0%				
	2030	31%	31%	38%	0%	44%	29%	28%	0%
	2050	62%	25%	13%	0%	64%	24%	12%	0%
Sri Lanka	2015	1%	35%	64%	0%				
	2030	30%	37%	33%	0%	41%	34%	25%	0%
	2050	58%	27%	15%	0%	59%	26%	14%	0%
Pacific Island State	2015	1%	35%	64%	0%				
	2030	29%	34%	37%	0%	39%	30%	30%	0%
	2050	55%	25%	20%	0%	55%	25%	20%	0%
Asia North West: Bangladesh, Myanmar, Thailand	2015	1%	35%	64%	0%				
	2030	23%	37%	40%	0%	33%	35%	32%	0%
	2050	48%	31%	21%	0%	50%	30%	20%	0%
Asia Central North: Viet Nam, Laos and Cambodia	2015	1%	35%	64%	0%				
	2030	27%	36%	36%	0%	38%	33%	29%	0%
	2050	53%	28%	20%	0%	56%	27%	17%	0%
Asia South West: Malaysia, Brunei	2015	1%	35%	64%	0%				
	2030	26%	40%	34%	0%	36%	37%	27%	0%
	2050	52%	29%	19%	0%	57%	28%	15%	0%

Indonesia	2015	1%	35%	64%	0%				
	2030	21%	34%	45%	0%	31%	35%	35%	0%
	2050	47%	30%	23%	0%	48%	30%	22%	0%
Philippines	2015	1%	35%	64%	0%				
	2030	34%	34%	32%	0%	48%	30%	22%	0%
	2050	63%	23%	13%	0%	65%	22%	13%	0%
Non-OECD Asia	2015	1%	35%	64%					
	2030	26%	35%	39%		36%	34%	30%	
	2050	52%	28%	19%		55%	28%	17%	

Table 8.74 Non-OECD Asia: capacity factors by generation type

Utilization of variable and dispatchable power generation:		2015	2020	2020	2030	2030	2040	2040	2050	2050
Non-OECD Asia			2.0 °C	1.5 °C	2.0 °C	1.5 °C	2.0 °C	1.5 °C	2.0 °C	1.5 °C
Capacity factor – average	[%/yr]	**55.4%**	**52%**	**53%**	**45%**	**42%**	**33%**	**33%**	**34%**	**32%**
Limited dispatchable: fossil and nuclear	[%/yr]	71.4%	52%	53%	44%	33%	31%	13%	25%	0%
Limited dispatchable: renewable	[%/yr]	40.5%	61%	61%	59%	56%	58%	53%	45%	49%
Dispatchable: fossil	[%/yr]	50.2%	32%	33%	23%	27%	37%	13%	28%	12%
Dispatchable: renewable	[%/yr]	34.4%	75%	75%	74%	69%	41%	58%	53%	51%
Variable: renewable	[%/yr]	13.1%	19%	19%	36%	35%	26%	31%	30%	29%

variable renewable generation. According to the Philippine Department of Energy, the peak demand in the Philippines in 2016 was 13.3 GW (PR-DoE 2016) (9.7 GW in Luzon, 1.9 GW in the Visayas, and 1.7 GW in Mindanao). The calculated load for the Philippines in 2020 was 16.3 GW, which seems realistic. The load will increase to 75.5 GW by 2050 under the 2.0 °C Scenario. The results for all Asian regions show a quadrupling of load by 2050.

The lack of interconnection potential between or even within most sub-regions will lead to some curtailment.

Table 8.76 shows that whereas countries on the Asian mainland will use and increase their capacity for hydro pump storage electricity, batteries will be used for most of the storage requirements of islands and island states. Where available, gas infrastructure must be converted to hydrogen-operated systems.

Table 8.75 Non-OECD Asia: load, generation, and residual load development—2.0 °C Scenario

Power generation structure		2.0 °C				1.5 °C			
		Max demand [GW]	Max generation [GW]	Max residual load [GW]	Max interconnection requirements [GW]	Max demand [GW]	Max generation [GW]	Max residual load [GW]	Max interconnection requirements [GW]
Asia West: Pakistan, Afghanistan, Nepal, Bhutan	2020	38.1	22.4	17.4		38.1	22.3	17.5	
	2030	65.1	58.2	44.4	0	67.1	64.2	47.9	0
	2050	145.6	194.0	117.5	0	137.5	185.6	112.2	0
Sri Lanka	2020	5.6	2.7	3.2		5.6	2.7	3.2	
	2030	9.0	8.6	6.3	0	9.2	10.5	6.5	0
	2050	19.7	31.1	15.2	0	18.2	29.7	14.1	0
Pacific Island State	2020	1.6	1.0	0.6		1.6	1.0	0.6	
	2030	2.6	2.3	1.8	0	2.7	2.6	1.9	0
	2050	5.6	8.2	4.2	0	5.5	7.9	4.2	0
Asia North West: Bangladesh, Myanmar, Thailand	2020	57.8	18.9	41.8		57.8	18.8	41.9	
	2030	97.4	88.8	67.1	0	99.6	101.2	71.5	0
	2050	218.9	321.0	171.4	0	198.1	306.3	155.8	0
Asia Central North: Viet Nam, Laos and Cambodia	2020	29.4	26.3	3.5		29.4	26.2	3.6	
	2030	47.0	44.4	29.8	0	47.9	61.2	32.2	0
	2050	109.6	191.0	83.1	0	93.7	182.6	70.2	19
Asia South West: Malaysia, Brunei	2020	38.2	16.0	25.0		38.2	15.1	25.5	
	2030	53.6	54.0	28.1	0	53.9	68.4	34.0	0
	2050	121.0	216.7	89.1	7	99.2	206.9	71.0	37

(continued)

Table 8.75 (continued)

Power generation structure		2.0 °C				1.5 °C			
		Max demand	Max generation	Max residual load	Max interconnection requirements	Max demand	Max generation	Max residual load	Max interconnection requirements
		[GW]	[GW]	[GW]	[GW]	[GW]	[GW]	[GW]	[GW]
Indonesia	2020	60.9	34.3	26.6		60.9	33.0	27.9	
	2030	106.7	99.8	59.9	0	108.7	114.5	77.4	0
	2050	239.4	363.8	188.3	0	218.6	348.2	173.2	0
Philippines	2020	16.3	13.7	3.9					
	2030	33.5	33.0	19.0	0	34.3	42.6	24.1	0
	2050	75.5	133.1	58.8	0	70.3	127.3	55.5	2

Table 8.76 Non-OECD Asia: storage and dispatch service requirements

Storage and dispatch		2.0 °C					1.5 °C				
		Required to avoid curtailment	Utilization battery -through-put-	Utilization PSH -through-put-	Total storage demand (incl. H2)	Dispatch hydrogen-based	Required to avoid curtailment	Utilization battery -through-put-	Utilization PSH -through-put-	Total storage demand (incl. H2)	Dispatch hydrogen-based
Non-OECD Asia		[GWh/year]	[GWh/year]	[GWh/year]	[GWh/year]	[GWh/year]	[GWh/year]	[GWh/year]	[GWh/year]	[GWh/year]	[GWh/year]
Asia West: Pakistan, Afghanistan, Nepal, Bhutan	2020	0	0	0	0	0	0	0	0	0	0
	2030	0	0	0	0	0	434	4	78	82	3356
	2050	36,251	767	716	1483	42,533	37,649	407	774	1181	44,157
Sri Lanka	2020	0	0	0	0	0	0	0	0	0	0
	2030	0	0	0	0	0	72	1	9	10	564
	2050	4755	135	125	260	7380	5471	74	144	218	7330
Pacific Island State	2020	0	0	0	0	0	0	0	0	0	0
	2030	12	0	2	2	0	183	1	14	14	142
	2050	2178	44	43	87	2101	1932	22	42	65	2211
Asia North West: Bangladesh, Myanmar, Thailand	2020	0	0	0	0	0	0	0	0	0	0
	2030	0	0	0	0	0	194	1	27	28	6617
	2050	19,992	1114	824	1938	93,720	29,141	657	1113	1770	92,309
Asia Central North: Viet Nam, Laos and Cambodia	2020	0	0	0	0	0	0	0	0	0	0
	2030	6	0	3	4	0	1031	5	121	126	3346
	2050	26,401	727	708	1435	49,483	40,048	416	919	1335	45,848

(continued)

Table 8.76 (continued)

Storage and dispatch		2.0 °C					1.5 °C				
		Required to avoid curtailment	Utilization battery -through-put-	Utilization PSH -through-put-	Total storage demand (incl. H2)	Dispatch hydrogen-based	Required to avoid curtailment	Utilization battery -through-put-	Utilization PSH -through-put-	Total storage demand (incl. H2)	Dispatch hydrogen-based
Non-OECD Asia		[GWh/year]	[GWh/year]	[GWh/year]	[GWh/ year]	[GWh/ year]	[GWh/ year]	[GWh/year]	[GWh/year]	[GWh/ year]	[GWh/year]
Asia South West: Malaysia, Brunei	2020	0	0	0	0	0	0	0	0	0	0
	2030	7	0	2	3	0	1036	5	120	125	4151
	2050	32,422	942	893	1835	59,371	55,862	610	1406	2016	51,750
Indonesia	2020	0	0	0	0	0	0	0	0	0	0
	2030	0	0	0	0	0	176	1	21	22	7391
	2050	11,890	720	530	1250	107,913	17,040	478	717	1195	107,330
Philippines	2020	0	0	0	0	0	0	0	0	0	0
	2030	112	3	22	25	0	3723	6	232	239	1917
	2050	38,084	507	670	1177	23,954	41,017	266	743	1009	24,126
Other Asia	2020	0	0	0	0	0	0	0	0	0	0
	2030	137	4	30	34	0	6848	23	622	646	27,484
	2050	171,973	4955	4510	9465	386,454	228,160	2930	5859	8789	375,061

8.12 India

8.12.1 India: Long-Term Energy Pathways

8.12.1.1 India: Final Energy Demand by Sector

The future development pathways for India's final energy demand when the assumptions on population growth, GDP growth, and energy intensity are combined are shown in Fig. 8.80 for the 5.0 °C, 2.0 °C, and 1.5 °C Scenarios. In the 5.0 °C Scenario, the total final energy demand will increase by 201% from the current 22,200 PJ/year to 66,800 PJ/year by 2050. In the 2.0 °C Scenario, the final energy demand will increase at a much slower rate by 57% compared with current consumption and will reach 34,900 PJ/year by 2050. The final energy demand in the 1.5 °C Scenario will reach 31,900 PJ, 44% above the 2015 level. In the 1.5 °C Scenario, the final energy demand in 2050 will be 9% lower than in the 2.0 °C Scenario. The electricity demand for 'classical' electrical devices (without power-to-heat or e-mobility) will increase from 750 TWh/year in 2015 to 3200 TWh/year in 2050 in both alternative scenarios. Compared with the 5.0 °C case (4720 TWh/year in 2050), efficiency measures in the 2.0 °C and 1.5 °C Scenarios will save around 1520 TWh/year by 2050.

Electrification will lead to a significant increase in the electricity demand by 2050. In the 2.0 °C Scenario, the electricity demand for heating will be approximately 1900 TWh/year due to electric heaters and heat pumps, and in the transport sector,

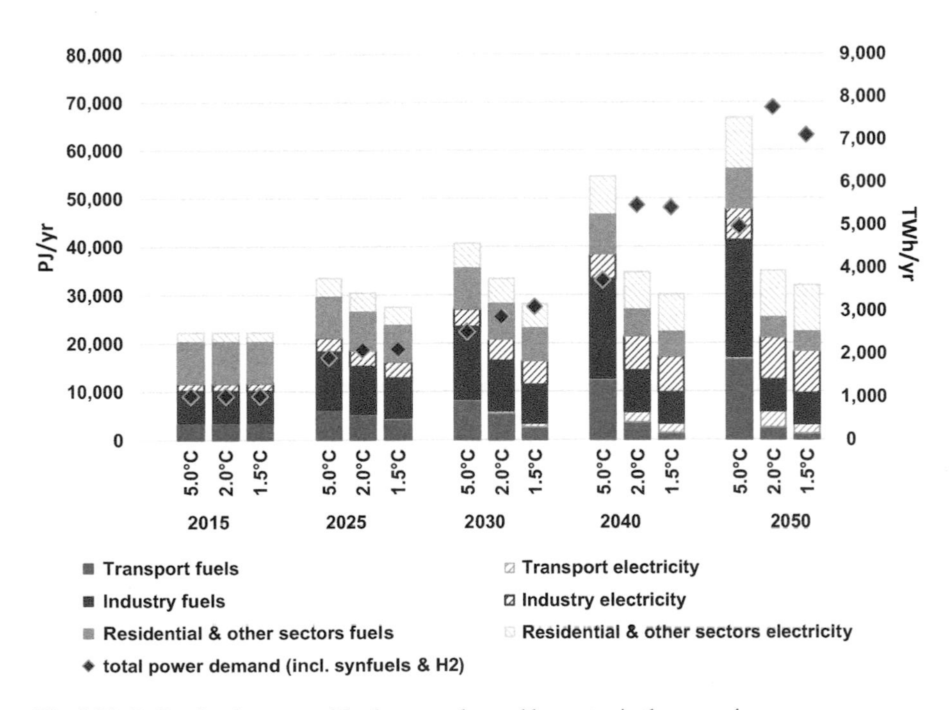

Fig. 8.80 India: development of final energy demand by sector in the scenarios

the electricity demand will be approximately 3400 TWh/year due to electric mobility. The generation of hydrogen (for transport and high-temperature process heat) and the manufacture of synthetic fuels (mainly for transport) will add an additional power demand of 1700 TWh/year. Therefore, the gross power demand will increase from 1400 TWh/year in 2015 to 8400 TWh/year in 2050 in the 2.0 °C Scenario, 31% higher than in the 5.0 °C case. In the 1.5 °C Scenario, the gross electricity demand will increases to a maximum of 7700 TWh/year in 2050.

Efficiency gains in the heating sector could be even larger than in the electricity sector. In the 2.0 °C and 1.5 °C Scenarios, a final energy consumption equivalent to about 9500 PJ/year and 9800 PJ/year, respectively, will be avoided through efficiency gains by 2050 compared with the 5.0 °C Scenario.

8.12.1.2 India: Electricity Generation

The development of the power system is characterized by a dynamically growing renewable energy market and an increasing proportion of total power from renewable sources. By 2050, 100% of the electricity produced in India will come from renewable energy sources in the 2.0 °C Scenario. 'New' renewables—mainly wind, solar, and geothermal energy—will contribute 90% of the total electricity generation. Renewable electricity's share of the total production will be 66% by 2030 and 89% by 2040. The installed capacity of renewables will reach about 1060 GW by 2030 and 3360 GW by 2050. The share of renewable electricity generation in 2030 in the 1.5 °C Scenario is assumed to be 77%. In the 1.5 °C Scenario, the generation capacity from renewable energy will be approximately 3040 GW in 2050.

Table 8.77 shows the development of different renewable technologies in India over time. Figure 8.81 provides an overview of the overall power-generation structure in India. From 2020 onwards, the continuing growth of wind and PV up to 1270 GW and 1570 GW, respectively, is complemented by up to 210 GW solar thermal generation, as well as limited biomass, geothermal, and ocean energy, in the 2.0 °C Scenario. Both the 2.0 °C Scenario and 1.5 °C Scenario will lead to a high proportion of variable power generation (PV, wind, and ocean) of 48% and 60%, respectively, by 2030, and 75% and 72%, respectively, by 2050.

8.12.1.3 India: Future Costs of Electricity Generation

Figure 8.82 shows the development of the electricity-generation and supply costs over time, including the CO_2 emission costs, in all scenarios. The calculated electricity generation costs in 2015 (referring to full costs) were around 5.4 ct/kWh. In the 5.0 °C case, the generation costs will increase until 2040, when they reach 11 ct/kWh, and then drop to 10.7 ct/kWh by 2050. The generation costs will increase in the 2.0 °C Scenario until 2030, when they reach 8.4 ct/kWh, and then drop to 5.7 ct/kWh by 2050. In the 1.5 °C Scenario, they will increase to 7.8 ct/kWh, and then drop to 5.8 ct/kWh by 2050. In the 2.0 °C Scenario, the generation costs in 2050 will be 5 ct/kWh lower than in the 5.0 °C case. In the 1.5 °C Scenario, the generation

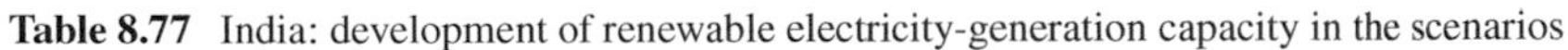

Table 8.77 India: development of renewable electricity-generation capacity in the scenarios

in GW	Case	2015	2025	2030	2040	2050
Hydro	5.0 °C	46	68	81	97	117
	2.0 °C	46	68	72	80	87
	1.5 °C	46	68	72	80	87
Biomass	5.0 °C	8	13	16	20	25
	2.0 °C	8	23	31	60	93
	1.5 °C	8	23	31	60	93
Wind	5.0 °C	25	82	119	185	246
	2.0 °C	25	200	421	938	1273
	1.5 °C	25	275	543	1002	1110
Geothermal	5.0 °C	0	0	0	0	0
	2.0 °C	0	3	8	42	68
	1.5 °C	0	3	8	42	68
PV	5.0 °C	5	115	198	345	545
	2.0 °C	5	230	469	1090	1572
	1.5 °C	5	365	648	1185	1412
CSP	5.0 °C	0	0	1	1	2
	2.0 °C	0	8	48	138	209
	1.5 °C	0	8	48	138	209
Ocean	5.0 °C	0	0	0	0	0
	2.0 °C	0	1	11	33	59
	1.5 °C	0	1	11	33	59
Total	5.0 °C	84	279	415	648	936
	2.0 °C	84	532	1061	2381	3360
	1.5 °C	84	742	1361	2540	3037

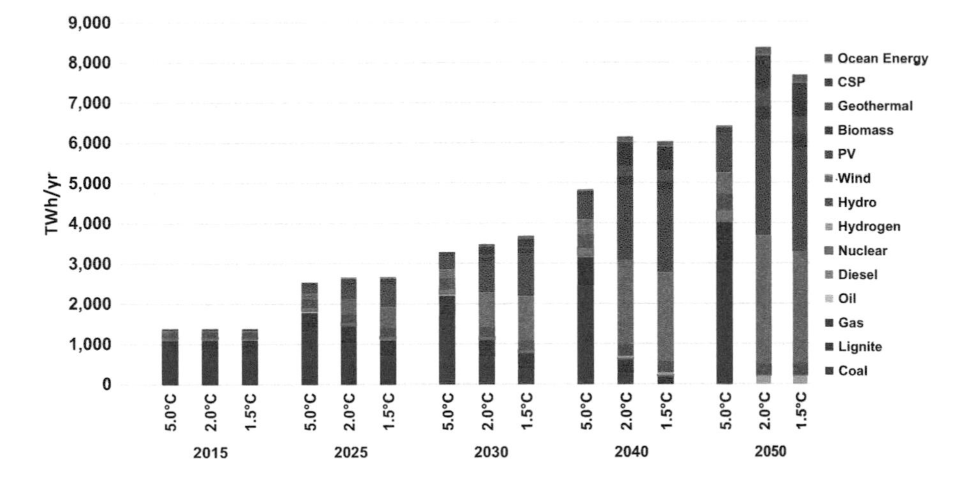

Fig. 8.81 India: development of electricity-generation structure in the scenarios

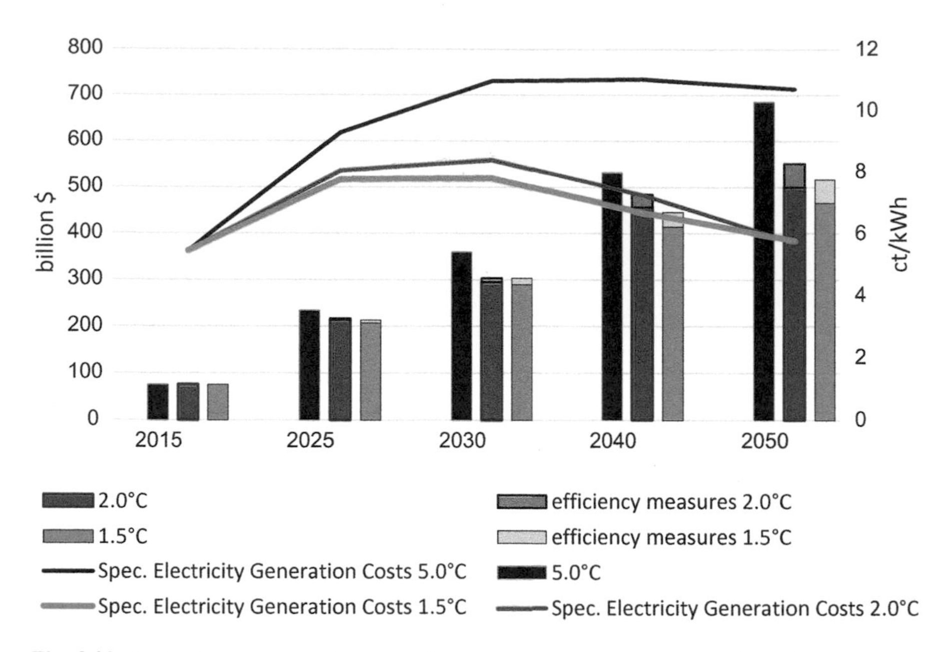

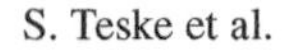

Fig. 8.82 India: development of total electricity supply costs and specific electricity generation costs in the scenarios

costs in 2050 will be 4.9 ct/kWh lower than in the 5.0 °C case. Note that these estimates of generation costs do not take into account integration costs such as power grid expansion, storage, or other load-balancing measures.

In the 5.0 °C case, the growth in demand and increasing fossil fuel prices will cause the total electricity supply costs to rise from today's $75 billion/year to more than $690 billion/year in 2050. In the 2.0 °C case, the total supply costs will be $500 billion/year and in the 1.5 °C Scenario, they will be $470 billion/year. The long-term costs for electricity supply will be more than 27% lower in the 2.0 °C Scenario than in the 5.0 °C Scenario as a result of the estimated generation costs and the electrification of heating and mobility. Further demand reductions in the 1.5 °C Scenario will result in total power generation costs that are 32% lower than in the 5.0 °C case.

Compared with these results, the generation costs, when the CO_2 emission costs are not considered, will increase in the 5.0 °C case to only 6.9 ct/kWh. In both alternative scenarios, they will still increase until 2030, when they reach 6.7 ct/kWh, and then drop to around 5.8 ct/kWh by 2050. The maximum difference in generation costs will be around 1 ct/kWh in 2050. If the CO_2 costs are not considered, the total electricity supply costs in the 5.0 °C case will rise to about $430 billion/year in 2050.

8.12.1.4 India: Future Investments in the Power Sector

An investment of around $5640 billion will be required for power generation between 2015 and 2050 in the 2.0 °C Scenario—including additional power plants for the production of hydrogen and synthetic fuels and investments in the replacement of plants after the end of their economic lifetimes. This value is equivalent to approximately $157

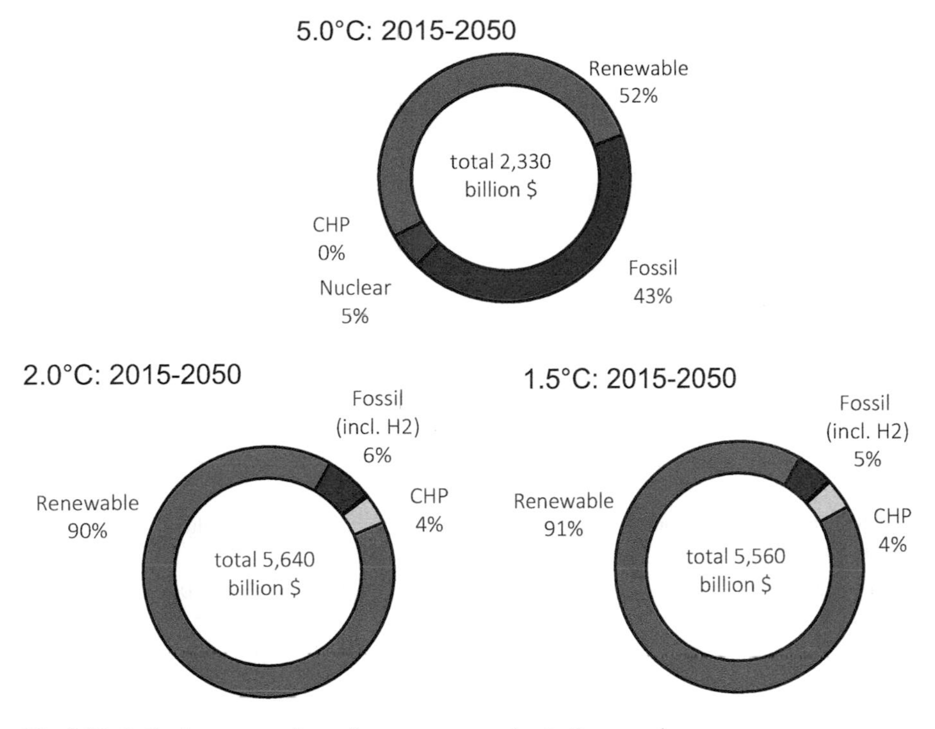

Fig. 8.83 India: investment shares for power generation in the scenarios

billion per year on average, and is \$3310 billion more than in the 5.0 °C case (\$2330 billion). An investment of around \$5560 billion for power generation will be required between 2015 and 2050 in the 1.5 °C Scenario. On average, this will be an investment of \$154 billion per year. In the 5.0 °C Scenario, the investment in conventional power plants will be around 48% of the total cumulative investments, whereas approximately 52% will be invested in renewable power generation and co-generation (Fig. 8.83).

However, in the 2.0 °C (1.5 °C) Scenario, India will shift almost 94% (95%) of its entire investment to renewables and co-generation. By 2030, the fossil fuel share of the power sector investment will predominantly focus on gas power plants that can also be operated with hydrogen.

Because renewable energy has no fuel costs, other than biomass, the cumulative fuel cost savings in the 2.0 °C Scenario will reach a total of \$3110 billion in 2050, equivalent to \$86 billion per year. Therefore, the total fuel cost savings will be equivalent to 90% of the total additional investments compared to the 5.0 °C Scenario. The fuel cost savings in the 1.5 °C Scenario will add up to \$3330 billion, or \$93 billion per year.

8.12.1.5 India: Energy Supply for Heating

The final energy demand for heating will increase in the 5.0 °C Scenario by 133%, from 11,900 PJ/year in 2015 to 27,800 PJ/year in 2050. Energy efficiency measures will help to reduce the energy demand for heating by 34% in 2050 in the 2.0 °C

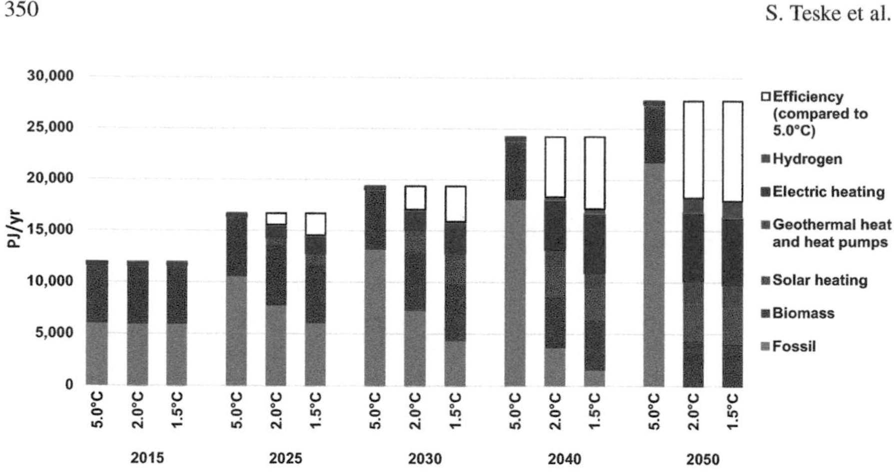

Fig. 8.84 India: development of heat supply by energy carrier in the scenarios

Table 8.78 India: development of renewable heat supply in the scenarios (excluding the direct use of electricity)

in PJ/year	Case	2015	2025	2030	2040	2050
Biomass	5.0 °C	5544	5633	5666	5595	5341
	2.0 °C	5544	5726	5600	4854	4366
	1.5 °C	5544	5600	5444	4758	4078
Solar heating	5.0 °C	28	77	115	200	310
	2.0 °C	28	589	1537	2964	3693
	1.5 °C	28	887	2271	3107	3626
Geothermal heat and heat pumps	5.0 °C	0	1	1	1	2
	2.0 °C	0	164	647	1627	2136
	1.5 °C	0	189	725	1497	2103
Hydrogen	5.0 °C	0	0	0	0	0
	2.0 °C	0	0	2	299	1409
	1.5 °C	0	0	2	437	1613
Total	5.0 °C	5572	5711	5781	5796	5653
	2.0 °C	5572	6478	7787	9743	11,603
	1.5 °C	5572	6675	8442	9800	11,420

Scenario, relative to the 5.0 °C case, and by 35% in the 1.5 °C Scenario. Today, renewables supply around 47% of India's final energy demand for heating, with the main contribution from biomass. Renewable energy will provide 53% of India's total heat demand in 2030 in the 2.0 °C Scenario and 68% in the 1.5 °C Scenario. In both scenarios, renewables will provide 100% of the total heat demand in 2050.

Figure 8.84 shows the development of different technologies for heating in India over time, and Table 8.78 provides the resulting renewable heat supply for all scenarios. Up to 2030, biomass will remain the main contributor. In the long term, the increasing use of solar, geothermal, and environmental heat will lead to a biomass share of 38% in the 2.0 °C Scenario and 36% in the 1.5 °C Scenario.

Heat from renewable hydrogen will further reduce the dependence on fossil fuels under both scenarios. Hydrogen consumption in 2050 will be around 1400 PJ/year in the 2.0 °C Scenario and 1600 PJ/year in the 1.5 °C Scenario. The direct use of electricity for heating will also increase by a factor of about 21 between 2015 and 2050, and the electricity for heating will have a final energy share of 36% in 2050 in both alternative scenarios.

8.12.1.6 India: Future Investments in the Heating Sector

The roughly estimated investments in renewable heating technologies up to 2050 amount to around $930 billion in the 2.0 °C Scenario (including investments for replacement after the economic lifetimes of the plants), or approximately $26 billion per year. The largest share of investment in India is assumed to be for solar collectors (around $490 billion), followed by heat pumps and biomass technologies. The 1.5 °C Scenario assumes an even faster expansion of renewable technologies and results in a higher average annual investment of around $29 billion per year (Table 8.79, Fig. 8.85).

8.12.1.7 India: Transport

The energy demand in the transport sector in India is expected to increase in the 5.0 °C Scenario by 377%, from around 3600 PJ/year in 2015 to 17,200 PJ/year in 2050. In the 2.0 °C Scenario, assumed technical, structural, and behavioural changes will save 66% (11,280 PJ/year) by 2050 compared to the 5.0 °C Scenario. Additional modal shifts, technology switches, and a reduction in the transport demand will lead to even higher energy savings in the 1.5 °C Scenario of 81% (or 13,930 PJ/year) in 2050 compared with the 5.0 °C case (Table 8.80, Fig. 8.86).

By 2030, electricity will provide 10% (160 TWh/year) of the transport sector's total energy demand in the 2.0 °C Scenario, whereas in 2050, the share will be 58%

Table 8.79 India: installed capacities for renewable heat generation in the scenarios

in GW	Case	2015	2025	2030	2040	2050
Biomass	5.0 °C	2049	1923	1836	1633	1432
	2.0 °C	2049	1954	1798	1311	856
	1.5 °C	2049	1916	1756	1276	785
Geothermal	5.0 °C	0	0	0	0	0
	2.0 °C	0	2	9	32	38
	1.5 °C	0	5	12	28	37
Solar heating	5.0 °C	6	17	25	43	67
	2.0 °C	6	126	327	619	777
	1.5 °C	6	191	486	653	763
Heat pumps	5.0 °C	0	0	0	0	0
	2.0 °C	0	12	42	90	131
	1.5 °C	0	11	46	82	129
Total[a]	5.0 °C	2055	1940	1861	1676	1499
	2.0 °C	2055	2094	2177	2052	1802
	1.5 °C	2055	2122	2300	2039	1715

[a] Excluding direct electric heating

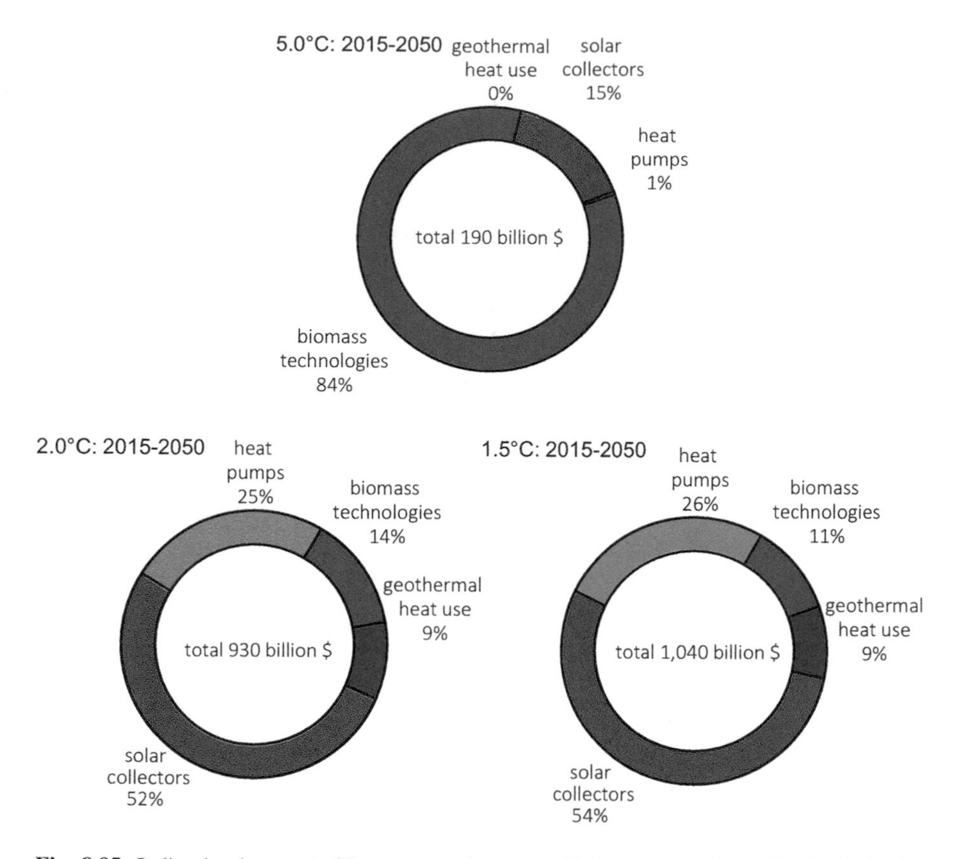

Fig. 8.85 India: development of investments for renewable heat-generation technologies in the scenarios

(950 TWh/year). In 2050, up to 860 PJ/year of hydrogen will be used in the transport sector as a complementary renewable option. In the 1.5 °C Scenario, the annual electricity demand will be 560 TWh in 2050. The 1.5 °C Scenario also assumes a hydrogen demand of 590 PJ/year by 2050.

Biofuel use is limited in the 2.0 °C Scenario to a maximum of around 1000 PJ/year. Therefore, around 2030, synthetic fuels based on power-to-liquid will be introduced, with a maximum amount of 610 PJ/year in 2050. Due to the lower overall energy demand in transport, biofuel use will be reduced in the 1.5 °C Scenario to a maximum of 510 PJ/year. The maximum synthetic fuel demand will amount to 310 PJ/year.

8.12.1.8 India: Development of CO_2 Emissions

In the 5.0 °C Scenario, India's annual CO_2 emissions will increase by 236%, from 2060 Mt. in 2015 to 6950 Mt. in 2050. The stringent mitigation measures in both alternative scenarios will cause the annual emissions to fall to 930 Mt. in 2040 in the 2.0 °C Scenario and to 200 Mt. in the 1.5 °C Scenario, with further reductions to

Table 8.80 India: projection of transport energy demand by mode in the scenarios

in PJ/year	Case	2015	2025	2030	2040	2050
Rail	5.0 °C	180	238	278	353	423
	2.0 °C	180	270	325	421	526
	1.5 °C	180	219	234	332	446
Road	5.0 °C	3294	5861	7880	12,152	16,455
	2.0 °C	3294	5017	5562	5301	5285
	1.5 °C	3294	4253	3125	2977	2730
Domestic aviation	5.0 °C	84	131	166	216	231
	2.0 °C	84	89	81	66	52
	1.5 °C	84	85	74	52	40
Domestic navigation	5.0 °C	29	34	36	40	52
	2.0 °C	29	34	36	40	52
	1.5 °C	29	34	36	40	52
Total	5.0 °C	3587	6263	8360	12,762	17,161
	2.0 °C	3587	5410	6006	5828	5914
	1.5 °C	3587	4590	3470	3401	3268

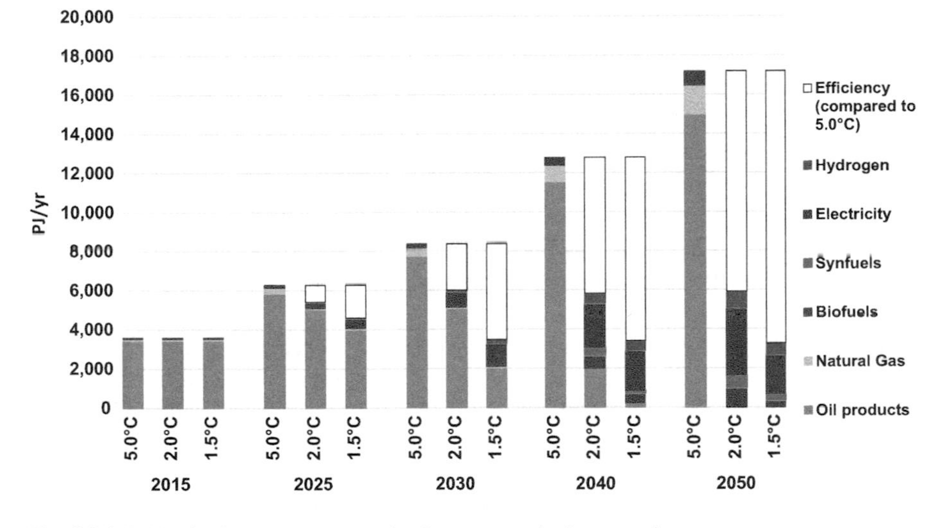

Fig. 8.86 India: final energy consumption by transport in the scenarios

almost zero by 2050. In the 5.0 °C case, the cumulative CO_2 emissions from 2015 until 2050 will add up to 169 Gt. In contrast, in the 2.0 °C and 1.5 °C Scenarios, the cumulative emissions for the period from 2015 until 2050 will be 55 Gt and 38 Gt, respectively.

Therefore, the cumulative CO_2 emissions will decrease by 67% in the 2.0 °C Scenario and by 78% in the 1.5 °C Scenario compared with the 5.0 °C case. A rapid reduction in the annual emissions will occur in both alternative scenarios. In the

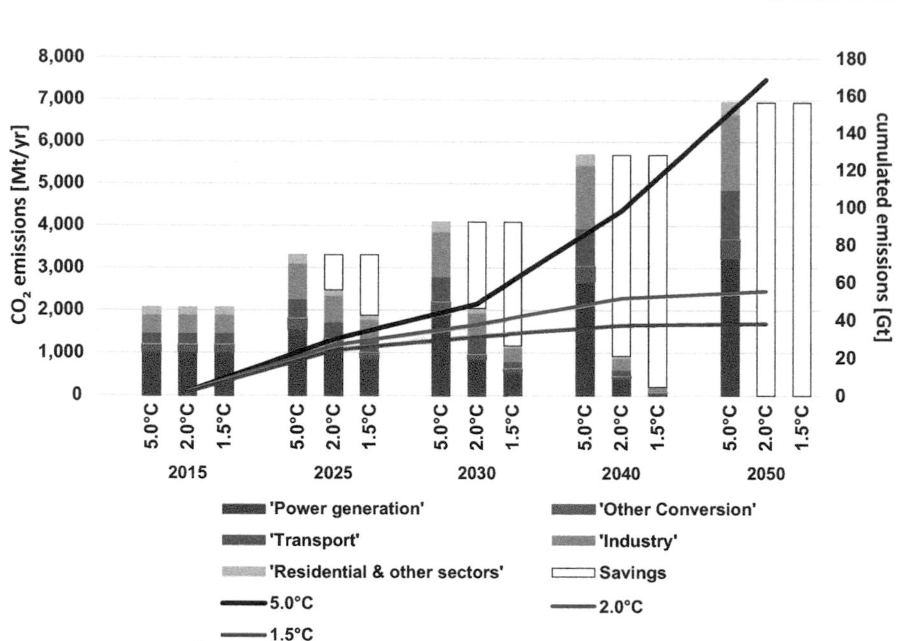

Fig. 8.87 India: development of CO_2 emissions by sector and cumulative CO_2 emissions (after 2015) in the scenarios ('Savings' = reduction compared with the 5.0 °C Scenario)

2.0 °C Scenario, the reduction will be greatest in the 'Residential and other' sector, followed by the 'Power generation' and 'Industry' sectors (Fig. 8.87).

8.12.1.9 India: Primary Energy Consumption

The levels of primary energy consumption in the three scenarios when the assumptions discussed above are taken into account are shown in Fig. 8.88. In the 2.0 °C Scenario, the primary energy demand will increase by 43%, from around 35,600 PJ/year in 2015 to 50,900 PJ/year in 2050. Compared with the 5.0 °C Scenario, the overall primary energy demand will decrease by 51% by 2050 in the 2.0 °C Scenario (5.0 °C: 104,800 PJ in 2050). In the 1.5 °C Scenario, the primary energy demand will be even lower (47,100 PJ in 2050) because the final energy demand and conversion losses will be lower.

Both the 2.0 °C and 1.5 °C Scenarios aim to rapidly phase-out coal and oil. This will cause renewable energy to have a primary energy share of 40% in 2030 and 94% in 2050 in the 2.0 °C Scenario. In the 1.5 °C Scenario, renewables will have a primary energy share of more than 94% in 2050 (including non-energy consumption, which will still include fossil fuels). Nuclear energy will be phased out by 2050 in both the 2.0 °C and 1.5 °C Scenarios. The cumulative primary energy consumption of natural gas in the 5.0 °C case will add up to 160 EJ, the cumulative coal consumption to about 1180 EJ, and the crude oil consumption to 570 EJ. In contrast,

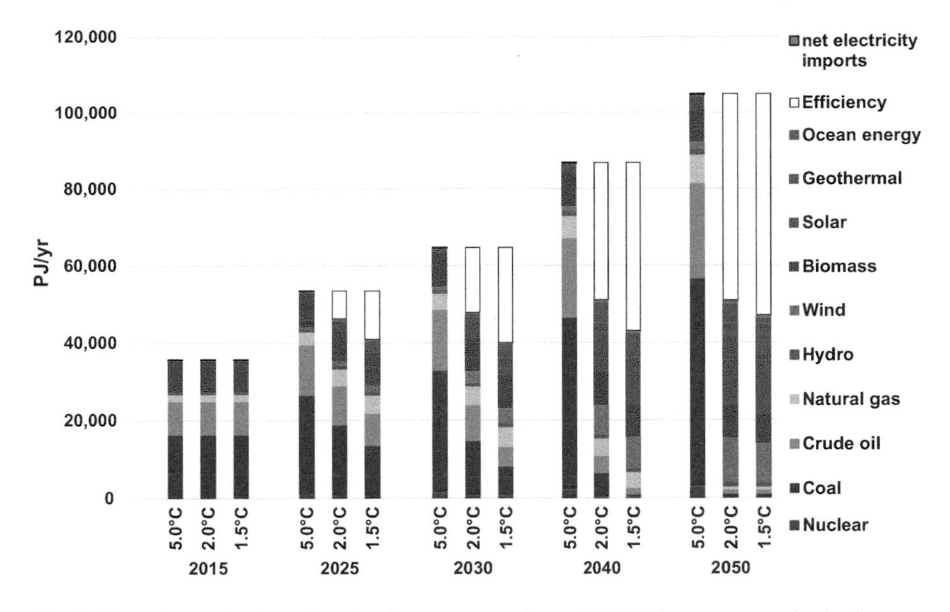

Fig. 8.88 India: projection of total primary energy demand (PED) by energy carrier in the scenarios (including electricity import balance)

in the 2.0 °C Scenario, the cumulative gas demand will amount to 120 EJ, the cumulative coal demand to 360 EJ, and the cumulative oil demand to 220 EJ. Even lower fossil fuel use will be achieved in the 1.5 °C Scenario: 130 EJ for natural gas, 220 EJ for coal, and 150 EJ for oil.

8.12.2 India: Power Sector Analysis

The electricity market in India is in dynamic development. The government of India is making great efforts to increase the reliability of the power supply and at the same time, it is developing universal access to electric power. In 2017, about 300 million Indians (RF 2018) had no power or inadequate power. In 2017, the Indian Government launched *The Third National Electricity Plan*, which covers two 5-year periods: 2017–2022 and 2022–2027. According to the International Energy Agency (IEA) Policies and Measures Database (IEA P + M DB 2018):

> […] "the plan covers short- and long-term demand forecasts in different regions and recommend areas for transmission and generation capacity additions … However, as India sets to meet its first nationally-determined contribution (NDC) under the Paris Agreement … Highlights of the plan include, that during the period 2017–22, no additional capacity of coal will be added – except for the coal power plants under construction […]".

In terms of renewable power generation, India aims to have a total capacity of 275 GW for solar and wind and 72 GW for hydro, with no further increase in the coal power plant capacity until at least 2027.

8.12.2.1 India: Development of Power Plant Capacities

The Third National Electricity Plan for India is an important foundation for strengthening India's renewable power market in order to achieve the levels envisaged in both alternative scenarios. Whereas the hydropower target is consistent with the 2.0 °C and 1.5 °C targets, the solar and wind capacity of 275 GW must be reached between 2020 and 2025 for both scenarios. The annual installation rates for solar PV installations must increase to around 50 GW—the market size in China in 2017—and remain at that level until 2040 to implement either the 2.0 °C or 1.5 °C Scenario. The installation rates for onshore wind must be equally high. In 2017, 4.15 GW of new wind turbines were installed, and significant growth is required. Offshore wind and concentrated solar power plants have significant potential for selected regions of India. Both technologies are vital to achieving the 2.0 °C or 1.5 °C targets (Table 8.81).

Table 8.81 India: average annual change in installed power plant capacity

India power generation: average annual change of installed capacity [GW/a]	2015–2025		2026–2035		2036–2050	
	2.0 °C	1.5 °C	2.0 °C	1.5 °C	2.0 °C	1.5 °C
Hard coal	7	−7	−6	−7	−15	−6
Lignite	0	−1	−1	−2	−2	−1
Gas	9	13	7	7	−14	17
Hydrogen-Gas	0	0	1	1	32	32
Oil/Diesel	0	−1	−1	−1	0	0
Nuclear	1	0	0	0	−1	−1
Biomass	2	2	2	2	4	4
Hydro	3	2	1	1	1	1
Wind (onshore)	20	55	54	59	44	21
Wind (offshore)	2	6	7	7	5	4
PV (roof top)	21	55	49	53	51	30
PV (utility scale)	7	18	16	18	17	10
Geothermal	0	1	3	3	4	4
Solar thermal power plants	1	6	11	11	10	10
Ocean energy	0	1	3	3	3	3
Renewable fuel based co-generation	0	1	2	2	3	3

8.12.2.2 India: Utilization of Power-Generation Capacities

The division of India into five sub-regions is intended to reflect the main grid zones and it is assumed that interconnection will continue to increase to 15% in 2030 and 20% in 2050. Both scenarios aim for an even distribution of variable power plant capacities across all Indian sub-regions. By 2030, the variable power generation will reach 40% in most regions, whereas dispatchable renewables will supply about one quarter of the demand by 2030 (Table 8.82).

India's average capacity factors for the entire power plant fleet remain at around 35% over the entire modelling period, as the calculation results in Table 8.83 show. Contributions from limited dispatchable fossil and nuclear power plants will remain high until 2030 and indicate that a significant replacement of coal for electricity must occur after 2030 in the 2.0 °C Scenario. In the 1.5 °C Scenario, coal will be phased-out just after 2035.

8.12.2.3 India: Development of Load, Generation, and Residual Load

Table 8.84 shows that India's load is predicted to quadruple in all five sub-regions between 2020 and 2050. Under the 2.0 °C Scenario, additional interconnection will increase—beyond the assumed 20% target—but may only be required for the western and southern sub-regions of India. However, for the 1.5 °C Scenario, interconnections must increase in four of the five regions. In the northern region, the calculated generation increases faster than the demand. This region has significant potential for concentrated solar power plants and could supply neighbouring regions.

Table 8.85 shows the storage and dispatch requirements under the 2.0 °C and 1.5 °C Scenarios. All the regions remain within the maximum curtailment target of 10%. Table 8.71 provides an overview of the calculated storage and dispatch power requirements by sub-region. Charging capacities are moderate compared with other world regions. Compared to all other world regions, hydrogen dispatch utilization is very low due to a relatively moderate increase in the gas and hydrogen capacities in India.

Table 8.82 India: power system shares by technology group

Power generation structure and interconnection		2.0 °C				1.5 °C			
		Variable RE	Dispatch RE	Dispatch Fossil	Inter-connection	Variable RE	Dispatch RE	Dispatch fossil	Inter-connection
India-Northern Region	2015	4%	32%	64%	10%				
	2030	41%	28%	31%	15%	56%	24%	20%	15%
	2050	60%	38%	2%	20%	48%	35%	17%	20%
India-North-Eastern Region	2015	4%	32%	64%	10%				
	2030	44%	26%	30%	15%	58%	21%	21%	15%
	2050	95%	5%	0%	20%	92%	5%	3%	20%
India-Eastern Region	2015	4%	32%	64%	10%				
	2030	51%	26%	23%	15%	68%	22%	10%	15%
	2050	73%	26%	1%	20%	69%	29%	2%	20%
India-Western Region	2015	4%	32%	64%	10%				
	2030	44%	26%	30%	15%	57%	21%	22%	15%
	2050	70%	29%	1%	20%	49%	24%	27%	20%
India-Southern Region	2015	4%	32%	64%	10%				
	2030	48%	23%	29%	15%	60%	18%	22%	15%
	2050	78%	21%	1%	20%	62%	19%	19%	20%
India	2015	4%	32%	64%					
	2030	45%	26%	29%		60%	21%	19%	
	2050	72%	27%	1%		58%	26%	16%	

Table 8.83 India: capacity factors by generation type

Utilization of variable and dispatchable power generation:		2015	2020	2020	2030	2030	2040	2040	2050	2050
India			2.0 °C	1.5 °C	2.0 °C	1.5 °C	2.0 °C	1.5 °C	2.0 °C	1.5 °C
Capacity factor – average	[%/yr]	**60.8%**	**53%**	**57%**	**35%**	**26%**	**33%**	**30%**	**37%**	**34%**
Limited dispatchable: fossil and nuclear	[%/yr]	67.7%	57%	61%	48%	38%	37%	27%	37%	12%
Limited dispatchable: renewable	[%/yr]	17.1%	24%	26%	38%	34%	58%	39%	44%	42%
Dispatchable: fossil	[%/yr]	44.7%	12%	19%	11%	12%	30%	29%	24%	29%
Dispatchable: renewable	[%/yr]	39.8%	60%	68%	57%	45%	40%	52%	65%	57%
Variable: renewable	[%/yr]	9.0%	8%	8%	19%	20%	27%	25%	29%	28%

Table 8.84 India: load, generation, and residual load development

Power generation structure		2.0 °C				1.5 °C			
		Max demand [GW]	Max generation [GW]	Max residual load [GW]	Max interconnection requirements [GW]	Max demand [GW]	Max generation [GW]	Max residual load [GW]	Max interconnection requirements [GW]
India-Northern Region	2020	87.8	85.6	11.1		87.8	78.2	19.9	
	2030	150.1	147.3	41.2	0	149.6	240.4	57.2	34
	2050	372.2	397.2	265.7	0	366.8	381.7	211.1	0
India-North-Eastern Region	2020	10.7	10.4	0.6		10.7	10.4	0.6	
	2030	18.3	21.7	2.7	1	18.3	30.4	2.7	9
	2050	45.4	69.1	32.7	0	44.8	223.9	9.3	170
India-Eastern Region	2020	64.5	47.5	25.3		64.5	38.5	34.2	
	2030	110.8	118.0	43.1	0	110.4	198.8	53.1	35
	2050	276.9	364.6	183.6	0	273.0	409.7	174.8	0
India-Western Region	2020	64.6	62.9	3.5		64.6	62.9	3.5	
	2030	111.0	173.5	19.4	43	110.6	196.4	20.0	66
	2050	277.4	542.0	207.2	57	273.4	401.3	86.4	42
India-Southern Region	2020	60.6	59.1	3.5		60.6	59.1	3.2	
	2030	103.0	163.4	5.2	55	102.6	195.0	15.2	77
	2050	252.8	507.5	164.8	90	249.1	448.0	76.7	122

Table 8.85 India: storage and dispatch service requirements

Storage and dispatch		2.0 °C					1.5 °C				
		Required to avoid curtailment	Utilization battery -through-put-	Utilization PSH -through-put-	Total storage demand (incl. H2)	Dispatch hydrogen-based	Required to avoid curtailment	Utilization battery -through-put-	Utilization PSH -through-put-	Total storage demand (incl. H2)	Dispatch hydrogen-based
India		[GWh/year]	[GWh/year]	[GWh/year]	[GWh/year]	[GWh/year]	[GWh/year]	[GWh/year]	[GWh/year]	[GWh/year]	[GWh/year]
India-Northern Region	2020	0	0	0	0	0	0	0	0	0	0
	2030	0	0	0	0	507	24,533	57	3063	3121	160
	2050	1244	42	51	93	9873	734	38	51	89	8647
India-North-Eastern Region	2020	1	0	1	1.1	0	1	0	1	1.1	0
	2030	307	1	65	66	0	3862	8	471	478	0
	2050	4923	126	332	457	1025	258,992	219	1896	2115	11
India-Eastern Region	2020	0	0	0	0	0	0	0	0	0	0
	2030	1657	10	427	437	476	54,903	95	4933	5028	156
	2050	27,180	729	2154	2884	6813	46,793	1519	3163	4682	5715
India-Western Region	2020	0	0	0	0	0	0	0	0	0	0
	2030	29,610	51	2978	3028	448	41,348	84	3928	4012	310
	2050	174,263	1709	5618	7327	5037	28,209	1228	2263	3491	2020
India-Southern Region	2020	0	0	0	0	0	0	0	0	0	0
	2030	27,824	42	2496	2537	328	57,916	88	4759	4847	144
	2050	165,200	1643	5274	6917	5365	103,156	1891	4931	6822	2066
India	2020	1	0	1	1	0	1	0	1	1	0
	2030	59,399	104	5966	6069	1759	182,561	333	17,154	17,487	769
	2050	372,809	4248	13,430	17,678	28,113	437,884	4895	12,304	17,199	18,459

8.13 China

8.13.1 China: Long-Term Energy Pathways

8.13.1.1 China: Final Energy Demand by Sector

The future development pathways for China's final energy demand when the assumptions on population growth, GDP growth and energy intensity are combined are shown in Fig. 8.89 for the 5.0 °C, 2.0 °C, and 1.5 °C Scenarios. In the 5.0 °C Scenario, the total final energy demand will increase by 56% from the current 73,600 PJ/year to 114,600 PJ/year in 2050. In the 2.0 °C Scenario, the final energy demand will decreases by 26% compared with current consumption and will reach 54,400 PJ/year by 2050. The final energy demand in the 1.5 °C Scenario will reach 49,200 PJ, 33% below the 2015 demand. In the 1.5 °C Scenario, the final energy demand in 2050 will be 10% lower than in the 2.0 °C Scenario. The electricity demand for 'classical' electrical devices (without power-to-heat or e-mobility) will increase from 3470 TWh/year in 2015 to around 5230 TWh/year in both alternative scenarios

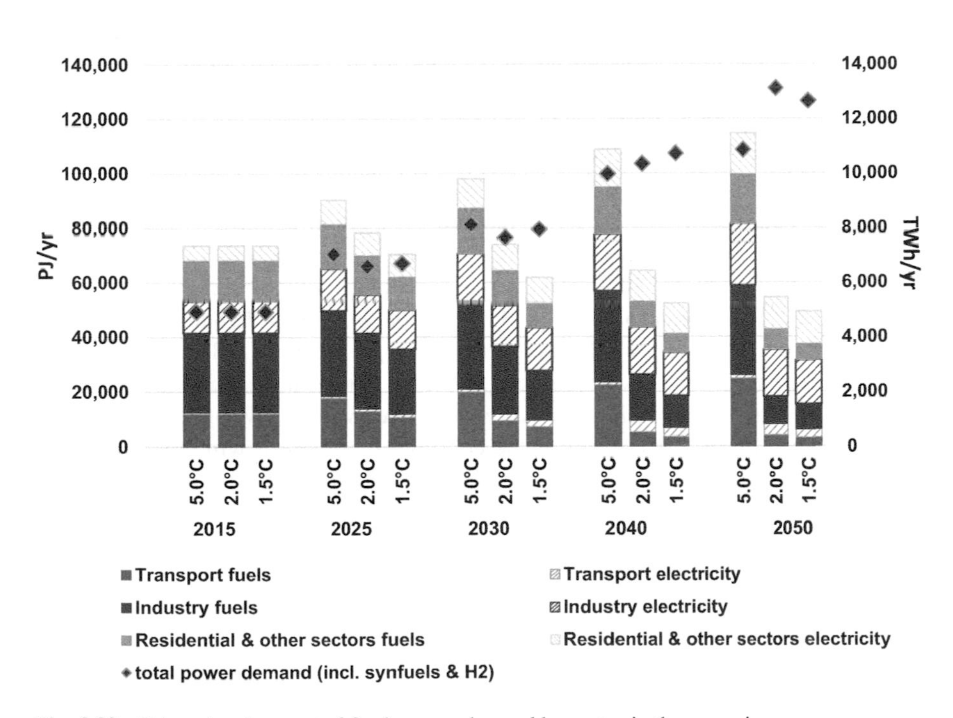

Fig. 8.89 China: development of final energy demand by sector in the scenarios

by 2050. Compared with the 5.0 °C case (9480 TWh/year in 2050), the efficiency measures in the 2.0 °C and 1.5 °C Scenarios save around 4250 TWh/year by 2050.

Electrification will lead to a significant increase in the electricity demand by 2050. In the 2.0 °C Scenario, the electricity demand for heating will be approximately 2800 TWh/year due to electric heaters and heat pumps and in the transport sector, the electricity demand will be approximately 4200 TWh/year due to electric mobility. The generation of hydrogen (for transport and high-temperature process heat) and the manufacture of synthetic fuels (mainly for transport) will add an additional power demand of 3900 TWh/year. Therefore, the gross power demand will rise from 5900 TWh/year in 2015 to 13,800 TWh/year in 2050 in the 2.0 °C Scenario, 11% higher than in the 5.0 °C case. In the 1.5 °C Scenario, the gross electricity demand will increase to a maximum of 13,300 TWh/year in 2050.

The efficiency gains in the heating sector could be even larger than in the electricity sector. In the 2.0 °C and 1.5 °C Scenarios, a final energy consumption equivalent to about 24,400 PJ/year and 27,600 PJ/year, respectively, will be avoided through efficiency gains by 2050 compared to the 5.0 °C Scenario.

8.13.1.2 China: Electricity Generation

The development of the power system is characterized by a dynamically growing renewable energy market and an increasing proportion of total power from renewable sources. By 2050, 100% of the electricity produced in China will come from renewable energy sources in the 2.0 °C Scenario. 'New' renewables—mainly wind, solar, and geothermal energy—will contribute 77% of the total electricity generation. Renewable electricity's share of the total production will be 54% by 2030 and 84% by 2040. The installed capacity of renewables will reach about 2170 GW by 2030 and 5420 GW by 2050. The share of renewable electricity generation in 2030 in the 1.5 °C Scenario is assumed to be 63%. In the 1.5 °C Scenario, the generation capacity from renewable energy will be approximately 5310 GW in 2050.

Table 8.86 shows the development of different renewable technologies in China over time. Figure 8.90 provides an overview of the overall power-generation structure in China. From 2020 onwards, the continuing growth of wind and PV, up to 1670 GW and 2220 GW, respectively, will be complemented by up to 680 GW solar thermal generation, as well as limited biomass, geothermal, and ocean energy, in the 2.0 °C Scenario. Both the 2.0 °C and 1.5 °C Scenarios will lead to a high proportion of variable power generation (PV, wind, and ocean) of 28% and 34%, respectively, by 2030, and 51% and 52%, respectively, by 2050.

Table 8.86 China: development of renewable electricity-generation capacity in the scenarios

in GW	Case	2015	2025	2030	2040	2050
Hydro	5.0 °C	320	395	424	477	525
	2.0 °C	320	383	396	420	450
	1.5 °C	320	383	396	420	450
Biomass	5.0 °C	11	24	29	39	48
	2.0 °C	11	57	101	158	195
	1.5 °C	11	72	106	160	195
Wind	5.0 °C	132	343	408	536	667
	2.0 °C	132	428	678	1299	1674
	1.5 °C	132	508	877	1460	1652
Geothermal	5.0 °C	0	0	0	1	3
	2.0 °C	0	4	19	77	134
	1.5 °C	0	7	29	77	119
PV	5.0 °C	43	265	330	430	565
	2.0 °C	43	504	889	1614	2218
	1.5 °C	43	604	1036	1781	2215
CSP	5.0 °C	0	3	5	7	11
	2.0 °C	0	11	84	413	677
	1.5 °C	0	16	103	391	614
Ocean	5.0 °C	0	0	0	1	1
	2.0 °C	0	1	7	33	74
	1.5 °C	0	1	7	33	62
Total	5.0 °C	505	1029	1196	1490	1819
	2.0 °C	505	1390	2175	4015	5421
	1.5 °C	505	1592	2555	4322	5307

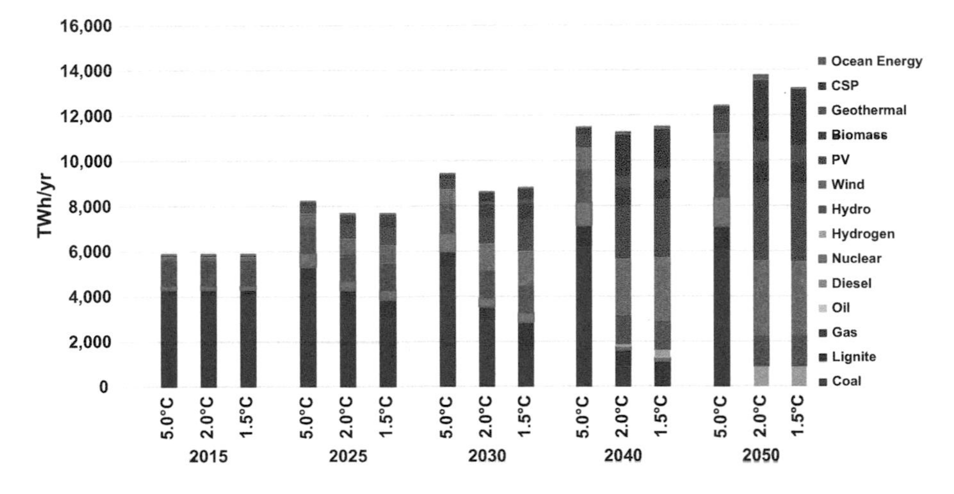

Fig. 8.90 China: development of electricity-generation structure in the scenarios

8.13.1.3 China: Future Costs of Electricity Generation

Figure 8.91 shows the development of the electricity-generation and supply costs over time, including the CO_2 emission costs, in all scenarios. The calculated electricity generation costs in 2015 (referring to full costs) were around 4.7 ct/kWh. In the 5.0 °C case, the generation costs will increase until 2030, when they reach 9.2 ct/kWh, and then drop to 8.8 ct/kWh by 2050. The generation costs will increase in the alternative scenarios until 2030, when they reach around 8 ct/kWh, and will then drop to 6.5 ct/kWh by 2050, 2.3 ct/kWh lower than in the 5.0 °C Scenario. Note that these estimates of generation costs do not take into account integration costs such as power grid expansion, storage, or other load-balancing measures.

In the 5.0 °C case, the growth in demand and increasing fossil fuel prices will cause total electricity supply costs to rise from today's $310 billion/year to more than $1230 billion/year in 2050. In the 2.0 °C case, the total supply costs will be $1030 billion/year and $1010 billion/year in the 1.5 °C Scenario. Therefore, the long-term costs for electricity supply will be more than 16% lower in the alternative scenarios than in the 5.0 °C case.

Compared with these results, the generation costs when the CO_2 emission costs are not considered will increase in the 5.0 °C case to 5.7 ct/kWh in 2030 and stabilize at 5.5 ct/kWh in 2050. In the 2.0 °C Scenario, they increase continuously until 2050, when they reach 6.6 ct/kWh. In the 1.5 °C Scenario, they will increase to 7 ct/kWh and then drop to 6.6 ct/kWh by 2050. In the 2.0 °C Scenario, the generation costs will be a maximum of 1 ct/kWh higher than in the 5.0 °C case, and this will

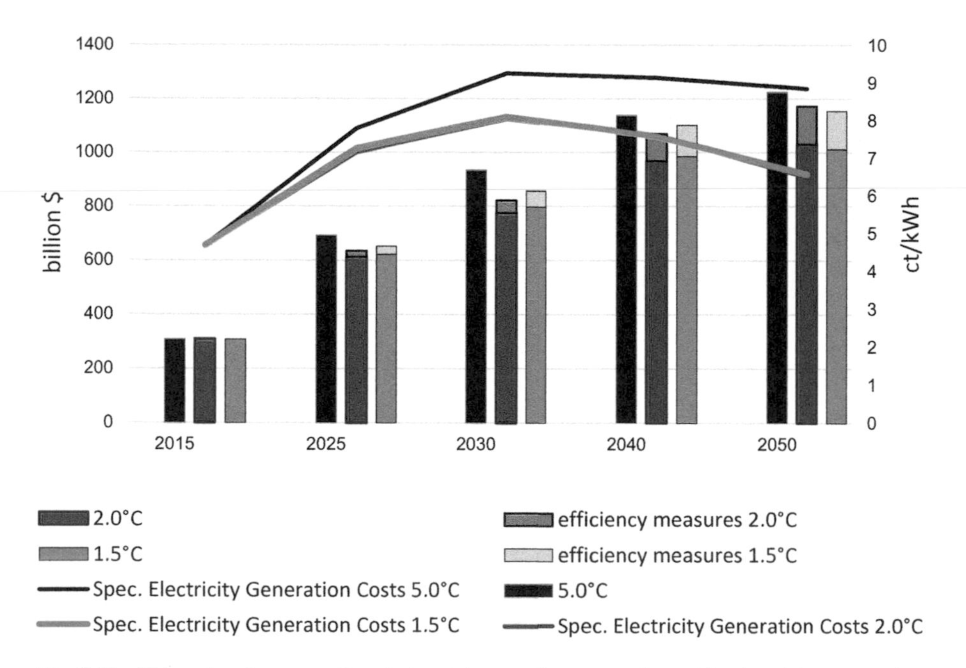

Fig. 8.91 China: development of total electricity supply costs and specific electricity-generation costs in the scenarios

occur in 2050. In the 1.5 °C Scenario, compared to the 5.0 °C Scenario, the maximum difference in generation costs will be 1.6 ct/kWh in 2040. The generation costs in 2050 will be 1.1 ct/kWh higher than in the 5.0 °C case. If the CO_2 costs are not considered, the total electricity supply costs in the 5.0 °C case will rise to about $810 billion/year in 2050.

8.13.1.4 China: Future Investments in the Power Sector

An investment of around $9740 billion will be required for power generation between 2015 and 2050 in the 2.0 °C Scenario—including additional power plants for the production of hydrogen and synthetic fuels and investments for plant replacement at the end of their economic lifetimes. This value will be equivalent to approximately $271 billion per year on average and will be $5680 billion more than in the 5.0 °C case ($4060 billion). An investment of around $9840 billion for power generation will be required between 2015 and 2050 in the 1.5 °C Scenario. On average, this will be an investment of $273 billion per year. In the 5.0 °C Scenario, the investment in conventional power plants will be around 29% of the total cumulative investments, whereas approximately 71% will be invested in renewable power generation and co-generation (Fig. 8.92).

However, in the 2.0 °C (1.5 °C) Scenario, China will shift almost 97% (98%) of its entire investment to renewables and co-generation. By 2030, the fossil fuel share of the power sector investment will predominantly focus on gas power plants that can also be operated with hydrogen.

Because renewable energy has no fuel costs, other than biomass, the cumulative fuel cost savings in both alternative scenarios will reach a total of more than $6200 billion in 2050, equivalent to $173 billion per year. Therefore, the total fuel cost savings will be equivalent to 110% of the total additional investments compared to the 5.0 °C Scenario.

8.13.1.5 China: Energy Supply for Heating

The final energy demand for heating will increase in the 5.0 °C Scenario by 38% from 42,300 PJ/year in 2015 to 58,200 PJ/year in 2050. Energy efficiency measures will help to reduce the energy demand for heating by 42% in 2050 in the 2.0 °C Scenario, relative to the 5.0 °C case, and by 47% in the 1.5 °C Scenario. Today, renewables supply around 11% of China's final energy demand for heating, with the main contribution from biomass. Renewable energy will provide 32% of China's total heat demand in 2030 in the 2.0 °C Scenario and 46% in the 1.5 °C Scenario. In both scenarios, renewables will provide 100% of the total heat demand in 2050.

Figure 8.93 shows the development of different technologies for heating in China over time, and Table 8.87 provides the resulting renewable heat supply for all scenarios. Up to 2030, biomass will remain the main contributor. In the long term, the

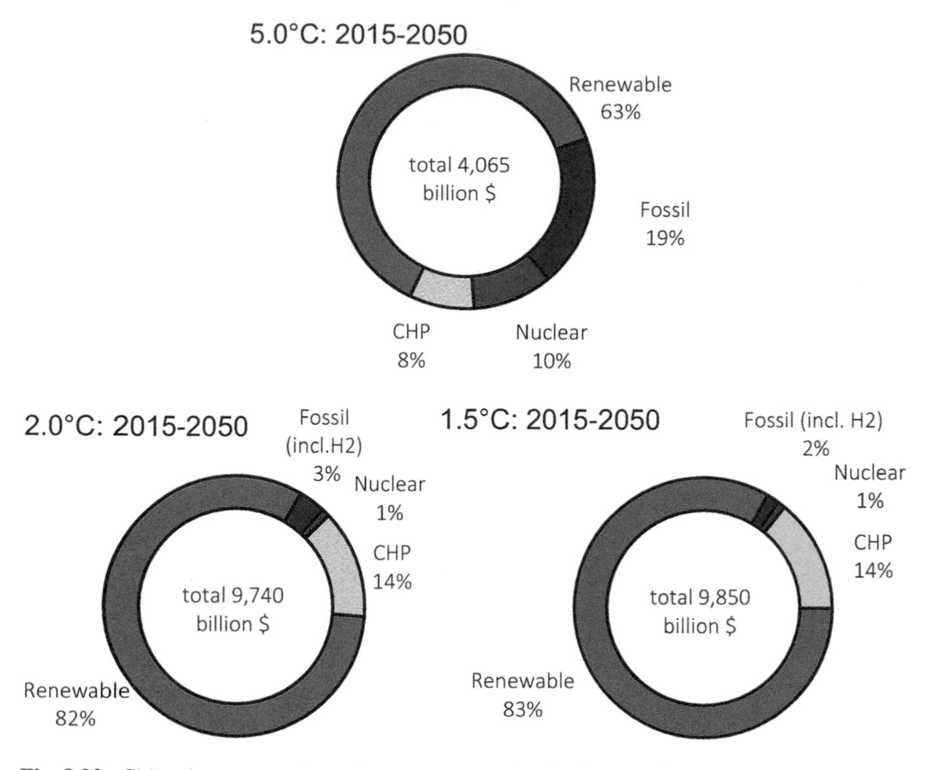

Fig. 8.92 China: investment shares for power generation in the scenarios

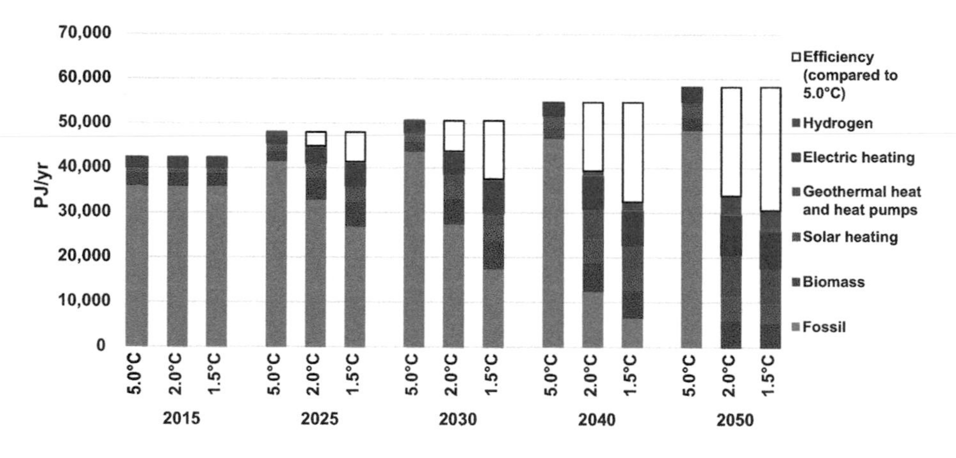

Fig. 8.93 China: development of heat supply by energy carrier in the scenarios

growing use of solar, geothermal, and environmental heat will lead to a biomass share of 24% in both alternative scenarios.

Heat from renewable hydrogen will further reduce the dependence on fossil fuels in both scenarios. Hydrogen consumption in 2050 will be around 4100 PJ/year in

Table 8.87 China: development of renewable heat supply in the scenarios (excluding the direct use of electricity)

in PJ/year	Case	2015	2025	2030	2040	2050
Biomass	5.0 °C	2776	2095	2079	2291	2877
	2.0 °C	2776	4609	5603	6254	5967
	1.5 °C	2776	5378	6263	6055	5385
Solar heating	5.0 °C	892	1297	1515	1962	2535
	2.0 °C	892	2066	2906	5454	5417
	1.5 °C	892	2364	3242	4381	4360
Geothermal heat and heat pumps	5.0 °C	306	452	526	743	1026
	2.0 °C	306	1304	2720	6690	9225
	1.5 °C	306	1269	2884	5706	7943
Hydrogen	5.0 °C	0	0	0	0	0
	2.0 °C	0	0	7	1020	4118
	1.5 °C	0	0	7	1890	4549
Total	5.0 °C	3974	3844	4120	4996	6438
	2.0 °C	3974	7978	11,237	19,417	24,727
	1.5 °C	3974	9011	12,396	18,031	22,237

the 2.0 °C Scenario and to 4500 PJ/year in the 1.5 °C Scenario. The direct use of electricity for heating will also increase by a factor of 3.7–4 between 2015 and 2050 and electricity for heating will have a final energy share of 27% in 2050 in both the 2.0 °C Scenario and 1.5 °C Scenario.

8.13.1.6 China: Future Investments in the Heating Sector

The roughly estimated investments in renewable heating technologies up to 2050 will amount to around $2780 billion in the 2.0 °C Scenario (including investments for the replacement of plants after their economic lifetimes), or approximately $77 billion per year. The largest share of investment in China is assumed to be for heat pumps (around $1200 billion), followed by solar collectors and geothermal heat use. The 1.5 °C Scenario assumes an even faster expansion of renewable technologies. However, the lower heat demand (compared with the 2.0 °C Scenario) will result in a lower average annual investment of around $67 billion per year (Table 8.88, Fig. 8.94).

8.13.1.7 China: Transport

The energy demand in the transport sector in China is expected to increase in the 5.0 °C Scenario by 107% from around 12,600 PJ/year in 2015 to 26,100 PJ/year in 2050. In the 2.0 °C Scenario, assumed technical, structural, and behavioural changes will save 68% (17,840 PJ/year) by 2050 compared with the 5.0 °C Scenario.

Table 8.88 China: installed capacities for renewable heat generation in the scenarios

in GW	Case	2015	2025	2030	2040	2050
Biomass	5.0 °C	1194	764	648	519	468
	2.0 °C	1194	1284	1214	921	578
	1.5 °C	1194	1267	1280	808	481
Geothermal	5.0 °C	0	0	0	0	0
	2.0 °C	0	20	46	187	272
	1.5 °C	0	20	42	139	161
Solar heating	5.0 °C	281	409	478	618	799
	2.0 °C	281	592	843	1546	1539
	1.5 °C	281	688	956	1252	1275
Heat pumps	5.0 °C	52	76	89	126	174
	2.0 °C	52	151	251	449	565
	1.5 °C	52	136	213	349	446
Total[a]	5.0 °C	1527	1250	1214	1263	1441
	2.0 °C	1527	2048	2355	3103	2954
	1.5 °C	1527	2111	2491	2549	2361

[a] Excluding direct electric heating

Additional modal shifts, technology switches, and a reduction in transport demand will lead to even higher energy savings in the 1.5 °C Scenario of 76% (or 19,900 PJ/year) in 2050 compared with the 5.0 °C case (Table 8.89, Fig. 8.95).

By 2030, electricity will provide 21% (680 TWh/year) of the transport sector's total energy demand in the 2.0 °C Scenario, whereas in 2050, the share will be 51% (1170 TWh/year). In 2050, up to 1600 PJ/year of hydrogen will be used in the transport sector as a complementary renewable option. In the 1.5 °C Scenario, the annual electricity demand is 860 TWh in 2050. The 1.5 °C Scenario also assumes a hydrogen demand of 1100 PJ/year by 2050.

Biofuel use is limited in the 2.0 °C Scenario to a maximum of 1900 PJ/year. Therefore, around 2030, synthetic fuels based on power-to-liquid will be introduced, with a maximum amount of 560 PJ/year in 2050. Due to the lower overall energy demand in transport, biofuel use will be reduced in the 1.5 °C Scenario to a maximum of around 1400 PJ/year. The maximum synthetic fuel demand will amount to 720 PJ/year.

8.13.1.8 China: Development of CO_2 Emissions

In the 5.0 °C Scenario, China's annual CO_2 emissions will increase by 25%, from 9060 Mt. in 2015 to 11,320 Mt. in 2050. The stringent mitigation measures in both alternative scenarios will cause annual emissions to fall to 1990 Mt. in 2040 in the 2.0 °C Scenario and to 760 Mt. in the 1.5 °C Scenario, with further reductions to almost zero by 2050. In the 5.0 °C case, the cumulative CO_2 emissions from 2015 until 2050 will add up to 392 Gt. In contrast, in the 2.0 °C and 1.5 °C Scenarios, the

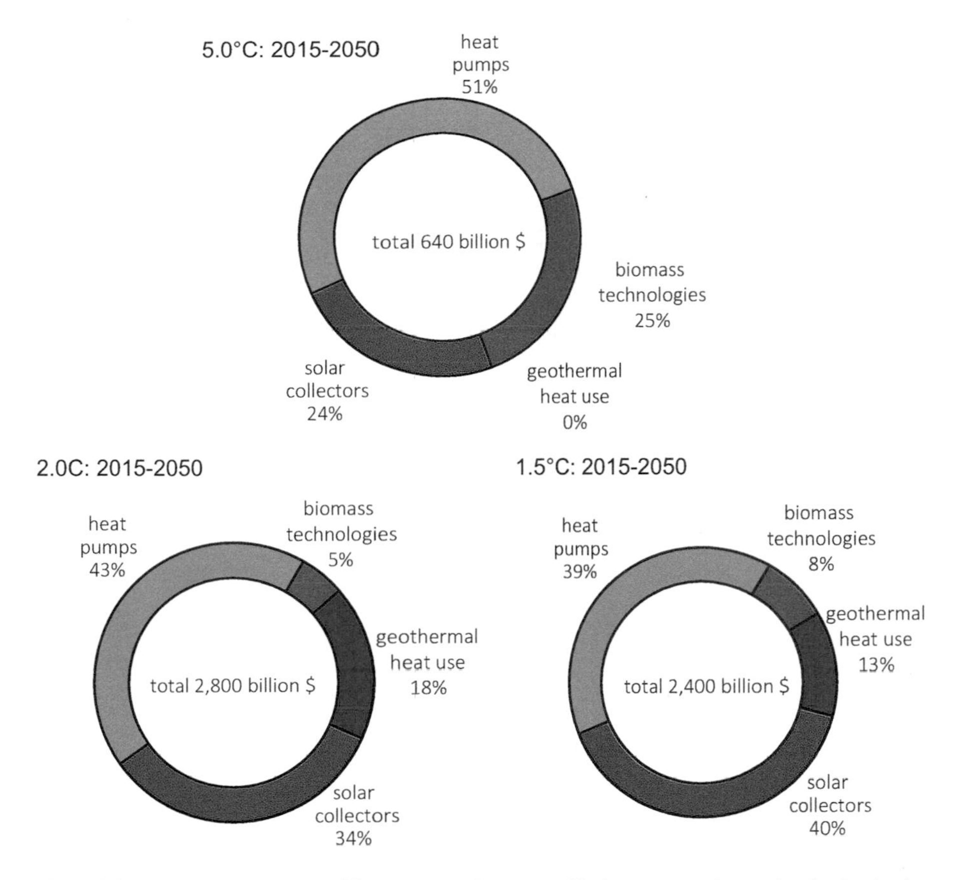

Fig. 8.94 China: development of investments for renewable heat-generation technologies in the scenarios

cumulative emissions for the period from 2015 until 2050 will be 174 Gt and 132 Gt, respectively.

Therefore, the cumulative CO_2 emissions will decrease by 56% in the 2.0 °C Scenario and by 66% in the 1.5 °C Scenario compared with the 5.0 °C case. A rapid reduction in annual emissions will occur in both alternative scenarios. In the 2.0 °C Scenario the reduction will be greatest in the 'Residential and other' sector, followed by 'Power generation' and 'Transport' sectors (Fig. 8.96).

8.13.1.9 China: Primary Energy Consumption

The levels of primary energy consumption in the three scenarios when the assumptions discussed above are taken into account are shown in Fig. 8.97. In the 2.0 °C Scenario, the primary energy demand will decrease by 30%, from around 125,000 PJ/year in 2015 to 87,800 PJ/year in 2050. Compared with the 5.0 °C Scenario, the

Table 8.89 China: projection of transport energy demand by mode in the scenarios

in PJ/year	Case	2015	2025	2030	2040	2050
Rail	5.0 °C	539	567	593	644	672
	2.0 °C	539	589	637	687	762
	1.5 °C	539	580	597	622	662
Road	5.0 °C	10,421	15,629	17,651	19,664	22,073
	2.0 °C	10,421	11,509	9395	7143	5894
	1.5 °C	10,421	9607	7372	4576	4020
Domestic aviation	5.0 °C	754	1234	1590	2070	2213
	2.0 °C	754	814	742	592	470
	1.5 °C	754	777	653	463	366
Domestic navigation	5.0 °C	877	984	1035	1113	1157
	2.0 °C	877	984	1035	1113	1157
	1.5 °C	877	984	1035	1113	1157
Total	5.0 °C	12,591	18,413	20,870	23,490	26,115
	2.0 °C	12,591	13,895	11,809	9535	8284
	1.5 °C	12,591	11,948	9657	6773	6206

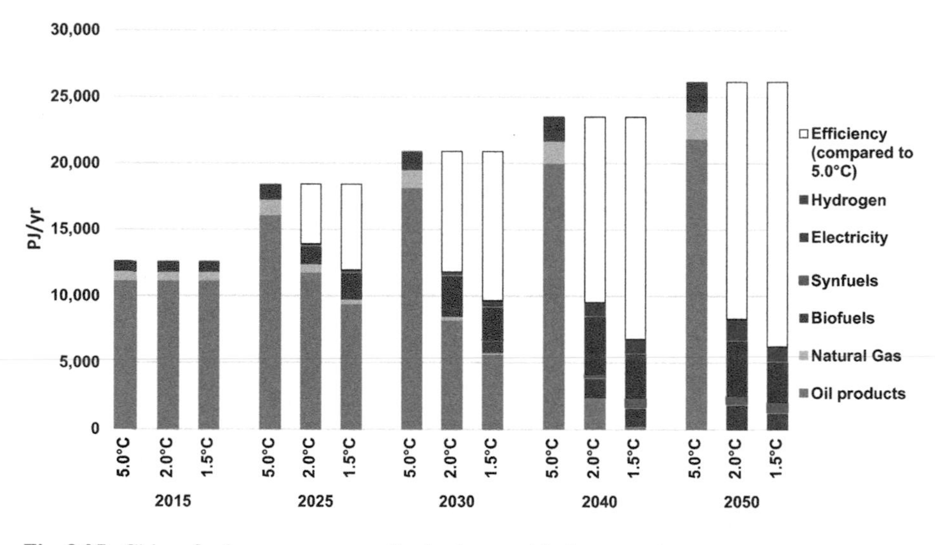

Fig. 8.95 China: final energy consumption by transport in the scenarios

overall primary energy demand will decrease by 54% by 2050 in the 2.0 °C Scenario (5.0 °C: 192,300 PJ in 2050). In the 1.5 °C Scenario, the primary energy demand will be even lower (80,700 PJ in 2050) because the final energy demand and conversion losses will be lower.

Both the 2.0 °C and 1.5 °C Scenarios aim to rapidly phase-out coal and oil. This will cause renewable energy to have a primary energy share of 28% in 2030 and 92% in 2050 in the 2.0 °C Scenario. In the 1.5 °C Scenario, renewables will have a primary energy share of more than 91% in 2050 (including non-energy consump-

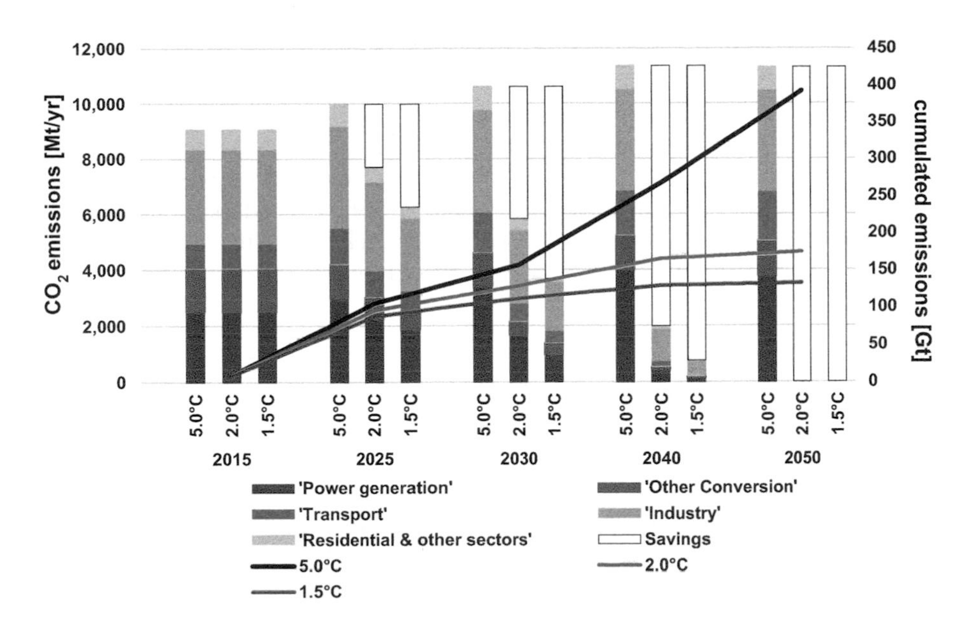

Fig. 8.96 China: development of CO_2 emissions by sector and cumulative CO_2 emissions (after 2015) in the scenarios ('Savings' = reduction compared with the 5.0 °C Scenario)

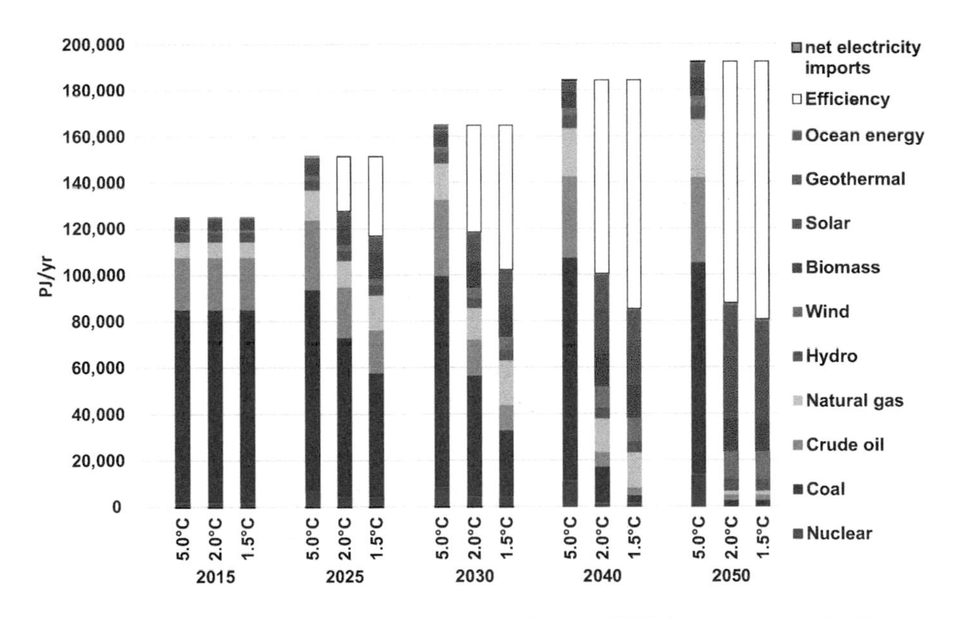

Fig. 8.97 China: projection of total primary energy demand (PED) by energy carrier in the scenarios (including electricity import balance)

tion, which will still include fossil fuels). Nuclear energy will be phased-out by 2050 in the 2.0 °C Scenario and by 2045 in the 1.5 °C Scenario. The cumulative primary energy consumption of natural gas in the 5.0 °C case will add up to 570 EJ, the cumulative coal consumption to about 3000 EJ, and the crude oil consumption to 1080 EJ. In contrast, in the 2.0 °C Scenario, the cumulative gas demand will amount to 360 EJ, the cumulative coal demand to 1360 EJ, and the cumulative oil demand to 430 EJ. Even lower fossil fuel use will be achieved in the 1.5 °C Scenario: 440 EJ for natural gas, 930 EJ for coal, and 340 EJ for oil.

8.13.2 China: Power Sector Analysis

China has by far the largest power sector of all world regions—about one quarter of the world's total electricity generation. China's National Energy Administration (NEA) released the *13th Energy Five-Year Plan (FYP)* in January 2016 (IEA RED 2016). The FYP that is in force from 2016 to 2020 introduces framework legislation that defines energy development for the next 5 years in China. In parallel to the main Energy FYP, there are 14 additional supporting FYPs, such as the Renewable Energy 13th FYP, the Wind FYP, and the Electricity FYP, which were all released at about the same time (GWEC-NL 2018). According to the Renewable Energy 13th FYP, by 2020, the total RE electricity installations will reach 680 GW, with electricity production of 1900 TWh/year This will account for 27% of electricity production. The wind power target is set to reach 210 GW by 2020, with electricity production of 420 TWh, supplying 6% of China's total electricity demand. The target for offshore wind is 5 GW by 2020 (GWEC-NL 2018). For other renewable power-generation technologies, the 2020 targets are 150 GW for solar PV, 10 GW for concentrated solar power (CSP), 15 GW for bioenergy, and 380 GW for hydropower, including 40 GW hydro pump storage (IEA-RED 2016). The renewable targets are consistent, to large extent, with both the 2.0 °C and 1.5 °C Scenarios. The onshore wind and solar PV capacities in both scenarios will increase to 50 GW and are within the current market size range. The targets for the 2.0 °C and 1.5 °C Scenarios for CSP, bioenergy, and offshore wind are slightly higher than current market volumes. However, the first decade of the 2.0 °C and 1.5 °C Scenarios will reflect the existing trends in China's power sector.

8.13.2.1 China: Development of Power Plant Capacities

China's solar PV and wind power markets are the largest in the world and represent about half the global annual market for solar PV (in 2017) and a third of the market for onshore wind. The continued growth of the annual renewable power market—for all technologies—for the Chinese market will continue to have a significant impact on other world regions. To implement the project's 2.0 °C Scenario, the current solar PV market in China must remain at the 2017 level, and to achieve the

Table 8.90 China: average annual change in installed power plant capacity

China power generation: average annual change of installed capacity [GW/a]	2015–2025		2026–2035		2036–2050	
	2.0 °C	1.5 °C	2.0 °C	1.5 °C	2.0 °C	1.5 °C
Hard coal	5	−51	−55	−81	−41	−5
Lignite	0	0	0	0	0	0
Gas	4	28	6	30	−16	−17
Hydrogen-Gas	0	0	1	3	24	38
Oil/Diesel	0	−1	0	−1	0	0
Nuclear	3	0	−2	0	−3	−4
Biomass	6	10	9	8	5	5
Hydro	8	5	3	3	3	3
Wind (onshore)	31	65	46	64	36	29
Wind (offshore)	2	12	20	22	11	9
PV (roof top)	41	77	69	76	62	50
PV (utility scale)	14	26	23	25	21	17
Geothermal	1	4	5	6	8	6
Solar thermal power plants	1	13	34	29	40	30
Ocean energy	0	1	2	2	5	4
Renewable fuel based co-generation	**4**	**9**	**10**	**9**	**8**	**8**

1.5 °C Scenario, it must double. The onshore wind market must increase by 50% compared with 2015 for the 2.0 °C Scenario and must triple to meet the 1.5 °C trajectory. All these annual market volumes must be maintained until 2035, before a moderate reduction in the annual market sizes can occur (Table 8.90).

8.13.2.2 China: Utilization of Power Generation Capacities

Across all regions, an interconnection capacity of 10% is assumed for the base year calculation. The interconnection capacity will increase to 20% by 2030, with no further increase thereafter. For the entire modelling period, it is assumed that Taiwan is not connected to any other region. Under the 2.0 °C Scenario, variable renewables will attain a share of around 30% in all sub-regions, whereas the 1.5 °C Scenario will lead to shares of over 40% in five of the seven sub-regions (Table 8.91).

Table 8.92 shows the results of the capacity factor calculations done under the assumption that variable and dispatchable power plants will have priority access to the grid and priority dispatch. The average capacity factors for limited dispatchable power plants will remain at around 30% until 2030 under the 2.0 °C Scenario. This relatively low factor indicates an overcapacity in China's power market. The curtailment rates of 20% (REW 1-2018) and more in 2017—mainly for wind farms—confirm this.

Table 8.91 China: power system shares by technology group

Power generation structure and interconnection		2.0 °C				1.5 °C			
China		Variable RE	Dispatch RE	Dispatch Fossil	Inter-connection	Variable RE	Dispatch RE	Dispatch fossil	Inter-connection
China-North	2015	7%	35%	58%	10%				
	2030	32%	21%	47%	20%	43%	29%	28%	20%
	2050	53%	43%	4%	20%	53%	37%	9%	20%
China-Northwest	2015	7%	35%	58%	10%				
	2030	29%	22%	49%	20%	40%	31%	29%	20%
	2050	49%	47%	3%	20%	54%	44%	2%	20%
China-Northeast	2015	6%	35%	60%	10%				
	2030	34%	24%	43%	20%	45%	31%	24%	20%
	2050	54%	43%	4%	20%	54%	45%	2%	20%
China-Tibet	2015	7%	35%	58%	10%				
	2030	37%	34%	29%	20%	49%	37%	14%	20%
	2050	43%	49%	7%	20%	42%	53%	5%	20%
China-Central	2015	6%	35%	60%	10%				
	2030	28%	26%	47%	20%	36%	32%	32%	20%
	2050	41%	52%	7%	20%	44%	48%	9%	20%
China-East	2015	6%	35%	60%	10%				
	2030	30%	25%	45%	20%	36%	29%	35%	20%
	2050	48%	47%	5%	20%	48%	38%	14%	20%

China-South	2015	6%	35%	60%	10%				
	2030	30%	28%	43%	20%	38%	31%	31%	20%
	2050	49%	47%	4%	20%	48%	46%	6%	20%
Taiwan	2015	7%	35%	59%	0%				
	2030	31%	24%	46%	0%	39%	29%	31%	0%
	2050	57%	40%	3%	0%	51%	37%	12%	0%
China	2015	6%	35%	59%					
	2030	30%	24%	46%		39%	30%	31%	
	2050	49%	47%	5%		49%	42%	9%	

Table 8.92 China: capacity factors by generation type

Utilization of variable and dispatchable power generation:		2015	2020	2020	2030	2030	2040	2040	2050	2050
China			2.0 °C	1.5 °C	2.0 °C	1.5 °C	2.0 °C	1.5 °C	2.0 °C	1.5 °C
Capacity factor – average	[%/yr]	**42.0%**	**30%**	**28%**	**26%**	**21%**	**37%**	**24%**	**37%**	**26%**
Limited dispatchable: fossil and nuclear	[%/yr]	39.2%	34%	29%	32%	25%	20%	17%	9%	16%
Limited dispatchable: renewable	[%/yr]	47.3%	20%	17%	21%	14%	68%	19%	47%	27%
Dispatchable: fossil	[%/yr]	30.7%	28%	40%	46%	34%	24%	37%	11%	37%
Dispatchable: renewable	[%/yr]	59.1%	27%	31%	28%	23%	47%	34%	62%	39%
Variable: renewable	[%/yr]	17.9%	15%	15%	17%	16%	22%	17%	22%	17%

8.13.2.3 China: Development of Load, Generation, and Residual Load

The load for China is calculated to continue to increase. Table 8.93 shows that the maximum load will double across all regions. However, the assumed interconnection rates of 20% are sufficient for the 2.0 °C Scenario, whereas significantly higher interconnection capacities will be required under the 1.5 °C Scenario. By 2050, all regions will have an oversupply under the 1.5 °C Scenario. This surplus electricity will be used to produce synthetic fuels and hydrogen. The [R]E 24/7 model does not interface with other world regions, so surplus generation will result in a negative residual load.

Finally, Table 8.94 provides an overview of the calculated storage and dispatch power requirements in the Chinese region. The calculated hydro pump storage increase by 2050 is consistent with the Thirteenth Five-Year Plan's requirement for 40 GW additional capacity. Furthermore, curtailment is within the acceptable range, at significantly below 10% in both scenarios by 2050. Battery capacities must increase significantly after 2030. The central, southern, and eastern sub-regions of mainland China have by far the highest storage requirements.

Table 8.93 China: load, generation, and residual load development

Power generation structure		2.0 °C				1.5 °C			
China		Max demand [GW]	Max generation [GW]	Max residual load [GW]	Max interconnection requirements [GW]	Max demand [GW]	Max generation [GW]	Max residual load [GW]	Max interconnection requirements [GW]
China-North	2020	168.7	168.7	3.6		167.9	167.9	3.6	
	2030	215.6	222.5	22.5	0	213.0	292.8	25.8	54
	2050	364.2	504.4	246.3	0	368.2	587.9	−133.4	353
China-Northwest	2020	77.4	80.5	6.1		77.1	82.4	6.1	
	2030	95.6	99.5	11.8	0	94.5	126.7	13.3	19
	2050	135.3	206.2	114.3	0	136.9	246.4	−48.1	158
China-Northeast	2020	67.8	67.7	1.9		67.4	67.3	1.9	
	2030	83.9	96.3	12.9	0	82.7	126.3	13.8	30
	2050	133.2	219.9	103.7	0	135.0	255.9	−22.8	144
China-Tibet	2020	0.8	0.8	0.0		0.8	0.8	0.0	
	2030	1.0	1.0	0.4	0	1.0	1.3	0.2	0
	2050	2.3	2.4	1.4	0	2.4	2.8	−0.9	1
China-Central	2020	208.7	208.7	5.9		207.2	207.2	5.9	
	2030	262.7	260.5	44.9	0	258.4	329.5	34.5	37
	2050	445.3	536.2	299.8	0	451.7	642.0	−218.4	409
China-East	2020	226.8	201.9	47.9		225.9	214.1	31.0	
	2030	286.3	284.3	40.1	0	283.6	372.4	41.5	47
	2050	454.4	633.5	320.4	0	458.5	739.3	−132.0	413

(continued)

Table 8.93 (continued)

Power generation structure		2.0 °C				1.5 °C			
		Max demand	Max generation	Max residual load	Max interconnection requirements	Max demand	Max generation	Max residual load	Max interconnection requirements
China		[GW]	[GW]	[GW]	[GW]	[GW]	[GW]	[GW]	[GW]
China-South	2020	173.6	173.6	9.0		173.6	173.6	9.0	
	2030	242.3	238.6	36.2	0	239.6	312.0	44.6	28
	2050	368.8	529.6	282.0	0	372.8	622.7	−49.1	299
Taiwan	2020	33.0	33.2	0.0					
	2030	46.0	45.9	3.8	0	45.7	52.5	5.9	1
	2050	63.7	92.0	47.1	0	64.1	105.7	−4.0	46

Table 8.94 China: storage and dispatch service requirements

Storage and dispatch		2.0 °C					1.5 °C				
		Required to avoid curtailment	Utilization battery -through-put-	Utilization PSH -through-put-	Total storage demand (incl. H2)	Dispatch hydrogen-based	Required to avoid curtailment	Utilization battery -through-put-	Utilization PSH -through-put-	Total storage demand (incl. H2)	Dispatch hydrogen-based
China		[GWh/year]	[GWh/year]	[GWh/year]	[GWh/year]	[GWh/year]	[GWh/year]	[GWh/year]	[GWh/year]	[GWh/year]	[GWh/year]
China-North	2020	0	0	0	0	0	0	0	0	0	0
	2030	45	3	38	41	11	6734	62	2363	2425	0
	2050	14,152	14,255	641	14,896	96,848	39,562	2958	6350	9308	17,528
China-Northwest	2020	158	2	302	304	0	326	3	547	550	0
	2030	7	1	9	10	1	3401	38	1240	1278	0
	2050	12,360	15,511	661	16,172	39,433	31,642	2171	4847	7018	10,080
China-Northeast	2020	0	0	0	0	0	0	0	0	0	0
	2030	912	22	563	585	143	11,430	57	2362	2418	1
	2050	24,955	22,345	1465	23,809	39,793	49,329	2238	5393	7631	10,012
China-Tibet	2020	0	0	0	0	0	0	0	0	0	0
	2030	0	0	0	0	4	43	0	15	15	0
	2050	0	0	0	0	754	3	1	1	3	230
China-Central Baltic	2020	0	0	0	0	0	0	0	0	0	0
	2030	6	1	10	11	576	6013	74	2305	2379	1
	2050	4763	7167	44	7211	167,132	23,175	2609	4372	6981	47,112
China-East	2020	0	0	0	0	0	0	0	0	0	0
	2030	59	4	79	83	797	8720	95	3042	3137	0
	2050	17,604	21,928	1036	22,964	148,351	50,402	3884	8341	12,225	18,866

(continued)

Table 8.94 (continued)

Storage and dispatch		2.0 °C					1.5 °C				
		Required to avoid curtailment	Utilization battery -through-put-	Utilization PSH -through-put-	Total storage demand (incl. H2)	Dispatch hydrogen-based	Required to avoid curtailment	Utilization battery -through-put-	Utilization PSH -through-put-	Total storage demand (incl. H2)	Dispatch hydrogen-based
China		[GWh/year]	[GWh/year]	[GWh/year]	[GWh/year]	[GWh/year]	[GWh/year]	[GWh/year]	[GWh/year]	[GWh/year]	[GWh/year]
China-South	2020	0	0	0	0	0	0	0	0	0	0
	2030	74	7	89	96	961	8676	93	3086	3179	0
	2050	21,703	28,028	1143	29,171	116,735	56,742	4139	9307	13,446	22,281
Taiwan	2020	0	0	0	0	0	0	0	0	0	0
	2030	0	0	0	0	89	202	5	121	126	0
	2050	6506	5734	943	6677	14,209	13,873	426	2985	3411	0
China	2020	158	2	302	304	0	326	3	547	550	0
	2030	1102	39	789	827	2582	45,217	424	14,533	14,957	2
	2050	102,042	114,967	5932	120,899	623,254	264,729	18,427	41,596	60,022	126,108

8.14 OECD Pacific

8.14.1 OECD Pacific: Long-Term Energy Pathways

8.14.1.1 OECD Pacific: Final Energy demand by Sector

The future development pathways for OECD Pacific's final energy demand when the assumptions on population growth, GDP growth, and energy intensity are combined are shown in Fig. 8.98 for the 5.0 °C, 2.0 °C, and 1.5 °C Scenarios. In the 5.0 °C Scenario, the total final energy demand will decrease by 2%, from the current 20,100 PJ/year to 19,600 PJ/year in 2050. In the 2.0 °C Scenario, the final energy demand will decrease by 46% compared with current consumption and will reach 10,800 PJ/year by 2050. The final energy demand in the 1.5 °C Scenario will reach 10,200 PJ, 49% below the 2015 demand. In the 1.5 °C Scenario, the final energy demand in 2050 will be 6% lower than in the 2.0 °C Scenario. The electricity demand for 'classical' electrical devices (without power-to-heat or e-mobility) will decrease from 1520 TWh/year in 2015 to 1150 TWh/year in 2050 in both alternative scenarios. Compared with the 5.0 °C case (1890 TWh/year in 2050), the efficiency measures in the 2.0 °C and 1.5 °C Scenarios will save 740 TWh/year in 2050.

Electrification will lead to a significant increase in the electricity demand by 2050. The 2.0 °C Scenario has an electricity demand for heating of approximately 400 TWh/year due to electric heaters and heat pumps, and in the transport sector,

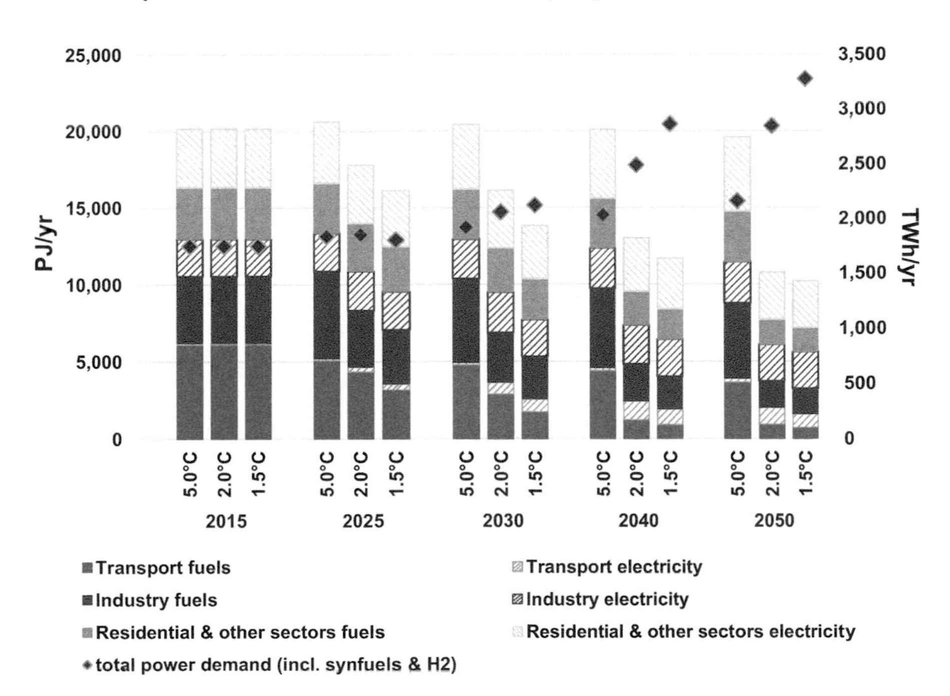

Fig. 8.98 OECD Pacific: development of final energy demand by sector in the scenarios

the electricity demand will be approximately 1100 TWh/year due to electric mobility. The generation of hydrogen (for transport and high-temperature process heat) and the manufacture of synthetic fuels (mainly for transport) will add an additional power demand of 1000 TWh/year. Therefore, the gross power demand will rise from 1900 TWh/year in 2015 to 3000 TWh/year in 2050 in the 2.0 °C Scenario, 25% higher than in the 5.0 °C case. In the 1.5 °C Scenario, the gross electricity demand will increase to a maximum of 3400 TWh/year in 2050.

The efficiency gains in the heating sector could be even larger than in the electricity sector. In the 2.0 °C and 1.5 °C Scenarios, a final energy consumption equivalent to about 3000 PJ/year and 3100 PJ/year, respectively, will be avoided by 2050 through efficiency gains compared with the 5.0 °C Scenario.

8.14.1.2 OECD Pacific: Electricity Generation

The development of the power system is characterized by a dynamically growing renewable energy market and an increasing proportion of total power coming from renewable sources. By 2050, 100% of the electricity produced in OECD Pacific will come from renewable energy sources in the 2.0 °C Scenario. 'New' renewables—mainly wind, solar, and geothermal energy—will contribute 82% of total electricity generation. Renewable electricity's share of the total production will be 60% by 2030 and 89% by 2040. The installed capacity of renewables will reach about 680 GW by 2030 and 1420 GW by 2050. The share of renewable electricity generation in 2030 in the 1.5 °C Scenario is assumed to be 68%. The 1.5 °C Scenario will have a generation capacity from renewable energy of approximately 1590 GW in 2050.

Table 8.95 shows the development of different renewable technologies in OECD Pacific over time. Figure 8.99 provides an overview of the overall power-generation structure in OECD Pacific. From 2020 onwards, the continuing growth of wind and PV, up to 320 GW and 830 GW, respectively, will complemented by up to 60 GW solar thermal generation, as well as limited biomass, geothermal, and ocean energy, in the 2.0 °C Scenario. Both the 2.0 °C and 1.5 °C Scenarios will lead to a high proportion of variable power generation (PV, wind, and ocean) of 40% and 47% by 2030, respectively, and of 68% in both scenarios by 2050.

8.14.1.3 OECD Pacific: Future Costs of Electricity Generation

Figure 8.100 shows the development of the electricity-generation and supply costs over time, including the CO_2 emission costs, in all scenarios. The calculated electricity-generation costs in 2015 (referring to full costs) were around 8 ct/kWh. In the 5.0 °C case, the generation costs will increase until 2030, when they reach 11.1 ct/kWh, and then drop to 10.9 ct/kWh by 2050. The generation costs will increase in the 2.0 °C Scenario until 2030, when they reach 10.5 ct/kWh, and then drop to 8.3 ct/kWh by 2050. In the 1.5 °C Scenario, they will increase to 10.7 ct/kWh, and then drop to 8.5 ct/kWh by 2050. In the 2.0 °C Scenario, the generation

Table 8.95 OECD Pacific: development of renewable electricity-generation capacity in the scenarios

in GW	Case	2015	2025	2030	2040	2050
Hydro	5.0 °C	69	73	76	78	78
	2.0 °C	69	76	78	82	84
	1.5 °C	69	76	78	82	84
Biomass	5.0 °C	9	13	15	16	18
	2.0 °C	9	23	26	35	43
	1.5 °C	9	23	29	42	47
Wind	5.0 °C	9	23	28	40	56
	2.0 °C	9	77	145	263	322
	1.5 °C	9	84	198	335	384
Geothermal	5.0 °C	2	4	5	7	11
	2.0 °C	2	4	14	27	37
	1.5 °C	2	4	14	27	37
PV	5.0 °C	43	84	96	102	107
	2.0 °C	43	225	394	701	831
	1.5 °C	43	253	427	782	932
CSP	5.0 °C	0	0	0	1	1
	2.0 °C	0	1	15	39	57
	1.5 °C	0	1	20	49	67
Ocean	5.0 °C	0	1	1	2	4
	2.0 °C	0	3	8	27	42
	1.5 °C	0	3	8	27	42
Total	5.0 °C	132	197	221	246	275
	2.0 °C	132	409	681	1176	1416
	1.5 °C	132	444	774	1345	1594

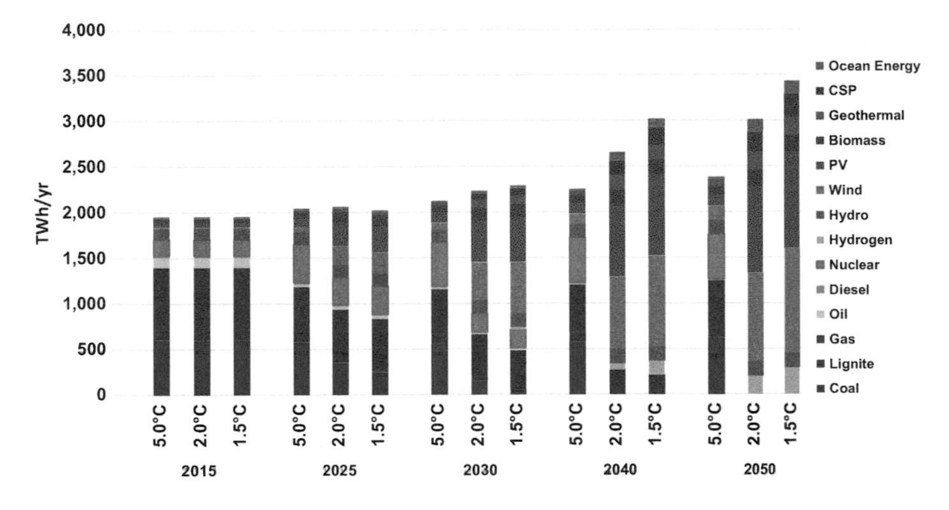

Fig. 8.99 OECD Pacific: development of electricity-generation structure in the scenarios

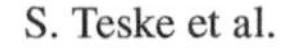

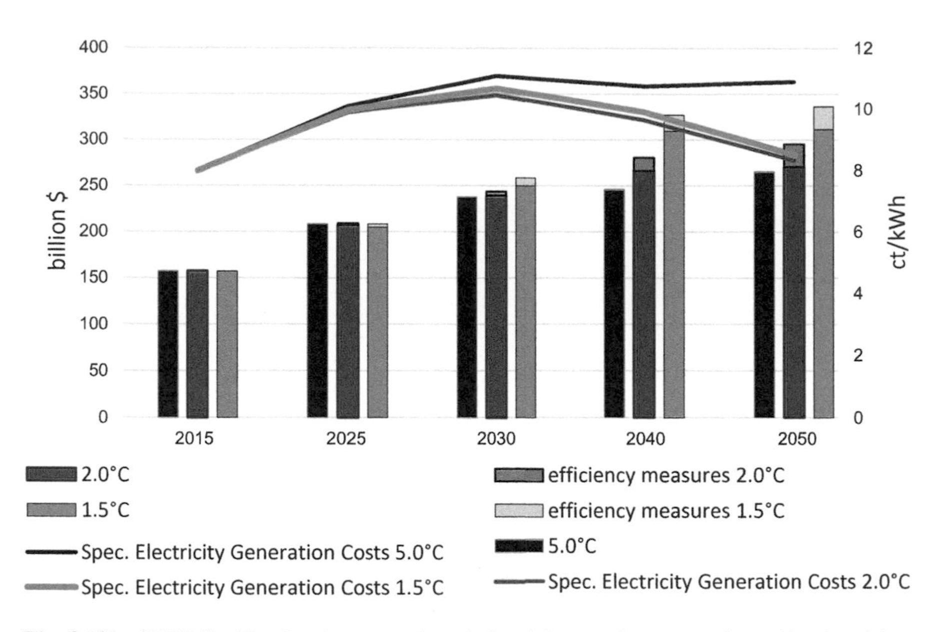

Fig. 8.100 OECD Pacific: development of total electricity supply costs and specific electricity-generation costs in the scenarios

costs in 2050 will be 2.6 ct/kWh lower than in the 5.0 °C case, and in the 1.5 °C Scenario, this difference will 2.4 ct/kWh. Note that these estimates of generation costs do not take into account integration costs such as power grid expansion, storage, or other load-balancing measures.

In the 5.0 ° C case, the growth in demand and increasing fossil fuel prices will cause the total electricity supply costs to rise from today's $160 billion/year to more than $270 billion/year in 2050. In the 2.0 °C Scenario, the total supply costs will be $270 billion/year, and in the 1.5 °C Scenario, they will be $310 billion/year. The long-term costs for electricity supply will be only 2% higher in the 2.0 °C Scenario than in the 5.0 °C Scenario as a result of the estimated generation costs and the electrification of heating and mobility. Further electrification and synthetic fuel generation in the 1.5 °C Scenario will result in total power generation costs that are 17% higher than in the 5.0 °C case.

Compared with these results, the generation costs when the CO_2 emission costs are not considered will increase in the 5.0 °C case to 8.3 ct/kWh in 2050. The generation costs in the 2.0 °C Scenario will increase until 2030, when they will reach 9.3 ct/kWh, and then drop to 8.3 ct/kWh by 2050. In the 1.5 °C Scenario, they will increase to 9.9 ct/kWh, and then drop to 8.5 ct/kWh by 2050. In the 2.0 °C Scenario, the generation costs will be a maximum of 1 ct/kWh higher than in the 5.0 °C case and this will occur in 2040. In the 1.5 °C Scenario, compared with the 5.0 °C Scenario, the maximum difference in the generation costs will be 1.4 ct/kWh, again in 2040. If the CO_2 costs are not considered, the total electricity supply costs in the 5.0 °C case will rise to about $200 billion/year in 2050.

8.14.1.4 OECD Pacific: Future Investments in the Power Sector

An investment of around $2780 billion will be required for power generation between 2015 and 2050 in the 2.0 °C Scenario—including additional power plants for the production of hydrogen and synthetic fuels and investments in the replacement of plants at the end of their economic lifetimes. This value will be equivalent to approximately $77 billion per year on average, and will be $1520 billion more than in the 5.0 °C case ($1260 billion). An investment of around $3100 billion for power generation will required between 2015 and 2050 in the 1.5 °C Scenario. On average, this is an investment of $86 billion per year. In the 5.0 °C Scenario, the investment in conventional power plants will be around 56% of the total cumulative investments, whereas approximately 44% will be invested in renewable power generation and co-generation (Fig. 8.101).

However, in the 2.0 °C (1.5 °C) Scenario, OECD Pacific will shift almost 93% (95%) of its entire investment to renewables and co-generation. By 2030, the fossil fuel share of the power sector investment will predominantly focused on gas power plants that can also be operated with hydrogen.

Because renewable energy has no fuel costs, other than biomass, the cumulative fuel cost savings in the 2.0 °C Scenario will reach a total of $1420 billion in 2050, equivalent to $39 billion per year. Therefore, the total fuel cost savings will be equivalent to 90% of the total additional investments compared to the 5.0 °C Scenario. The fuel cost savings in the 1.5 °C Scenario will add up to $1510 billion, or $42 billion per year.

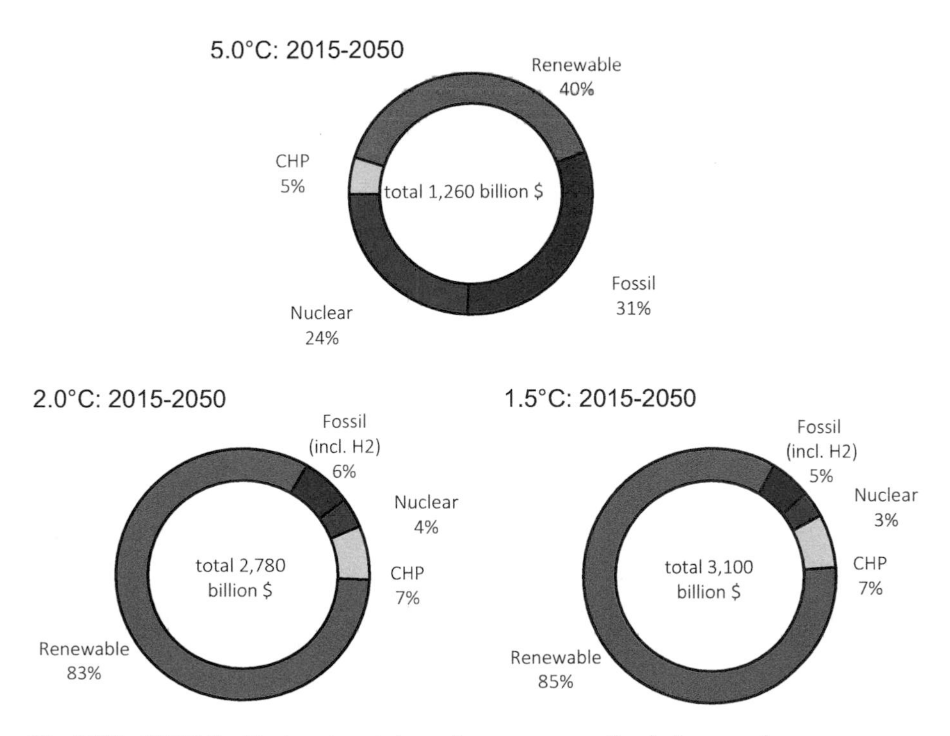

Fig. 8.101 OECD Pacific: investment shares for power generation in the scenarios

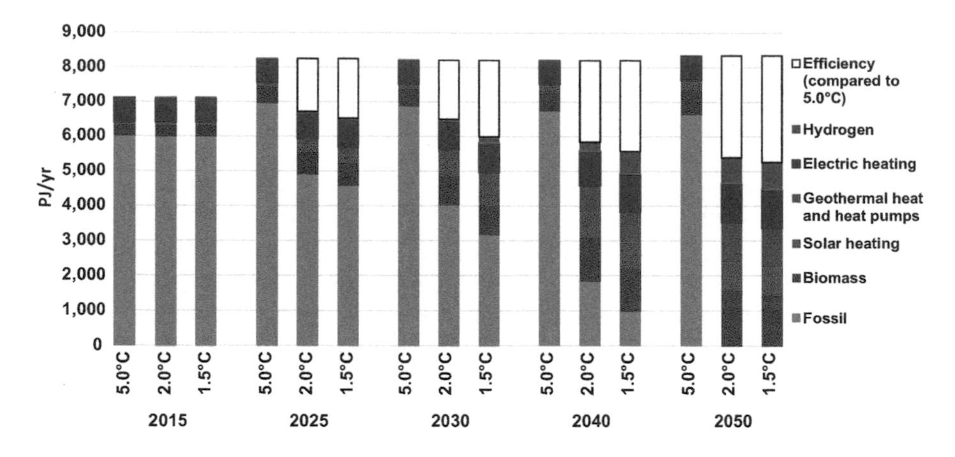

Fig. 8.102 OECD Pacific: development of heat supply by energy carrier in the scenarios

8.14.1.5 OECD Pacific: Energy Supply for Heating

The final energy demand for heating will increase in the 5.0 °C Scenario by 17%, from 7100 PJ/year in 2015 to 8300 PJ/year in 2050. Energy efficiency measures will help to reduce the energy demand for heating by 35% in 2050 in the 2.0 °C Scenario, relative to the 5.0 °C case, and by 37% in the 1.5 °C Scenario. Today, renewables supply around 7% of OECD Pacific's final energy demand for heating, with the main contribution from biomass. Renewable energy will provide 33% of OECD Pacific's total heat demand in 2030 in the 2.0 °C Scenario and 42% in the 1.5 °C Scenario. In both scenarios, renewables will provide 100% of the total heat demand in 2050.

Figure 8.102 shows the development of different technologies for heating in OECD Pacific over time, and Table 8.96 provides the resulting renewable heat supply for all scenarios. Up to 2030, biomass will remain the main contributor. The growing use of solar, geothermal, and environmental heat will lead, in the long term, to a biomass share of 37% in the 2.0 °C Scenario and of 35% in the 1.5 °C Scenario.

Heat from renewable hydrogen will further reduce the dependence on fossil fuels in both scenarios. The hydrogen consumption in 2050 will be around 700 PJ/year in the 2.0 °C Scenario and 800 PJ/year in the 1.5 °C Scenario. The direct use of electricity for heating will also increases by a factor of 1.6 between 2015 and 2050, and will achieves a final energy share of 21% in 2050 in the 2.0 °C Scenario and 22% in the 1.5 °C Scenario.

8.14.1.6 OECD Pacific: Future Investments in the Heating Sector

The roughly estimated investments in renewable heating technologies up to 2050 will amount to around $530 billion in the 2.0 °C Scenario (including the investments for the replacement of plants after their economic lifetimes), or approximately $15 billion per year. The largest share of the investment in OECD Pacific is

Table 8.96 OECD Pacific: development of renewable heat supply in the scenarios (excluding the direct use of electricity)

in PJ/year	Case	2015	2025	2030	2040	2050
Biomass	5.0 °C	314	471	504	584	714
	2.0 °C	314	633	815	1250	1579
	1.5 °C	314	650	823	1229	1463
Solar heating	5.0 °C	45	76	92	150	236
	2.0 °C	45	221	452	737	819
	1.5 °C	45	252	543	772	795
Geothermal heat and heat pumps	5.0 °C	30	33	34	36	38
	2.0 °C	30	157	307	737	1119
	1.5 °C	30	197	420	830	1094
Hydrogen	5.0 °C	0	0	0	0	0
	2.0 °C	0	6	16	251	728
	1.5 °C	0	9	160	642	772
Total	5.0 °C	390	580	629	769	988
	2.0 °C	390	1017	1591	2975	4245
	1.5 °C	390	1107	1946	3473	4124

Table 8.97 OECD Pacific: installed capacities for renewable heat generation in the scenarios

in GW	Case	2015	2025	2030	2040	2050
Biomass	5.0 °C	44	60	63	69	75
	2.0 °C	44	77	92	117	94
	1.5 °C	44	79	91	113	80
Geothermal	5.0 °C	0	0	0	0	0
	2.0 °C	0	3	8	20	28
	1.5 °C	0	3	7	22	26
Solar heating	5.0 °C	13	22	27	43	69
	2.0 °C	13	64	128	207	230
	1.5 °C	13	73	152	215	224
Heat pumps	5.0 °C	5	5	5	5	6
	2.0 °C	5	11	23	54	74
	1.5 °C	5	16	36	63	71
Total[a]	5.0 °C	62	87	95	117	150
	2.0 °C	62	156	250	397	426
	1.5 °C	62	171	287	413	401

[a] Excluding direct electric heating

assumed to be for solar collectors (around $240 billion), followed by heat pumps and biomass technologies. The 1.5 °C Scenario assumes an even faster expansion of renewable technologies, but with a similar average annual investment of around $15 billion per year (Table 8.97, Fig. 8.103).

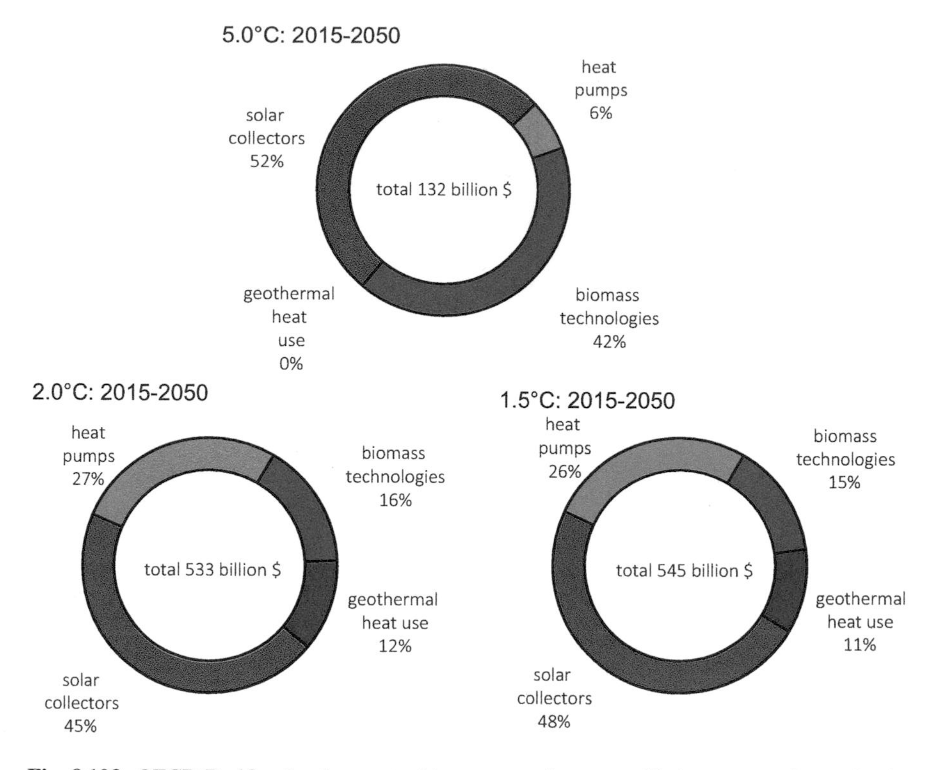

Fig. 8.103 OECD Pacific: development of investments for renewable heat-generation technologies in the scenarios

8.14.1.7 OECD Pacific: Transport

Energy demand in the transport sector in OECD Pacific is expected to decrease by 37% in the 5.0 °C Scenario, from around 6200 PJ/year in 2015 to 3900 PJ/year in 2050. In the 2.0 °C Scenario, assumed technical, structural, and behavioural changes will save 49% (around 1900 PJ/year) by 2050 compared with the 5.0 °C Scenario. Additional modal shifts, technology switches, and a reduction in the transport demand will lead to even higher energy savings in the 1.5 °C Scenario of 59% (or 2300 PJ/year) in 2050 compared with the 5.0 °C case (Table 8.98, Fig. 8.104).

By 2030, electricity will provide 20% (200 TWh/year) of the transport sector's total energy demand in the 2.0 °C Scenario, whereas in 2050, the share will be 53% (300 TWh/year). In 2050, up to 480 PJ/year of hydrogen will be used in the transport sector as a complementary renewable option. In the 1.5 °C Scenario, the annual electricity demand will be 240 TWh in 2050. The 1.5 °C Scenario also assumes a hydrogen demand of 360 PJ/year by 2050.

Biofuel use is limited in the 2.0 °C Scenario and the 1.5 °C Scenario to a maximum of approximately 200 PJ/year. Therefore, around 2030, synthetic fuels based on power-to-liquid will be introduced, with a maximum amount of 270 PJ/year in 2050 in the 2.0 °C Scenario. Due to the lower overall energy demand in transport, the maximum synthetic fuel demand will amount to 210 PJ/year in the 1.5 °C Scenario.

Table 8.98 OECD Pacific: projection of transport energy demand by mode in the scenarios

in PJ/year	Case	2015	2025	2030	2040	2050
Rail	5.0 °C	158	162	163	162	161
	2.0 °C	158	154	156	154	159
	1.5 °C	158	156	156	162	161
Road	5.0 °C	5515	4317	3902	3365	2614
	2.0 °C	5515	3961	2979	1837	1456
	1.5 °C	5515	2891	1975	1399	1123
Domestic aviation	5.0 °C	331	524	663	863	922
	2.0 °C	331	338	308	242	194
	1.5 °C	331	307	240	147	109
Domestic navigation	5.0 °C	173	178	181	186	193
	2.0 °C	173	178	181	186	193
	1.5 °C	173	178	181	186	193
Total	5.0 °C	6176	5182	4908	4576	3890
	2.0 °C	6176	4631	3624	2419	2002
	1.5 °C	6176	3533	2551	1893	1586

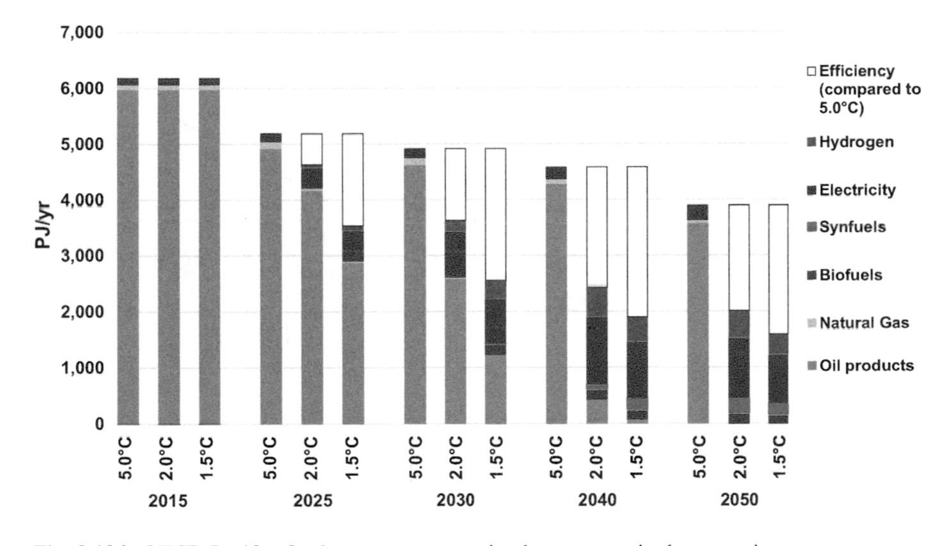

Fig. 8.104 OECD Pacific: final energy consumption by transport in the scenarios

8.14.1.8 OECD Pacific: Development of CO_2 Emissions

In the 5.0 °C Scenario, OECD Pacific's annual CO_2 emissions will decrease by 21%, from 2080 Mt in 2015 to 1640 Mt in 2050. The stringent mitigation measures in both alternative scenarios will cause the annual emissions to fall to 280 Mt in 2040 in the 2.0 °C Scenario and to 160 Mt. in the 1.5 °C Scenario, with further reductions to almost zero by 2050. In the 5.0 °C case, the cumulative CO_2 emissions from 2015 until 2050 will add up to 67 Gt. In contrast, in the 2.0 °C and 1.5 °C Scenarios, the cumulative emissions for the period from 2015 until 2050 will be 31 Gt and 26 Gt, respectively.

Therefore, the cumulative CO_2 emissions will decrease by 54% in the 2.0 °C Scenario and by 61% in the 1.5 °C Scenario compared with the 5.0 °C case. A rapid reduction in the annual emissions will occur under both alternative scenarios. In the 2.0 °C Scenario, this reduction will be greatest in 'Power generation', followed by 'Transport' and 'Industry' (Fig. 8.105).

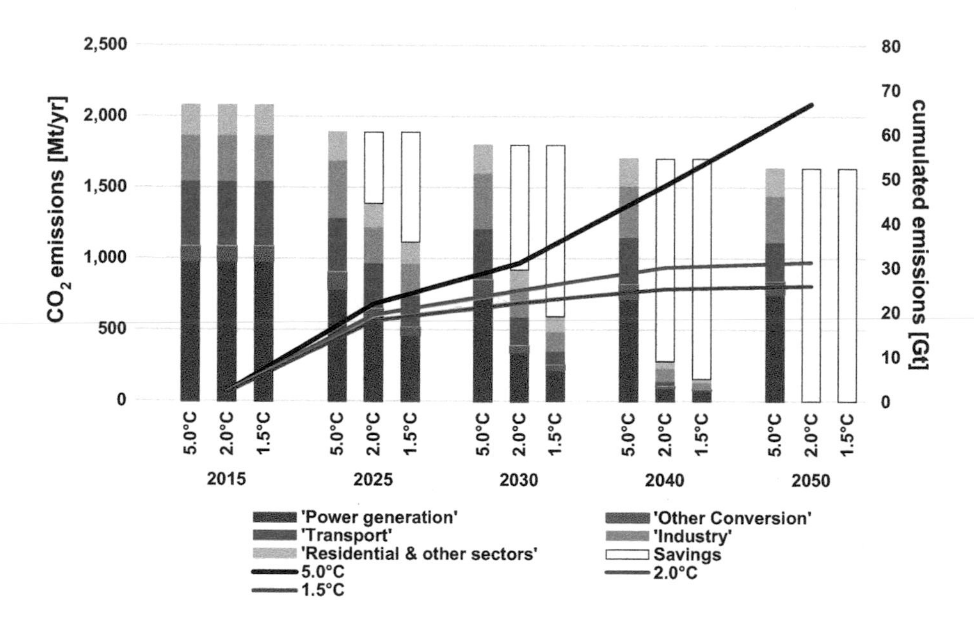

Fig. 8.105 OECD Pacific: development of CO_2 emissions by sector and cumulative CO_2 emissions (after 2015) in the scenarios ('Savings' = reduction compared with the 5.0 °C Scenario)

8.14.1.9 OECD Pacific: Primary Energy Consumption

The levels of primary energy consumption in the three scenarios when the assumptions discussed above are taken into account are shown in Fig. 8.106. In the 2.0 °C Scenario, the primary energy demand will decrease by 48%, from around 36,300 PJ/year in 2015 to 18,900 PJ/year in 2050. Compared with the 5.0 °C Scenario, the overall primary energy demand will decrease by 45% by 2050 in the 2.0 °C Scenario (5.0 °C: 34,700 PJ in 2050). In the 1.5 °C Scenario, the primary energy demand will be even lower (19,900 PJ in 2050) because the final energy demand and conversion losses will be lower.

Both the 2.0 °C Scenario and 1.5 °C Scenario aim to rapidly phase-out coal and oil. This will cause renewable energy to have a primary energy share of 33% in 2030 and 88% in 2050 in the 2.0 °C Scenario. In the 1.5 °C Scenario, renewables will have a primary energy share of more than 89% in 2050 (including non-energy consumption, which will still include fossil fuels). Nuclear energy will be phased-out in 2040 in both the 2.0 °C and 1.5 °C Scenarios. The cumulative primary energy consumption of natural gas in the 5.0 °C case will add up to 230 EJ, the cumulative coal consumption to about 300 EJ, and the crude oil consumption to 380 EJ. In contrast, in the 2.0 °C Scenario, the cumulative gas demand will amount to 150 EJ, the cumulative coal demand to 100 EJ, and the cumulative oil demand to 230 EJ. Even lower

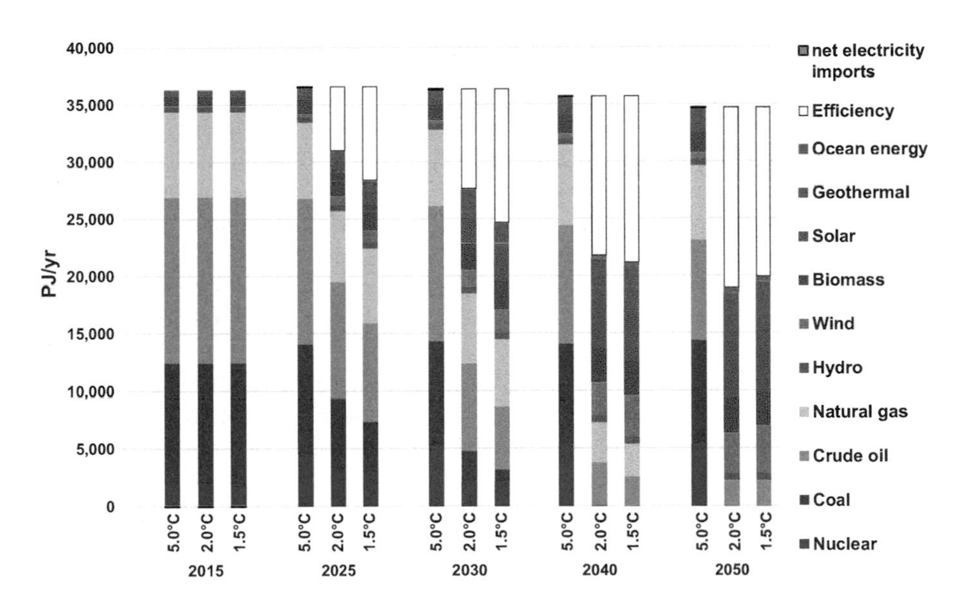

Fig. 8.106 OECD Pacific: projection of total primary energy demand (PED) by energy carrier in the scenarios (including electricity import balance)

fossil fuel use will be achieved in the 1.5 °C Scenario: 150 EJ for natural gas, 70 EJ for coal, and 190 EJ for oil.

8.14.2 *OECD Pacific: Power Sector Analysis*

South Korea, Japan, Australia, and New Zealand form the OECD Pacific region (also referred to as OECD Asia Pacific or OECD Asia Oceania). Like Non-OECD Asia, a regional interconnected power market with regular electricity exchange is unlikely. Therefore, the region is broken down into seven sub-regions: (1) South Korea; (2) the north of Japan; (3) the south of Japan; (4) Australia's National Electricity Market (NEM) (covering the entire east coast); (5) the SWIS-NT grid region (comprising Western Australia and the Northern Territory); (6) the North Island of New Zealand; and (7) the South Island of New Zealand. The sub-regions have very different electricity policies, power-generation structures, and demand patterns. In this analysis, simplifications that may not reflect the local conditions are made to ensure that the results comparable on a global level. Therefore, the results for specific countries are only estimates.

8.14.2.1 OECD Pacific: Development of Power Plant Capacities

The region has significant potential for all renewables, including the dominant renewable power technologies of solar PV and onshore wind. Japan has significant geothermal power resources, and offshore wind potentials are substantial across the region. There is also potential for ocean energy across the region, although it is currently a niche technology. Australia has one of the best solar resources in the world, so concentrated solar power plants will be an important part of both scenarios in Australia. Coal and nuclear capacities will be phased-out as plants come to the end of their lifetimes. In the 1.5 °C Scenario, the last coal power plant will be phased out just after 2030.

The solar PV market will reach 8 GW in 2020 under the 2.0 °C Scenario—the same level as the actual regional market of 8.3 GW (REN21-GSR 2018) in 2017—and increase rapidly to 43 GW by 2030. The 1.5 °C Scenario requires that solar PV will achieve an equal market size by 2030 and remain at this level until 2040.

However, the onshore market must increase significantly compared with the market in 2017, which was only 0.54 GW (GWEC-NL 2018). By 2025, 12 GW of onshore wind capacity must be installed annually across the region under the 2.0 °C

Table 8.99 OECD Pacific: average annual change in installed power plant capacity

OECD Pacific power generation: average annual change of installed capacity [GW/a]	2015–2025		2026–2035		2036–2050	
	2.0 °C	1.5C°	2.0 °C	1.5 °C	2.0 °C	1.5 °C
Hard coal	−4	−9	−5	−4	−1	0
Lignite	0	−1	−2	−2	0	0
Gas	2	−2	−1	−3	−14	0
Hydrogen-gas	0	1	1	5	12	12
Oil/diesel	−3	−2	−2	−2	−1	−1
Nuclear	0	−5	−3	−3	−2	−2
Biomass	2	1	1	1	1	1
Hydro	2	1	0	1	0	0
Wind (onshore)	7	18	12	17	7	6
Wind (offshore)	1	4	5	5	2	2
PV (roof top)	17	33	33	33	16	21
PV (utility scale)	6	11	11	11	5	7
Geothermal	0	2	2	2	2	2
Solar thermal power plants	1	2	3	4	2	3
Ocean energy	0	1	2	2	2	2
Renewable fuel based co-generation	**1**	**1**	**1**	**2**	**1**	**1**

Scenario, and 17 GW under the 1.5 °C Scenario. By 2030, geothermal, concentrated solar power, and ocean energy must increase by around 2 GW each (Table 8.99).

8.14.2.2 OECD Pacific: Utilization of Power Generation Capacities

The very different developments of variable and dispatch power plants in all sub-regions reflect the diversity the Pacific region. Table 8.100 shows that because there is no interconnection between the northern and southern parts of Japan, we assume that even within Japan, the separate electricity markets of the 50 Hz and 60 Hz regions will remain as they are. For Australia, it is assumed that the east- and west-coast electricity markets will have limited interconnection capacities by 2030. The North and South Islands of New Zealand are calculated to have an increased inter-connection capacity by 2050.

Table 8.101 shows that for the region as a whole, the limited dispatchable power plants will retain a relatively high capacity factor, compared with other regions, until after 2020 and decrease thereafter. The average capacity factors from 2030 onwards will be consistent with all other regions.

Table 8.100 OECD Pacific: power system shares by technology group

Power generation structure and interconnection		2.0 °C				1.5 °C			
OECD Pacific		Variable RE	Dispatch RE	Dispatch Fossil	Inter-connection	Variable RE	Dispatch RE	Dispatch Fossil	Inter-connection
South Korea	2015	4%	34%	61%	5%				
	2030	40%	18%	43%	20%	46%	17%	38%	0%
	2050	70%	28%	2%	25%	62%	24%	14%	0%
Japan – North (50 Hz)	2015	4%	34%	61%	5%				
	2030	46%	26%	28%	20%	51%	23%	26%	0%
	2050	72%	26%	2%	25%	64%	21%	15%	0%
Japan – South (60 Hz)	2015	4%	34%	61%	5%				
	2030	36%	38%	26%	20%	41%	35%	24%	0%
	2050	72%	25%	2%	25%	64%	20%	15%	0%
Australia – East and South (NEM)	2015	5%	34%	61%	5%				
	2030	17%	82%	0%	20%	17%	83%	0%	10%
	2050	73%	26%	2%	25%	67%	21%	12%	20%
Australia West and North (SWIS + NT)	2015	5%	34%	61%	5%				
	2030	41%	37%	22%	20%	46%	33%	21%	10%
	2050	73%	25%	2%	25%	67%	21%	12%	20%

New Zealand – North Island	2015	5%	34%	61%	5%				
	2030	39%	61%	0%	20%	45%	55%	0%	10%
	2050	77%	22%	2%	25%	70%	18%	12%	20%
New Zealand – South Island	2015	5%	34%	61%	5%				
	2030	39%	61%	0%	20%	45%	55%	0%	10%
	2050	77%	22%	2%	25%	70%	18%	12%	20%
OECD Pacific	2015	4%	34%	61%					
	2030	40%	31%	30%		45%	29%	27%	
	2050	71%	26%	2%		64%	22%	14%	

Table 8.101 OECD Pacific: capacity factors by generation type

Utilization of variable and dispatchable power generation:		2015	2020	2020	2030	2030	2040	2040	2050	2050
OECD Pacific			2.0 °C	1.5 °C	2.0 °C	1.5 °C	2.0 °C	1.5 °C	2.0 °C	1.5 °C
Capacity factor – average	[%/yr]	54.8%	55%	55%	29%	29%	29%	29%	34%	31%
Limited dispatchable: fossil and nuclear	[%/yr]	65.1%	54%	54%	26%	31%	19%	29%	25%	32%
Limited dispatchable: renewable	[%/yr]	42.7%	63%	54%	29%	30%	54%	25%	27%	25%
Dispatchable: fossil	[%/yr]	48.6%	48%	50%	20%	23%	35%	21%	19%	26%
Dispatchable: renewable	[%/yr]	43.1%	73%	73%	50%	52%	37%	46%	49%	46%
Variable: renewable	[%/yr]	23.2%	17%	17%	20%	20%	27%	27%	31%	28%

8.14.2.3 OECD Pacific: Development of Load, Generation, and Residual Load

Table 8.102 shows the development of the maximum load, generation, and resulting residual load in the Pacific region. To verify the calculation results, we compared the peak demands in Australia and Japan.

The peak load for Australia's NEM was calculated to be 32.6 GW in 2020, which corresponds to the reported summer peak of 32.5 GW in the summer of 2017/2018 (AER 2018). Japan's peak demand was 152 GW in 2015 according to the Tokyo Electric Power Company (TEPCO -2018) and TEPCO predicts that it will be 136 GW in 2020, which is 11% lower.

In the long term, the Pacific region will be a renewable fuel producer for the export market. Therefore, the calculated increased interconnection capacities indicate overproduction, which will be used for international bunker fuels.

The storage and dispatch requirements for all sub-regions are shown in Table 8.103. The Pacific region has vast solar and wind resources and will therefore be one of the production hubs for synthetic fuels and hydrogen, which may be used for industrial processes, for bunker fuels, or to replace natural gas. Therefore, the storage and dispatch demand may vary significantly because they depend on the extent to which renewable fuel production is integrated into the national power sectors or used for dispatch and demand-side management. The more integrated the fuel production is, the lower the overall requirement for battery or hydro pump storage technologies. Further research is required to develop a dedicated plan to produce renewable bunker fuels in Australia.

Table 8.102 OECD Pacific: load, generation, and residual load development

Power generation structure		2.0 °C				1.5 °C			
		Max demand	Max generation	Max Residual Load	Max interconnection requirements	Max demand	Max generation	Max residual load	Max interconnection requirements
OECD Pacific		[GW]	[GW]	[GW]	[GW]	[GW]	[GW]	[GW]	[GW]
South Korea	2020	86.8	86.8	1.6		86.8	86.8	1.6	
	2030	92.3	145.0	6.6	46	94.5	167.5	14.9	58
	2050	116.4	298.0	58.3	123	134.7	339.9	62.4	143
Japan – North (50 Hz)	2020	130.6	97.9	35.1		130.6	97.4	36.0	
	2030	79.1	125.0	4.1	42	81.6	145.4	4.1	60
	2050	106.0	252.2	49.4	97	120.1	288.5	33.9	134
Japan – South (60 Hz)	2020	83.6	83.6	3.5		83.7	83.6	3.5	
	2030	87.5	126.1	4.8	34	90.4	148.0	11.3	46
	2050	116.2	287.9	51.2	121	132.6	329.3	38.8	158
Australia – East and South (NEM)	2020	6.7	7.0	1.2		6.7	7.0	1.2	
	2030	4.3	6.4	3.8	0	4.4	6.6	3.9	0
	2050	5.7	12.6	2.6	4	6.4	14.4	2.5	5
Australia West and North (SWIS + NT)	2020	32.6	32.6	1.1		32.6	32.6	1.1	
	2030	33.9	49.6	1.1	15	34.8	58.3	1.1	22
	2050	44.7	111.5	21.3	45	51.4	127.4	22.5	53
New Zealand – North Island	2020	5.5	5.5	3.9		5.5	5.5	3.9	
	2030	5.0	6.9	0.2	2	5.1	8.2	0.2	3
	2050	6.5	15.9	3.0	6	7.5	18.1	2.1	9
New Zealand – South Island	2020	1.3	4.4	0.0					
	2030	1.5	2.1	0.0	1	1.5	2.4	0.0	1
	2050	2.0	4.8	0.9	2	2.2	5.4	0.6	3

Table 8.103 OECD Pacific: storage and dispatch service requirements

Storage and dispatch		2.0 °C					1.5 °C				
		Required to avoid curtailment	Utilization battery -through-put-	Utilization PSH -through-put-	Total Storage demand (incl. H2)	Dispatch Hydrogen-based	Required to avoid curtailment	Utilization battery -through-put-	Utilization PSH -through-put-	Total storage demand (incl. H2)	Dispatch hydrogen-based
OECD Pacific		[GWh/year]	[GWh/year]	[GWh/year]	[GWh/year]	[GWh/year]	[GWh/year]	[GWh/year]	[GWh/year]	[GWh/year]	[GWh/year]
South Korea	2020	0	0	0	0	70	0	0	0	0	51
	2030	13,803	275	1968	2242	241	27,635	400	3197	3596	890
	2050	156,658	46,248	8717	54,965	22,747	176,909	45,195	8599	53,793	16,906
Japan – North (50 Hz)	2020	0	0	0	0	85	0	0	0	0	62
	2030	25,236	357	2820	3177	200	44,298	418	3819	4238	131
	2050	156,580	32,626	6784	39,411	21,744	185,902	32,676	6917	39,594	15,831
Japan – South (60 Hz)	2020	0	0	0	0	121	0	0	0	0	88
	2030	21,734	343	2320	2663	303	37,937	439	3490	3929	774
	2050	199,561	38,310	8309	46,618	24,062	233,815	38,207	8381	46,588	17,382
Australia – East and South (NEM)	2020	114	0	0	0	15	202	0	0	0	11
	2030	850	0	55	55	0	1696	0	86	86	0
	2050	9375	1983	457	2440	924	11,304	1981	472	2453	538
Australia West and North (SWIS + NT)	2020	4	0	0	0	0	49	0	0	0	0
	2030	11,311	219	1621	1840	87	19,866	255	2289	2544	79
	2050	90,062	18,266	4625	22,891	8053	103,839	18,187	4637	22,824	5140

New Zealand – North Island	2020	0	0	0	0	4	0	0	0	0	3
	2030	1223	20	142	162	0	2165	26	221	247	0
	2050	12,361	2316	546	2862	1090	14,474	2304	548	2852	779
New Zealand – South Island	2020	0	0	0	0	0	0	0	0	0	0
	2030	374	6	43	49	0	658	8	67	74	0
	2050	3733	695	164	859	328	4371	691	164	855	235
OECD Pacific	2020	118	0	0	0	295	251	0	0	0	215
	2030	84,079	1246	9157	10,403	831	146,440	1564	13,290	14,855	1874
	2050	654,287	140,807	29,623	170,431	81,215	760,962	139,369	29,724	169,093	59,243

References

AER (2018), Australian Energy Regulator (AER), Seasonal peak demand (NEM), website, viewed in September 2018, https://www.aer.gov.au/wholesale-markets/wholesale-statistics/seasonal-peak-demand-nem

ASEAN-CE (2018) ASEAN Centre for Energy, ASEAN POWER GRID initiative, http://www.aseanenergy.org/programme-area/apg/ (viewed September 2018)

ASEAN (2018), The Association of Southeast Asian Nations, or ASEAN, was established on 8 August 1967 in Bangkok, Thailand, with the signing of the ASEAN Declaration (Bangkok Declaration) by the Founding Fathers of ASEAN, namely Indonesia, Malaysia, Philippines, Singapore and Thailand. Brunei Darussalam then joined on 7 January 1984, Viet Nam on 28 July 1995, Lao PDR and Myanmar on 23 July 1997, and Cambodia on 30 April 1999, making up what is today the ten Member States of ASEAN. (Source: https://asean.org/asean/about-asean/)

C2ES (2017), INTERCONNECTED: CANADIAN AND U.S. ELECTRICITY, March 2017, Doug Vine, C2ES, Center for Climate and Energy Solutions, 2101 WILSON BLVD. SUITE 550 ARLINGTON, VA 22201 703-516-4146, https://www.c2es.org/site/assets/uploads/2017/05/canada-interconnected.pdf - total cumulative current interconnection capacity between Canada and the USA kV4,606

ENTSO-E (2018), Statistical Factsheet 2017, Electricity system data of member TSO countries, https://docstore.entsoe.eu/Documents/Publications/Statistics/Factsheet/entsoe_sfs_2017.pdf

EU-EG (2017), Towards a sustainable and integrated Europe, Report of the Commission Expert Group on electricity interconnection targets, November 2017, Page 25, https://ec.europa.eu/energy/sites/ener/files/documents/report_of_the_commission_expert_group_on_electricity_interconnection_targets.pdf

GWEC (2018), Global Wind Report: Annual Market Update 2017, Global Wind Energy Council, (GWEC), Rue d'Arlon 80, 1040 Brussels, Belgium, http://files.gwec.net/files/GWR2017.pdf?ref=PR

GWEC-NL (2018), GWEC 2016, Newsletter – November 2016, China's new Five-Year-Plan, http://gwec.net/chinas-new-five-year-energy-plan/

Hohmeyer (2015) A 100% renewable Barbados and lower energy bills: A plan to change Barbados' power supply to 100% renewables and its possible benefits, January, 2015, Prof. Dr. Olav Hohmeyer, Europa University Flensburg, Discussion Papers, CENTER FOR SUSTAINABLE ENERGY SYSTEMS (CSES/ZNES); System Integration Department, ISSN: 2192–4597(Internet Version), https://www.uni-flensburg.de/fileadmin/content/abteilungen/industrial/dokumente/downloads/veroeffentlichungen/diskussionsbeitraege/znes-discussions-papers-005-barbados.pdf

IDB (2013) Uruguay – Rapid Assessment and Gap Analysis, UNDP, Inter-American Development Bank (IDB), 2013, Sustainable Energy For All, https://www.seforall.org/sites/default/files/Uruguay_RAGA_EN_Released.pdf

IEA RED (2016) – Renewable Policy Updated, Issue 11, 17 November 2016, https://www.iea.org/media/topics/renewables/repolicyupdate/REDRenewablePolicyUpdateNo11FINAL20161117.pdf

IEA P+M DB (2018), International Energy Agency, Policies and Measure Database, viewed September 2018, http://www.iea.org/policiesandmeasures/pams/india/name-168047-en.php?s=dHlwZT1jYyZzdGF0dXM9T2s,&return=PG5hdiBpZD0iYnJlYWRjcnVtYiI-PGEgaHJlZj0iLyI-SG9tZTwvYT4gJnJhcXVvOyA8YSBocmVmPSIvcG9saWNpZXNhbmRtZW-FzdXJlcy8iPlBvbGljaWVzIGFuZCBNZWFzdXJlczwvYT4gJnJhcXVvOyA8YSBocmVmP-SIvcG9saWNpZXNhbmRtZWFzdXJlcy9jbGltYXRlY2hhbmdlLyI-Q2xpbWF0ZSBDaGFu-Z2U8L2E-PC9uYXY

IRENA (2014), African Clean Energy Corridor – internet page viewed September 2018, http://www.irena.org/cleanenergycorridors/Africa-Clean-Energy-Corridor

PVM-(3-2018), PV-Magazine, Kazahkstan opened a tender for 290 MW solar and 620 MW wind power plants in March 2018, Kazahkstan tenders 290 MW of solar, 19th March 2018, Emilliano Bellini, PV-Magazine, https://www.pv-magazine.com/2018/03/09/kazahkstan-tenders-290-mw-of-solar/

PR-DoE (2016), Republic of the Philippines, Department of Energy, government website, viewed in September 2018, 2016 Philippines Power Situation Report, https://asean.org/asean/about-asean/

REN21-GSR (2018). 2018; Renewables 2018 Global Status Report, (Paris: REN21 Secretariat), ISBN 978-3-9818911-3-3, http://www.ren21.net/status-of-renewables/global-status-report/

REW (1-2018), Renewable Energy World, Wind Power Curtailment in China on the Mend, January 26, 2018, Liu Yuanyuan, https://www.renewableenergyworld.com/articles/2018/01/wind-power-curtailment-in-china-on-the-mend.html

RF (2018), Rockefeller Foundation, 24x7 Power is About "Access", Not "Electrification", Jaideep Mukherji, 22nd January 2018, https://www.rockefellerfoundation.org/blog/24x7-power-access-not-electrification/

TEPCO (2018) Tokyo Electric Power Company (TEPCO), companies website, viewed September 2018, https://www4.tepco.co.jp/en/corpinfo/illustrated/power-demand/peak-demand-international-e.html

TYNDP (2016) ENTSO-E, TYNDP 2016 (published) and TYNDP 2018 (in consultation) are published online only; https://tyndp.entsoe.eu/

WB-DB (2018), World Bank – Database, World Development Indicators, The per capita electricity demand of Turkey has been 2850 kWh (2014), compared to 6352 kWh in the Euro area, http://databank.worldbank.org/data/reports.aspx?source=2&series=EG.USE.ELEC.KH.PC&country=

WPM (3-2018), Wind Power monthly, The first commercial wind farm in Russia opened in 2018, and a project pipeline of more than 1000 MW is reported in Wind Power monthly; Russian partners plan Leningrad wind farm, 27th March 2018, https://www.windpowermonthly.com/article/1460599/russian-partners-plan-leningrad-wind-farm

Chapter 9
Trajectories for a Just Transition of the Fossil Fuel Industry

Sven Teske

Abstract This section provides historical production data for coal, oil and gas between 1980 and 2015. The 2.0 °C and 1.5 °C scenario lead to specific phase-out pathways for each of the fossil fuel types. Current regional production volumes are compared with future demands. The results provide the input for the employment analysis in the following chapter for the fossil fuel sector. This section discusses the need to shift the current political debate about coal, oil and gas which is focused on security of supply and price security towards an open debate about an orderly withdrawal from coal, oil and gas extraction industries.

The implementation of the 2.0 °C and 1.5 °C climate mitigation pathways presented here will have a significant impact on the global fossil fuel industry. Although this may appear to be stating the obvious, current climate debates have not yet involved open discussion of the orderly withdrawal from the coal, oil, and gas extraction industries. Instead, the political debate about coal, oil, and gas has focused on the security of supply and price security. However, mitigating climate change is only possible when fossil fuels are phased-out. This section provides an overview of the time-frame of this phase-out under the 2.0 °C and 1.5 °C Scenarios compared with the 5.0 °C pathway.

9.1 Fossil Energy Resources—The Sky Is the Limit

An unrelenting increase in fossil fuel extraction conflicts with the finite nature of these resources. At the same time, the global distribution of oil and gas resources does not match the distribution of demand. Therefore, some countries currently rely almost entirely on imported fossil fuels. Therefore, is the relative scarcity of fossil fuels an additional reason an energy transition? The Global Energy Assessment

S. Teske (✉)
Institute for Sustainable Futures, University of Technology Sydney, Sydney, NSW, Australia
e-mail: sven.teske@uts.edu.au

S. Teske (ed.), *Achieving the Paris Climate Agreement Goals*,
https://doi.org/10.1007/978-3-030-05843-2_9

Table 9.1 Fossil reserves, resources, and additional occurrences

Energy carrier	Reserves [EJ/year]	Resources [EJ/year]	Demand in 2015 [EJ/year]
Conventional oil	4900–7610	4170–6150	41.9
Unconventional oil	3750–5600	11,280–14,800	
Conventional gas	5000–7100	7200–8900	33.8
Unconventional gas	20,100–67,100	40,200–121,900	
Coal	17,300–21,000	291,000–435,000	16.5

(GEA 2012), an integrated assessment of the global energy system, has published a comprehensive overview of estimated available fossil fuel reserves and resources. Table 9.1 shows the estimates for conventional and unconventional coal, oil, and gas reserves and resources. The distinction between reserves and resources is based on the current technology (exploration and production) and market conditions. The resource data are not cumulative and do not include reserves (GEA 2012).

The assessment shows that there is no shortage of fossil fuels. There might be a shortage of conventional oil and gas, but unconventional resources are still significantly larger than our climate can cope with. Reducing global fossil fuel consumption for reasons of resource scarcity alone is not essential, even though there may be substantial price fluctuations and regional or structural shortages, as we have seen in the past (Teske and Pregger 2015).

9.2 Coal—Past Production and Future Trajectories Under Three Scenarios

Global coal production is dominated by China, which in 2017, produced over 3.5 billion tonnes of coal, 45% of the world volume, followed by India with 716 million tonnes, the USA with 702 million tons, and Australia with 481 million tons. The top 10 producers, in order of annual production, are China, India, USA, Australia, Indonesia, the Russian Federation, South Africa, Germany (mainly lignite), Poland, and Kazakhstan. These countries account for 90% of the global coal production.

Figure 9.1 shows the historical time series for global coal production. The data are based on the BP Statistical Review 2018 (BP 2018), as are the following overviews of oil and gas. Production volumes have declined in recent years, mainly due to changes in demand in China, but they rose again in 2017.

Under the 5.0 °C scenario, the required production of thermal coal (excluding coal for non-energy uses, such as steel production) will remain at the 2015 level, with an annual increase of around 1% per year until 2050. As shown in Fig. 9.2, under the 2.0 °C Scenario, coal production will decline sharply between 2020 and 2030 at a rate of around 6% per year. By 2030, the global coal production will be equal to China's annual production in 2017, at 3.7 billion tons, whereas that volume will be reached in 2025 under the 1.5 °C Scenario.

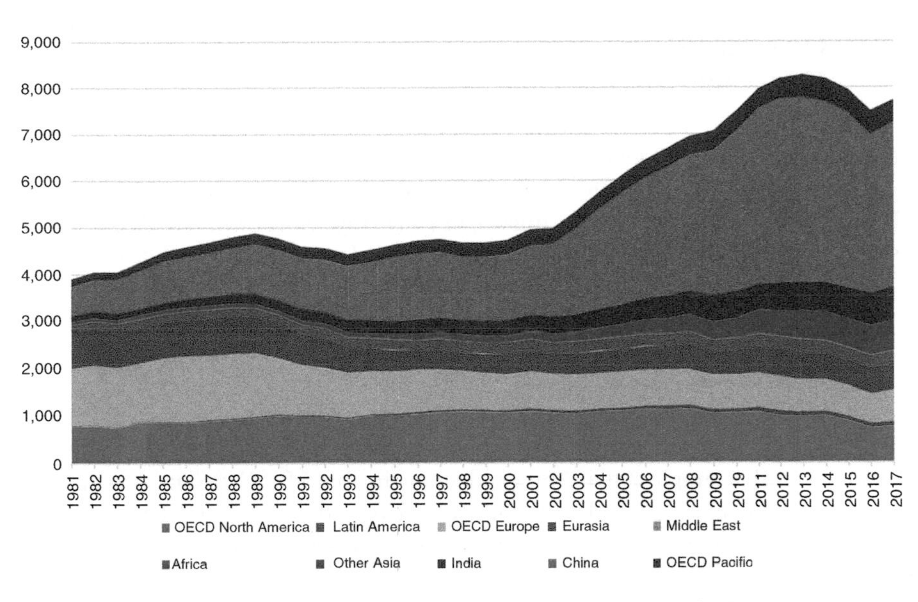

Fig. 9.1 Global coal production in 1981–2017 (BP 2018—Statistical Review)

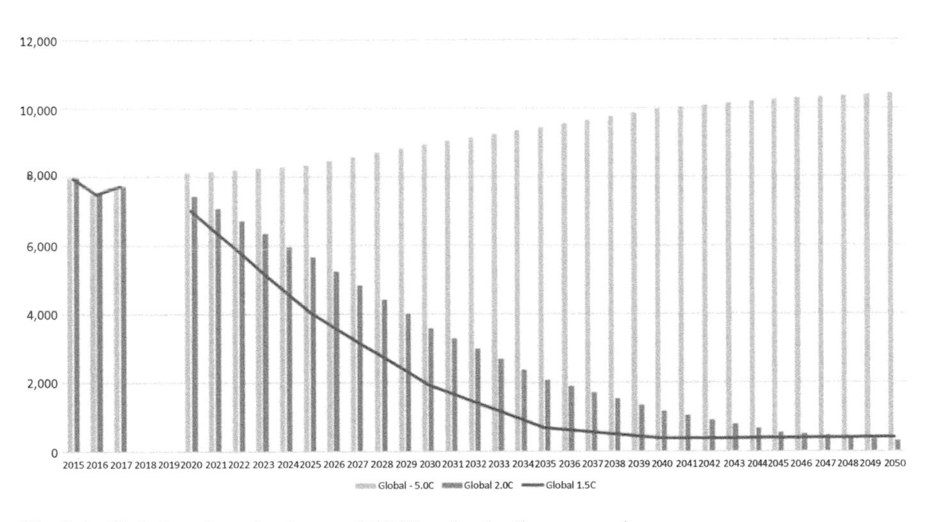

Fig. 9.2 Global coal production until 2050 under the three scenarios

9.3 Oil—Past Production and Future Trajectories Under the Three Scenarios

Oil production almost doubled between 1965 and 1975. After the early 1990s, it grew almost constantly and by 2017, the production volume was about three times higher than in 1965 and twice as high as in 1985 (Fig. 9.3). Unlike coal, there is no sign of a decline in oil production in response to reduced demand. Oil production is more widely distributed than coal production. Three countries, the USA, Russia, and Saudi Arabia, have global market shares of around 12–14% each, whereas four countries, Canada, Iran, Iraq, and China, produce around 5% each. The other oil-producing countries have significantly lower market shares.

Figure 9.4 shows the global oil production levels required by the three calculated scenarios. Oil for non-energy uses, such as the petrochemical industry, is not included in this graph. Again, oil production in the 5.0 °C Scenario will grow steadily by 1% until the end of the modelling period in 2050. Under the 2.0 °C Scenario, oil production will decline by 5% per year until 2030 and by 3% annually until 2025. After 2030, production will decline by around 7% per year, on average, until oil production for energy is phased-out entirely. The oil production capacity of the USA, Saudi Arabia, and Russia in 2017 would be sufficient to supply the global demand calculated for the 2.0 °C Scenario in 2035. The 1.5 °C Scenario will cut the required production volumes in half by 2030, reducing them further to the equivalent of the production volume of just one of the three largest oil producers (the USA, Saudi Arabia, or Russia) by 2040.

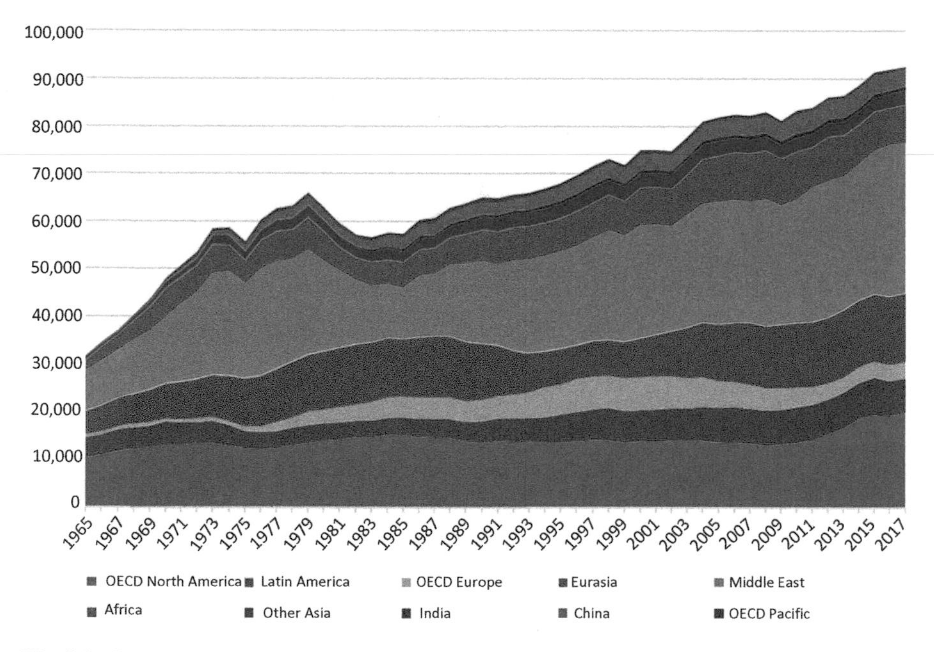

Fig. 9.3 Global oil production in 1965–2017 (BP 2018—Statistical Review)

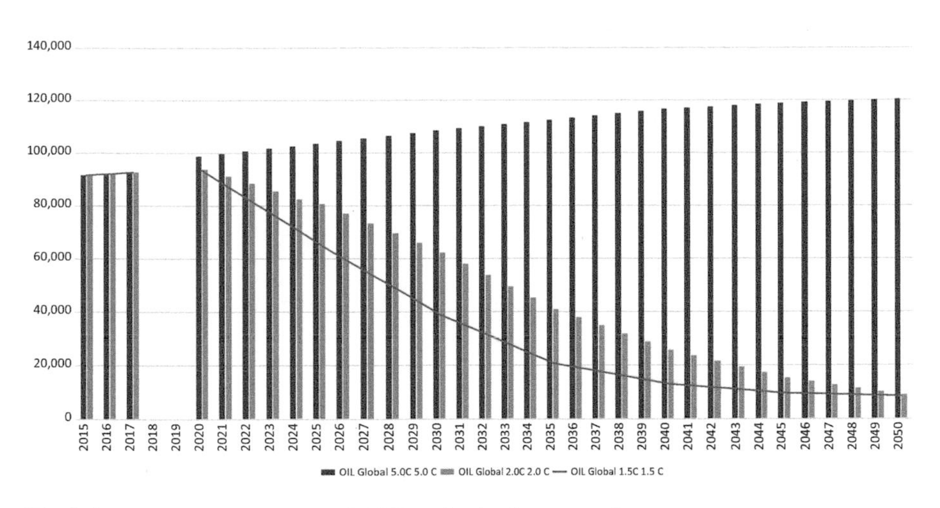

Fig. 9.4 Global oil production until 2050 under the three scenarios

9.4 Gas—Past Production and Future Trajectories Under the Three Scenarios

Gas production has grown steadily over the past four decades, leading to an overall production of 3500 billion cubic meters—3.5 times higher than in 1970. The production of natural gas is even more widely distributed than oil production. According to 2017 figures, by far the largest producers are the USA, with 20% of the global volume, and Russia with 17%. Four countries have market share of around 5% each: Canada (4.8%), Iran (6.1%), Qatar (4.8%), and China (4.1%). The remaining 43% of global gas production is distributed over 42 countries (Fig. 9.5).

In the 5.0 °C Scenario, gas production will increase steadily by 2% a year for the next two decades, leading to an overall production increase of about 50% by 2050. Compared with coal and oil, the gas phase-out will be significantly slower in the 2.0 °C and 1.5 °C Scenarios. Furthermore, these scenarios assume that infrastructure, such as gas pipelines and power plants, will be used after this phase-out for hydrogen and/or renewable methane produced with electricity from renewable sources (see Chap. 5, Sect 5.2). Under the 2.0 °C Scenario, gas production will only decrease by 0.2% per year until 2025, then by 1% until 2030, and on average by 4% annually until 2040. This represents a rather slow phase-out and will allow the gas industry to gradually transfer to hydrogen. The phase-out under the 1.5 °C Scenario will be equally slow, and a 4%/year reduction will occur after 2025 (Fig. 9.6).

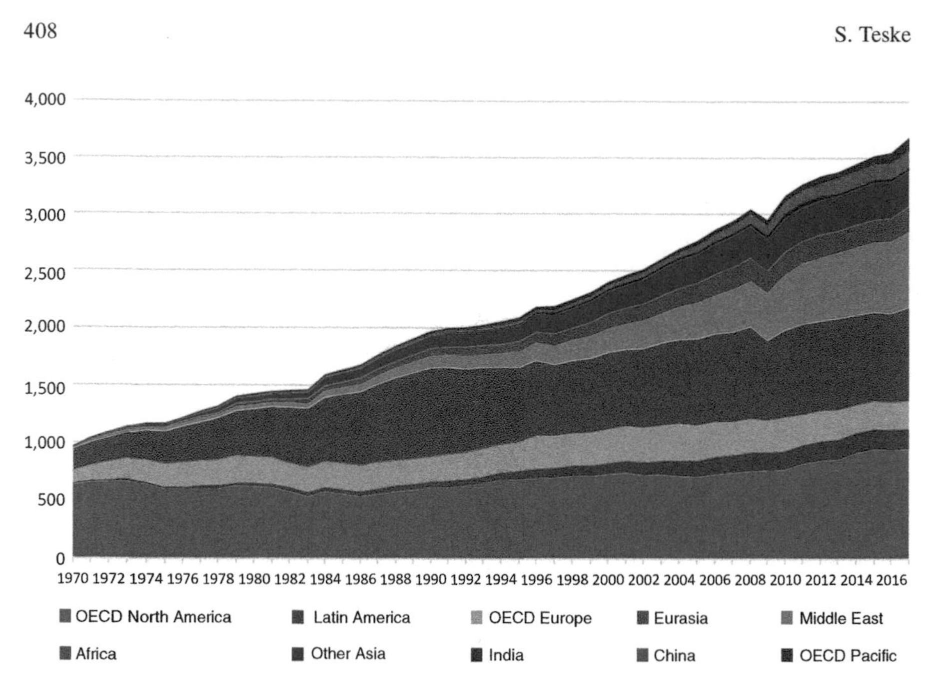

Fig. 9.5 Global gas production in 1970–2017 (BP 2018—Statistical Review)

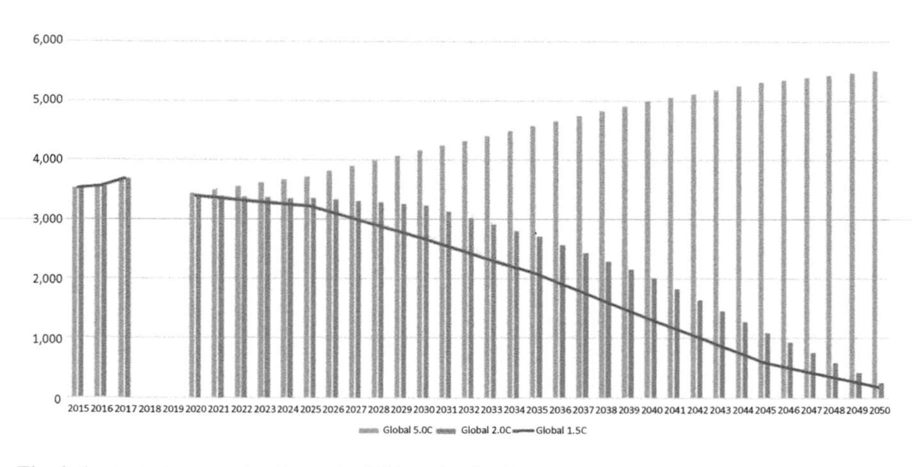

Fig. 9.6 Global gas production until 2050 under the three scenarios

9.5 Overview: Required Fossil Fuel Resources Under the 5.0 °C, 2.0 °C, and 1.5 °C Trajectories

In summary, the global fossil fuel extraction industry must reduce production at a rate of 2% per annum under the 2.0 °C Scenario and 3% per annum under the 1.5 °C Scenario. A constant reduction in production seems unlikely if no international measures are taken to organize the economic and social transitions in the producing countries, and for the communities and workers involved. The idea of a 'just transition' is well documented in the international literature. According to the International Labour Organization (ILO 2015), the concept was first mentioned in the 1990s, when North American unions began developing the concept of just transition.

Initially, trade unionists understood 'just transition' to be a program of support for workers who lost their jobs due to environmental protection policies. Since then, several UNFCCC Climate Conferences have referred to the 'just transition' concept. The Paris Climate Agreement 2015, during the 21st session of the Conference of the Parties (COP 21) "*decided to continue and improve the forum on the impact of the implementation of response measures (hereinafter referred to as the improved forum), and adopted the work programme, comprising two areas: (1) economic diversification and transformation; and (2) just transition of the workforce, and the creation of decent work and quality jobs*" (UNFCCC-JT 2016).

Table 9.2 provides possible trajectories for global coal, oil, and gas production, consistent with the Paris Agreement targets. These trajectories are the results of the 2.0 °C and 1.5 °C Scenarios, documented in detail over the previous six chapters of this book. Chapter 10 uses these trajectories to calculate possible employment effects, both in terms of job losses in the fossil fuel industry, job gains in the renewable energy industry, and options for transitioning the gas industry towards a renewably produced hydrogen industry.

Table 9.2 Summary—coal, oil, and gas trajectories for a just transition under the 5.0 °C, 2.0 °C, and 1.5 °C Scenarios

	2015	2020	2025	2030	2035	2040	2045	2050
5.0 °C: Primary energy [PJ/a]								
Coal	140,895	147,324	153,529	167,795	179,666	191,078	196,692	200,680
Lignite	19,835	18,836	18,550	18,573	18,597	19,028	19,546	19,562
Natural gas	123,673	133,732	145,075	162,132	178,213	194,467	207,273	214,702
Crude oil	166,465	173,082	181,520	190,294	197,300	204,563	208,561	210,970
5.0 °C production units								
Coal [million tons per year]	5871	6138	6397	6991	7486	7962	8196	8362
Lignite [million tons per year]	2088	1983	1953	1955	1958	2003	2058	2059
Natural gas [billion cubic meters]	3171	3429	3720	4157	4570	4986	5315	5505
Oil [thousand barrels per day]	94,836	98,606	103,414	108,412	112,403	116,541	118,819	120,191
2.0 °C—primary energy [PJ/a]								
Coal	140,624	136,111	114,647	77,766	45,445	25,594	12,480	7568
Lignite	19,835	16,779	8333	3203	1630	912	288	0
Natural gas	123,770	132,209	130,797	126,054	105,321	78,390	42,535	9949
Crude oil	166,472	164,438	141,523	109,213	71,812	45,013	26,649	15,461
2.0 °C—production units								
Coal [million tons per year]	5859	5671	4777	3240	1894	1066	520	315
Lignite [million tons per year]	2088	1766	877	337	172	96	30	0
Natural gas [billion cubic meters]	3174	3390	3354	3232	2701	2010	1091	255
Oil [thousand barrels per day]	94,840	93,682	80,627	62,219	40,912	25,644	15,182	8808
1.5 °C— primary energy [PJ/a]								
Coal	141,275	125,431	84,267	41,360	14,243	9134	9363	9759
Lignite	19,835	16,956	5006	2056	777	0	0	0
Natural gas	123,426	132,241	125,494	104,349	80,940	50,883	23,202	7315
Crude oil	166,472	163,957	114,986	68,449	36,541	22,923	16,772	14,794
1.5 °C production units								
Coal [million tons per year]	5886	5226	3511	1723	593	381	390	407
Lignite [million tons per year]	2088	1785	527	216	82	0	0	0
Natural gas [billion cubic meters]	3165	3391	3218	2676	2075	1305	595	188
Oil [thousand barrels per day]	94,840	93,408	65,509	38,996	20,818	13,059	9555	8428

References

BP (2018), British Petrol, Statistical review, website with statistical data for download, downloaded in September 2018, https://www.bp.com/en/global/corporate/energy-economics/statistical-review-of-world-energy/downloads.html

GEA (2012): Global Energy Assessment - Toward a Sustainable Future, Cambridge University Press, Cambridge, UK and New York, NY, USA and the International Institute for Applied Systems Analysis, Laxenburg, Austria; www.globalenergyassessment.org

ILO (2015), International Labour Organization, Just Transition – A report for the OECD, May 2017, Just Transition Centre, www.ituc-csi.org/just-transition-centre

Teske, Pregger (2015), Teske, S, Pregger, T., Naegler, T., Simon, S., Energy [R]evolution - A sustainable World Energy Outlook 2015, Greenpeace International with the German Aerospace Centre (DLR), Institute of Engineering Thermodynamics, System Analysis and Technology Assessment, Stuttgart, Germany https://www.scribd.com/document/333565532/Energy-Revolution-2015-Full

UNFCCC-JT (2016) Just Transition of the Workforce, and the Creation of Decent Work and Quality Jobs – Technical paper, United Nations – Framework Convention on Climate Change (UNFCCC) 20 https://unfccc.int/sites/default/files/resource/Just%20transition.pdf

Chapter 10
Just Transition: Employment Projections for the 2.0 °C and 1.5 °C Scenarios

Elsa Dominish, Chris Briggs, Sven Teske, and Franziska Mey

Abstract This section provides the input data for two different employment development calculation methods: The quantitative analysis, which looks into the overall number of jobs in renewable and fossil fuel industries and the occupational analysis which looks into specific job categories required for the solar and wind sector as well as the oil, gas, and coal industry. Results are given with various figures and tables.

10.1 Introduction: Employment Modelling for a Just Transition

The transition to a 100% renewable energy system is not just a technical task. It is also a socially and economically challenging process, and it is imperative that the transition is managed in a fair and equitable way. One of the key concerns is the employment of workers in the affected industries (UNFCCC 2016; ILO 2015). However, it should be noted that the 'just transition' concept is concerned not only with workers' rights, but also with the well-being of the broader community (Smith 2017; Jenkins et al. 2016; Sovacool and Dworkin 2014). This includes, for example, community participation in decision-making processes, public dialogue, and policy mechanisms to create an enabling environment for new industries, to ensure local economic development.

Although it is acknowledged that a just transition is important, there are limited data on the impacts that the transition will have on employment. There is even less information on the types of occupations that will be affected by the transition, either by project growth or a decline in employment. This study provides projections for jobs in construction, manufacturing, operations and maintenance (O&M), and fuel and heat supply across 12 technologies and 10 world regions, based on the energy

E. Dominish (✉) · C. Briggs · S. Teske · F. Mey
Institute for Sustainable Futures, University of Technology Sydney, Sydney, NSW, Australia
e-mail: elsa.dominish@uts.edu.au; chris.briggs@uts.edu.au; sven.teske@uts.edu.au; franziska.mey@uts.edu.au

S. Teske (ed.), *Achieving the Paris Climate Agreement Goals*,
https://doi.org/10.1007/978-3-030-05843-2_10

scenario from the Leonardo Di Caprio project (see Chap. 3 ff. This study is funded by the German Greenpeace Foundation and builds on the methodology developed by UTS/ISF (Rutovitz et al. 2015), with an updated framework that disaggregates jobs by specific occupations. Projected employment is calculated regionally, but in this chapter, we present an overview of the global data, which are an aggregate of the results for the 10 world regions. Further details, including a further regional breakdown of employment data, are provided in the full report (Dominish et al. 2018).

10.2 Quantitative Employment Modelling

This section discusses the calculation factors used for the quantitative employment modelling (an overview of the methodology is given in Sect. 3.6 of Chap. 3). The factors were analysed on a regional basis where possible, to take into account the significant economic differences between world regions. The results are then presented in the following section.

10.2.1 Employment Factors

Employment factors were used to calculate the number of jobs required per unit of electrical or heating capacity, or per unit of fuel. The employment factors differ depending on whether they involve manufacturing, construction, operation and maintenance, or fuel supply. Information about these factors usually comes from OECD countries because that is where most data are collected, although local data were used wherever possible. For job calculations in non-OECD regions, regional adjustments were made when a local factor was not available (see Sect. 10.2.2). The employment factors used in the calculations are shown in Table 10.1.

The employment factors were based on coal supplies, because employment per tonne varies significantly across the world regions and because coal plays a significant role in energy production in many countries. In Australia and the USA, coal is extracted at an average rate of more than 9000 tonnes per person per year, whereas in Europe, the average coal miner is responsible for less than 1000 tonnes per year. China has relatively low per capita productivity at present, with 650 tonnes per worker per year, but the annual increases in productivity are very high. India and Eurasia have significantly increased their productivity since a similar analysis was performed in 2015. Local data were also used for gas extraction in every region except India, the Middle East, and Non-OECD-Asia. The calculation of coal and gas employment per petajoule (PJ) drew on data from national statistics and company reports, combined with production figures from the BP Statistical Review of World Energy 2018 (BP-SR 2018) or other sources. Data were collected for as many major coal-producing countries as possible, and coverage was obtained for 90% of the world coal production (Table 10.2).

Table 10.1 Summary of employment factors used in a global analysis in 2012

	Construction/installation	Manufacturing	Operations & maintenance	Fuel – PRIMARY energy demand
	Job years/ MW	Job years/ MW	Jobs/MW	
Coal	11.4	5.1	0.14	Regional
Gas	1.8	2.9	0.14	Regional
Nuclear	11.8	1.3	0.6	0.001 jobs per GWh final energy demand
Biomass	14.0	2.9	1.5	29.9 jobs/PJ
Hydro-large	7.5	3.9	0.2	
Hydro-small	15.8	11.1	4.9	
Wind onshore	3.0	3.4	0.3	
Wind offshore	6.5	13.6	0.15	
PV	13.0	6.7	0.7	
Geothermal	6.8	3.9	0.4	
Solar thermal	8.9	4.0	0.7	
Ocean	10.3	10.3	0.6	
Geothermal – heat	6.9 jobs/ MW (construction and manufacturing)			
Solar – heat	8.4 jobs/ MW (construction and manufacturing)			
Nuclear decommissioning	0.95 jobs per MW decommissioned			
Combined heat and power	CHP technologies use the factor for the technology, i.e. coal, gas, biomass, geothermal, etc., increased by a factor of 1.5 for O&M only.			

Note: For details of sources and derivation of factors see Dominish et al. (2018)

Table 10.2 Employment factors used for coal fuel supply (mining and associated jobs)

	Employment factor Jobs per PJ	Tonnes per person per year (coal equivalent)
World average	36.2	943
OECD North America	3.5	9613
OECD Europe	36.2	942
OECD Asia-Oceania	3.6	9455
India	33.6	1016
China	52.9	645
Africa	13.7	2482
Eastern Europe/Eurasia	36.0	948
Developing Asia	6.5	5273
Latin America	12.5	2725
Middle East	Used world average because no employment data were available	

10.2.2 Regional Adjustments

The employment factors used in this model for energy technologies other than coal mining were usually for OECD regions, which are typically wealthier than other regions. A regional multiplier was applied to make the jobs per MW more realistic for other parts of the world. In developing countries, there are generally more jobs per unit of electricity because those countries have more labour-intensive practices. The multipliers change over the study period, consistent with the projections for GDP per worker. This reflects the fact that as prosperity increases, labour intensity tends to fall. The multipliers are shown in Table 10.3.

10.2.2.1 Local Employment Factors

Local employment factors were used where possible. These region-specific factors were:

- *OECD Americas*—gas and coal fuel, photovoltaics (PV) and offshore wind (all factors), and solar thermal power (construction and operation and maintenance (O&M)
- *OECD Europe*—gas and coal fuel, offshore wind (all factors), solar thermal power (construction and O&M), and solar heating
- *OECD Pacific*—gas and coal fuel
- *Africa*—gas, coal, and biomass fuel
- *China*—gas and coal fuel, and solar heating
- *Eastern Europe/Eurasia*—gas and coal fuel
- *Developing Asia*—coal fuel
- *India* – coal fuel and solar heating
- *Latin America*—coal and biomass fuels, onshore wind (all factors), nuclear (construction and O&M), large hydro (O&M), and small hydro (construction and O&M).

Table 10.3 Regional multipliers used for the quantitative calculation of employment

	2015	2020	2030	2040	2050
OECD (North America, Europe, Pacific)	1.0	1.0	1.0	1.0	1.0
Latin America	3.4	3.4	3.4	3.1	2.9
Africa	5.7	5.7	5.6	5.2	4.9
Middle East	1.4	1.5	1.5	1.4	1.3
Eastern Europe/Eurasia	2.4	2.4	2.2	2.0	1.8
India	7.0	5.6	3.7	2.7	2.0
Developing Asia	6.1	5.3	4.2	3.5	2.9
China	2.6	2.2	1.6	1.3	1.1

Source: Derived from ILO (2012) Key Indicators of the Labour Market, eighth edition software, with growth in GDP per capita derived from IEA World Energy Outlook 2018 and World Bank data

10.2.2.2 Local Manufacturing and Fuel Production

Some regions do not manufacture the equipment (e.g., wind turbines or solar PV panels) required for the introduction of renewable technologies. This model includes estimates of the percentages of renewable technology that are made locally and assumes that the percentage of local manufacturing will increase over time as the industry matures. Based on this, the jobs involving the manufacture of components for export were calculated for the region in which the manufacturing occurs. The same applies to coal and gas, because they are traded internationally, so the jobs in fuel supply were calculated regionally, based on historical data.

10.2.2.3 Learning Adjustments or 'Decline Factors'

Learning adjustments are used to account for the projected reductions in the costs of renewables over time, as technologies and companies become more efficient and production processes are scaled up. Generally, jobs per MW are projected to fall in parallel with this trend. The cost projections for each of the calculated energy scenario regions (see Sect. 5.3 of Chap. 5) were used to derive these factors.

10.2.3 Results of Quantitative Employment Modelling

The 2.0 °C and 1.5 °C Scenarios will result in an increase in energy-sector jobs in the world as a whole at every stage of the projection. The 1.5 °C Scenario will increase the renewable energy capacities faster, so employment will increase faster than in the 2.0 °C Scenario. By 2050, employment in te energy sector will be within the same range in both scenarios, at around 48–50 million jobs.

- In 2025, there will be 29.6 million energy-sector jobs in the 5.0 °C Scenario, 42.3 million in the 2.0 °C Scenario, and 48.1 million in the 1.5 °C Scenario.
- In 2030, there will be 30.3 million energy-sector jobs in the 5.0 °C Scenario, 49.2 million in the 2.0 °C Scenario, and 53.8 million in the 1.5 °C Scenario.
- In 2050, there will be 29.6 million energy-sector jobs in the 5.0 °C Scenario, 50.4 million in the 2.0 °C Scenario, and 47.8 million in the 1.5 °C Scenario.

Figure 10.1 shows the changes in job numbers under the 5.0 °C, 2.0 °C, and 1.5 °C Scenarios for each technology between 2015 and 2030. the 5.0 °C Scenario, jobs will drop to 4% below the 2015 levels by 2020 and then remain quite stable until 2030. Strong growth in renewable energy will lead to an increase of 44% in the total energy-sector jobs in the 2.0 °C Scenario and 66% in the 1.5 °C Scenario by 2025. In the 2.0 °C (1.5 °C) Scenario the renewable energy sector will account for 81% (86%) in 2025 and 87% (89%) in 2030, with PV having the greatest share of 24% (26%), followed by biomass, wind, and solar heating.

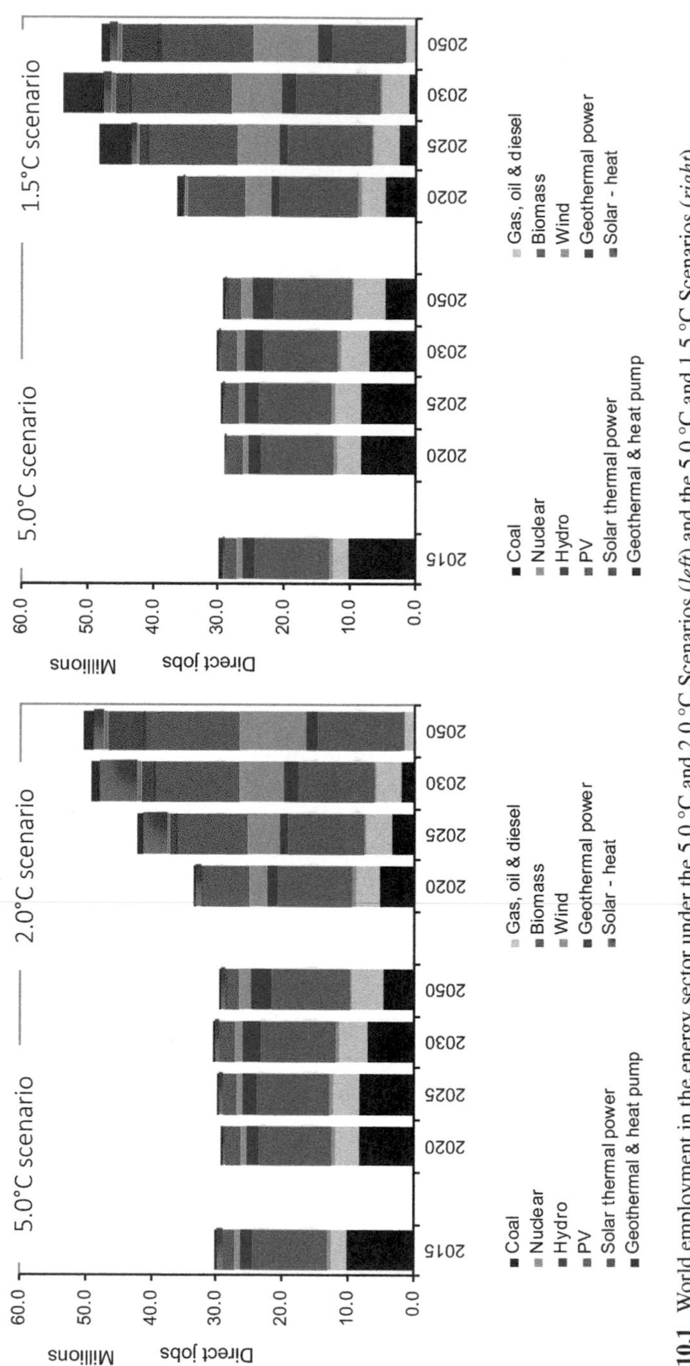

Fig. 10.1 World employment in the energy sector under the 5.0 °C and 2.0 °C Scenarios (*left*) and the 5.0 °C and 1.5 °C Scenarios (*right*)

10.3 Occupational Employment Modelling

To plan for a just transition, it is important to understand the occupations and locations at which jobs are likely to be lost or created. The modelling of employment by type of occupation is based on a new framework developed by UTS/ISF and financed by the German Greenpeace Foundation. The framework is applied to the results of the employment modelling discussed in Sect. 10.2. This information can be used to attempt to understand where labour is likely to be required in the renewable energy transition, and where job losses are likely to occur.

10.3.1 Background: Development of Occupational Employment Modelling

The occupational employment modelling framework used in this study was developed for renewable energy (solar PV, onshore wind, offshore wind) and fossil fuels (coal and gas). The three primary studies that classified and measured the occupational composition of renewable energy industries were conducted by the International Renewable Energy Agency (IRENA). Through surveys of around 45 industry participants across a range of developed and developing nations, IRENA estimated the percentages of person–days for the various occupations across the solar PV and onshore and offshore wind farm supply chains (IRENA 2017a, b, 2018). Figure 10.2 is an example (in this case, for solar PV manufacturing).

IRENA's studies are the most detailed estimates of the occupational compositions of the solar PV and onshore wind industries to date. ISF has extended the application of IRENA's work in two key ways:

1. **Mapping IRENA's job categories against the International Standard Classification of Occupations (ISCO):** IRENA uses its own occupational classification system, which does not match the ISCO, which is the basis for national statistical agency data. For example, 'regulation and standardization experts' is not a category in the ISCO. Consequently, the IRENA job categories have been mapped and translated across to the ISCO to facilitate comparisons between renewable energy technologies and fossil fuel sectors. The best fit for each of the occupations in the IRENA studies has been identified at one-digit, two-digit, three-digit, and four-digit levels of the ISCO.
2. **Unpacking mid- and low-skill job categories in IRENA's study**: Some of the categories in the IRENA studies containe jobs that are of interest from a just transition perspective. Specifically, IRENA combines:
    - 'Factory workers' for solar PV and onshore and offshore wind manufacturing
    - 'Ship crews' for offshore wind construction and operation and maintenance
    - 'Construction workers' for solar PV, onshore wind farm construction, and operation and maintenance.

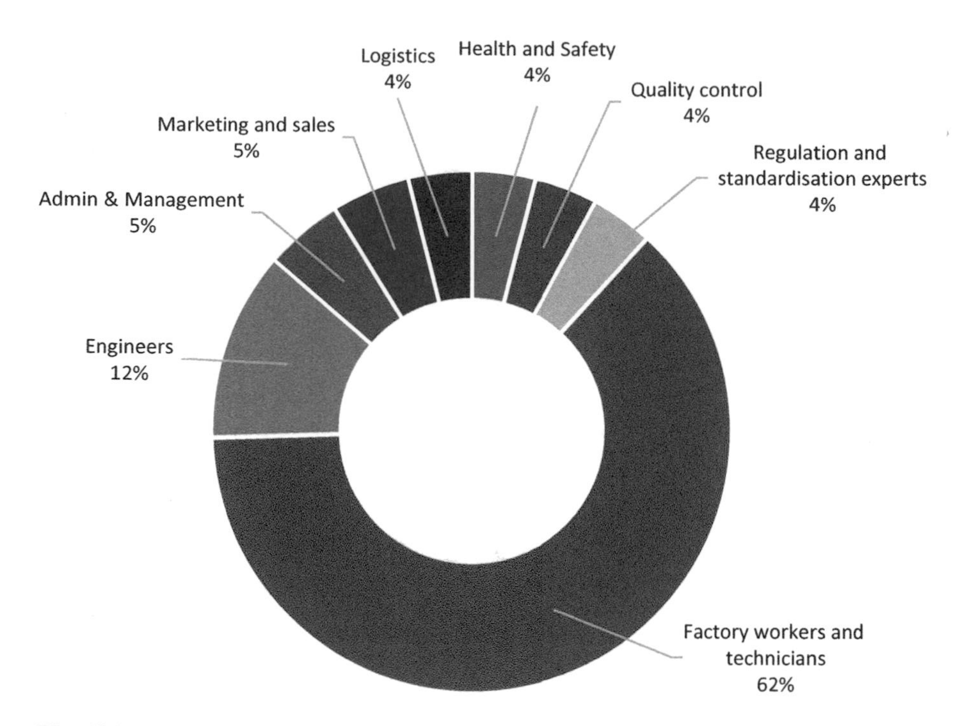

Fig. 10.2 Distribution of human resources required to manufacture the main components of a 50 MW solar photovoltaic power plant. (IRENA 2017a)

These categories combine a range of technicians, trades, machinery operators, drivers, and assemblers, and labourers.

For factory workers, industry data from the *Benchmarks of Global Clean Energy Manufacturing* study by the Clean Energy Manufacturing Analysis Centre was combined with the occupational framework from the Australian census. The international standard classification of industries used for the Benchmarks of Global Clean Energy Manufacturing study were translated across to the Australian New Zealand Standard Classification of Industry framework (which is based on the international classification standard) (Australian Bureau of Statistics & Statistics New Zealand 2013).

Data from the Australian census on the occupational composition of these manufacturing sectors were then used to derive the breakdown of employment (Australian Bureau of Statistics 2017). The census includes a comprehensive stocktake of employment, with data at one-, two-, three- and four-digit levels for each industry. The Australian–New Zealand Standard Classification of Occupations is based on the ISCO. Clean Energy Manufacturing Analysis data on the relative share of the value added by component was used to weight the shares of employment. The Australian manufacturing sectors used are not wind or solar PV manufacturing activities, but as the Clean Energy Manufacturing Analysis Centre notes, 'Large portions of the wind energy supply chain connect well to core manufacturing indus-

Table 10.4 Wind and solar PV manufacturing–study methodology

Technology	Component	I-O Industry Category	Equivalent ANZSIC Classification
Wind	Nascelles	Machinery & equipment manufacturing, not elsewhere classified	22 fabricated metal product manufacturing 222 structural metal product manufacturing 2221 structural steel fabricating
	Blades	Manufacturing not elsewhere classified	22 fabricated metal product manufacturing 222 structural metal product manufacturing 2221 structural steel fabricating
	Towers	Fabricated metal product manufacturing	22 fabricated metal product manufacturing 222 structural metal product manufacturing 2221 structural steel fabricating
	Steel	Basic metal manufacturing	21 primary metal and metal product manufacturing 211 basic ferrous metal manufacturing 2110 Iron Smelting & Steel Manufacturing
	Generators	Electrical machinery & apparatus manufacturing,	24 Machinery & Equipment Manufacturing 243 electrical equipment manufacturing 2439 other electrical equipment manufacturing
Solar PV	Modules Cells Wafers	Computers, electronic and optical equipment manufacturing	24 Machinery & Equipment Manufacturing 243 electrical equipment manufacturing 2439 other electrical equipment manufacturing

tries' (CEMAC 2017). To clarify construction worker categories for onshore wind and PV construction, and for the operation and maintenance of solar PV, interviews were conducted with project managers who were currently overseeing or had recently completed construction projects (Table 10.4).

The IRENA studies are the richest data source on employment in solar PV and onshore wind projects, but further work is required to directly match renewable energy job levels against existing fossil fuel sectors and to generate data on mid- and low-skill jobs, which are of primary interest from a just transition perspective.

As an example, Table 10.5 shows the occupational hierarchy for solar PV construction and how it is matched against ISCO. ISCO classifies occupations from a one-digit level (left) to a four-digit level (right). Each level is more detailed than the previous one in terms of the labour force required for the type of work. This methodology has been transferred to an occupational hierarchy that has been constructed for solar PV, onshore wind, and offshore wind using IRENA and other data sources to map jobs against the ISCO. The result is a matrix with percentages allocated to each occupation at the one-, two-, three-, and four-digit levels of aggregation.

The framework for fossil fuels was derived from labour statistics from the Australian 2016 national census for coal mining, gas supply, and coal and gas generation (Australian Bureau of Statistics 2017). Although these data are specific to Australia, these statistics provide the best source of data, and regional multipliers in the quantitative modelling can adjust the results to account for economic differences between regions.

Table 10.5 Occupational hierarchy, solar PV construction

ILO 1-digit		ILO 2-digit		ILO 3-digit		ILO 4-digit	
1 MANAGERS	1.7%	Production and specialised service MANAGERS (13)	1.7%	Manufacturing, mining, construction and distribution MANAGERS (132)	1.7%	Supply, distribution and related MANAGERS (1324)	1.7%
2 PROFESSIONALS	12.0%	Science & engineering PROFESSIONALS (21)	4.4%	Physical, life & earth science PROFESSIONALS (211)	0.1%	Geotechnical experts (2114)	0.1%
				Life science PROFESSIONALS (213)	2.1%	Environmental protection professionals (2133)	2.1%
				Engineering PROFESSIONALS (214)	1.0%	Mechanical engineers (2144)	0.8%
		Health professionals (22)	4.4%	Electrotechnology engineers (215)	0.8%	Civil engineers (2142)	0.3%
				Other health professionals (226)	4.4%	Electrical engineers (2151)	0.8%
						Environmental and occupational health and hygiene professionals (2263)	4.4%
		Business & administrative PROFESSIONALS (24)	3.4%	Finance PROFESSIONALS (241)	2.6%	Financial analysts (2413)	1.9%
				Administration PROFESSIONALS (242)	0.8%	Accountants (2411)	0.7%
						Policy administration professionals (2422)	0.8%
3 TECHNICIANS and Associate professionals	27.8%	Business and administration Associate professionals (33)	0.6%	Business services agents (333)	0.7%	Real estate agents and property managers (3334)	0.7%
		Science and engineering TECHNICIANS and supervisors (31)	24.9%	Physical and science engineering TECHNICIANS (311)	11.9%	Civil engineering technicians (3112)	7.0%

		Information & communications technicians (35)	2.1%			Electrical engineering technicians (3113)	4.9%
				Mining, manufacturing & construction supervisors (312)	13.0%	Construction supervisors (3123)	13.0%
				ICT Operations & Support Technicians (351)	2.1%	ICT operations technicians (3511)	2.1%
4 Clerical Support workers	0.3%	Numerical and material recording clerks (43)	0.3%	Numerical clerks (431)	0.3%	Accounting and bookkeeping clerks (4311)	0.2%
						Payroll clerks (4313)	0.2%
7 Craft and related TRADES workers	31.6%	Electrical and electronics TRADES workers (74)	31.6%	Electrical equipment installers and repairers (741)	31.6%	Building Frame & Finisher Trades (711 & 712)	9.9%
						Sheet & Structural Metal Workers (721)	7.9%
						Electricial equipment installers (741)	13.8%
8 Plant and machine operators and assemblers	22.0%	Assemblers 821	9.8%	Assemblers 821	9.8%	Mechanical machinery assemblers (8211)	4.2%
		Drivers and mobile plant OPERATORS (83)	12.1%	Heavy truck and bus drivers (833)	4.3%	Electrical assemblers (8212)	5.6%
						Truck and lorry drivers (8332)	4.3%
				Mobile plant OPERATORS (834)	7.8%	Earthmoving plant operators (8342)	3.5%
						Crane, hoist and related plant operators (8343)	4.3%
9 Elementary occupations	4.3%	Labourers in mining, construction, manufacturing and transport (93)	4.3%	Mining & construction labourers (931)	4.3%	Civil engineering Labourers (9312)	4.3%

For each modelling run, the results for the installed capacity of the renewable energy technologies (MW) and full-time-equivalent jobs (FTE/MW) were used to generate an aggregate level of employment for construction, manufacturing, and operation and maintenance. The matrix was then used to calculate the number of jobs at different levels of disaggregation (Sect. 3.6.1.)

The final framework is shown in Table 10.4. This is based on a composite profile for each technology using a mix of one-, two-, three-, and four-digit levels of occupation, depending on which best illustrates the breakdown of jobs and allows comparison to be made across technologies. Choices have been made based on the proportion of jobs and the labels that are most readily understandable to readers (noting that the categories used in the ISCO do not always correspond to popularly used titles). Note, for example, that 'Managers' is a one-digit category, but trades are broken down into construction trades, metal trades, and electricians because this provides more meaningful descriptions.

In the example shown in Table 10.6, it is notable that wind and solar farms employ higher proportions of professionals and technicians for their operations and maintenance than coal mining, and similar or higher proportions of elementary occupations, but much lower proportions of machinery operators and drivers.

10.3.1.1 Methodology and Limitations

- At the aggregate level, it is assumed that rising labour productivity over time will reduce the labour intensity (i.e., less FTE/MW) that is applied in the construction, manufacture, and operation and maintenance of each renewable energy technology. No assumptions have been made about changes to the relative labour intensity between occupations. Over time, we would expect that the proportion of less-skilled jobs would fall as a result of mechanization. Therefore, the share of less-skilled jobs is likely to be overestimated.
- IRENA estimates a single global figure for each occupation, averaged from surveys of industry participants across different global markets. In practice, there are variations in labour intensity and the compositions of jobs across supply chains between different regions (broadly speaking, supply chains in lower-wage nations are more labour intensive). ISF takes account of regional conditions in the job factors applied at the level of major sub-sectors (construction, manufacturing, operation and maintenance), but not at the disaggregated level. Therefore, it is likely that the proportion of less-skilled jobs is overestimated for rich economies and underestimated for less-developed economies.
- The breakdown of the category of 'construction workers' is based on interviews with some Australian solar and wind project managers. The project managers had overseen recent projects and provided detailed estimates of the contributions of different jobs. Nonetheless, the breakdowns are based on a limited sample, and further research is required to generate more-accurate estimates.

Table 10.6 Occupational compositions for renewable and fossil fuel technologies

		Solar PV			Onshore Wind			Offshore wind			Fossil fuels		
ISCO category	Name	Construction	Manufacturing	O&M	Construction	Manufacturing	O&M	Construction	Manufacturing	O&M	Coal mining	Gas supply	Coal and gas generation
1	Managers	1.0%	4.2%	6.3%	1.7%	7.6%	1.5%	2.2%	4.6%	3.0%	6.7%	16.8%	13.0%
2	Other professionals (Legal, finance, scientific)	5.0%	12.7%	4.4%	10.6%	11.3%	11.6%	7.0%	26.0%	15.8%	10.0%	14.7%	6.7%
2	Engineers (Industrial, electrical & civil)	3.8%	14.3%	14.7%	1.8%	8.7%	27.0%	6.2%	6.3%	7.7%	0.0%	5.6%	8.5%
3	Technicians & associate professionals	7.2%	6.3%	26.2%	27.8%	6.5%	46.9%	0.1%	3.5%	25.1%	7.5%	10.5%	22.5%
4	Clerical support workers	3.3%	4.9%	1.3%	0.3%	4.6%	4.7%	0.2%	9.2%	8.4%	5.1%	18.8%	12.1%
7	Construction trades	0.8%	0.0%	0.0%	9.9%	2.5%	0.0%	0.0%	2.0%	5.5%	0.0%	13.9%	1.6%
7	Metal trades	1.8%	7.9%	0.0%	7.9%	28.4%	0.0%	0.0%	23.3%	0.0%	16.3%	0.0%	12.1%
7	Electricians	14.2%	21.6%	32.3%	13.8%	4.0%	4.1%	0.2%	3.3%	5.5%	5.5%	0.0%	11.2%
8	Plant & machine operators & assemblers	55.6%	10.6%	0.0%	21.9%	18.3%	0.0%	12.9%	15.0%	20.0%	46.4%	13.2%	6.0%
9	Elementary occupations (Labourers)	7.4%	17.5%	14.7%	4.3%	8.2%	4.1%	0.8%	6.7%	8.9%	2.6%	6.5%	6.4%
	Ship crew							70.3%					

10.3.2 Results of Occupational Employment Modelling

There will be an increase in jobs in the 1.5 °C Scenario across all occupations between 2015 and 2025, except in metal trades, which will display a minor decline of 2%, as shown in Fig. 10.3.

There will be an increase in jobs across all occupations between 2015 and 2025 in the 2.0 °C Scenario, as shown in Fig. 10.4. The occupations with the highest number of jobs will be plant and machine operators and assemblers, followed by technicians (including electrical, mechanical, civil, and IT technicians) and electricians. The occupations that will have the largest percentage increase in jobs from 2015 to 2025 will be labourers, engineers, electricians, and construction trades. The results are similar in the 1.5 °C Scenario, except for managers and metal trades, which will experience minor reductions in overall jobs (3% each) (Table 10.7).

However, the results are not uniform across regions. For example, China and India will both experience a reduction in the number of jobs for managers and clerical and administrative workers between 2015 and 2025, as shown in Table 10.8.

Table 10.8 and Fig. 10.6 show the employment changes between 2015 and 2025 under the 1.5 °C Scenario. Across all eight employment groups, the net effect of the energy transition will positive or stable (Fig. 10.5).

However, the results are not uniform across regions. For example, China and India both foresee a reduction in the number of jobs for managers and clerical and administrative workers between 2015 and 2025, as shown in Table 10.9.

Table 10.10 and Fig. 10.5 show the employment changes between 2015 and 2025 under the 2.0 °C Scenario. Across all eight employment groups, the net effect of the energy transition is positive. Further research is required to identify the training needs for all employment groups.

10.4 Conclusions

Under both the 1.5 °C and 2.0 °C Scenarios, the renewable energy transition is projected to increase employment. Importantly, this analysis has reviewed the locations and types of occupations and found that the jobs created in wind and solar PV alone are enough to replace the jobs lost in the fossil fuel industry across all occupation types. Further research is required to identify the training needs and supportive policies needed to ensure a just transition for all employment groups.

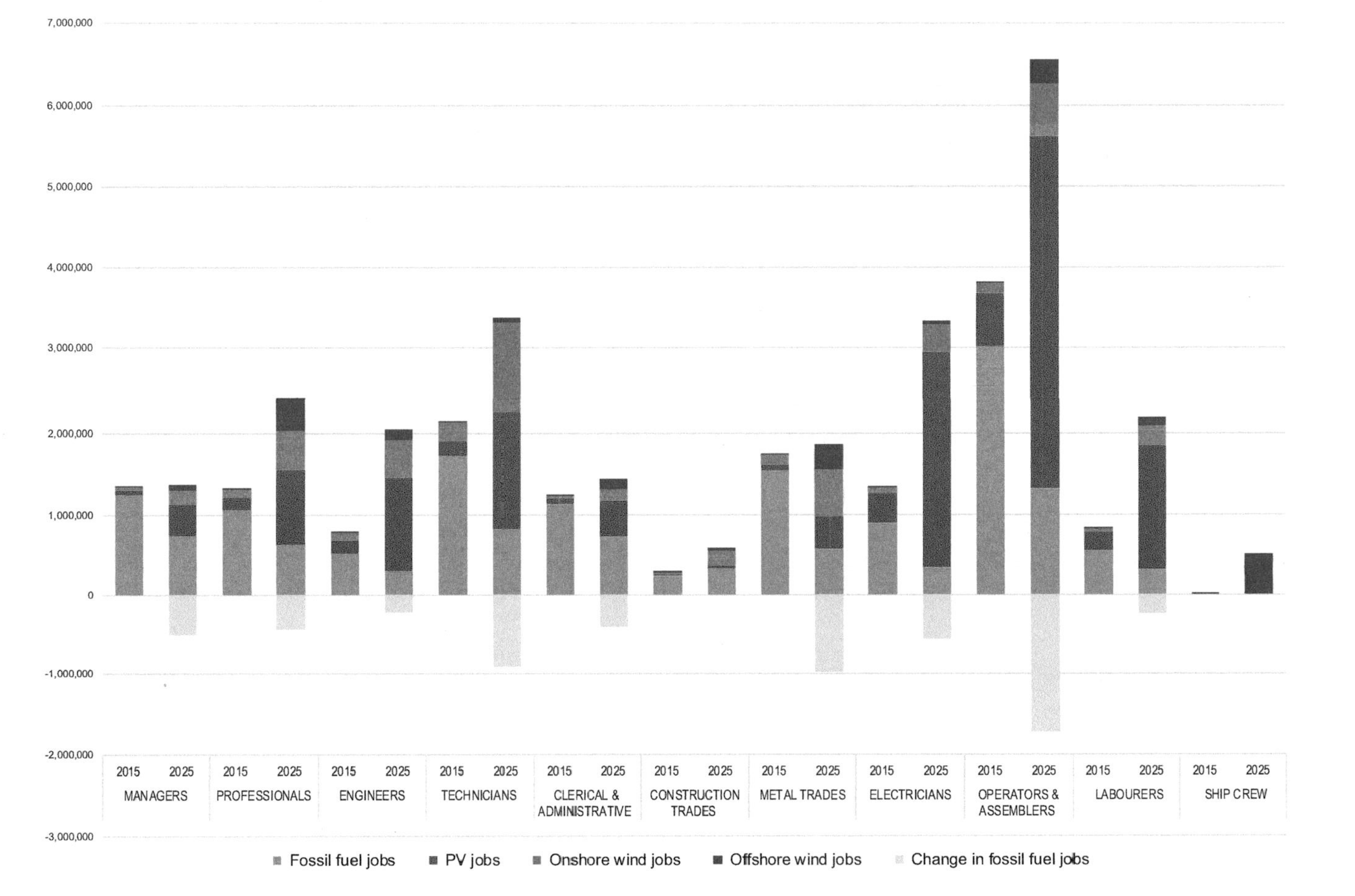

Fig. 10.3 Division of occupations between fossil fuels and renewable energy in 2015 and 2025 under the 1.5 °C Scenario

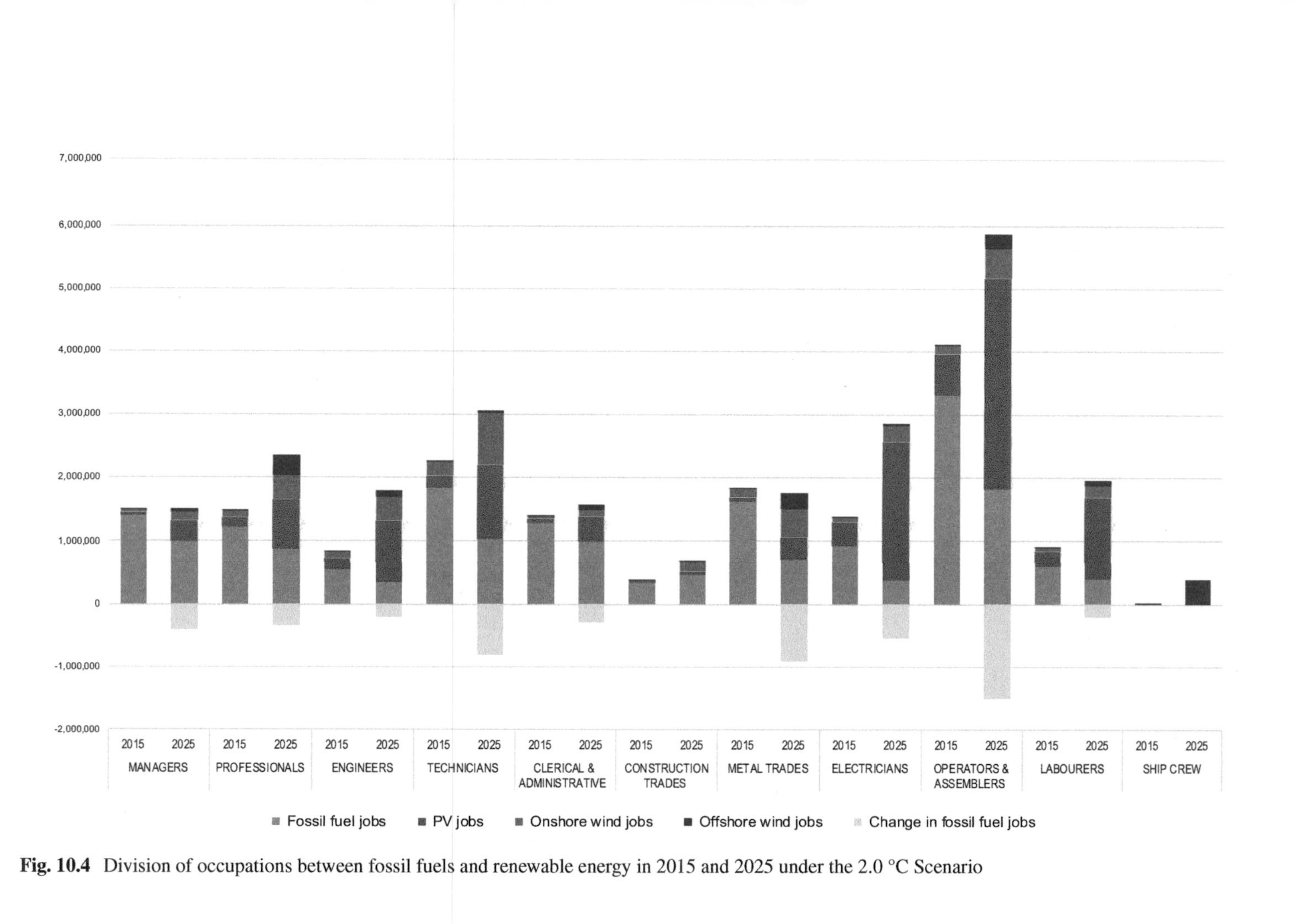

Fig. 10.4 Division of occupations between fossil fuels and renewable energy in 2015 and 2025 under the 2.0 °C Scenario

Table 10.7 Jobs created and lost between 2015 and 2025 under the 1.5 °C Scenario

	Jobs created or lost				Total jobs		Difference in jobs	
	Fossil fuels	PV	Wind – onshore	Wind – offshore	Total jobs in 2015	Total jobs in 2025	Total difference	% difference
Managers	−505,000	355,000	130,000	75,000	1,345,000	1,380,000	35,000	3%
Professionals (Social, Legal, finance, scientific)	−430,000	770,000	375,000	385,000	1,330,000	2 430,000	1,100,000	83%
Engineers (Industrial, electrical & civil)	−225,000	990,000	365,000	125,000	790,000	2,050,000	1,260,000	159%
Technicians (Electrical, mechanical, civil & IT)	−900,000	1,240,000	840,000	55,000	2 130,000	3 365,000	1 235,000	58%
Clerical & administrative workers	−395,000	375,000	100,000	120,000	1 235,000	1,440,000	200,000	16%
Construction trades	80,000	45,000	155,000	30,000	290,000	600,000	310,000	107%
Metal trades	−970,000	335,000	450,000	295,000	1,750,000	1,860,000	110,000	6%
Electricians	−560,000	2,250,000	260,000	45,000	1 335,000	3 330,000	1,995,000	150%
Plant & machine operators & assemblers	−1,700,000	3,360,000	505,000	290,000	3,820,000	6 560,000	2,740,000	72%
Labourers (Manufacturing, construction & transport)	−235,000	1,300,000	190,000	95,000	835,000	2,185,000	1,350,000	162%
Total	**−5 845,000**	**11,300,000**	**3,370,000**	**1,510,000**	**14 900,000**	**25 195,000**	**10 335,000**	**70%**
Ship crew	–	–	–	490,000	12,000	500,000	490,000	4000%

Table 10.8 Jobs created or lost between 2015 and 2025 by region under the 1.5 °C Scenario

	OECD North America	Latin America	OECD Europe	Africa	Middle East	Eastern Europe/ Eurasia	India	Developing Asia	China	OECD Pacific	Global
Managers	138%	57%	26%	13%	42%	42%	−14%	36%	−24%	40%	3%
Professionals (Social, legal, finance, scientific)	383%	185%	113%	186%	119%	119%	133%	194%	1%	172%	83%
Engineers (Industrial, electrical & civil)	414%	306%	199%	201%	179%	179%	140%	200%	82%	214%	159%
Technicians (Electrical, mechanical, civil & IT)	236%	189%	65%	140%	153%	153%	46%	125%	−5%	138%	58%
Clerical & administrative workers	165%	63%	50%	24%	47%	47%	−7%	53%	−13%	52%	16%
Construction trades	145%	75%	137%	202%	64%	64%	107%	93%	124%	115%	107%
Metal trades	258%	76%	44%	50%	112%	112%	33%	83%	−36%	76%	6%
Electricians	489%	556%	165%	237%	500%	500%	157%	346%	24%	206%	150%
Plant & machine operators & assemblers	390%	456%	53%	326%	502%	502%	166%	649%	−29%	178%	72%
Labourers (Manufacturing, construction & transport)	475%	395%	209%	220%	221%	221%	181%	266%	48%	210%	162%
Total	**327%**	**237%**	**87%**	**159%**	**168%**	**168%**	**90%**	**199%**	**−10%**	**152%**	**70%**
Ship crew	4150%		731%					68,465%	1462%	8742%	4000%

Table 10.9 Jobs created and lost between 2015 and 2025 under the 2.0 °C Scenario

	Jobs created or dlost				Total jobs		Difference in jobs	
	Fossil fuels	PV	Wind – onshore	Wind – offshore	Total jobs in 2015	Total jobs in 2025	Total difference	% difference
Managers	−382,067	255,887	89,967	56,041	1,444,621	1,464,449	19,828	1%
Professionals (Social, legal, finance, scientific)	−305,186	609,373	260,542	288,834	1,421,732	2 275,295	853,563	60%
Engineers (Industrial, electrical & civil)	−191,753	764,217	250,117	94,195	830,073	1,746,850	916,776	110%
Technicians (Electrical, mechanical, civil & IT)	−778,902	945,091	580,639	41,848	2,195,839	2,984,516	788,677	36%
Clerical & administrative workers	−278,592	297,860	69,929	91,610	1,350,727	1,531,534	180,807	13%
Construction trades	129,278	37,123	109,843	21,101	371,818	669,163	297,345	80%
Metal trades	−844,156	267,032	315,134	222,511	1,752,952	1,713,473	−39,479	−2%
Electricians	−504,687	1,740,194	182,757	34,307	1,342,336	2,794,906	1,452,570	108%
Plant & machine operators & assemblers	−1,350,554	2,949,489	354,152	215,137	3,903,199	6,071,423	2,168,224	56%
Labourers (Manufacturing, construction & transport)	−186,954	1,013,583	133,271	71,034	881,189	1,912,124	1,030,935	117%
Total	−4,693,573	8,879,851	2,346,350	1,136,618	15,494,487	23,163,733	7,669,246	49%
Ship crew	–	–	–	355,791	12,199	367,990	355,791	2917%

Table 10.10 Jobs created or lost between 2015 and 2025 by region under the 2.0 °C Scenario

	OECD North America	Latin America	OECD Europe	Africa	Middle East	Eastern Europe/ Eurasia	India	Developing Asia	China	OECD Pacific	Global
Managers	132%	35%	14%	16%	40%	40%	−20%	36%	−26%	26%	1%
Professionals (Social, legal, finance, scientific)	308%	106%	80%	151%	82%	82%	102%	120%	−5%	137%	60%
Engineers (Industrial, electrical & civil)	369%	187%	134%	144%	126%	126%	90%	124%	53%	172%	110%
Technicians (Electrical, mechanical, civil & it)	223%	111%	39%	100%	114%	114%	18%	82%	−13%	105%	36%
Clerical & administrative workers	154%	38%	33%	29%	43%	43%	−11%	46%	−16%	38%	13%
Construction Trades	140%	50%	100%	163%	52%	52%	87%	72%	101%	54%	80%
Metal trades	202%	26%	29%	48%	101%	101%	20%	56%	−38%	61%	−2%
Electricians	456%	378%	100%	203%	416%	416%	102%	239%	11%	180%	108%
Plant & machine operators & assemblers	369%	258%	30%	317%	289%	289%	130%	398%	−27%	164%	56%
Labourers (Manufacturing, construction & transport)	436%	230%	136%	191%	143%	143%	125%	167%	31%	176%	117%
Total	**298%**	**134%**	**55%**	**138%**	**113%**	**113%**	**60%**	**129%**	**−14%**	**127%**	**49%**
Ship crew	1724%		644%					43,668%	1098%	8610%	2917%

2 degree scenario

Managers

Fossil fuels 2015: 1,251,000
Renewable energy 2015: 93,000
Fossil fuels 2025: 810,000
Renewable energy 2025: 493,000
Jobs lost: 41,000

Engineers (industrial, electrical & civil)

Fossil fuels 2015: 515,000
Renewable energy 2015: 275,000
Jobs created: 885,000
Fossil fuels 2025: 305,000
Renewable energy 2025: 1,370,000

Professionals (social, legal, finance & scientific)

Fossil fuels 2015: 1,070,000
Renewable energy 2015: 260,000
Jobs created: 790,000
Fossil fuels 2025: 715,000
Renewable energy 2025: 1,405,000

Construction trades

Fossil fuels 2015: 240,000
Jobs created: 250,000
Renewable energy 2015: 48,000
Fossil fuels 2025: 325,000
Renewable energy 2025: 213,000

Metal trades

Fossil fuels 2015: 1,595,000
Renewable energy 2015: 205,000
Fossil fuels 2025: 695,000
Renewable energy 2025: 1,055,000
Jobs lost: 50,000

Electricians

Fossil fuels 2015: 900,000
Renewable energy 2015: 440,000
Jobs created: 1,425,000
Fossil fuels 2025: 390,000
Renewable energy 2025: 2,375,000

Technicians (electrical, mechanical, civil & IT)

Fossil fuels 2015: 1,730,000
Renewable energy 2015: 400,000
Jobs created: 725,000
Fossil fuels 2025: 910,000
Renewable energy 2025: 1,945,000

Clerical & administrative workers

Fossil fuels 2015: 1,130,000
Renewable energy 2015: 105,000
Jobs created: 110,000
Fossil fuels 2025: 790,000
Renewable energy 2025: 555,000

Plant & machine operators

Fossil fuels 2015: 3,025,000
Renewable energy 2015: 800,000
Jobs created: 2,075,000
Fossil fuels 2025: 1,630,000
Renewable energy 2025: 4,270,000

Labourers (manufacturing, construction & drivers)

Fossil fuels 2015: 550,000
Renewable energy 2015: 285,000
Jobs created: 995,000
Fossil fuels 2025: 340,000
Renewable energy 2025: 1,490,000

Ship crew

Renewable energy 2015: 12,000
Jobs created 2015: 350,000
Renewable energy 2025: 362,000

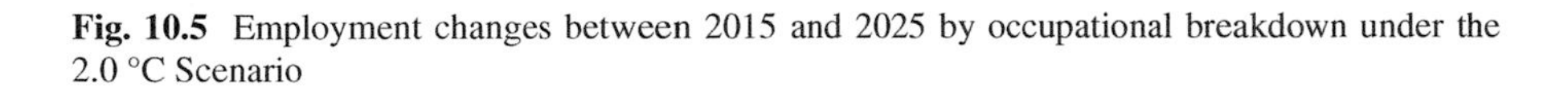

Fig. 10.5 Employment changes between 2015 and 2025 by occupational breakdown under the 2.0 °C Scenario

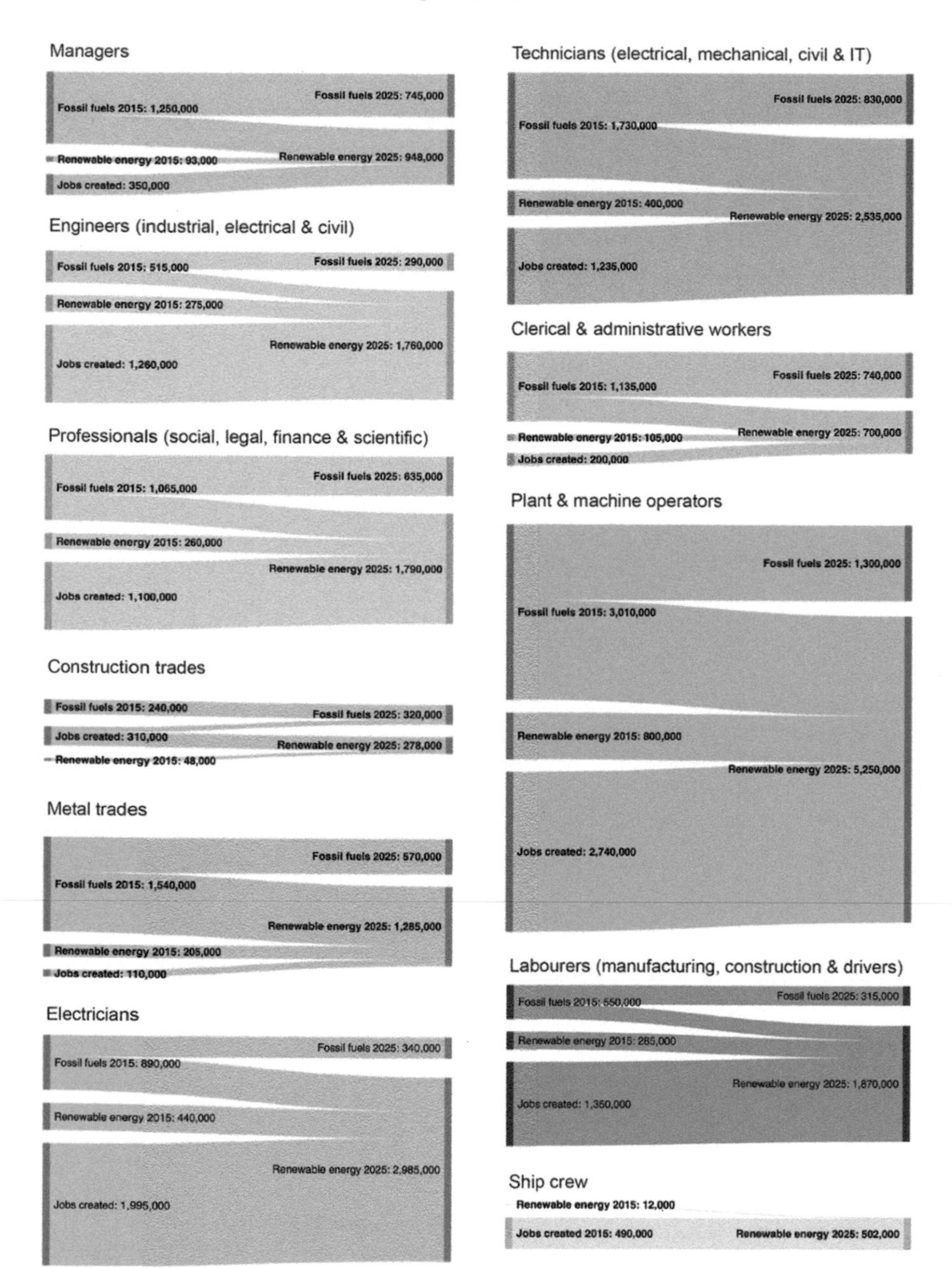

Fig. 10.6 Employment changes between 2015 and 2025 by occupational breakdown under the 1.5 °C Scenario

References

Australian Bureau of Statistics (2017), Employment in Renewable Energy Activities – Explanatory Notes. http://www.abs.gov.au/ausstats/abs@.nsf/Lookup/4631.0Explanatory+Notes12015-16. Accessed September 16 2018.

Australian Bureau of Statistics & Statistics New Zealand (2013), Australian and New Zealand Standard Classification of Occupations, http://www.abs.gov.au/ANZSCO, Accessed September 16 2018

Clean Energy Manufacturing Analysis Center (CEMAC), 2017. Benchmarks of global clean energy manufacturing. Available at: https://www.nrel.gov/docs/fy17osti/65619.pdf

Dominish, E., Teske S., Briggs, C., Mey, F., and Rutovitz, J. (2018). Just Transition: A global social plan for the fossil fuel industry. Report prepared by ISF for German Greenpeace Foundation, November 2018.

International Labour Office (2012), International Standard Classification of Occupations, Geneva, ILO

International Labour Office. (2015). *Guidelines for a just transition towards environmentally sustainable economies and societies for all.* Geneva

IRENA (2017a) *Renewable Energy Benefits: Leveraging Local Capacity for Onshore Wind, IRENA, Abu Dhabi*;

IRENA (2017b) *Renewable Energy Benefits: Leveraging Local Capacity for Solar PV, IRENA, Abu Dhabi.*

IRENA (2018) *Renewable Energy Benefits: Leveraging Local Capacity for Offshore Wind, IRENA, Abu Dhabi.*

Jenkins, K., McCauley, D., Heffron, R., Stephan, H., & Rehner, R. (2016). Energy justice: A conceptual review. *Energy Research and Social Science*, *11*, 174–182. https://doi.org/10.1016/j.erss.2015.10.004

Rutovitz, J., Dominish, E., & Downes, J. (2015). *Calculating global energy sector jobs: 2015 methodology*. Prepared for Greenpeace International by the Institute for Sustainable Futures, University of Technology Sydney.

Smith, S. (2017). Just Transition – A Report for the OECD. Brussels. Retrieved from https://www.oecd.org/environment/cc/g20-climate/collapsecontents/Just-Transition-Centre-report-just-transition.pdf

Sovacool, B. K., & Dworkin, M. H. (2014). *Global energy justice: Problems, principles, and practices. Global energy justice: Problems, principles, and practices*. https://doi.org/10.1017/CBO9781107323605

UNFCCC. (2016). *Just transition of the workforce, and the creation of decent work and quality jobs: Technical paper by the Secretariat.* Paris. https://doi.org/10.1186/1750-9378-6-S2-S6

Chapter 11
Requirements for Minerals and Metals for 100% Renewable Scenarios

Damien Giurco, Elsa Dominish, Nick Florin, Takuma Watari, and Benjamin McLellan

Abstract This chapter explores the magnitude of the changes in patterns of material use that will be associated with the increasing deployment of renewable energy and discusses the implications for sustainable development. In particular, this chapter focuses on the increased use of lithium and cobalt, metals which are used extensively in battery technologies, and silver used in solar cells. Consistent with the strong growth in renewable energy and electrification of the transport system required in a 1.5°C scenario, the material requirements also rise dramatically, particularly for cobalt and lithium. Scenarios developed for this study show that increasing recycling rates and material efficiency can significantly reduce primary demand for metals.

11.1 Introduction

Globally, recent investments in new renewable energy infrastructure have been double the investments in new energy from fossil fuels and nuclear power (REN21 2018). This is strong evidence of the increasing momentum of the energy transition away from fossil fuels. A rapid transition to 100% renewables offers hope for reducing carbon emissions and increasing the chance that global warming will be maintained to below 2.0 °C. However, the transition to 100% renewables also comes with requirements for new patterns of material use to support the renewable energy infrastructure, including wind turbines, solar cells, batteries, and other technologies.

This chapter explores the magnitude of the changes in patterns of material use that will be associated with the increasing deployment of renewable energy, and discusses the implications for sustainable development. In particular, this chapter focuses on the increased use of lithium (used extensively in battery technology),

D. Giurco (✉) · E. Dominish · N. Florin
Institute for Sustainable Futures, University of Technology Sydney, Sydney, NSW, Australia
e-mail: damien.giurco@uts.edu.au; elsa.dominish@uts.edu.au; nick.florin@uts.edu.au

T. Watari · B. McLellan
Graduate School of Energy Science, Kyoto University, Kyoto, Japan

S. Teske (ed.), *Achieving the Paris Climate Agreement Goals*,
https://doi.org/10.1007/978-3-030-05843-2_11

cobalt (again used in batteries, particularly for vehicles), and silver (used in solar cells because it is an excellent conductor of electricity).

The importance of understanding how material use is affected by renewable energy technologies is already established (Ali et al. 2017; Valero et al. 2018). A range of studies have identified different aspects of the challenges associated with the supply of the materials needed for renewable energy. The first aspect is the availability of mineral supplies. For example, Mohr et al. (2012) developed commodity-focused supply projections for lithium, with an emphasis on the origins and locations of available lithium deposits over time, which are contrasted with a simple demand function. Other studies have placed more emphasis on demand scenarios, either for a specific geographic location, such as Germany (Viebahn et al. 2015) or globally (Valero et al. 2018; Watari et al. 2018), and some have explored the role of technology mixes and material substitution in detail (Månberger and Stenqvist 2018).

In addition to issues of resource availability, environmental and social issues have also been explored (Giurco et al. 2014; Florin and Dominish 2017) and the need for improved resource governance has been highlighted (Prior et al. 2013; Ali et al. 2017). This chapter evaluates the total demand to determine which metals may present bottlenecks to the supply of renewable energy technologies. The potential to offset primary demand is explored through a range of scenarios for technology type, material efficiency, and recycling rates. This chapter draws on the results of a larger study funded by Earthworks (see Dominish et al. 2019).

This chapter has five sections. Following this introductory section, the second section outlines the material requirements for key renewable energy infrastructure technologies, namely solar photovoltaics (PV), wind turbines, and batteries. The key assumptions, and the energy and resource scenarios are described in Sect. 11.3, together with a summary of the methodology, which is described in detail in Chap. 3. The requirements for selected materials (lithium, cobalt, and silver) are presented in Sect. 11.4. Following the results, a discussion is presented in Sect. 11.5.

11.2 Overview of Metal Requirements for each Technology

Renewable energy and storage technologies typically have high and diverse metal requirements. Moreover, there are often competing technologies or component technologies, which add to the complexity of material considerations. The key metals used for solar PV, wind power, batteries, and EV are discussed below.

11.2.1 Solar PV

A typical crystalline silicon (c-Si) PV panel, which is currently the dominant technology, with over 95% of the global market, contains about 76% glass (panel surface), 10% polymer (encapsulant and back-sheet foil), 8% aluminium (frame), 5% silicon (solar cells), 1% copper (interconnectors), and less than 0.1% silver (contact lines) and other metals (e.g., tin and lead). Thin film technologies,

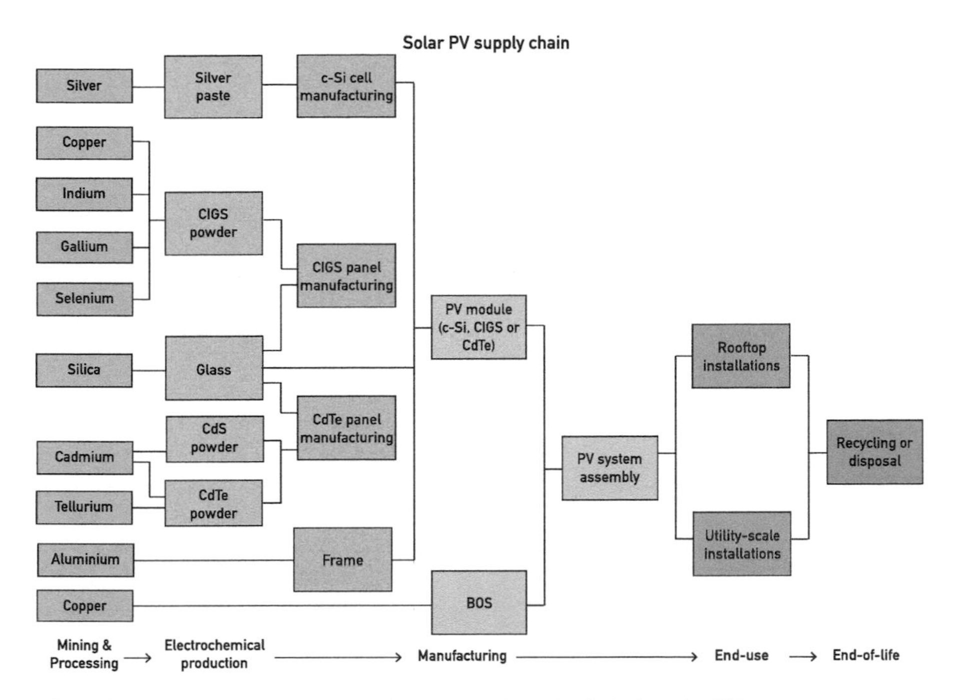

Fig. 11.1 Overview of key metal requirements and supply chain for solar PV

copper–indium–gallium–(di)selenide (CIGS) and cadmium telluride (CdTe), make up the remainder of the market. These technologies require less material overall than crystalline silicon. For CdTe panels, the composition is 96–97% glass, 3%–4% polymer, and less than 1% semi-conductor materials (CdTe) and other metals (e.g., nickel, zinc, tin). CIGS contain about 88%–89% glass, 7% aluminium, 4% polymer, and less than 1% semi-conductor material (indium, gallium, selenium) and other metals (e.g., copper) (Weckend et al. 2016). Figure 11.1 provides a simplified diagram of the PV supply chain, including key materials and sub-components.

11.2.2 Wind

The major raw materials required for the manufacture of wind turbine components are bulk commodities: iron ore, copper, aluminium, limestone, and carbon. Wind turbines use steel for the towers, nacelle structural components, and the drivetrain, accounting for about 80% of the total weight. Some turbine generator designs use direct-drive magnetics, which contain the rare earth metals neodymium and dysprosium (Fig. 11.2). The development of direct-drive permanent magnet generators (PMG) by major producers (e.g., Siemens and General Electric) simplifies the design by eliminating the gearbox, and this is attractive for offshore applications because it reduces maintenance (Zimmermann et al. 2013). It is estimated that about 20% of all installed wind turbines (both onshore and offshore) use rare earth magnets (CEMAC 2017).

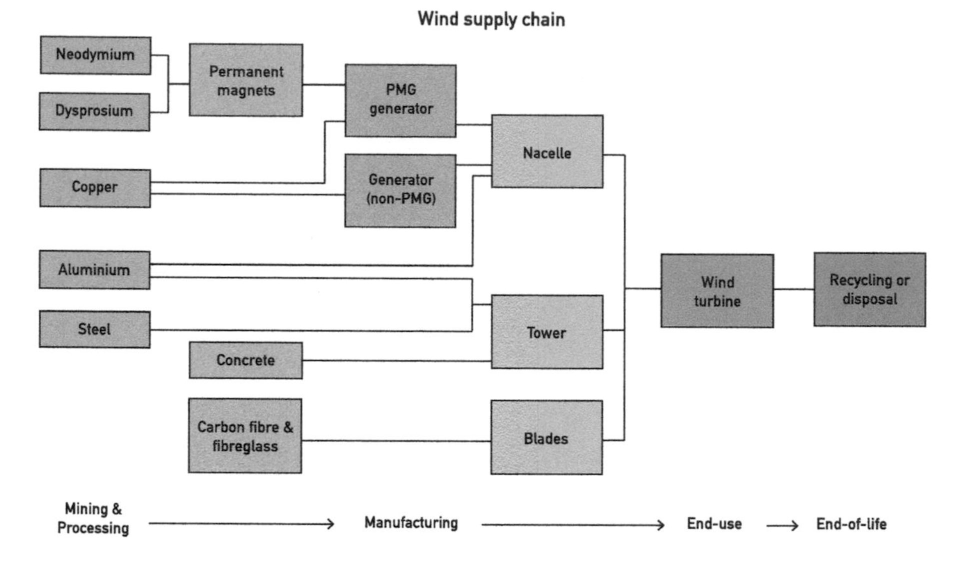

Fig. 11.2 Overview of key metal requirements and supply chain for wind power

11.2.3 Batteries and Electric Vehicles

This study focuses on lithium ion batteries (LIBs), which power almost all electric vehicles (EVs) on the market today and are also an important technology for stationary energy-storage applications. LIBs are made of two electrodes (anode and cathode), current collectors, a separator, electrolyte, a container, and sealing parts. The anode is typically made of graphite, with a copper foil current collector. The cathode is typically a layered transition metal oxide, with an aluminium foil current collector. In between the electrodes is a porous separator and electrolyte. All of these components are typically housed in an aluminium container. LIBs are generally referred to by the material content of the cathode, which accounts for 90% of the material value and about 25% of the total weight (Gratz et al. 2014).

The size and type of the LIB has the greatest impact on the material requirements. Since commercialization in the 1990s, a range of different types ('chemistries') have been developed for different applications, named according to the metals in the cathode. The most common LIB types for EV applications are nickel–manganese–cobalt (NMC), lithium–iron phosphate (LFP), nickel–cobalt–aluminium (NCA), and lithium–manganese oxide (LMO) (Vaalma et al. 2018). NMC is the most common battery type for passenger vehicles, and NCA is also common, with a small share for LMO. However, in China, LFP is the dominant chemistry. Electric buses have traditionally used LFP batteries (BNEF 2018) and lead–acid batteries are most commonly used for two-wheel vehicles. However, the application of LIBs in this market sector is growing (Yan et al. 2018).

For energy storage, NMC and NCA are the most commonly used chemistries. A simplified overview of the lithium-ion battery supply chain, including its key metals

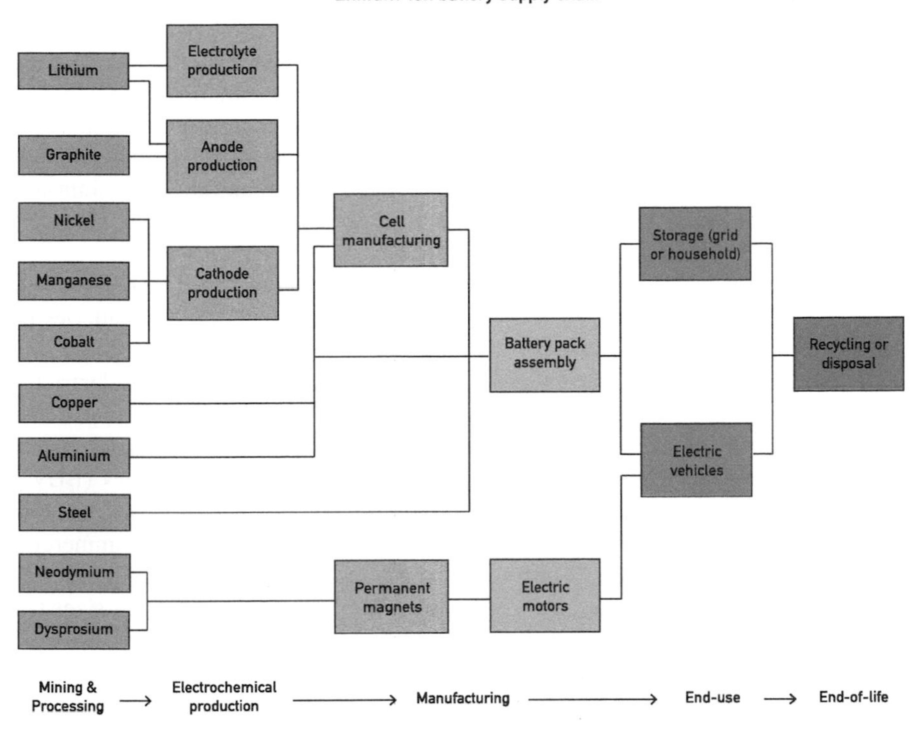

Fig. 11.3 Overview of key metal requirements and supply chain for LIB and EV

(for the NMC chemistry) and sub-components, is shown in Fig. 11.3. Rare earth permanent magnets using neodymium and dysprosium are common in most electric vehicle motors. Other motor technologies, or those that replace rare earths with lower-cost materials, are under development and are already used in some vehicles. However, rare earth magnets are expected to remain the standard for the foreseeable future because of their higher performance characteristics (Widmer et al. 2015).

11.3 Scenarios and Key Assumptions

11.3.1 Electricity and Transport Scenarios

The energy scenarios presented here were developed to achieve the climate target in the 2015 Paris Climate Agreement of limiting anthropogenic climate change to a maximum of 1.5 °C above pre-industrial levels. This projection is more ambitious than the IEA's annual World Energy Outlook (WEO) scenarios, which project that current policies, and therefore the global development of renewable power and electric mobility, will not change. In contrast, many scenarios have been proposed in the academic literature that extend as far as zero emissions, whereas the IEA's own "Energy Technology Perspectives" (ETP) scenarios include both a 2.0 °C scenario (2DS) and a more ambitious "Beyond 2.0 °C" scenario (B2DS), which aim to achieve the Paris targets. The scenario proposed here includes both a shift to 100% renewable electricity and a shift to renewable electricity and fuels in the transport sector.

In this scenario, solar PV will account for more than one-third of the installed capacity by 2050, with the remainder from wind and other renewables. Lithium-ion batteries will account for approximately 6% of energy storage (which will be dominated by pumped hydro and hydrogen).

In the transport system, we focus on the material requirements for the batteries used in road transport, because other types of transport do not require batteries to power drivetrains or are assumed to rely on other forms of energy (e.g., biofuels for aviation). In 2050, most of the energy for road transport will come from electricity (55%) and hydrogen (22%), and the remainder will be from biofuels and synfuels. In the 1.5 °C Scenario, the batteries required to electrify road transport are specified for electric buses and passenger cars, including battery electric vehicles (BEV), plug-in hybrid electric vehicles (PHEV), and commercial vehicles. Passenger cars will account for 92% of vehicles and 51% of battery capacity, whereas commercial vehicles are projected to account for 48% of battery capacity, although they will make up only 8% of the total fleet of vehicles. This is because the battery sizes for commercial vehicles (assumed to be 250 kWh in 2015 and rising to 600 kWh in 2050) are larger than those for passenger vehicles (5–15 kWh for PHEV and 38–62 kWh for BEV). Buses will account for a small percentage (1%) of both vehicles and batteries. Electric bikes and scooters have been excluded, because although they are currently a growing market in Asia, by 2050, their share of electricity consumption will be negligible compared with the predicted uptake of electric passenger and commercial vehicles. Lead–acid batteries are also the main type of battery for electric bikes and scooters, although lithium-ion may replace these in future.

11.3.2 Resource Scenario Development

Five resource scenarios were developed to estimate the metal demand for 100% renewables, as shown in Table 11.1. The scenarios were developed based on the current market trends and the likelihood of changes in material efficiency or technology. Our aim is to understand how primary demand can be offset through changes in technology or recycling rates.

11.3.3 Technology Assumptions

The total metal demand for renewable energy and battery technology each year is estimated based on the metal intensity of a specific technology and the capacity of each technology introduced in a specific year. This introduced stock will accounts

Table 11.1 Summary of metal scenarios

Scenario name	Market share/ materials efficiency	Recycling	Colour in figures
Total demand	Current materials intensity and current market share	No recycling	Red
Current recycling	Current materials intensity and current market share	Current recycling rates	Pink
Potential recycling	Current materials intensity and current market share	Improved recycling rates	Orange
Future technology	Improved materials efficiency for PV and current market share Technology shift for batteries	No recycling	Dark blue
Future technology & potential recycling	Improved materials efficiency for PV and current market share Technology shift for batteries	Improved recycling rates	Light blue

for the new capacity and the replacement of technologies at the end of their lives, based on a lifetime distribution curve for the average lifetime.

The values for metal intensity are given in tonnes/GW (for solar PV) or tonnes/GWh (for batteries) of capacity, and two values are given for each metal to evaluate the impact of improving the material efficiency (for solar PV) and technology shifts (for batteries). To evaluate the impact of recycling, the primary demand is estimated by multiplying the discarded products by the recycling rate. The recycling rate is obtained by multiplying the collection rate by the recovery efficiency of a metal from a specific technology. This recycling rate is also varied to obtain a current rate and a potential rate. The potential recycling rate is technologically possible, but is not currently applicable because it is not economic. The detailed assumptions for batteries and solar PV are explained in detail in the following section.

11.3.4 Batteries

The 'current materials intensity' for LIB for EVs and storage (Table 11.2) is estimated based on the assumed market shares of the LIB technologies: NMC (60%), LMO (20%), NCA (15%), and LFP (5%) (Vaalma et al. 2018). The dominant battery technologies in the future are not likely to be the same as those commercialized today. Therefore, for the 'future technology scenario', we assume that lithium–sulfur batteries will replace LIB for EVs (Cano et al. 2018). We have modelled a future market (Table 11.3) in which Li–S will achieve a 50% market share for EVs by 2050, with deployment scaling up at a linear rate, assuming the first commercialization in 2030. In this scenario, the technology does not change for storage batteries.

We have assumed a collection efficiency of 100% for all batteries and a recovery rate of 90% for Co and Ni (Georgi-Maschler et al. 2012). A 10% recovery rate is assumed for Li, acknowledging that pyrometallurgical processing routes account for most of the current global capacity, and Li recovery may not be possible by this

Table 11.2 Material intensity and recycling rates

	Solar PV	Batteries	
Materials	Silver	Lithium	Cobalt
Current materials intensity	20 t/GW	113 t/GWh	124 t/GWh
Future technology	4 t/GW	411 t/GWh	
Current recycling rate [%]	0%	0%	90%
Potential recycling rate [%]	81%	95%	95%

Table 11.3 Market share

	Solar PV	Batteries	
Technology	c-Si	Li-ion	Li-S
Current market share	95.8%	100%	
Future market share		Decreases to 50% by 2050	50% by 2050, beginning from 2030

route (King et al. 2018). For scenarios that assume a 'potential future recycling' rate, we have assumed 95% recovery for all metals. This is reasonable given that 100% recovery has been reported in the laboratory (Gratz et al. 2014).[1] However, some losses during processing seem unavoidable.[2]

11.3.5 Solar PV

For solar PV, we assume that the technology types do not change until 2050, and that they retain their current market shares, so that crystalline silicon will remain the dominant technology. We have modelled the potential to offset demand through increases in material efficiency and increases in recycling.

A high and a low value are given for silver to show the impact of material efficiency on silver demand. The current data on silver intensity are from a survey of the PV industry (ITRPV 2018), and the future material efficiency is based on an assumed minimum amount of silver (Kavlak et al. 2015). The 'current recycling rate' scenarios assume a current collection rate of 85% for all panels, consistent with the target of the EU WEEE Directive.[3] This should be considered an average rate, noting that remote and/or distributed roof-top systems will be more costly to collect and transport than large utility-scale PV. We assume that no recycling is currently occur-

[1] See: https://americanmanganeseinc.com/investor-info-3/investment-proposition/

[2] Pers comms Boxall, N.

[3] More details available here: http://ec. europa.eu/environment/waste/weee/index_en.htm

Table 11.4 Metal assumptions

Metal	Production 2017 (tonnes/year)	Reserve (tonnes)	Resources (tonnes)
Cobalt	110,000	7,100,000	25,000,000
Lithium	46,500	16,000,000	53,000,000
Silver	25,000	530,000	N/A

ring for silver from PV, but that 95% recovery may be possible. Therefore, for the 'potential recycling rate' scenarios, we assume a 95% recovery efficiency and an 85% collection rate, which will result in an 81% metal recycling rate.

11.3.6 Metal Assumptions

For each scenario, the annual primary demand is compared with the current production (2017 data), and the cumulative demand to 2050 is compared with current reserves. The data presented (Table 11.4) highlight the annual production, total resources, and reserves. 'Reserves' are the subset of the total resources that can be economically mined under current conditions. They are dependent on a multitude of factors and can change over time. By contrast, resources are less certain economically and there may be no firm plan to mine them. Over time, new resources can be discovered and as economic conditions change, resources may be upgraded to reserves (for example, where the price for the metal increases, thus making lower-grade or more-challenging ores profitable or where a new technology for extraction allows lower cost processing). In contrast, reserve estimates can also be downgraded over time (as occurred with coal reserve estimates in UK and Germany).

11.4 Results for Lithium, Cobalt, and Silver

The cumulative demand from renewable energy technologies for cobalt, lithium, and silver by 2050 has been modelled, and is compared to current reserves in Fig. 11.4. The cumulative demand for cobalt from renewable energy and transport exceeds the current reserves in all scenarios, and for lithium, the cumulative demand is exceeded in all scenarios, except the 'potential recycling scenario'. For silver, the total demand for silver from renewable energy will reach around 50% of current reserves.

The annual demand in 2050 is compared with the current rates of production (based on 2017 data). Both cobalt and lithium have annual demands that far exceed the current rates of production—particularly lithium in the 'future technology' scenario. However, the annual demand for silver will remain below current production levels (Fig. 11.5).

The detailed results for each metal are shown in the following section.

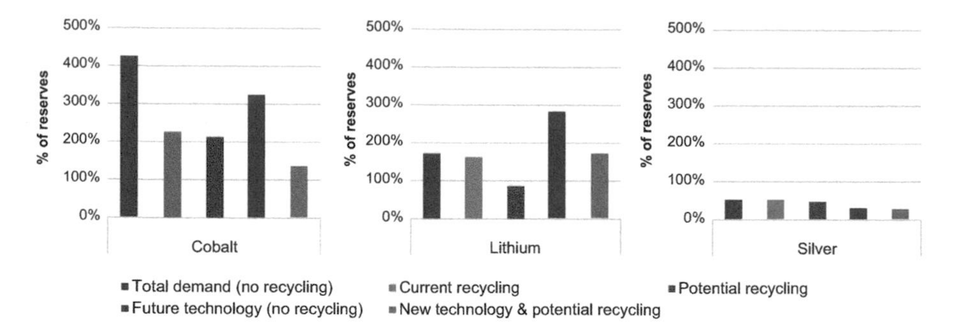

Fig. 11.4 Cumulative demand from renewable energy and transport technologies to 2050 compared with reserves

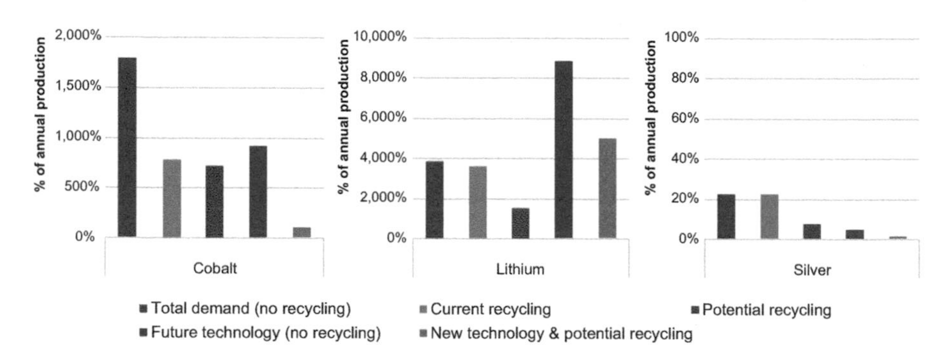

Fig. 11.5 Annual demand from renewable energy and storage technologies in 2050 compared with current production rates (note that scale varies across the metals)

11.4.1 Cobalt

The annual demand for cobalt from EVs and storage could exceed the current production rates in around 2023 (in all scenarios). In the 'future technology' scenario, shifting to Li–S instead of LIB will reduce the demand for cobalt. However, recycling, rather than shifting technologies, will have the greatest impact on reducing the primary demand in both the current technology and future technology scenarios.

The cumulative total demand to 2050 (with current technology and no recycling) could exceed current reserves by 400%, and exceed current resources by 20%. Even with recycling and a shift to technologies that use less cobalt, the cumulative demand will still exceed reserves. However, in these scenarios, the demand will remain below the resource levels (Figs. 11.6 and 11.7).

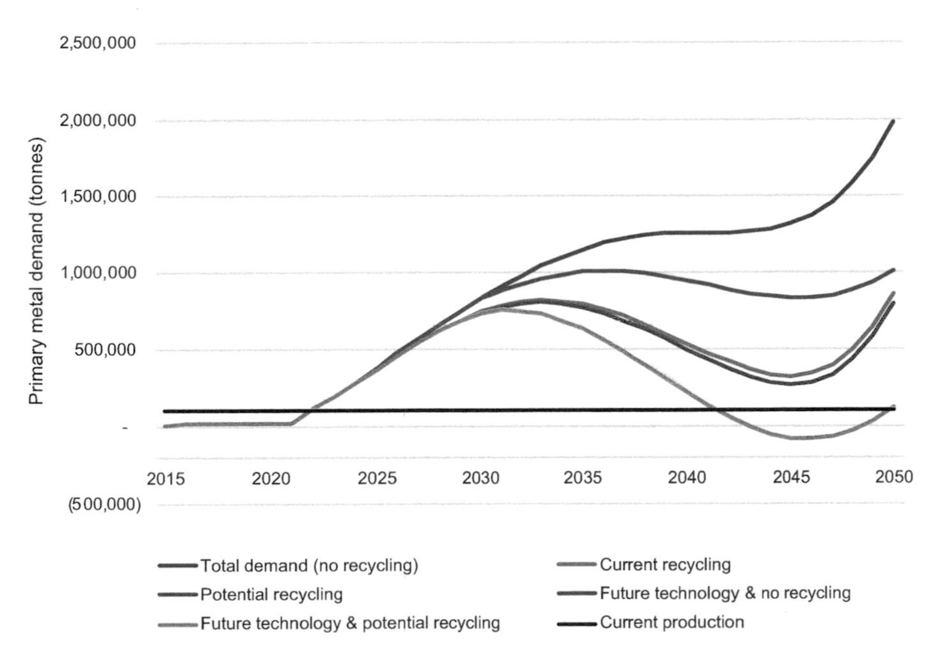

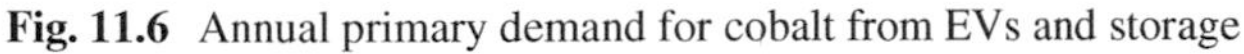

Fig. 11.6 Annual primary demand for cobalt from EVs and storage

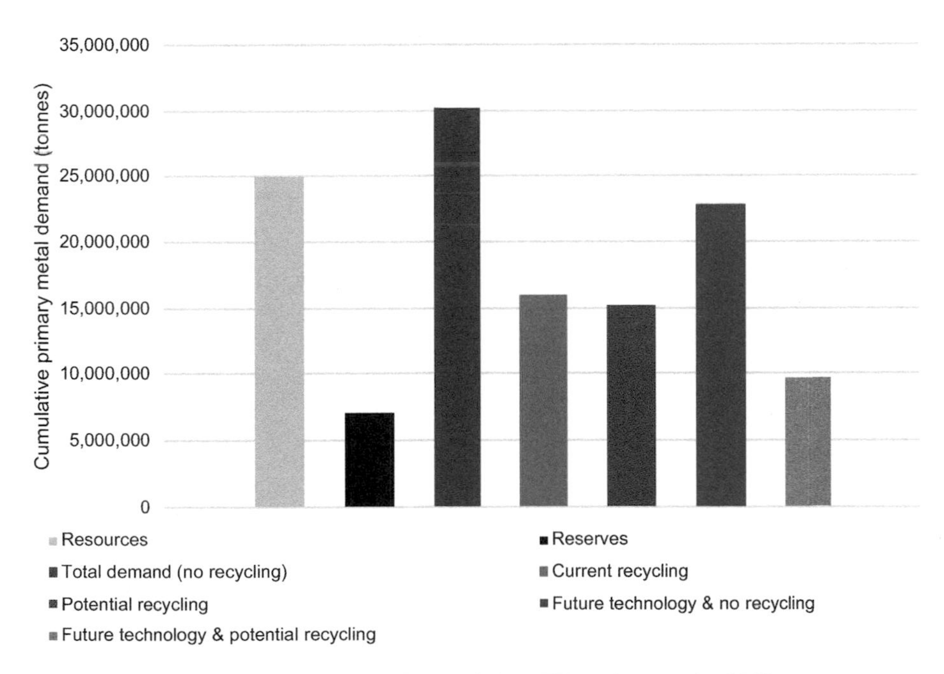

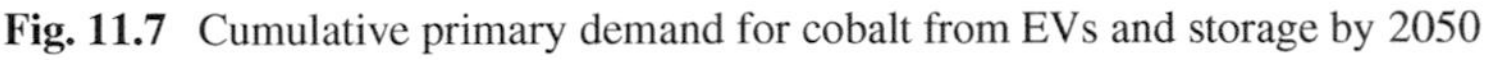

Fig. 11.7 Cumulative primary demand for cobalt from EVs and storage by 2050

11.4.2 Lithium

The annual demand for lithium for EVs and storage could exceed the current production rates by around 2022 (in all scenarios). In the 'future technology and no recycling' scenario, a shift to Li–S will increase the demand for lithium, because these batteries have a higher amount of lithium. Increasing recycling from its current low levels (which are assumed to be 10%) will offer the greatest potential to offset the primary demand for lithium.

The cumulative demand for lithium by 2050 will be below the resource levels for all scenarios, but will exceed the reserves unless there is a shift to a high recycling rate. The cumulative demand could be as high as 170% of the current reserves with the current technology, and could be 280% of current reserves with a switch to Li–S batteries (Figs. 11.8 and 11.9).

11.4.3 Silver

The total annual demand for silver could reach more than 40% of the current production rates by 2050, assuming no recycling and that the materials efficiency does not change (Fig. 11.10). The cumulative demand to 2050 could reach around half the current reserves with the current technology, and around one-quarter if the technology improves (Fig. 11.11). The reduction in material intensity in the 'future technology' scenario, in which silver use decreases from 20 to 4 tonnes/GW, has the greatest potential to reduce demand.

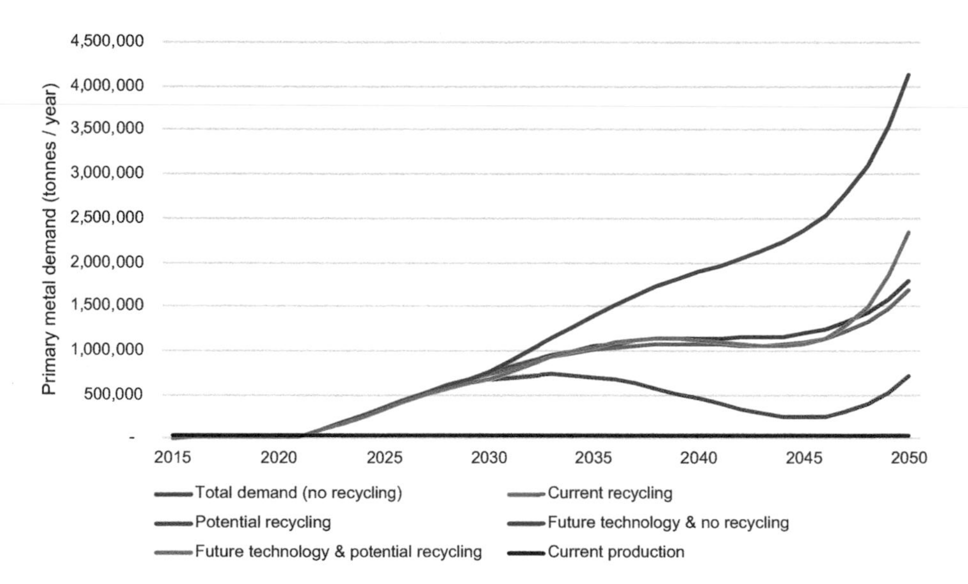

Fig. 11.8 Annual primary demand for lithium from EVs and storage

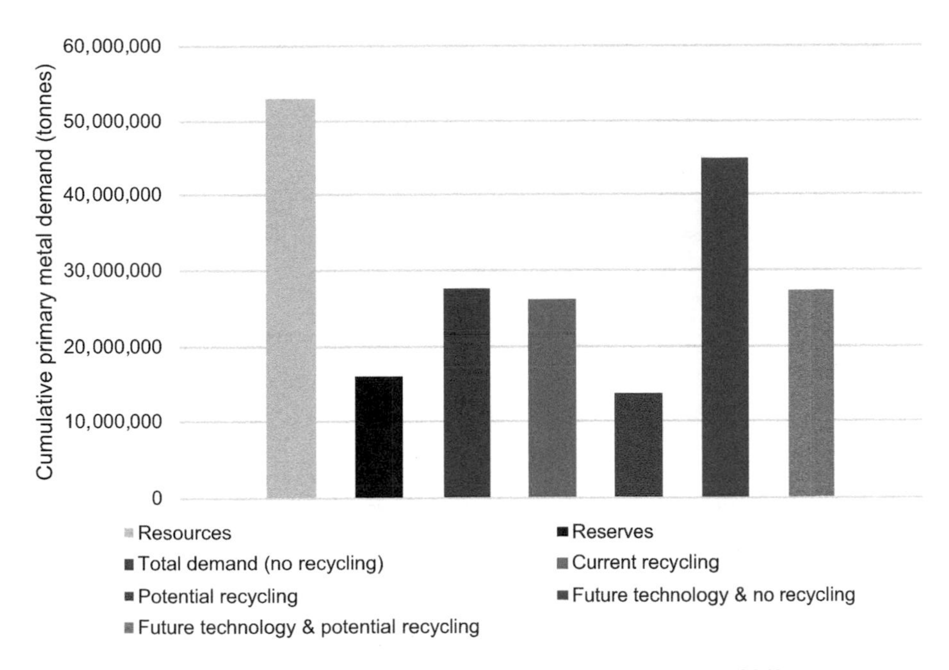

Fig. 11.9 Cumulative primary demand for lithium from EVs and storage by 2050

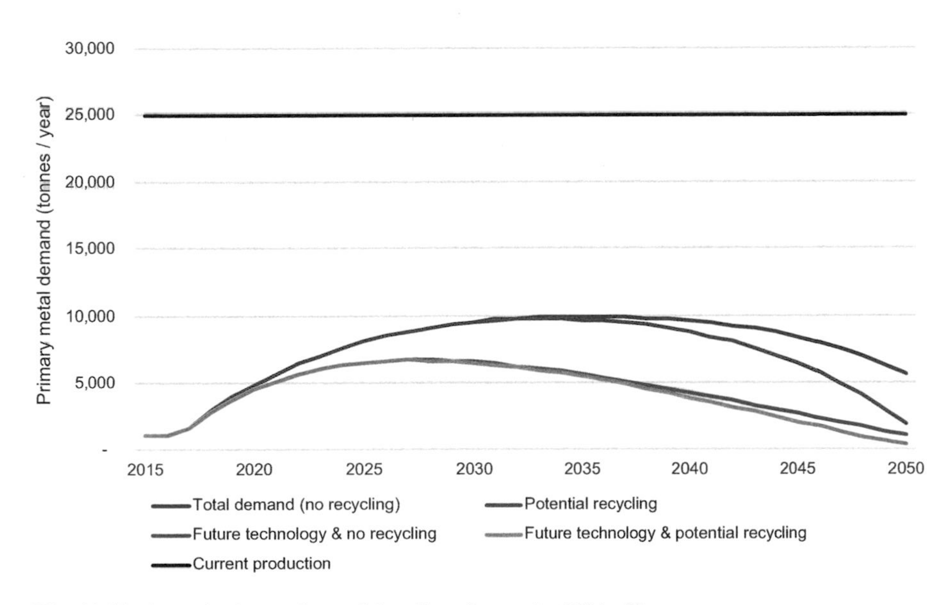

Fig. 11.10 Annual primary demand for silver from solar PV (c-Si)

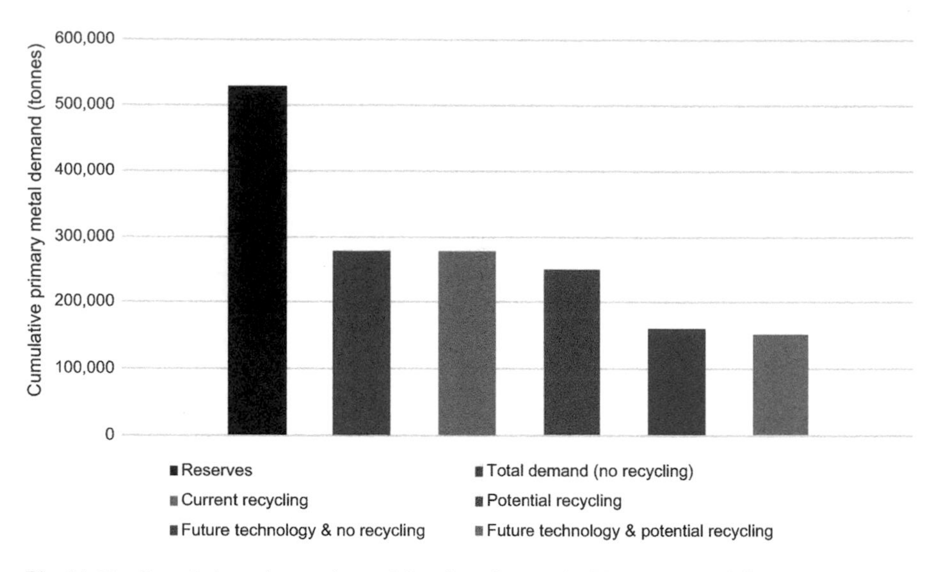

Fig. 11.11 Cumulative primary demand for silver from solar PV (c-Si) by 2050

11.5 Discussion

Within the context of the increasing metal resource requirements for the renewable energy and storage technologies, the rapid increase in the demand for both cobalt and lithium is of greatest concern, and the demand for both metals will exceed the current production rates by 2023 and 2022, respectively. The demand for these metals will increase more rapidly than it does for silver, partly because solar PV is a more established technology and silver use has become very efficient, whereas the electrification of the transport system and the rapid expansion of lithium battery usage has only begun to accelerate in the last few years.

The potential to offset the primary demand differs, depending on the technology. Offsetting demand through secondary sources of cobalt and lithium has the most potential to reduce total primary demand for these metals, as batteries have a comparatively short lifetime of approximately 10 years. The cumulative demand for both metals will exceed current reserves, but with high recycling rates, they can remain below resource levels. However, there is a delay until the period in which recycling will offset demand, because there must be enough batteries in use and reaching the ends of their lives to be collected and recycled. This delay could be further extended by strategies for the reuse of vehicular batteries as stationary storage, which might save costs in the short term and increase the uptake of PV. The efficiency of cobalt use in batteries also significantly reduces demand, and this is already happening as manufacturers shift towards lower-cobalt chemistries (Energy Insights by McKinsey 2018). However, this is also likely to lead to an increase in the demand for lithium.

Increasing the efficiency of material use has the biggest potential utility in offsetting the demand for PV metals, whereas recycling will have a smaller impact on demand. This is attributable to the long life-span of solar PV panels, and their lower potential for recycling. Although not as extreme as the growth of cobalt and lithium, the growth of silver and its demand in 2050 are still considerable. This is important, especially when considering that solar PV currently consumes approximately 9% of end-use silver (The Silver Institute & Thomson Reuters 2018), and although it is possible to create silver-less solar panels, these panels are not expected to be on the market in the near future (ITRPV 2018).

11.5.1 Limitations

This study only focuses on the metal demand for renewable energy and transport. It does not take into account other demands for these metals. It is expected that with the increase in renewable energy, renewable energy technologies will consume a greater share of these metals. In our modelling, the recycled content used in technologies comes only from metals from the same technologies at the end of their lives. However, the demand could potentially be offset by accessing other secondary sources of the metal. At the same time, these technologies are only one end-use within the overall economic demand for these metals. The expansion of renewable energy and storage technologies could have a significant effect on the overall market dynamics, including influencing prices, which may feed back to the efforts to reduce material intensity. These results only focus on cobalt, lithium, and silver because of their importance. However, further analysis is required of other metals that will be important in the renewable energy transition.

Another important limitation is that this analysis does not consider the impact of the demand for a mineral on the mining of that mineral and therefore on the amount of energy required for this mining and processing activity. In particular, as the potential ore grade declines, and polymetallic ore processing and the mining of deeper ore bodies increase, it is possible that this feedback loop could have a more-than-marginal influence on the overall sectoral energy consumption. To examine this is an area accurately will require much more-complex modelling.

11.5.2 Comparison with Other Studies

A large number of studies have examined scenarios for renewable energy and storage technologies that will mitigate climate change. In recent years, there has also been an upsurge in studies of mineral 'criticality', which have paralleled the present study in terms of the high penetration of renewable and storage technologies and the potential constraints that certain minerals may impose. This increased interest has been prompted to some extent by China's rare earth export restrictions of

Table 11.5 Comparison of results with other studies

Study	Energy Scenario	Metals (% of reserves up to 2050)		
		Cobalt	Lithium	Silver
This study	1.5 °C100% renewable energy scenario	420%	170%	50%
Watari et al. (2018)	Beyond 2 degree scenario (IEA)	180%	130%	78%
Månberger and Stenqvist (2018)	Beyond 2 degree scenario (IEA)	440%	100%	18%
Valero et al. (2018)	2 degree scenario	70%	40%	70%

2009–2011, which reflect the sense that mineral supply chains are still quite insecure. Most of these studies have addressed specific technologies or specific countries or regions, rather than global climate targets. A number of studies have specifically and directly addressed the Paris Agreement targets (ensuring that the temperature rise does not exceed 2.0 °C), although the modelling frameworks have been slightly different. Some of the authors of the present chapter have been involved in these studies (Tokimatsu et al. 2017; Watari et al. 2018).

A range of variables affect the results, including the installed capacity and technology type in the energy scenario; the assumed market demand of technology types (e.g., type of battery or solar panel); and the material intensity assumptions. The projected future demand for lithium is higher than in previous studies, and that for cobalt is similar or higher (Månberger and Stenqvist 2018; Watari et al. 2018; Valero et al. 2018; Watari et al. 2018), as shown in Table 11.5. This is primarily attributable to the ambitious renewable energy scenario used in this study, which includes achieving a 100% renewable transport system by 2050, whereas the other studies have still included a large share of gasoline-powered cars in 2050. This study also includes batteries for stationary energy storage, whereas the other studies have only included batteries for road transport. Moreover, this study assumes a shorter battery life of 10 years, based on current warranties, whereas some scenarios assume a longer life.

The results for silver are in the middle of the range of results given by other studies. Our scenario includes a higher installed capacity of PV in 2050 than the other studies. However, this study also assumes a lower metal intensity than previous studies, because new data have been published based on the current material use by the PV industry (ITRPV 2018).

11.6 Supply Impacts and Challenges

In addition to the material requirements for renewable technologies explored earlier in this chapter, it is important to understand the changes in the available supply, the geopolitical landscape, and the associated social and environmental impacts, which are outlined below.

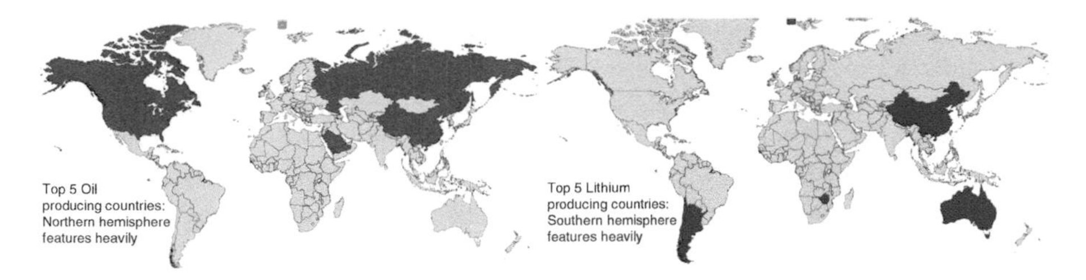

Fig. 11.12 Top five oil-producing countries (left) versus lithium-producing countries (right)

11.6.1 Geopolitical Landscape

The geopolitical shift underway in the supply of the resources required for the globe's future energy mix is clearly illustrated in Fig. 11.12. Whereas the value of the lithium industry is much less than the value of the oil industry, this comparison highlights a distinct shift in the energy commodities that society values. Oil's rate of use is projected to decline somewhat in the decade ahead (Mohr et al. 2015), whereas lithium's production is expected to grow rapidly (Mohr et al. 2012).

Although it is not a key focus of this particular chapter, the dominance of China in the supply of rare earths has encouraged manufacturing countries to look at diversifying their supply sources, including through the recovery of rare earths from recycled material. A similar situation of heavily concentrated supply occurs for cobalt, where the Democratic Republic of the Congo is the largest supplier, at around 66,000 tonnes per year. The next four largest cobalt-producing countries (China, Canada, Russia, and Australia) only produce 5–7000 tonnes per year each. The supply chain is also concentrated downstream, with around 50% of cobalt smelted and refined in China.

11.6.2 Social and Environmental Impacts

The mining and supply chain for these metals can have adverse social and environmental consequences for workers, local communities, and the environment. These impacts are most significant for the cobalt mined in the Democratic Republic of Congo, where there are human rights violations, child labour, and severe environmental pollution affecting health (Florin and Dominish 2017).

These types of impacts at mine sites and along the supply chain also influence the availability of primary resources. For example, whereas Australia and Chile are large producers of lithium, large deposits remain undeveloped in Bolivia, due in part to local concerns over the social and environmental impacts. The global silver market receives less media attention than the market for lithium, but the world's second-largest silver mine (Escobal) in Guatemala is currently closed by a constitutional court ruling that the Xinca Indigenous peoples were not adequately consulted before a mine licence was granted (Jamasmie 2018).

The increased use of materials such as lithium, cobalt, and silver has economic implications for the future of battery manufacture. For example, the cost of cobalt has risen dramatically from US$20000/t in 2016 to US$80000 in 2018 (Tchetvertakov 2018). This is prompting manufacturers to look for alternatives, such as nickel, vanadium, and zinc (Tchetvertakov 2018). At the same time, a significant proportion of these price fluctuations are attributable to non-industrial factors, such as investment and speculation in metal markets, which may also require greater regulation to avoid unnecessary restrictions on renewable energy and storage technologies.

11.6.2.1 Recycling Challenges

Whilst recycling can help to offset primary material demand through recycled sources, there are technological, social and environmental challenges to increasing recycling. The collection systems and infrastructure required to recycle metals from renewable energy technologies are not well established. For example, although silver has an overall recycling rate of 30–50% (Graedel et al. 2011), almost no recycling of silver from PV panels occurs, because most recycling of PV panels focuses on recycling the glass, aluminium, and copper. Most of the processes used to recycle lithium-ion batteries focus on the recovery of cobalt and nickel, because of their higher price, so that lithium is downcycled into less valuable products, such as cement. It should also be noted that the recycling of some key energy materials, such as rare earths in magnets, does not offer significant cost savings or environmental benefits over their extraction from primary resources (McLellan et al. 2013).

The establishment of effective collection networks is important for recycling. Collection networks must be easily understood and must provide easily accessible deposit mechanisms. Recycling can be both informal and formal, and whereas in some cases, informal collection networks offer greater rates of recycling, the social and environmental impacts can be higher. For example, recycling electronic waste on open fires creates hazardous fumes for informal recyclers and has detrimental effects on the environment. There is also a potential tension between reducing material intensity and the ability to effectively separate, recover, and recycle materials. Increasingly, thin layers of material are being utilized because of improved manufacturing processes, which can reduce the costs of products in their first lifetimes. However, as the material content decreases, the value of the secondary production decreases, making it a less attractive option for investment. Again, the use of advanced, complex materials can sometimes make valuable materials difficult to separate from each other, and this again increases the costs and reduces the profits for investors in recovery processes.

To ensure that social and environmental issues will be addressed, several initiatives across supply chains are being developed, including the China-led Responsible Cobalt Initiative, the Cobalt Institute, the Initiative for Responsible Mining

Assurance, Solving the Ewaste Problem (StEP), and the R2 standard for sustainable electronics recycling. In the USA, the Dodd-Frank Act has mandated the traceability of the gold, tin, tungsten, and tantalum from the Democratic Republic of Congo. This has arisen from the international concern over the social impacts and implications of sourcing metals from that country, and in particular, to avoid supporting conflict and environmental damage. However, cobalt is not explicitly mentioned in the Dodd-Frank Act.

International resource governance that looks beyond renewable energy and beyond a single commodity focus is increasingly recognised as a missing area in environmental policy (Ali et al. 2017). The transition to renewable energy and the associated requirements for resources to support this transition could be a catalyst for advancing such policies.

11.7 Concluding Remarks

This chapter has given an overview of the material requirements for key renewable energy technologies and the projected demands for lithium, cobalt, and silver. Consistent with the strong growth in renewable energy and the electrification of the transport system required by the 1.5 °C scenario, these material requirements will also increase dramatically, particularly those for cobalt and lithium. The high total demand requirements for these metals emphasize the importance of recycling, alternative chemistries, and transport planning and practices (including city design to encourage active and public transport, as well as car sharing and pooling). The key messages of this chapter are: it is important to design for both renewable energy and resource cycles; and it is important to adopt a systems view that considers the available supply and also social and environmental factors. To support sustainable development goals, both the primary and secondary sources of the resources required to underpin the renewable energy transformation must be stewarded effectively as the supply chains develop.

References

Ali, S. H., D. Giurco, N. Arndt, E. Nickless, G. Brown, A. Demetriades, R. Durrheim, M. A. Enriquez, J. Kinnaird, A. Littleboy, L. D. Meinert, R. Oberhänsli, J. Salem, R. Schodde, G. Schneider, O. Vidal and N. Yakovleva (2017). "Mineral supply for sustainable development requires resource governance." Nature 543: 367.

Bloomberg New Energy Finance, 2018., Electric Buses in Cities. Available at: http://www.ourenergypolicy.org/wp-content/uploads/2018/04/1726_BNEF_C40_Electric_buses_in_cities_FINAL_APPROVED_2.original.pdf

Cano, Z.P., Banham, D., Ye, S., Hintennach, A., Lu, J., Fowler, M. and Chen, Z., 2018. Batteries and fuel cells for emerging electric vehicle markets. Nature Energy, 3(4), p.279.

Clean Energy Manufacturing Analysis Center (CEMAC), 2017. Benchmarks of global clean energy manufacturing. Available at: https://www.nrel.gov/docs/fy17osti/65619.pdf

Energy Insights by McKinsey, 2018. Metal mining constraints on the electric mobility horizon. Available at: https://www.mckinseyenergyinsights.com/insights/metal-mining-constraints-on-the-electric-mobility-horizon/

Florin, N., Dominish, E. (2017). Sustainability evaluation of energy storage technologies, Institute for Sustainable Futures for the Australian Council of Learned Academies.

Georgi-Maschler, T., Friedrich, B., Weyhe, R., Heegn, H. and Rutz, M., 2012. Development of a recycling process for Li-ion batteries. Journal of power sources, 207, pp.173–182.

Giurco, D., B. McLellan, D. M. Franks, K. Nansai and T. Prior (2014). "Responsible mineral and energy futures: views at the nexus." Journal of Cleaner Production 84: 322–338.

Graedel, T.E., Allwood, J., Birat, J.P., Buchert, M., Hagelüken, C., Reck, B.K., Sibley, S.F. and Sonnemann, G. (2011). What do we know about metal recycling rates?. Journal of Industrial Ecology, 15(3), 355–366.

Gratz, E., Sa, Q., Apelian, D. and Wang, Y., 2014. A closed loop process for recycling spent lithium ion batteries. Journal of Power Sources, 262, pp.255–262.

International Technology Roadmap for Photovoltaic (ITRPV), 2018, International Technology Roadmap for Photovoltaic Results 2017, Ninth Edition. Available at: http://www.itrpv.net/Reports/Downloads/

Jamasmie, C. (2018) Guatemala delays ruling on Tahoe's Escobal mine reopening. Mining.com Accessed 26 September 2018, http://www.mining.com/guatemala-delays-ruling-tahoes-escobal-mine-reopening/

Kavlak, G., McNerney, J., Jaffe, R.L. and Trancik, J.E., 2015. Metal production requirements for rapid photovoltaics deployment. Energy & Environmental Science, 8(6), pp.1651–1659.

King S, Boxall NJ, Bhatt AI (2018) Australian Status and Opportunities for Lithium Battery Recycling. CSIRO, Australia

Månberger, A. and B. Stenqvist (2018). "Global metal flows in the renewable energy transition: Exploring the effects of substitutes, technological mix and development." Energy Policy 119: 226–241.

McLellan, B. C., G. D. Corder and S. H. Ali (2013). "Sustainability of rare earths—an overview of the state of knowledge." Minerals 3(3): 304–317.

Mohr, S. H., G. M. Mudd and D. Giurco (2012). "Lithium Resources and Production: Critical Assessment and Global Projections." Minerals 2(1): 65.

Mohr, S. H., Wang, J., Ellem, G., Ward, J., & Giurco, D. (2015). Projection of world fossil fuels by country. Fuel, 141, 120–135.

Prior, T., P. A. Wäger, A. Stamp, R. Widmer and D. Giurco (2013). "Sustainable governance of scarce metals: The case of lithium." Science of The Total Environment 461–462: 785–791.

REN21 (2018) Renewables 2018 Global Status Report, REN21 Secretariat, Paris

Tchetvertakov, G. (2018). "Panasonic and Tesla seek to remove cobalt from electric car batteries." Retrieved 15 September 2018, from https://smallcaps.com.au/panasonic-tesla-remove-cobalt-electric-car-battery/.

The Silver Institute and Thomson Reuters, 2018. World Silver Survey 2018. Available at: https://www.silverinstitute.org/wp-content/uploads/2018/04/WSS-2018.pdf

Tokimatsu, K., H. Wachtmeister, B. McLellan, S. Davidsson, S. Murakami, M. Höök, R. Yasuoka and M. Nishio (2017). "Energy modeling approach to the global energy-mineral nexus: A first look at metal requirements and the 2 °C target." Applied Energy 207: 494–509.

Vaalma, C., Buchholz, D., Weil, M. and Passerini, S., 2018. A cost and resource analysis of sodium-ion batteries. Nature Reviews Materials, 3, p.18013

Valero, A., A. Valero, G. Calvo and A. Ortego (2018). "Material bottlenecks in the future development of green technologies." Renewable and Sustainable Energy Reviews 93: 178–200.

Viebahn, P., O. Soukup, S. Samadi, J. Teubler, K. Wiesen and M. Ritthoff (2015). "Assessing the need for critical minerals to shift the German energy system towards a high proportion of renewables." Renewable and Sustainable Energy Reviews 49: 655–671.

Watari, T., B. McLellan, S. Ogata and T. Tezuka (2018). "Analysis of Potential for Critical Metal Resource Constraints in the International Energy Agency's Long-Term Low-Carbon Energy Scenarios." Minerals 8(4): 156.

Weckend, S.;Wade, A.; Heath, G. 2016 End-of-Life Management Solar Photovoltaic Panels; International Renewable Energy Agency and International Energy Agency Photovoltaic Power Systems: Paris, France

Widmer, J.D., Martin, R. and Kimiabeigi, M., 2015. Electric vehicle traction motors without rare earth magnets. Sustainable Materials and Technologies, 3, pp.7–13.

Yan, X., He, J., King, M., Hang, W. and Zhou, B., 2018. Electric bicycle cost calculation models and analysis based on the social perspective in China. Environmental Science and Pollution Research, pp.1–13.

Zimmermann, T., Rehberger, M. and Gößling-Reisemann, S., 2013. Material flows resulting from large scale deployment of wind energy in Germany. Resources, 2(3), pp.303–334.

Chapter 12
Implications of the Developed Scenarios for Climate Change

Malte Meinshausen

Abstract This section provides a summary of the implications of the developed 2.0 °C and 1.5 °C scenarios for global mean climate change. Specifically, we consider atmospheric CO_2 concentrations, radiative forcing, global-mean surface air temperatures and sea level rise.

The question addressed in this section is what the implications are for future climate change if the world were to follow the energy-related CO_2 emissions developed here, complemented by land-use CO_2 emissions and those of other greenhouses gases (GHGs). According to the high-emission scenario with unabated fossil fuel use, the world could experience 1400 ppm CO_2 concentrations by the end of 2100, which is five times higher than the pre-industrial background concentration of 278 ppm. Our ice-core records have shown that over the last 850,000 years, the CO_2 concentrations have only oscillated between approximately 180 ppm and 280 ppm. In fact, for the last 10 million years on this planet, the CO_2 concentrations have probably not exceeded the CO_2 concentrations that our thirst for fossil fuels would propel the world into in just two centuries.

That is a dramatic change to the thin layer of atmosphere that wraps our planet. Even if any climate change consequences that follow from this dramatic change in CO_2 concentrations are disregarded, the consequences will be dramatic. At atmospheric CO_2 concentrations above 900 ppm, the acidity level in the oceans will drop below the so-called 'aragonite saturation level', which is the level required for coral reefs and other organisms with calciferous shells to sustain their structures (Ricke et al. 2013). Therefore, without even considering ocean warming and the associated bleaching, this would mean the end of much marine life as we know it.

M. Meinshausen (✉)
Australian-German Climate and Energy College, University of Melbourne, Parkville, Victoria, Australia
e-mail: malte.meinshausen@unimelb.edu.au

S. Teske (ed.), *Achieving the Paris Climate Agreement Goals*,
https://doi.org/10.1007/978-3-030-05843-2_12

12.1 Background on the Investigated Scenarios

The international community uses various scenarios to explore these future changes to CO_2, involving other GHGs and ultimately, temperature, precipitation changes, and changes involving extreme events. Here, we compare the scenarios developed in this study with the standard scenarios developed for the forthcoming Intergovernmental Panel on Climate Change (IPCC) Sixth Assessment Report. The scenarios used in the forthcoming IPCC Sixth Assessment Report are the so-called 'SSP scenarios'. The full range of those nine SSP scenarios spans cases of unabated fossil fuel use at the higher-emissions end to two scenarios at the lower-emission end that are considered to be consistent with a 2 °C and 1.5 °C warming target. These two scenarios are called 'SSP1_26' and 'SSP1_19', respectively.

Both of those lower-emissions scenarios imply a substantial amount of biomass use for energy, combined with CCS, to draw down atmospheric CO_2 levels. This is the key difference from the scenarios developed in the present study. Whereas we use a substantial amount of negative CO_2 emissions in the reforestation and forest restoration options, our analysis does not depend on the assumption that biomass, combined with CCS, must be used on large scale to achieve either the 2 °C or 1.5 °C target.

If we do not rely on some negative CO_2 emission options, in particular biomass and CCS, the questions are: (1) to what extent are those negative emissions created by other means (e.g., reforestation); or (2) whether CO_2 emissions can be reduced in the first place so that we do not need to rely on negative emissions options; or (3) whether other non-CO_2 gases can be reduced even further to make negative CO_2 emissions unnecessary. The scenarios developed in this study use all three options, as outlined in the previous chapters. Not only will energy-related CO_2 emissions be radically reduced, CO_2 uptake via reforestation and forest restoration will also play an important role. This study takes a more conservative approach to non-CO_2 gases by being consistent with other stringent mitigation scenarios. Thereby, we ensure that the feasibility constraints implicitly or explicitly set by other modelling frameworks are not violated.

12.2 Comparison of Atmospheric CO_2 Concentrations and Radiative Forcing

As mentioned above, unabated fossil fuel use over the last century and the twenty-first century will dramatically change the oceans, simply by creating an atmospheric CO_2 concentration beyond that present on the Earth for the last 10 million years. Currently, we just exceeded the historical maximum level of 400 ppm atmospheric CO_2 and are continuing to add 2–3 ppm a year. If we consider the most stringent mitigation scenario used in the preparation of the forthcoming IPCC Sixth Assessment Report, SSP1_19, then we will reach 400 ppm concentrations again in the latter half of the century. Staying below 400 ppm is a prerequisite in the long term for remaining below 2 °C or reaching 1.5 °C warming.

Figure 12.1 shows the global CO_2, CH_4, and N_2O concentrations under the key RCP and SSP scenarios and the three scenarios developed as part of this study. Our 1.5 °C scenario is clearly lower in terms of its CO_2 concentrations than the lowest SSP scenario, SSP1_19, for practically the entire twenty-first century (upper panel). Only towards the end of the twenty-first century do the strongly negative CO_2 emissions in the SSP1_19 scenario bring the CO_2 concentration closer towards our 1.5 °C scenario.

Aggregating all the greenhouse gas and aerosol emissions by their radiative forcing, and expressing the resulting radiative forcing again as if it were only caused by CO_2 yields the so-called 'CO_2 equivalence concentrations'. In Fig. 12.2, the CO_2 equivalence concentrations for the four RCP scenarios, the nine SSP scenarios, and three scenarios developed in this study are shown. The reference scenario in this study is quite similar to both the RCP6.0 and SSP4_60 scenarios, providing a medium-high reference case. The radiative forcing and CO_2 equivalence concentration of the 2 °C scenario of this study is actually quite closely aligned with the lower SSP1_19 scenario, at least initially until the middle of the second half of this century. Thereafter, the net negative CO_2 emissions implied by the SSP1_19 scenario lead to stronger reductions in radiative forcing than the 2.0 °C scenario of our study. However, our lower 1.5 °C scenario first undercuts the radiative forcing trajectory of the SSP1_19 scenario, but then ends up at a similar radiative forcing level by 2100.

Figure 12.2 shows CO_2 equivalence concentrations (upper panel) and radiative forcing (lower panel) of the main scenarios used in IPCC Assessment Reports and this study's scenarios. The RCP scenarios (shown in thin dotted lines) underlie the IPCC Fifth Assessment Report and the so-called 'SSP scenarios' provide the main basis for the scenarios considered in the IPCC Sixth Assessment Report. The three scenarios developed in this study are shown in thick blue lines.

12.3 Comparison of Cumulative CO_2 Emissions

Since the IPCC Fifth Assessment Report, cumulative CO_2 emissions have been introduced as a key metric into the international climate debate. Every tonne of additional CO_2 emitted will add to that cumulative burden and warm the planet for the next hundreds and in fact thousands of years. The only way to halt further global warming is to halt cumulative CO_2 emissions, which means bringing the annual CO_2 emissions to basically zero levels. Only with net negative emissions can the temperature thermostat of the Earth be dialled back again. Therefore, while achieving such net negative emissions is tremendously challenging, it is the only way to maintain long-term climate change at levels close to those of today or well below 1.5 °C. Few scenarios include very strong near-term reductions and no substantially negative CO_2 emissions and can limit the temperature increase to 1.5 °C, with no or only a slight overshoot (IPCC Special Report on 1.5 °C). In this study, as well as very strong near-term reductions in energy-related CO_2 emissions, we have used a range of land-use-based sequestration options. These are not unambitious, as

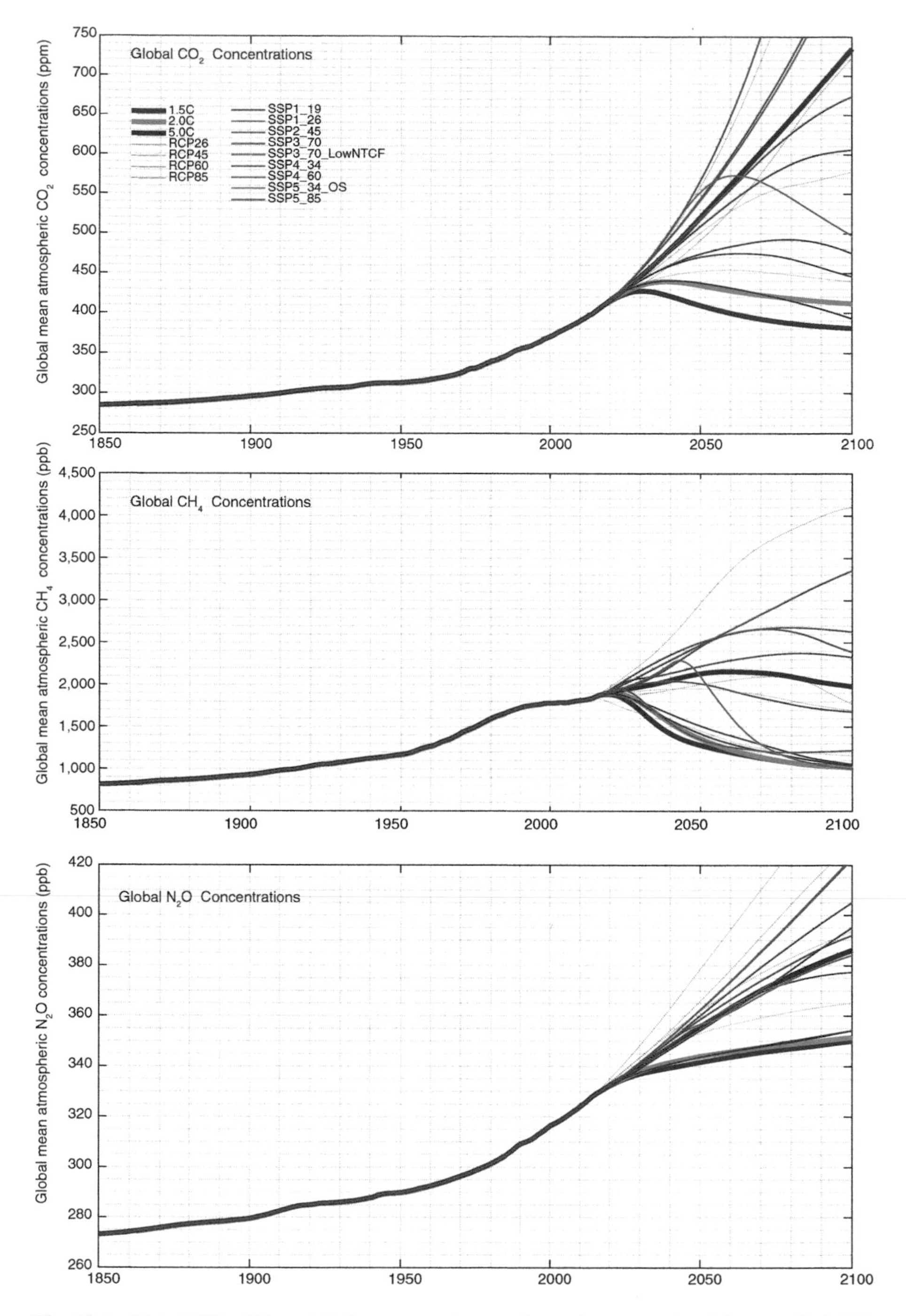

Fig. 12.1 Global CO_2, CH_4 and N_2O concentrations under various scenarios. The so-called SSP scenarios are going to inform the Sixth Assessment Report by the IPCC, the RCP scenarios are the previous generation of scenarios and the LDF scenarios are those developed in this study

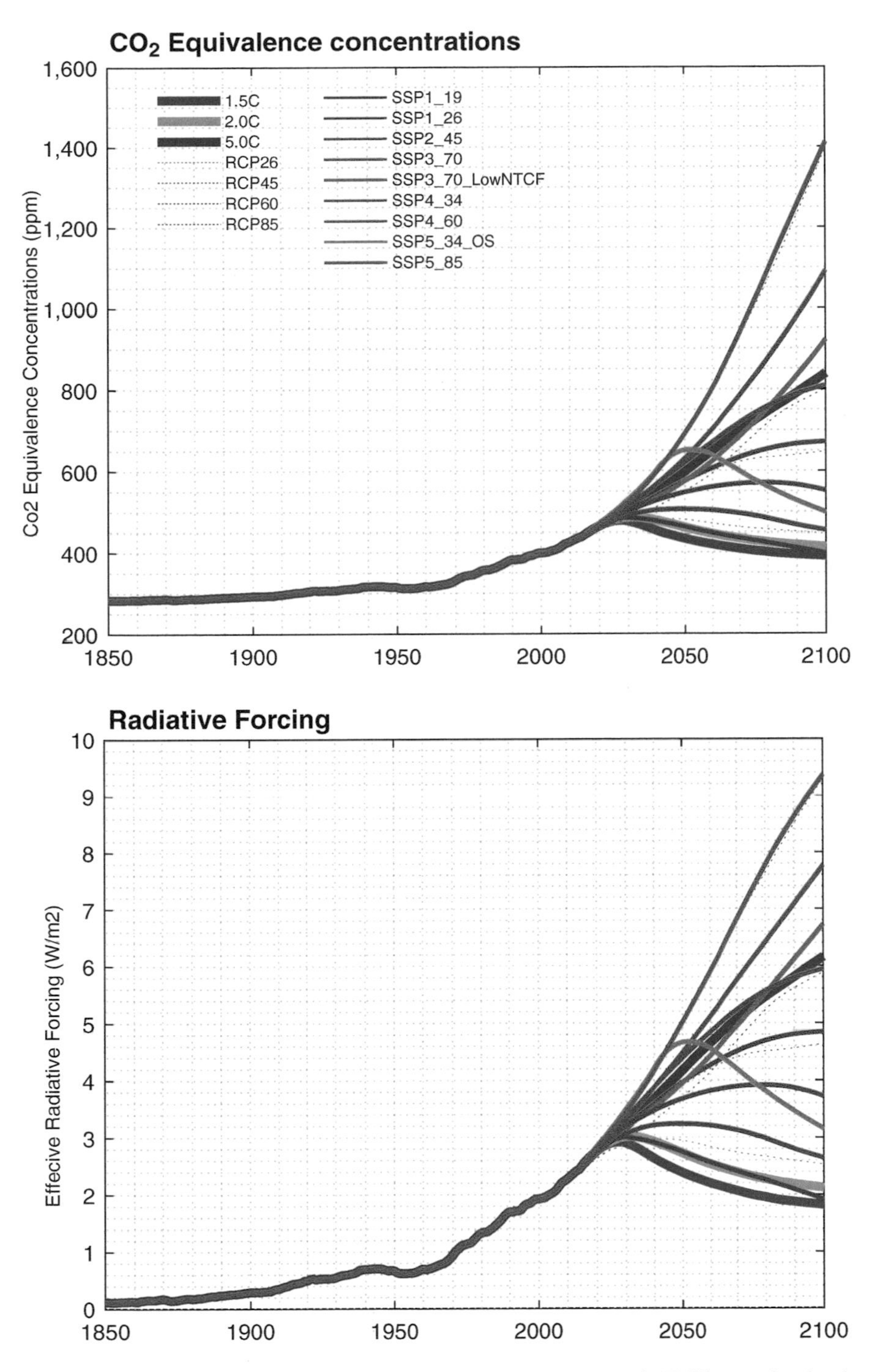

Fig. 12.2 CO_2 equivalence concentrations and radiative forcing of main IPCC scenarios for the forthcoming Sixth Assessment (so-called SSP scenarios), the RCP scenarios underlying the Fifth IPCC Assessment Report and the LDF scenarios developed in this study

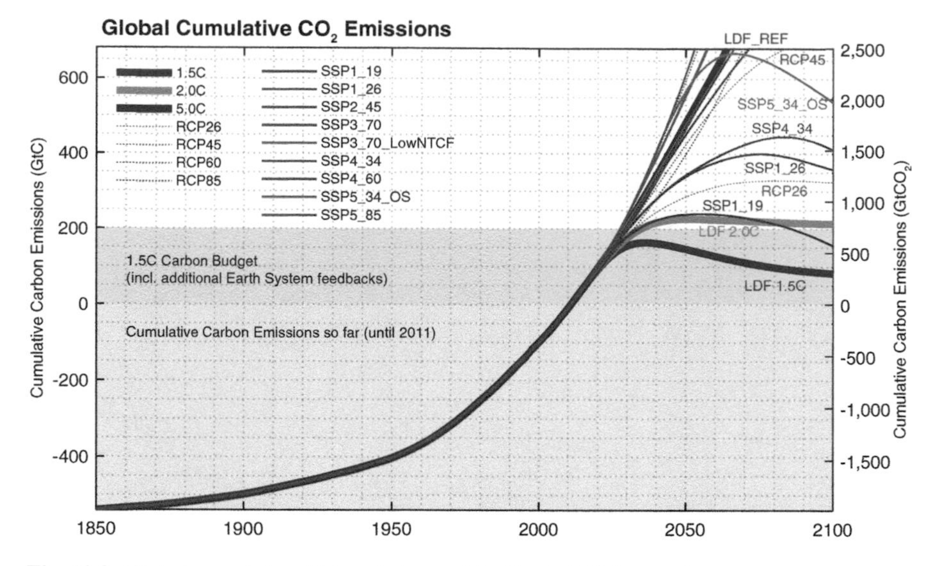

Fig. 12.3 Global cumulative CO2 emissions – 2.0 °C and 1.5 °C scenarios

outlined in previous chapters. In fact, in the case of the 1.5 °C scenario, they practically require that the Earth's forests be returned to their pre-industrial coverage—a substantial challenge given the widespread deforestation that has occurred over the last 100 years. By the end of the century, the cumulative CO_2 emissions beyond levels will constitute around 300 $GtCO_2$ (see Fig. 12.3).

Without going into too much detail about why cumulative emission numbers differ between various studies, the key point is that CO_2 emissions must be reduced to zero. Otherwise, cumulative emissions and global warming will continue to increase. Therefore, there is no way to avoid emissions-free electricity, transport, and industrial systems, that either learn to completely get rid themselves of fossil fuels or to compensates for any remaining use, either by CCS or land-based sequestration. Because the latter negative emissions option must be used to offset some remaining agricultural emissions, the story becomes simpler. There is no way around the fast and complete cessation of all fossil fuel use. This is the simple and powerful logic of carbon budgets.

Global cumulative CO_2 emissions for the scenarios developed in this study (thick blue lines) and other literature-reported scenarios from the RCP and SSP sets. It is clear that this study's 2.0 °C pathway initially reaches a similar cumulative emissions level as the SSP1_19 scenario, before the cumulative CO_2 emissions are reduced again in SSP1_19 with large-scale net negative CO_2 emissions via bioenergy with carbon capture and storage. In contrast, the reduction of cumulative CO_2 emissions in this study's 1.5 °C scenarios starts to plateau much earlier (by 2035) and are then reduced by land -based sequestration options, such as reforestation and forest restoration (thick blue line at the lower end). The stated blue 1.5 °C carbon budget range is the effective central value presented in the recent IPCC Special Report on 1.5 °C warming.[1]

12.4 Implications for Temperature and Sea-Level Rise

This section examines the implications of the three scenarios developed in this study for the probabilistic global mean temperature and sea level rise. Based on the latest version 7 of the reduced complexity carbon cycle and climate model MAGICC, we can derive a range of projections for every scenarios that provides a good measure of the projection uncertainties over the twenty-first century. We ran this model 600 times for each scenario, varying wide ranges of feedback and forcing parameters. For the sea-level rise projections, we used the new sea-level rise module of MAGICC7, as described by Nauels et al. (2017) and drove that with our probabilistic temperature projections. We did not include the recent finding that possible Antarctic ice sheet instability could lead to a much greater sea-level rise this century. These extra contributions to sea-level rise are also assumed to affect the higher end of the projections for high-emission scenarios, and do not therefore change the sea-level rise projections of the lower-emissions scenarios.

A key question regarding future temperature projections is whether the scenarios will stay below the envisaged 1.5 °C and 2.0 °C warming levels relative to pre-industrial levels. When answering this question in terms of the consistency of the 1.5 °C and 2.0 °C warming levels, the uncertainty in historical reconstructions of temperatures must play a role. The latest IPCC Special Report on 1.5 °C warming estimated from an average of four studies that the difference between early-industrial (1850–1900) and recent surface air temperature levels (2006–2015) was 0.97 °C. If we accept here the slightly oversimplified assumption that the 1850–1900 temperature levels can be equated with pre-industrial temperatures, we can evaluate the difference towards a 1.5 °C warming level, namely 0.53 °C. Note that this distance could be shorter—a matter of ongoing scientific debate.

Anyway, the median temperatures for the (lowest) 1.5 °C Scenario in this study do indeed—somewhat by design—reach the 1.5 °C target level by 2100. This means that after a slight overshoot, there will be a 50% chance by 2100 that the global mean temperature is at or below 1.5 °C warming—and without the widespread deployment of bioenergy with carbon capture and storage (BECCS).

[1] The IPCC Special Report estimates for a 0.53 °C distance of the 1.5 °C degree target from the 2006–2015 level, a carbon budget of 560 $GtCO_2$ from 1 January 2018 onwards. Reduced by 100 $GtCO_2$ for additional Earth System Feedbacks (see Table 2.2 in the IPCC Special Report on 1.5 °C) and adding approximately 270 $GtCO_2$ emissions for the period from 2011 to 2017, the central estimate for a 50% chance of staying below 1.5 °C warming as stated by IPCC is approximately 730 $GtCO_2$ from 2011 onwards. This is for a definition of historical warming that is based on a consistent surface air temperature estimate over the land and oceans. A further reduction of approximately 150 $GtCO_2$ would result, if we take into account that the IPCC carbon budget estimate refer to a 1850–1900 early industrial reference period for the 1.5 °C warming, rather than a 1750 pre-industrial reference period (which makes approximately a 0.1 °C difference, with +– 0.1 °C uncertainty).

In the case of our 2 °C Scenario, the centric 66% range is almost entirely below the 2 °C warming level, which means that the chance of maintaining temperature change below 2 °C is 80%–85%, in the modelling framework used. This warming level and likelihood fits well with the adapted Paris Agreement target, which shifted from a 'below 2 °C' to a 'well below 2°C' formulation.

Of course, there are several uncertainties that are not addressed in those probabilistic temperature projections. Therefore, future investigations might shift the estimates for a 50%, lower, or higher chance of staying below 1.5 °C warming level relative to the pre-industrial level.

Figure 12.4 shows the global mean surface air temperature projections, and their 66% and 90% ranges, for the three scenarios developed in this study. The reference scenario is shown in red, the 2.0 °C pathway is shown in blue, and the 1.5 °C pathway is shown in green. For historical temperatures, a mixture between ocean-surface and land-surface air temperatures is shown, namely the HadCRUT4 dataset. The most recent estimate by the IPCC of the pre-industrial air surface temperature levels is that they were around 0.97 °C below the 2006–2015 levels. The median projections of the three scenarios are shown in thick solid lines.

As a first approximation, the global mean temperature is proportional to the sum of all historical CO_2 emissions. In a similar vein, sea-level rise is the sum over all past temperature rises relative to the pre-industrial level. Combining these two approximations yields the rule of thumb that sea-level rise is proportional to the double integral of CO_2 emissions. Therefore, whereas temperatures are relatively agnostic about when CO_2 emissions occurred, sea-level rise will be higher the longer ago the CO_2 emission occurred.

This proportionality has implications for our scenarios. As we saw previously, our scenarios are relatively low in terms of radiative forcing early in the twenty-first century compared with the lower scenarios of the SSP and RCP sets. Towards the end of this century, the strongly net negative emissions of the SSP1_19 scenario will cause radiative forcing (and approximate temperature) to be similar under SSP1_19 and our 1.5 °C scenario. That our scenarios will first entail lower and then similar forcing and temperature levels suggests that the implied sea-level rise will be lower for the second half of the twenty-first century and beyond, even though the 2100 temperature level might be similar. This is a clear benefit of our scenarios and should be investigated further in the future.

Thus, there is clearly an advantage in undertaking concerted early action rather a slower decline in emissions followed by strong negative emissions. This is important. However, in the larger scheme of things, it is clearly a second-order effect. Even under the strongest mitigation scenarios, we cannot expect sea-level rise to stop any time soon. To halt sea-level rise in the 21st or even the twenty-second century, we will require massively negative CO_2 emissions, drawing back out of the atmosphere a lot of the CO_2 that we emitted this century. Therefore, even under our low 1.5 °C and 2.0 °C scenarios, the expected sea-level rise by 2100 will be above 30 cm relative to the 2010 levels—and will continue to rise to 2100 (Fig. 12.5).

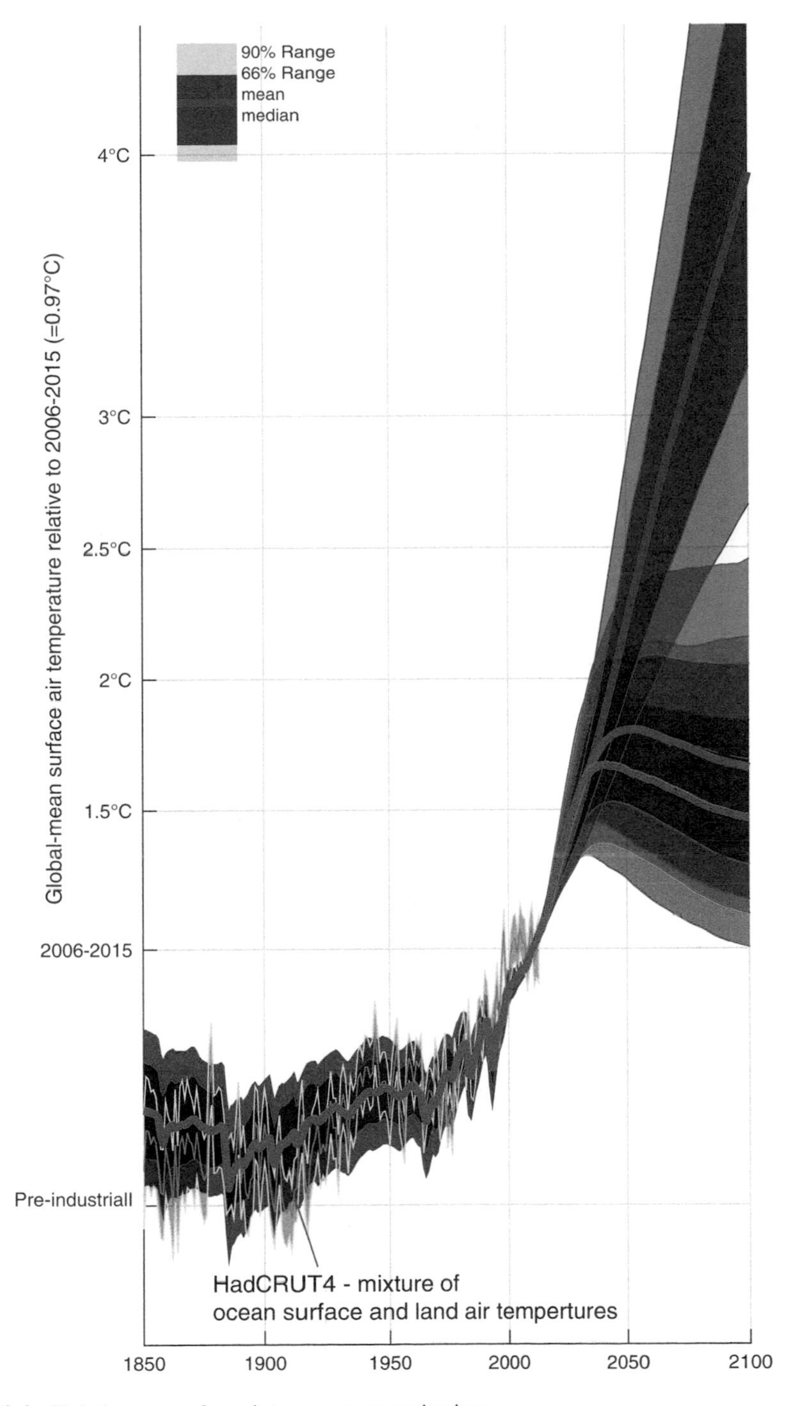

Fig. 12.4 Global-mean surface air temperature projections

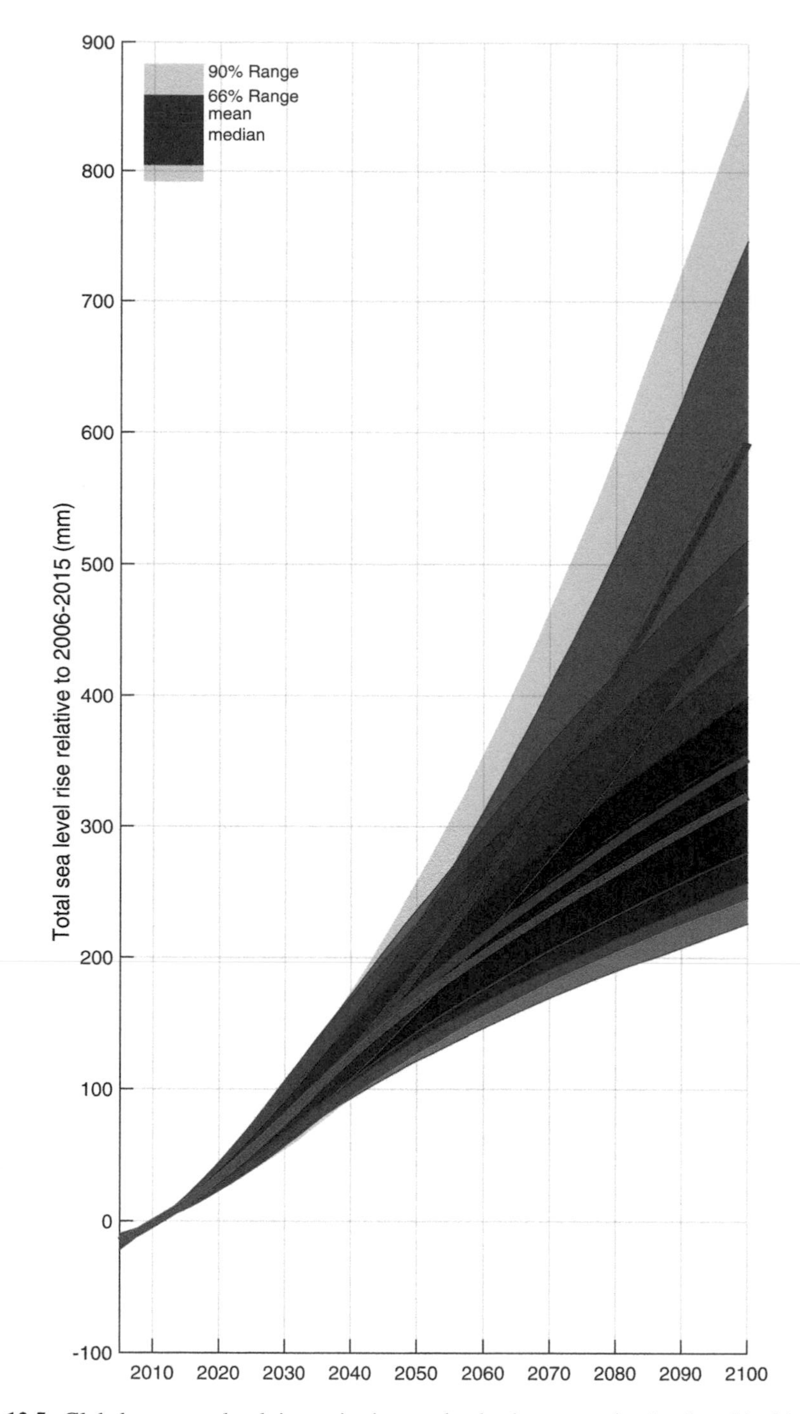

Fig. 12.5 Global-mean sea level rise projections under the three scenarios developed in this study

References

Ricke, K.L., Orr, J.C., Schneider, K. and Caldeira, K., 2013. Risks to coral reefs from ocean carbonate chemistry changes in recent earth system model projections. Environmental Research Letters, 8(3), p.034003.

Nauels, A., Meinshausen, M., Mengel, M., Lorbacher, K. and Wigley, T.M., 2017. Synthesizing long-term sea level rise projections-the MAGICC sea level model v2. 0. Geoscientific Model Development, 10(6).

Chapter 13
Discussion, Conclusions and Recommendations

Sven Teske, Thomas Pregger, Johannes Pagenkopf, Bent van den Adel, Özcan Deniz, Malte Meinshausen, and Damien Giurco

> *The Paris Agreement central aim is to strengthen the global response to the threat of climate change by keeping a global temperature rise this century well below 2 °C above pre-industrial levels and to pursue efforts to limit the temperature increase even further to 1.5 °C.*
>
> UNFCCC (2015)

Abstract The following section focuses on the main findings in all parts of the research, with priority given to high-level lessons, to avoid the repetition of previous chapters. The key findings as well as the research limitations and further research requirements are given for following topics:

Renewable energy potential mapping, Transport scenario and long-term energy scenario development, power sector analysis, employment and mineral resource implications for the 2.0C and 1.5C scenarios and non-energy GHG scenarios,

Policy recommendations for the energy sector with a focus on policies for buildings sector decarbonisation, for the transport and industry sector as well as a recommended political framework for power markets are provided.

S. Teske (✉) · D. Giurco
Institute for Sustainable Futures, University of Technology Sydney, Sydney, NSW, Australia
e-mail: sven.teske@uts.edu.au; damien.giurco@uts.edu.au

T. Pregger
Department of Energy Systems Analysis, German Aerospace Center (DLR), Institute for Engineering Thermodynamics (TT), Pfaffenwaldring, Germany
e-mail: thomas.pregger@dlr.de

J. Pagenkopf · B. van den Adel · Ö. Deniz
Department of Vehicle Systems and Technology Assessment, German Aerospace Center (DLR), Institute of Vehicle Concepts (FK), Pfaffenwaldring, Germany
e-mail: johannes.pagenkopf@dlr.de; Bent.vandenAdel@dlr.de; oezcan.deniz@dlr.de

M. Meinshausen
Australian-German Climate and Energy College, University of Melbourne, Parkville, Victoria, Australia
e-mail: malte.meinshausen@unimelb.edu.au

S. Teske (ed.), *Achieving the Paris Climate Agreement Goals*,
https://doi.org/10.1007/978-3-030-05843-2_13

The Paris Agreement's goals require significant change in how we use and produce energy on a global level. This energy transition must be started immediately and without further delay, and concerns energy-producing industry and utilities as well as every energy consumer—from the industry level down to the residential sector. A combination of energy efficiency and the use of renewable energies will involve new business concepts for the energy sector, which will require entirely new policies that provide a new market framework. The implementation of new technologies and business concepts will include policy changes on the community level as well as by national governments and international organizations. While the 2.0 °C Scenario allows for a 3–5 years transition period in which to implement policy measures, the 1.5 °C Scenario allows no further delay but requires an immediate start, at the latest in 2020. Therefore, the 1.5 °C Scenario presented in this book provides a technical pathway that assumes and allows no political delay, and may therefore be seen as a technical benchmark scenario.

However, this energy transition cannot be seen in isolation. To stay within the Paris Agreement Goal requires reductions in greenhouse gas (GHG) emission that are greater than can be delivered by the energy transition alone (see Chap. 4). However, there are technical and logistical limits to how fast new technologies can be implemented. The establishment of efficient transport systems and industry processes, or simply the provision of required renewable energy and storage technologies will require a minimum time. Production capacities must increase significantly, people must be trained, and existing buildings must be retro-fitted. Infrastructural changes, such as smart power grids, must be planed, construction permits must be issued. The energy transition will require political determination and public acceptance. Therefore, the planning process will require resolute stakeholder involvement. Furthermore, a fundamental shift in today's resource-intensive lifestyles seems unavoidable if we are to limit global warming. In particular, the way consumption and mobility are organized in developed countries today will challenge planetary boundaries to the extreme if they remain at their current levels.

During the development of the energy scenario pathways, it became clear that even the most ambitious 'man-on-the-moon' program—if focussed only on the energy sector—would not be enough. Therefore, the political framework required to implement the Paris Agreement on national levels must also take the land-use sector, as well as the main GHGs, methane, N_2O, and fluorinated gases, into account.

The following sections present the main findings and lessons of this research project, and highlight its limitations and further research requirements. Finally, we provide policy recommendations for the energy sector in order to implement the -energy scenarios.

13.1 Findings and Limitations—Modelling

The following section focuses on the main findings in all parts of the research, with priority given to high-level lessons, to avoid the repetition of previous chapters.

13.1.1 Key Findings—Renewable Energy Potential Mapping

Various research projects have analysed renewable energy potentials and all have in common that the renewable energy potential exceeded the current and projected energy demands over the next decades by an order of magnitude. However, regional distributions are uneven, and especially in densely populated regions, such as urban areas, the local renewable energy resource might be unable to supply the demand. Therefore, renewable energies might have to be transported by power lines or in the form of gasified or liquid renewable fuels to the demand centre. However, all 10 regions and 75 sub-regions examined had sufficient renewable energy resources to meet the regional needs.

13.1.2 Limitations and Further Research—Renewable Energy Potential

The quality of data varies significantly across the regions and especially detailed high-resolution surveys have their limits. Due to the limited available data, detailed mapping of the Eurasia region—especially for Russia and parts of central Asian countries—was not possible. Therefore, in this case, we relied on research published in the scientific literature. Onshore wind data were generally only available for 80 m above the ground, whereas no data for 100–120 m were available in most cases. However, modern wind turbines operate at those higher levels and wind resources are generally better at these elevations. Therefore, further research is required using open source data for onshore and offshore wind data at the 100 m level.

13.1.3 Key Findings—Transport Scenario

Transport modelling has shown that the 2.0 °C and 1.5 °C pathways can be met when strong and determined measures are taken, starting immediately. They include rapid electrification across passenger and freight transport modes, a shift towards more energy-efficient transport modes, and a build-up of biomass and synfuel capacities for the transport modes that are less inclined towards electrification due to range or constructional constraints, as is the case in aviation. In the road sub-sector, which is the most relevant emission source in the transport area, battery electric and fuel-cell electric vehicles must be widely introduced, which will also require a stringent parallel build-up of recharging and refuelling infrastructures.

In general, we found that beyond pure technical measures with regard to powertrain shifts and overall efficiency enhancements, fundamental changes in today's mobility patterns will also be required to meet the 2.0 °C Scenario, and even more so for the 1.5 °C pathway. This will apply particularly to the car use habits in the OECD countries. In essence, definite limitations on transport activities and modal shifts towards mainly buses and railways in some world regions and sub-sectors will be required to meet the Paris goals. However, the Non-OECD world regions will mainly increase their overall transport activity until 2050. The 2.0 °C and 1.5 °C transport scenario pathways will not be achieved automatically, but will require long-sighted infrastructural and transport policy framework settings on both inter- and intra-governmental levels.

13.1.4 Limitations and Further Research Requirement—Transport

The statistical databases for several world regions on transport activities and fleet and powertrain shares are limited, and in those cases, projections, conclusions by analogy, and estimations were required in our modelling. Therefore, further studies should focus on enhancing these databases and specify the modelling in more detail, which could also include case studies of countries instead of regions, to better address spatial particularities in the transport models. Detailed investigations of mode shift potentials, based on infrastructure capacity constraints, were considered to some extent, but deserve more in-depth modelling in future works. Further research is required to refine the coupling of renewable energy potentials, transport infrastructure upgrades, and the expansion of on-board energy storage usage.

13.1.5 Key Findings—Long-Term Energy Scenario

The 2.0 °C and the 1.5 °C scenarios both represent ambitious pathways that require fundamental changes in current energy consumption and supply. The key strategies of these pathways are the implementation of renewable energy technologies and efficiency improvements in all sectors. The electrification of the transport and heating sectors and the diversification of supply technologies are core elements of both alternative scenarios. Besides numerous technical and structural improvements, behavioural changes among end-users and major changes in investment activities and strategies must be achieved. This applies, for example, to the per capita electricity consumption (electric appliances without heating) of 'residential and other sectors', which will decrease in OECD countries by one third between 2015 and 2050 in the 1.5 °C Scenario, but will grows in the non-OECD regions by only 70% in the same period. This could imply limitations on personal consumption compared with today's standards, particularly in OECD countries—at least for as long as fossil fuels still play a significant role. Another example is the final energy demand per $GDP in the 'Industry' sector, which will decrease by 65% in OECD regions and by 80% in non-OECD regions, an ambitious pathway that will require stringent technology change and replacement strategies and supporting regulatory and governance measures to trigger huge investments for its realization.

On the energy supply side, considerable contributions in the future are assumed from wind and solar power, electrification and synthetic fuel use in transport and heating, sustainable biomass use, especially for co-generation and biofuels, and district heating systems that integrate solar, geothermal, and ambient heat potentials. The exploitation of renewable energy potentials depends strongly on regional conditions. In regions such as India and China, a 100% renewable electricity supply will require the extensive use of existing potentials. The global installed capacities of renewable power generation technologies will increase by a factor of more than 12, from 2000 GW in 2015 to more than 25,000 GW in 2050. Cumulative investments for power generation are estimated to increase between 2015 and 2050 by up to around US$ 50,000 billion compared with around US$20000 billion in the reference case. Fuel cost savings could offset around 90% of the additional investment costs as consequence of the fossil fuel phase-out and reductions in demand, but without considering the additional infrastructure demands of the transition arising from grid expansion, storage, and other flexibility demands. In the scenarios, large-scale and long-range electricity transport between Europe and the MENA countries is assumed to be a possible and promising example of supra-regional exchange between regions of production and regions of demand. Many of these import/export relationships must also be realized among countries *within* individual world regions (which are thus not resolved in our model) to increase the security and cost-efficiency of the energy supply. Decentralization and digitalization, but also the efficient implementation of new respectively the expansion of existing central infrastructures, are other implicit core elements of the scenario narratives. The large-scale generation and use of synthetic fuels are expected to play key roles in the deep-decarbonisation scenario, at least if the intensity of today's freight transport and air traffic is to be maintained, despite the huge energy losses this option will have. Both alternative scenarios, especially the 1.5 °C Scenario, require a rapid reduction in the final energy demand and, as far as possible, stagnation in the strong global growth in the demand for energy services, at least for as long as fossil energies dominate the energy supply structures.

Besides the structural similarities between both the 2.0 °C and 1.5 °C Scenarios, one main difference between them is the rate of transformation: To maintain the average global temperature increase due to climate change below 1.5 °C, the transformation must be accomplished as fast as technically possible. The trend in increasing global energy-related CO_2 emissions must be reversed as soon as possible (in the early 2020s, at the latest) and emissions must be reduced by 70% in 2035 and by 85% in 2040 (compared with emission levels in 1990). Every single year without significant emission reductions on the global level will dramatically reduce our chance to confine global warming to 1.5 °C. In contrast, in the 2.0 °C Scenario, emission reductions in 2035 may be in the order of 40% and 65% in 2040, leaving a little more time for the transition process.

13.1.6 Limitations and Further Research Requirement—Long-Term Energy Scenario

The 1.5 °C Scenario may seem more difficult than similarly ambitious scenarios that were made 10 to even 20 years ago. However, the opportunities to respond to climate change have been largely wasted in the last two decades, and all transition processes have faced huge obstacles in the past due to the inertia and conflicting aims of societies, governments, and most relevant stakeholders. Too often, more attention has been paid to doubters than to facts. Therefore, the scenarios also show that the longer the governments wait, the more difficult it will be to prevent severe climate damage and the greater will be the technical and economic challenges that will be encountered in the energy system transformation.

The coarse regional resolution of such global scenarios does not allow sufficient account to be taken of sub-regional differences in energy demand and the characteristic and favourable possibilities of sustainable supply. However, it can provide rather fundamental insight into basic technical and structural possibilities and requirements of a target-oriented pathway. Our results clearly reveal and quantitatively describe that the coming years will be most critical regarding a successful energy transition because for both parts of the energy transition—efficiency improvement/demand reduction and the implementation of new technologies—huge investments and fundamental changes in producing, distributing, and consuming energy will be needed. Such transformation processes must be analysed and planned carefully under the complex economic and societal framework conditions of each region, down to the country, sub-country, and community levels. Such analyses can then form the basis for the further investigation of the economic implications of these pathways.

Another limitation of this approach is that the economic, technical, and market assumptions made probably have limited consistency. Carbon, fuel, and technology costs are assumed independently of the assumptions regarding overall economic development and the final energy demand. It also remains unclear to what extent the energy transition will change the overall material demand and activity of the manufacturing industry. Furthermore, which economic framework conditions and market mechanisms will be necessary for rapid decarbonisation remain largely unclear, as is whether the current market mechanisms are capable of supporting the fundamental paradigm shift of this target-oriented energy transition.

13.1.7 Key Findings—Power Sector Analysis

Although there are significant differences across all the regions and sub-regions analysed, there are some similarities:

Increasing loads: The loads in all regions will increase significantly between 2020 and 2050, due to increased electrification of the transport and, to some extent, the heating sectors. Higher loads will require the adaptation of power-lines and transformer stations, especially on distribution grids where electric vehicles are most likely to be charged.

Increasing reverse power flows: Almost all regions will have periods of negative residual load, when generation is higher than the required load within a specific period of time. This leads to reversed power flow, and the generated electricity must be transported to other regions. The dominant power-generation technologies are wind and solar photovoltaic (PV). Solar PV is mainly connected to the distribution grid and export requires that the electricity must be able to flow from low- to medium-voltage levels, which requires adaptation at transformer stations.

Ratio between maximum variable generation capacity and maximum load: The results presented in Chap. 8 suggest that a ratio of 180% variable renewable capacity to the maximum load represents the optimum relationship. If the capacity were higher, short-term peaks would appear more frequently, which would lead to higher curtailment, or high storage and transmission demands. Example: The maximum load of a region is 100 GW and the installed capacity of wind and solar PV combined is 180 GW. Short-term peaks are relatively rare, and curtailment is below 10% of the total annual generation potential. If the variable power generation were above 200 GW, while the load remained at 100 GW, the curtailment rates would increase significantly, to 15% and higher.

Technology variety reduces storage demand: The combination of solar PV and wind leads to a lower storage demand than does a solar- or wind-dominated supply scenario. The relative interaction between the available wind and solar resources, not just in regard to day and night, but also seasonally, will reduce the storage demand in all regions. Therefore, the levelized cost of electricity generation cannot be seen in isolation. Even if, for example, wind is more expensive than solar PV in one region, a combination of wind and solar will lead to a reduction in storage demand and lower systems costs.

Variety in storage technology and sector coupling: The combination of battery and hydro pump storage technology and the conversion of the gas sector to hydrogen and synthetic fuels will be beneficial. The hydrogen produced can be used for the management of demand, and hydrogen-fuelled power plants can provide valuable dispatch services.

13.1.8 Limitations and Further Research Requirement—Power Sector Analysis

Measured load curves for regions, countries, or states/provinces are often unavailable and, in some countries, are even classified information. Therefore, it was not always possible to compare the calculated loads with actual current actual. However, the comparison of 2020 calculations with current maximum loads for regions and countries with published data showed that the calculations were within a ± 10% range. However, verification was not possible for several regions. Therefore, our analysis may over- or underestimate the current loads and therefore the future projections of loads as well. Therefore, the optimal ratio of maximum load to installed variable capacity requires further research.

More research is also required to verify the thesis of an optimal ratio between the variable generation and maximum load, because the sample size of this study was not sufficient to ensure validity. Furthermore, the optimal mix of solar and wind requires better meteorological data and actual measured load profiles. Access to more detailed data to calculate more case studies is vital to determining the possible optimal combination of wind and solar.

13.1.9 Key Findings—Non-energy Scenarios

The key finding of the land-use-related emission scenarios is that a dedicated concerted effort to sequester carbon by reforestation and forest restoration could re-establish the terrestrial carbon stock of pre-industrial times. That would undoubtedly come with multiple co-benefits, but would not be without challenges. After all, there is a reason why humans in the various corners of the planet pursued deforestation, whether for short-term and short-sighted gains or to establish agricultural areas that fed an increasing population. Therefore, land-use conflicts and trade offs are an inherent part of future mitigation actions, whether CO_2 sequestration is pursued by reforestation sequestration or by some biomass and CCS use. Nevertheless, the important result of this study is that the addition of land-use CO_2 and other GHG emission pathways to energy-related scenarios yields scenarios that stay below or get below 1.5 °C warming without a reliance on massive net negative CO_2 emission potentials towards the second half of this century.

Going beyond the land-use CO_2 emission pathways that we sketched for a series of sequestration options, we also designed trajectories for all the other GHGs and aerosols. An unprecedented wealth of scenario information is now available thanks to the recent concerted efforts of the larger integrated assessment community. Designing a novel method here, the Generalized Quantile Walk method, we were able to distil non-CO_2 pathways from this rich scenario database—in a way that respects the correlations and dependencies between energy-related CO_2 and other gas emissions. This is not only a new methodological advance in scenario research, but also key to the proper estimation of the climate effects of the energy-related CO_2 scenarios designed in the main part of this study.

13.1.10 Limitations and Further Research Requirement—Non-energy Scenarios

There are a number of limitations associated with the derived non-energy-related emission trajectories. Possibly the most important opportunity for future research will involve a more fine-grained look at land-use-based sequestration options in various countries and biomes. This study assumed only a rather coarse approximation of the available land areas, sequestration rates, and cumulative changes in land carbon stocks to estimate the potential and time trajectories of those reforestation, forest restoration, agroforestry, and other land-based sequestration options.

In terms of the non-CO_2 emission trajectories, this study relied heavily on the collective wisdom embodied within a large set of literature-reported scenarios. Although we have designed probably the most advanced method to distil that knowledge into emission trajectories that are consistent with our energy-related pathways, this meta-analytical approach is not without its limitations. In particular, a bottom-up energy-system and land-use/agricultural model must be able to estimate methane and N_2O emissions from various agricultural activities in a more coherent way, which could provide results on a regional level. Such regionally and sectorally specific information would, in turn, allow the examination of various mitigation options for non-CO_2 emissions. This bottom-up modelling capacity is missing from our meta-analytical approach.

13.1.11 Key Findings—Employment Analysis

The occupational employment analysis was developed in 2018 to improve the database for the 'just transition' concept. Not only is the number of jobs that will be created or lost as a result of a global or regional energy transition important, but also the specific occupations that will be to develop a socially sound transition. This analysis breaks new grounds because very little information has been available. However, the results indicate that even within the seven occupation types, job losses are the exception and almost all trades will gain more jobs.

Very specialized jobs, such as machine operators in coal mines, will be lost and there will be no replacement. Therefore, a detailed analysis of all sectors is required to identify those highly specialized tasks and to develop re-training possibilities.

13.1.12 Limitations and Further Research Requirement—Employment Analysis

The data available on the detailed employment requirements for renewable energies are very limited. Although there are some data for solar PV and onshore and offshore wind, there are almost none for concentrated solar power plants or geothermal energy. Furthermore, occupational surveys of the heating and energy efficiency sectors are required.

13.1.13 Key Findings—Mineral Resource Analysis

Lithium sees the highest projected increase from mined ore of around 40 times current production to above 80 times current production should future technologies be introduced without recycling.

Cumulative primary lithium demand by 2050 for the majority of scenarios is above current reserves of lithium except for the "potential recycling" scenario but less than known resources. The scenarios anticipate a scale up in resource exploration, discovery and production for primary resources to meet demand—assuming lithium-ion batteries continue to dominate as the chemistry of choice. It is important that future mines be responsibly developed and that battery designs be compatible with circular-economy thinking. Significant infrastructure for reuse and recycling will need to be developed to achieve high rates of lithium recycling.

For cobalt, future scenarios exceed currently known reserves and approach currently known resources in 2050. Given the concentrated supply source from the Democratic Republic of Congo, this will continue to keep pressure on exploring alternative battery chemistries and on increasing cobalt recycling. Attention must continue to be paid to reducing the social and environmental impacts of supply whilst supporting development noting the significant adverse impacts on human and environmental health associated with cobalt mining. While the value of cobalt in EV battery recycling is already an important component of the recycling economics because of supply limitations, the social and environmental challenges provide a further driver for increasing recycling.

For silver, the potential of materials efficiency (using less silver per GW solar PV panel) has potential to reduce demand owing to the long lifetimes of PV panels. Under some scenarios using future technology with recycling, the levels of silver demand are similar to current production.

13.1.14 Limitations and Further Research Requirement—Mineral Resource Analysis

This study focuses only on the metal demand for renewable energy (generation and storage) and transport and does not consider other demands for these metals. However, it is expected that with the increase in renewable energy, renewable energy technologies will consume a greater share of these metals and it is anticipated that this growth will have significant influence on overall market dynamics, including influencing prices, which may feedback to efforts to reduce material intensity and invest in reuse and recycling infrastructure.

Promoting the transition to circular economy for both renewable energy and resource cycles; and adopting a systems view that considers available supply as well as social and environmental factors is critically important. To support sustainable development goals, both the primary and secondary sources of the resources required to underpin this renewable energy transformation needs to be stewarded as the supply chains develop. The high total demand requirements for energy metals, demonstrates the importance of redesigning technologies and systems to eliminate the adverse social and environmental supply chain impacts, to promote long-life products, and to actively encourage efficient material use in both energy and transport sectors.

13.2 Policy Recommendations by Sector

To implement the 2.0 °C or 1.5 °C Scenario will require a significant shift in current policies. This section documents the policy measures that have been assumed for the scenarios presented in the previous chapters, as well as policy measures known to be successful. The global legal frameworks and regulations differ significantly on national and community levels. Therefore, only the functions and aims of suggested policy measures can be discussed, but not how they can be integrated into current jurisdictions. That said, it is assumed that the Paris Agreement will be the global basis.

13.2.1 Energy

The energy sector is not a homogeneous sector but is highly diverse. Therefore, policy measures can follow different strategies and can address different aspects and stakeholders. The supply side can be divided into the main sub-sectors of electricity, heating, and fuel supply. The demand side can be broken down further into buildings, industry, and transport. However, because all these sub-sectors are directly or indirectly connected via resources and energy markets, policy interventions can have different effects.

The basic principles for the development of the 2.0 °C and 1.5 °C Scenarios derive from the long-term experiences with scenario development of the authorship team, and have led to a 'seven-step logic'.

This logic extends from the definition of the final state of the energy systems in the long-term future to the key drivers of the energy demand and the energy efficiency potentials, a technological analysis of supply and demand and the market development potential, and the specific policy measures required to implement a theoretical concept in the real market-place.

The seven steps are:

1. Define the maximum carbon budget and other targets, milestones, and constraints to achieve the climate goal;
2. Define the renewable energy resource potentials and limits within a space-constrained environment;

3. Identify the economic and societal drivers of demand;
4. Define the efficiency potentials and energy intensities by energy service and sector;
5. Establish time lines and narratives for the technology implementation on the end-user and supply sides;
6. Estimate the infrastructure needs, generation costs, and other effects;
7. Identify the required policies and discuss the policy options.

13.2.1.1 General Energy Policies

The energy sector requires both very specific measures, such as grid codes and efficiency standards, and overarching measures. International collaboration and co-operation are required to define and implement mandatory standards and to effectively develop energy policy and its regulative interventions. The most important interventions to accelerate the energy transition are:

- Renewable energy targets and incentives for their deployment and expansion;
- Internalization of external costs by carbon tax or surcharge;
- Phase-out of fossil fuel subsidies;
- Accelerated replacement of fossil and inefficient technologies.

Renewable energy targets are vital to accelerate the deployment of renewable energy. Experiences of the past two decades clearly show the effectiveness of renewable energy policy development. The Renewable Policy Network for the 21st Century (REN21) states in their annual market analysis, Renewables 2018, that "Targets remain one of the primary means for policy makers to express their commitment to renewable energy deployment. Targets are enacted for economy-wide energy development as well as for specific sectors" (REN21-GSR-2018). To achieve these goals, innovation processes must be initiated, markets developed, and investment stimulated. For the latter, auctions and feed-in tariffs have proven suitable. It is important in this context to guarantee investment security and to enable long-term but appropriate revenues.

Climate change leads to a number of types of environmental damage. Carbon emissions lead to climate change. Therefore, it is vital to put a price on carbon to internalize the external costs. Carbon-pricing schemes can be established as *cap-and-trade* schemes or taxes. Carbon pricing is not sufficient on its own to achieve the objective of the Paris Agreement, and many leading international agencies and institutions argue that a much more concerted and widespread global take-up of carbon pricing will be necessary (Carbon Tracker 2018). To make carbon pricing an efficient measure, the price of carbon must be sufficient to reflect the environmental damage it causes and it must be reliable. Therefore, a minimum price should be implemented to provide planning security.

Subsidies of fossil fuels counteract any efforts to make energy efficiency and renewable energy competitive. According to the International Energy Agency, the total amount of global fossil fuel subsidies was estimated to be around US$260 billion in 2016 (IEA-DB 2018). The governments of the G20 and the Asia-Pacific Economic Cooperation (APEC) reached an agreement to "*rationalize and phase out over the medium term inefficient fossil fuel subsidies that encourage wasteful consumption*" (OECD-IEA 2018).

Setting legally binding national targets for 100% renewable energy pathways will lead to an orderly phase-out of fossil fuels. This is vital in planning the socio-economic effects of the energy transition (see Chap. 10). Supporting measures to achieve the replacement of fossil and inefficient technologies and energy sources will be necessary because targets and economic incentives may not be sufficient in all areas. Regulatory interventions for the decommissioning and replacement of facilities are a way to stop opposing business scenarios and to overcome the inertia of consumers and investors. Moreover, economic policy measures and the clearer definitions and stringent enforcement of international standards will accelerate the implementation of the best available technologies in the industry.

Both a minimum price on carbon and the immediate phase-out of fossil fuel subsidies must be implemented in order to support the global energy transition. Policy support for technology innovation is another important measure to create the basis for the energy transition processes.

13.2.1.2 Policies for Buildings Sector Decarbonisation

To reduce the energy demands of existing building stock and new buildings, constantly most stringent energy demand standards (= *building energy codes*) are required for all building types and across all countries. The goal must be to achieve (near) zero-energy buildings, so that each building reduces its heating and cooling demand to the lowest possible level and aims to supply the remaining energy with on-site renewable energy technologies, such as solar collectors, electric heaters, advanced bioenergy and heat pumps, or with low-temperature heating networks. Mandatory municipal heating plans are an appropriate way to define an efficient strategy that balances insulation, local heating systems, and the grid-connected heat supply in regions with a significant space-heating demand. An analysis of building energy codes in 15 countries (Young 2014) distinguished between residential and commercial buildings and listed six technical requirement categories, ranging from heating and cooling requirements and the insulation of the building envelope to the building design. However, building energy codes must be mandatory and include existing building stock as well as new buildings. The most urgent need for action is in developing countries, where the rapid growth in building construction can be expected over the next decades. Economic measures to increase heating/cooling costs, e.g., by introducing fossil energy taxes or surcharges or by phasing-out fossil fuel subsidies, could support efforts to save energy. However, they must be accompanied by social policies.

13.2.1.3 Policies for more-Efficient Electrical Appliances

To reduce the electricity demand of consumer goods in households and equipment in buildings, the efficiency standards for electrical appliances, all forms of information and communication technologies (computers, smart devices, screens, televisions, etc.), white goods (washing machines, dryers, dishwasher, fridges, and freezers), electrical building equipment for thermal comfort, and lighting technologies are required. These efficiency standards must be dynamic and designed to support competition for the most efficient design. The Japanese front runner system (IEA-PM 2018) is a positive example of dynamic efficiency standards. Labelling programmes and purchase subsidies for the best available technologies can support the replacement of old devices with the most efficient technologies. Measures for training and capacity building are also essential, most importantly in non-OECD countries.

13.2.1.4 Policies for the Transport Sector

As well as resolute electrification and further technical advances in all transport modes, decision-makers in politics and the urban planning context must carefully steer transport habits and infrastructure development toward a climate-friendly transport system. Their aim should be to promote the use of less ecologically problematic transport modes. This can be done by, for example, the introduction of fiscal and regulatory measures that effectively reduce the subsidization of currently untaxed and internalisation of external costs.

In parallel, environmentally less harmful transportation modes should be incentivized. Investments must also be channelled towards highly productive and energy-efficient passenger and freight railway systems and towards a dense network of battery recharging and hydrogen refuelling infrastructures for road vehicles. In the passenger car and truck context, direct subsidies or tax incentives for electric vehicles will speed up the electrification of fleets. CO_2 taxation, road tolls, and congestion charges could be applied, in addition to parking-space management schemes to reduce road traffic and thus internal combustion engines in a transition to car-reduced cities. The assignment of parking lots and driving lanes exclusively to electric cars will speed the phase-out of internal combustion engines.

In aviation, measures could include the taxation of jet fuel and CO_2, the application of an emission trading scheme on the direct and indirect climate effects of flight at high altitudes. Direct and indirect public subsidies for carriers and airports should be abolished (investment and operational grants should be reduced and funding should be allocated to a competitive and attractive rail system).

All measures curtailing the use of individual passenger transport should be accompanied by the promotion of ubiquitous, fast, comfortable, and price-competitive public transport systems, ride and car sharing and on-demand services (especially for less densely populated semi-urban and rural areas). Last but not least, an attractive and safe infrastructure for bicycles and e-bikes will help to reduce emissions and other unwanted side-effects of transport. In this arena, Copenhagen and Amsterdam are at the cycling forefront and inspiring more and more cities in following their path. Cities must also curtail tendencies to urban sprawl and 'reinvent' the compact city ideal, which means becoming pedestrian-friendly cities, thus reducing the need for motorized individual mobility and freeing up space for recreation and green spaces.

Cities in developed countries should aim to transform their transport systems (often) from passenger-car-centred urban structures and policies towards pedestrian-, bike-, and mass-transport-friendly environments. The often densely populated emerging megacities in the upcoming economic powerhouses of Africa, Latin America, and Asia should invest, right from the start, in resilient public-transport-oriented urban structures instead of relying too strongly on individual passenger car traffic, as the OECD countries have done in the past.

13.2.1.5 Policies for the Industry Sector

Policies to achieve the implementation of new highly efficient technologies and to replace fossil fuel use in industry must be defined region-wide or even on the global level, and will require stringent and regulatory implementation. Economic incentives, national initiatives, and voluntary agreements with industrial branches will most probably not, by themselves, see the achievement of a rapid technological change. Concrete standards and requirements must be defined at a very detailed level, covering as far as possible all technologies and their areas of application. The systematic implementation of already-identified best-available technologies could begin in the next few years. Mandatory energy management systems should be introduced to identify efficiency potentials and to monitor efficiency progress. The sustainability features along process chains and material flows must also be taken into account when designing political measures. Particular attention must be paid to the material efficiency of both production processes and their products, because this can open up major energy efficiency potentials and reduce other environmental effects. Public procurement policies and guidelines can help to establish new markets and to demonstrate new more-efficient products and opportunities. The effectiveness of policy interventions must be assessed by independent experts and the further development of efficiency programs and measures will require ongoing coordination by independent executive agencies. The public provision of low-interest loans, investment risk management, and tax exemptions for energy-efficient technologies and processes will significantly support technological changes and incentivize the huge investments required. Knowledge transfer between sectors and countries can be achieved through networks initiated and coordinated by governments. Public funding for research and development activities with regard to technological innovation, low-carbon solutions, and their process integration will be vital to push the technological limits further. Innovative approaches to the realization of material cycles and recycling options, the recovery of industrial waste heat, and low-carbon raw materials and process routes in industry must also be identified and implemented.

13.2.1.6 Political Framework for Power Markets

The 2.0 °C and 1.5 °C Scenarios will lead to 100% renewable electricity supply, with significant shares of variable power generation. The traditional electricity market framework has been developed for central suppliers operating dispatchable and limited dispatchable ('*base load*') thermal power plants. The electricity markets of the future will be dominated by variable generation without marginal/fuel costs. The power system will also require the built-up and economic operation of a combination of dispatch generation, storage, and other system services whose operation will be conditioned by renewable electricity feed-ins. For both reasons, a significantly different market framework is urgently needed, in which the technologies can be operated economically and refinanced. Renewable electricity should be guaranteed priority access to the grid. Access to the exchange capacity available at any given moment should be fully transparent and the transmission of renewable electricity must always have preference. Furthermore, the design of distribution and transmission networks, particularly for interconnections and transformer stations, should be guided by the objective of facilitating the integration of renewables and to achieve a 100% renewable electricity system.

To establish fair and equal market conditions, the ownership of electrical grids should be completely disengaged from the ownership of power-generation and supply companies. To encourage new businesses, relevant grid data must be made available from transmission and distribution system operators. This will require establishing communication standards and data protection guidelines for smart grids. Legislation to support and expand demand-side management is required to create new markets for the flexibility services for renewable electricity integration.

Public funding for research and development is required to further develop and implement technologies that allow variable power integration, such as smart grid technology, virtual power stations, low-cost storage solutions, and responsive demand-side management. Finally, a policy framework that supports the electrification and sector coupling of the heating and transport sectors is urgently needed for a successful and cost-efficient transition process.

References

Carbon Tracker (2018), Closing the Gap to a Paris-compliant EU-ETS, 25th April 2018, website, viewed October 2018, https://www.carbontracker.org/reports/carbon-clampdown/

IEA-DB (2018), International Energy Agency Agency Database 2018, Energy Subsidies, webbased database, viewed October 2018, https://www.iea.org/statistics/resources/energysubsidies/

IEA-PM (2018), International Energy Agency – Policy and Measures database, Energy Conservation Frontrunner Plan-Japan, website, viewed October 2018, https://www.iea.org/policiesandmeasures/pams/japan/name-22959-en.php
OECD-IEA (2018), OECD website, OECD-IEA analysis of fossil fuels and other support, viewed October 2018, http://www.oecd.org/site/tadffss/
UNFCCC (2015), The Paris Agreement, website, viewed October 2018, https://unfccc.int/process-and-meetings/the-paris-agreement/the-paris-agreement
REN21-GSR (2018), REN21, 2018, Renewables 2018, Global Status Report, Paris/France, http://www.ren21.net/status-of-renewables/global-status-report/ page 20
Young, R. (2014), Young, Rachel, Global Approaches: A Comparison of Building Energy Codes in 15 Countries, American Council for an Energy-Efficient Economy, 2014 ACEEE Summer Study on Energy Efficiency in Buildings, https://aceee.org/files/proceedings/2014/data/papers/3-606.pdf

Annex

Glossary, regions, and countries included in the scenarios, further detailed descriptions of the assumptions (tables) of all models and methodologies, are detailed results tables.

Acronyms

2DS	2.0 °C Scenario – published in the IEA ETP
ACEC	Africa Clean Energy Corridor
ACOLA	Australian Council of Learned Academies
B2DS	beyond 2.0 °C Scenario – published in the IEA ETP
BC (emissions)	Black carbon emission
BECCS	Bioenergy with carbon capture and storage
BEV	Battery electric vehicles
BP	British Petrol
CCGT	Closed-Cycle Gas Turbine (gas power plant technology)
CCS	Carbon Capture and Sequestration
CdTd	Cadmium Telluride (Solar photovoltaic cell technology)
CEDS	Community Emissions Data System (for Historical Climate relevant Emissions)
CHP	Combine Heat and Power (=co-generation)
CIGS	Copper Indium Gallium Di-Selenide (Solar photovoltaic cell technology)
CMIP6	Coupled Model Inter-comparison Project Phase 6 (computer climate model)
CO2	Carbon Dioxide
CSP	Concentrated Solar Power
DBFZ	Deutsches Biomasse Forschungs-Zentrum (German Bio Mass Research Centre)

S. Teske (ed.), *Achieving the Paris Climate Agreement Goals*,
https://doi.org/10.1007/978-3-030-05843-2

DLR	Deutsche Luft und Raumfahrt, German Aero Space Centre
DREA	Distributed renewables for energy access
EAPP	East Africa Power Pool
EJ/year	Exa-Joule per year
EJ	Exa-Joule
ENTSO-E	European Network of Transmission System Operators for Electricity
ESM	Earth System Model (computer climate model)
ETP	Energy Technology Perspectives (publication of the International Energy Agency)
EU-WEEE	Waste electronic equipment – Environment, Directive of the European Commission
EV	Electric Vehicles
FAO	Food and Agriculture Organization of the United Nations
FCEV	Fuel-Cell-Electric-Vehicle
FYP	Five-Year-Plan
G20	Group of 20 (The G20 is an international forum for the governments and central bank governors from Argentina, Australia, Brazil, Canada, China, the European Union, France, Germany, India, Indonesia, Italy, Japan, Mexico, Russia, Saudi Arabia, South Africa, South Korea, Turkey, the United Kingdom, and the United States.)
GDP	Gross Domestic Product
GEA	Global Energy Assessment
GIS	Global Information System
GLOBIOM	GLObal BIOsphere Model (computer model for land-use)
GW	Gigawatt
GWh/year	Gigawatthours per year
HDV	High Duty Vehicle (= truck)
HOV	High Occupancy Vehicle lanes (transport modelling)
HST	High Speed Trains
HWP	Harvested Wood Product (computer model for carbon flows)
IAM	Integrated Assessment Modelling
ICAO	International Civil Aviation Organization
IEA	International Energy Agency
ILO	International Labour Organization
IMAGE	Integrated Model to Assess the Global Environment (computer climate model)
IMO	International Maritime Organization
INDC	Intended Nationally Determined Contributions
IRENA	International Renewable Energy Agency
ITRPV	International Technology Roadmap for Photovoltaic
LCOE	Levelized Cost of Electricity
LDF	Leonardo DiCaprio Foundation
LDV	Light Duty Vehicle (= passenger car)

LIB	lithium Ion Batteries
MAGICC	Model for the Assessment of Greenhouse-gas Induced Climate Change
MEPC	Marine Environment Protection Committee
MESSAGE	Model for Energy Supply Strategy Alternatives and their General Environmental Impact (computer energy model)
MJ/year	Mega-Joule per year
MW	Megawatt
MWh/year	Megawatthours per year
NDC	Nationally Determined Contributions
NEA	National Energy Administration (China)
NGV	Natural gas vehicles
OC (emissions)	Organic Carbon emission
OD	Origin Destination (transport modelling)
OECD	Organization for Economic Co-operation and Development
PJ/year	Peta-Joule per year
PMG	Permanent magnet generators
PPMC	Paris Process on Mobility and Climate
PPP	Purchasing Power Parity (economic term)
PRIMAP	Potsdam Real-time Integrated Model for probabilistic Assessment of emissions Paths (computer climate model)
PV	Photovoltaic
QGIS	Quantum - Global Information System (open source software)
RCP	Representative Concentration Pathway
REF	Reference scenario
REN21	Renewable Energy Network for the 21st Century
SAPP	South Africa Power Pool
SF6	Sulphur Hexafluoride
SHS	solar home systems
SOx	Sulphur Oxides
SSP	Shared Socio Economic Pathways (for climate scenarios)
TWh/year	Terawatthours per year
TYNDP	Ten-Year-Network Development Plan
UIC	Union Internationale des Chemins de fer; International union of raiways
UNFCCC	United Nations Framework Convention on Climate Change
USA-EPA	United States of America – Environmental Protection Agency
USD	US Dollar
UTCE	Union for the Coordination of the Transmission of Electricity
UTS-ISF	University of Technology Sydney – Institute for Sustainable Futures
VRE	Variable Renewable Energy
WEO	World Energy Outlook (publication of the International Energy Agency)

9783030058425